Theory & Application of External Thermal Insulation Technology

(2nd Edition)

Editors-in-chief:
Beijing Building Energy-Efficiency & Environment Engineering Association
Technology and Industrialization Development Center of the Ministry of Housing and Urban-Rural Development
Beijing Building Research Institute of CSCEC
Shandong Academy of Building Research
Beijing Zhenli Energy Conservation & Environmental Protection Technology Co., Ltd.

China Architecture & Building Press
中国建筑工业出版社

图书在版编目（CIP）数据

外保温技术理论与应用（第二版）Theory & Application of External Thermal Insulation Technology（2nd Edition）：英文/北京建筑节能与环境工程协会主编. —北京：中国建筑工业出版社，2018.9
ISBN 978-7-112-22510-1

Ⅰ.①外… Ⅱ.①北… Ⅲ.①建筑物-外墙-保温-英文 Ⅳ.①TU111.4

中国版本图书馆 CIP 数据核字（2018）第 179480 号

The *Theory & Application of External Thermal Insulation Technology*（*2nd Edition*）is compiled based on the practice and development of the past four years. It is a version of collation and revision of the first edition，based on the achievement and experience summary completed by experts and scholars under organization of the editors-in-chief. This book further enriches the practical technology and theory of the first edition，and involves new contents of external thermal insulation development. This classic book for building energy conservation in China is suitable for both building energy conservation technicians and college students.

Editor-in-charge：Zhang Lei and Qu Ruduo

Theory & Application of External Thermal Insulation Technology（2nd Edition）

Editors-in-chief：
Beijing Building Energy-Efficiency & Environment Engineering Association
Technology and Industrialization Development Center of the Ministry of Housing and Urban-Rural Development
Beijing Building Research Institute of CSCEC
Shandong Academy of Building Research
Beijing Zhenli Energy Conservation & Environmental Protection Technology Co.，Ltd.

*
中国建筑工业出版社出版、发行（北京海淀三里河路 9 号）
各地新华书店、建筑书店经销
北京佳捷真科技发展有限公司制版
天津翔远印刷有限公司印刷
*
开本：880×1230 毫米 1/16 印张：27 字数：832 千字
2018 年 10 月第一版 2018 年 10 月第一次印刷
定价：**85.00** 元
ISBN 978-7-112-22510-1
（32327）

Editors-in-chief of the 2nd edition:

Beijing Building Energy-Efficiency & Environment Engineering Association

Technology and Industrialization Development Center of the Ministry of Housing and Urban-Rural Development

Beijing Building Research Institute of CSCEC

Shandong Academy of Building Research

Beijing Zhenli Energy Conservation & Environmental Protection Technology Co., Ltd.

Co-editors of the 2nd edition:

Beijing Building Research Institute of CSCEC

Tianjin Fire Research Institute of MPS

Tsinghua University

China Institute of Building Standard Design Research

Harbin Institute of Technology

Civil Engineering Research Institute of Jiaxing University

Beijing Building Materials Testing Academy Co., Ltd.

Beijing Institute of Real Estate Science and Technology

China Green Building Industry Technology Innovation Strategy Alliance

Beijing Building Technology Development Co., Ltd.

China Building Material Enterprise Management Association

Building and Building Material Expert Committee of China Association for Promotion of Private Sci-tech Enterprises

Editing Committee of 2nd Edition

Editor-in-chief: Huang Zhenli

Associate editors: Tu Fengxiang, Liang Junqiang, Qu Hongle and Sun Hongming

Members: (listed according to the stroke sequence of Chinese family names)

Wang Chuan, Wang Yongkui, Wang Junsheng, Wang Mansheng, Zuo Qing, Shi Shuli, Fu Haiming, Zhu Qing, Zhu Chunling, Xiang Lina, Liu Bo, Ren Lin, Yan Chunyan, Sun Jiajin, Sun Guifang, Sun Huiqin, Li Peng, Li Wenbo, Li Xiaoliang, Li Xichao, Yang Xingming, Qiu Junfu, He Xiaoyan, Zou Haimin, Zhang Yong, Zhang Jun, Zhang Yuxiang, Zhang Libo, Zhang Shujun, Zhang Xiaoying, Chen Zhihao, Lin Yancheng, Luo Shuxiang, Ji Guangqi, Zhou Hongyan, Zheng Jinli, Ju Shibao, Zhong Shengjun, Hong Mei, Xu Xin, Gao Wei, Guo Qing, Huang Kai, Cao Dejun and Xie Fuzhou

Review Committee of 2nd Edition

Chief reviewer: Jin Hongxiang

Associate reviewers: Yang Xiwei and Wang Lichen

Members: (listed according to the stroke sequence of Chinese family names)

Wang Zheng, Wang Qingsheng, Wang Guohui, Fang Zhanhe, Ye Jincheng, Tian Lingjiang, Feng Jinqiu, Liu Xiaojun, Liu Huaiyu, Liu Xiaozhong, Liu Jingjiang, An Yanhua, Xu Jinfeng, Sun Sihai, Sun Kefang, Li Dongyi, Li Jinbao, Li Xiaoming, Song Bo, Chen Danlin, Chen Dianying, Lin Guohai, Zheng Xiangqin, Zhu Genli, Jia Dongmei, Gu Taichang, Tao Siji, Cui Qi, Peng Canyun and Tan Chunli

Editing Committee of 1st Edition

Editor-in-chief: Huang Zhenli

Associate editors: Tu Fengxiang and Liang Junqiang

Members: (listed according to the stroke sequence of Chinese family names)

Wang Chuan, Wang Liqian, Wang Mansheng, Fu Haiming, Zhu Chunling, Ren Lin, Liu Feng, Liu Xiangzhi, Sun Xiaoli, Li Wenbo, Li Zhiguo, Li Chunlei, Song Changyou, Wu Xihui, He Liudong, Zou Haimin, Zhang Jun, Zhang Ming, Zhang Lei, Zhang Leilei, Chen Zhihao, Lin Yancheng, Ji Guangqi, Zheng Jinli, Ju Shibao, Hu Yongteng, Gao Hanzhang, Tu Fengxiang, Cao Dejun and Xu Junfeng

Review Committee of 1st Edition

Chief reviewer: Yang Rong

Associate reviewers: Jin Hongxiang, Zhu Qing and Yang Xiwei

Members: (listed according to the stroke sequence of Chinese family names)

Ding Wanjun, Yu Qingshan, Ma Heng, Ma Yunyu, Wang Jian, Wang Gongshan, Wang Hanyi, Wang Qingsheng, Wang Guojun, Wang Guohui, Wang Chuntang, Wang Shaojun, Wang Zhenqing, Wang Mansheng, You Xiaofei, Fang Ming, Fang Zhanhe, Ye Jincheng, Tian Lingjiang, Feng Ya, Feng Jinqiu, Feng Baochun, Qu Zhenbin, Zhu Chuansheng, Zhu Yingbao, Liu Xiaojun, Liu Jiaping, Liu Younong, Liu Huaiyu, Liu Nianxiong, Liu Zhendong, Liu Zhenxian, Liu Xiaozhong, Liu Jingjiang, An Yanhua, Jiang Chenggui, Xu Jinfeng, Sun Sihai, Sun Kefang, Sun Hongming, Su Dan, Su Xianghui, Li Dongyi, Li Jinbao, Li Xiaoming, Li Haojie, Li Xikuan, Yang Honghai, Yang Huizhong, Wu Gang, Wu Xihui, Song Bo, Zhang Jusong, Zhang Guoxiang, Zhang Shujun, Zhang Jianfeng, Zhang Ruijing, Lu Shanhou, Chen Xing, Chen Danlin, Chen Dianying, Lin Guohai, Lin Haiyan, Zheng Yibo, Zheng Dejin, Zheng Xiangqin, Zhao Shiqi, Zhao Lihua, Zhao Chenggang, Hao Bin, Hu Xiaoyuan, Zhu Genli, Qin Gang, Qin Zheng, Qin Youguo, Jia Dongmei, Gu Taichang, Xu Weidong, Guo Li, Luan Jingyang, Tang Liang, Tu Fengxiang, Tao Siji, Mei Yingting, Cao Yongmin, Cui Ronghua, Liang Xiaonong, Liang Junqiang, Jiang Wei, Su Dongqing, Han Hongwei, You Guangcai, Liao Libing and Wei Yongqi

Preface I of 1st Edition

In the past two decades, China has seen great progress in social practice in building energy conservation. Among all wall insulation technologies, the external thermal insulation has developed fastest. It has entered a mature technology stage and played a leading role in the field of building wall energy conservation in China. Technical standards of design, construction, material, acceptance and approval have been formulated. In addition, the external thermal insulation is an important technical guarantee for the basic national policy of energy conservation and emission reduction.

The publication of the*Theory & Application of External Thermal Insulation Technology* shows that the social practice in building energy conservation has spawned the development of its theory. The external insulation technology is a discipline focusing on the impact of natural forces (namely thermal stress, water, fire, wind and earthquake) on the external insulation system. It is the progress from the practice to theory and from the realm of necessity to that of freedom to study the motions of the above five natural destructive forces, analyze their inherent laws, and propose appropriate technical measures.

The external thermal insulation structure can provide a reasonable temperature field for a building, minimize the impact of the ambient temperature on buildings and extend their service life. It is a main method of wall energy conservation.

It is another development stage when the research focus turns to how to extend the service life of the external thermal insulation and synchronize it with the building life. This book is a good start in studying the impact of the above five natural forces on the external thermal insulation.

There is still a large space forthe development of wall insulation and the exploration of external thermal insulation. To summarize the practical experience is helpful to develop the theory and better guide the practice. Combining the practice and theory, we can continuously promote the external thermal insulation technology and broaden its applications to make greater contributions to building energy conservation.

Building Energy Conservatoin & Technology Department
Ministry of Housing and Urban-Rural Development
March 2011

Preface II of 1st Edition

The publication of the *Theory & Application of External Thermal Insulation Technology* is a remarkable event. The external thermal insulation has been applied for more than two decades in the field of building energy conservation in China. We continuously strive to overcome difficulties and make achievements in this unprecedented social practice. Two "reductions" in this book, namely "reduction of energy consumption and reduction of waste generation" reflect that this basic national policy of the Party has has been rooted in the hearts of the people. Along with the social practice in energy conservation and emission reduction, various supporting technical standards have been formulated in the field of building energy conservation, in which the external thermal insulation plays an important role. Such standards for different functions are powerful for comprehensive, extensive and in-depth implementation of the government-led social practice in energy conservation and emission reduction. They are derived from the practice but can guide, standardize, improve and further develop the practice.

During the writing process, we explored the regular practical experiences by trial and error and summed up those mature practical experiences to form these standards. The standards are applied to guide the practice but also unceasingly revised in the practice. In the process of constant negation and exploration of new laws, the internal scientific laws contained in the standards are developed into theories. Therefore, the social practice has spawned the energy conservation theory.

The theory is a product of leaps in practice, revealing and refining internal laws. It is repeatedly verified in dynamic development, so its trajectory can be calculated by mathematical formulas.

In recent years, the standardization has continued to promote the in-depth research of the theory of external thermal insulation. In particular, the study of basic theories has become a focus of dominant enterprises and research institutes. This will make the standards as scientific and rigorous application tools and also an excellent element of advanced productivity.

This book presenting some technical theories in the field of external thermal insulation in China is a stage symbol of building energy conservation. Without technical theories, the related field will not develop constantly. However, this book involving scientific research results is still immature in the aspect of technical theories. Further explorations should be made in the social practice for maturity of such theories.

Theoretical research is needed for the successful implementation of the basic national policy of energy conservation and emission reduction. They are enable dominant enterprises to win out, small- and medium-sized enterprises to follow leading enterprises in the industry, and administrative managers to make correct decisions and perform scientific administration. One important feature of an innovation-oriented country is that stage theories increase in the social practice. The concern about theoretical research is undoubtedly essential in the Scientific Outlook on Development.

Standard Rating Department
Ministry of Housing and Urban-Rural Development
March 2011

Preface of 2nd Edition

The *Theory & Application of External Thermal Insulation Technology* is popular with readers in the industry since its publication. It is used as a training book for cadre re-education of local construction organizations and a supporting teaching material for building energy conservation courses in colleges and universities. This best-seller has been reprinted several times in the four years after its publication. It is encouraging that the insights in this book are widely circulated.

In the second year after its publication, ASTM vice president paid a special visit to Beijing to survey the impact of five natural destructive forces on the external thermal insulation in this book, and visited the sites of large-scale weathering test of the exterior insulation and finish system (EIFS) and research on the structure, window and corner resistance to fire. Later, President of the ASTM granted the excellent thesis award to the author of this book, which is an affirmation of relevant research results of China.

The history of Chinese and world's architectures proves that the soul and core of architectural technology research is to extend the building life to one hundred years or more. Certainly, buildings of long service life have unmatched effects of energy conservation and emission reduction.

As the basic contents and core ideas of this book, whether the study on the building temperature field or the analysis of the life extension via external thermal insulation, are for energy conservation and long life (or 100 years) of buildings.

Among wall insulation technologies, the external thermal insulation is the most efficient, popular and extensive. The qualified external thermal insulation will prolong the building life. Deservedly, it has become a leading technology in the field of building energy conservation in China. Technical standards for the whole process from the design, construction and material to acceptance and assessment have been formulated as an important technical pillar for energy conservation and emission reduction. China has the world's largest area of construction, most ambitious market of building energy conservation, and most energy-saving application technologies. The technological essence and theory derived scientifically from the world's richest practical experience in energy conservation will not only promote China's technologies of building energy conservation, but also benefit the world and the mankind as a whole.

At present, there is emerging money worship and eager for quick success and instant benefits, and the market economy isfar from mature. Some construction companies pursuing profit maximization may resort to unscrupulous means, such as competition with low quality and price and jerry building after bid awarding, resulting in the inferior quality projects. Some enterprises only pay attention to making money instead of investment in technology R&D. They use "copycat" materials and just apply one thin plastering method to different materials. Some government administrative departments may abuse their power, issue documents without rigorous examination of the laws of technology, permit or prohibit non-conforming products, materials and technologies and even immature techniques, exaggerate common technological problems or grossly negate technologies involving common problems. These market chaos lead to engineering quality accidents and shorten the project life. Because of insulation accidents and consequent poor reputation, some local administrative departments curtly make a wrong decision to use the internal thermal insulation or sandwich insulation instead of external thermal insulation, without thorough analysis, investigation and research, which also causes irreparable losses.

Errors and setbacks have warned us that how important and necessary scientific energy-saving ideas are. Some experts suggest publishing the second edition of the *Theory & Application of External Thermal Insulation Technology*, in order to broadly publicize the mechanism and practical technology of thermal insulation for building energy conservation, achieve scientific administration, reduce the misleading and misjudgment of administrative departments to energy-saving technologies in the field of building energy conservation and emission reduction, control the quality problems of thermal insulation arising from low-cost competition, and strengthen the research on the technologies and theories of building energy conservation.

In order to add recent achievementsto this book, the second edition is compiled in one and a half years, involving the efforts of experts as well as scientific and technical personnel.

The following contents are mainly added based on the theory in the first edition: the influence of the temperature stress and temperature field on the motion of the building structure, researched by Professor Zhang Jun from Tsinghua University; the analysis of 48 large-scale weathering tests of the external thermal insulation system, conducted by the R&D department of Beijing Zhenli; the mechanism of preventing the external thermal insulation structure from damage caused by moisture and phase change; the destructive effect of negative wind pressure on the internal cavity of the external thermal insulation layer; and the verification of damage of the low-elastic-modulus wall in the earthquake.

At the same time, the contents on external thermal insulation are increased in the second edition. Engineering cases of external thermal insulation in Chapter 11 are seven representative works selected from 120 essays. They describe external thermal insulation accidents in recent years, analyze the causes of differences in the structural design, material temperature and construction method based on cases, and also provide solutions related to the design ideas, structural practice adjustment and construction method improvement.

The new Chapter 9 mainly introduces the technical connotation of the mineral binder and expanded polystyrene granule insulation and the structural design principle of the mineral binder and expanded polystyrene granule composite insulation board system, and presents the technical means to control cracking in the external thermal insulation layer and crack-resistant layer. This chapter, with a close combination of the technical theory and application, is also a highlight of the second edition.

Main personnel engaged in compilation of the second edition:

Chapter 1: written by Lin Yancheng, and revised by Huang Zhenli, Tu Fengxiang and Jin Hongxiang;

Chapter 2: written by Wang Chuan and Zhang Jun, and revised by Huang Zhenli and Jin Hongxiang;

Chapter 3: written by Wang Chuan and Zhang Jun, and revised by Huang Zhenli and Jin Hongxiang;

Chapter 4: written by Wang Chuan and Zhang Jun, and revised by Huang Zhenli and Jin Hongxiang;

Chapter 5: written by Wang Chuan, and revised by Huang Zhenli, Jin Hongxiang, Zhu Chunling and Ji Guangqi;

Chapter 6: written by Wang Chuan, and revised by Huang Zhenli and Jin Hongxiang;

Chapter 7: written by Wang Chuan, and revised by Huang Zhenli and Jin Hongxiang;

Chapter 8: not amended;

Chapter 9: written by Fu Haiming, and revised by Huang Zhenli and Jin Hongxiang;

Chapter 10: not amended;

Section 11.1: written by Hong Mei, and revised by Huang Zhenli, Fang Zhanhe, Wang Qingsheng, Jin Hongxiang and Tu Fengxiang;

Section 11.2: written by Gao Wei and Lin Yancheng, and revised by Huang Zhenli, Fang Zhanhe, Wang Qingsheng, Jin Hongxiang and Tu Fengxiang;

Section 11.3: written by Lin Yancheng, and revised by Huang Zhenli and Jin Hongxiang;

Section 11.4: written by Lin Yancheng, Luo Shuxiang, Sun Guifang, Qiu Junfu and Wang Yongkui, and revised by Huang Zhenli, Fang Zhanhe, Wang Qingsheng and Jin Hongxiang;

Section 11.5: written by Huang Kai and Lin Yancheng; revised by Huang Zhenli, Feng Jinqiu, Li Dongyi and Jin Hongxiang;

Section 11.6: written by Zhou Hongyan and Lin Yancheng, and revised by Huang Zhenli and Jin Hongxiang;

Section 11.7: written by Sun Jiajin and Lin Yancheng, and revised by Huang Zhenli and Jin Hongxiang;

Chapter 12: written by Tu Fengxiang, and revised by Jin Hongxiang.

Author
April 2015

Preface of 1st Edition

Buildings consume about one third of the world's natural resources and energy. With the improvement of people's living conditions, the proportion of China's building energy consumption to its total social energy consumption is increasing year by year. This proportion has exceeded 1/4 at present, and will reach a higher level in the near future. Therefore, it is imperative to promote building energy conservation. External thermal insulation is an essential part of building energy conservation, so the research on external thermal insulation will certainly make useful contributions to the progress of building energy-saving technologies.

Over 20 years of development, China has made great achievements in external thermal insulation. Under the guidance of China's basic national policy for energy conservation and emission reduction, great importance must be attached to the quality of external thermal insulation to ensure its safety and service life, in addition to the conformity to thermal insulation, energy conservation, environmental protection and low-carbon economy. A number of experts have made unremitting efforts in this respect.

The Theory & Application of External Thermal Insulation Technology is a product of the development of external thermal insulation technology. This book, based on over ten years of engineering practice and a large amount of experimental data, studies the impact of five natural destructive forces (thermal stress, fire, wind, water/vapor and seismic force) on the wall subject to external thermal insulation, and summarizes the laws of development of the external thermal insulation technology. In combination with engineering practices, this book also studies and discusses the technical theory of external thermal insulation, making people pay more attention to the performance indexes (safety, durability, etc.) and overall service life of external thermal insulation instead of the thermal resistance of walls.

In this book, finite element method and finite difference method are applied to establish a mathematical model of temperature field thermal stress for numerical simulations, and the impact of the insulation structure and measures on the building stability and safety is studied. Among the five natural destructive forces, the hygrothermal stress and fire hazard have the most severe impact on and damage to the external thermal insulation of the building. Starting from the fire protection with the external thermal insulation system as the focus, this book accumulates and analyzes 1.2 million data collected in combustion testing, and provides technical data of the fire resistance of the external thermal insulation system as well as the applicable building height. The impact of and damage caused by hygrothermal stress should not be underestimated. Hygrothermal stress may not only cause wall cracking and bulging, but also finish falling to result in damage to properties or injuries of people on the ground. This book, combining the theory and practice, studies and discusses the exterior finish tile practice with the greatest impact. In addition, this book analyzes the comprehensive utilization of resources in external thermal insulation, and reflects the idea of integrating the external thermal insulation with low-carbon economy, as the low-carbon green technology is helpful for building energy conservation.

The research on and application of building energy conservation and external thermal insulation require the attention of people from all walks of life, and the industry-university-research support and cooperation. We should make endeavors to achieve faster and better development of building energy conservation and external thermal insulation on the path of independent innovation.

Main personnel engaged in the writing and review of this book:

Chapter 1: written by Lin Yancheng, Zhang Jun, Ren Lin and Song Changyou; and edited by Liang Junqiang, Tu Fengxiang, Jin Hongxiang, Yang Xiwei, Wang Qingsheng, Feng Baochun, Sun Kefang, Gu Taichang, Zhu Genli, Liu Xiaojun, Fang Zhanhe, You Guangcai, Zheng Xiangqin and Wang Guojun.

Chapter 2: written by Zhang Jun, Wang-Mansheng, Liu Feng and Wu Xihui; and edited by Lin Haiyan, Jin Hongxiang, Tu Fengxiang, Yang Xiwei, Xu Jinfeng, Feng Ya, Tian Lingjiang, Liu Younong, Zhang Jusong, Zhang Shujun and Sun Sihai.

Chapter 3: written by Ju Shibao and Liu Feng; and edited by Liu Nianxiong, Qin Youguo, Liu Jiaping, Zhu Yingbao, Jin Hongxiang, Tu Fengxiang, Yang Xiwei, Su Xianghui, Zhao Lihua, Hao Bin, Liu Jingjiang, Tang Liang, Han Hongwei, Yang Honghai and Chen Danlin.

Chapter 4: written by Hu Yongteng, Liu Feng and Zheng Jinli; and edited by Song Bo, Feng Jinqiu, Yang Xiwei, Tu Fengxiang, Jin Hongxiang, Sun Hongming, Fang Ming, Zhang Guoxiang, Wei Yongqi, Zhang Jianfeng, Liu Huaiyu and Xu Weidong.

Chapter 5: written by Zhu Chunling, Ji Guangqi, Zhang Ming, Hu Yongteng, Zhang Leilei and Cao Dejun; and edited by Liang Junqiang, Ma Heng, Cui Ronghua, Wang Guohui, Zhao Chenggang, Jin Hongxiang, Tu Fengxiang, Yang Xiwei, Ding Wanjun, Liu Zhendong, Jia Dongmei, Wang Shaojun, Chen Xing, Su Dan, Mei Yingting, Cao Yongmin, Liang Xiaonong, Wu Gang, Wang Jian, Wu Xihui and You Xiaofei.

Chapter 6: written by Liu Feng and Zou-Haimin; and edited by Jin Hongxiang, Tu Fengxiang, Yang Xiwei, Wang Chuntang, Li Dongyi, Li Xiaoming, Li Xikuan, Jiang Wei, Tao Siji, Li Jinbao, Qu Zhenbin, Qin Zheng, An Yanhua, Guo Li and Lin Guohai.

Chapter 7: written by Li Wenbo, Liu Feng and Sun Xiaoli; edited by Jin Hongxiang, Tu Fengxiang, Yang Xiwei, Wang Mansheng, Qin Gang, Jiang Chenggui, Zhu Chuansheng, Li Haojie and Ma Yunyu.

Chapter 8: written by Fu Haiming, Zhang Lei, Xu Junfeng, Liu Feng, Liu Xiangzhi, Chen Zhihao and Li Chunlei; and edited by Jin Hongxiang, Tu Fengxiang, Yang Xiwei, Yang Huizhong, Lu Shanhou, Liu Xiaozhong, Zheng Dejin, Chen Dianying, Zhang Ruijing, Wang Hanyi, Wang Gongshan, Wang Zhenqing and Su Dongqing.

Chapter 9: written by Wang Chuan, He Liudong and Sun Xiaoli; and edited by Jin Hongxiang, Tu Fengxiang, Yang Xiwei, Luan Jingyang, Hu Xiaoyuan, Liu Zhenxian, Liao Libing, Yu Qingshan and Zheng Yibo.

Chapter 10: written by Tu Fengxiang; and edited by Jin Hongxiang, Liang Junqiang, Ye Jincheng and Zhao Shiqi.

This book can be used as a reference for researchers, designers and construction technicians engaged in external wall insulation.

There are still many problems to be solved despite rapid development of the external thermal insulation technology. All criticism and correction of inadequacies and errors in this book will be appreciated.

Author
March 2011

Contents

1. Overview

1.1 Development of External Thermal Insulation Technology

1.1.1 Status Quo of Buildings Energy Consumption in China

The relevant national statistics show that over 90% of China's fossil energy resources are coal, of which the per capita reserve is 1/2 of the world's average. The per capita reserve of oil is 11% of the world's average, and that of natural gas is only 4.5%. China's coal consumption accounts for 40% of the world's total, and the oil consumption ranks only second to the United States. China's dependence on overseas energy is more than 50%.

Despite its energy storage, China's energy consumption and greenhouse gas (CO_2, etc.) emissions rank first in the world along with its sustained and rapid economic development, resulting in severe environmental pollution and ecological damage. The global warming and gradual depletion of fossil energy have aroused increasingly acute contradictions of the world's economic and social sustainable development and the desires and efforts to achieve modernization in developing countries (including China) against the sustainability of the ecological environment and the ability to supply natural resources on Earth. It is of great importance for the mankind to reduce greenhouse gas emissions, alleviate global warning, protect the living environment, reduce the ecological damage, and save and effectively use fossil fuels. It is proposed in the *China-U.S. Joint Statement on Climate Change* signed in the Beijing APEC Summit on December 12, 2014 that China plans to achieve and will make efforts to achieve in advance the peak of CO_2 emissions approximately in 2030, and also plans to increase the proportion of non-fossil energy consumption to primary energy consumption to roughly 20%. China's commitment is encouraging to the world. This is also an international obligation that China should undertake.

The "energy conservation and emission reduction" has become one basic national policy of China. It is cold in winter and hot in summer in China. Due to large-scale construction and high energy consumption of heating and air-conditioning, the building energy consumption account for about 30% of the total energy consumption. If that of building materials of material production and building construction, the energy consumption of the building field has accounted for 46% of the social terminal energy consumption and exceeded the industrial energy consumption. With the rapid development of China's urbanization and continuous improvement of people's living standards, the total building energy consumption will still increase, which will not only need a lot of coal but also lead to air pollution and ecological destruction. We must work together to vigorously reduce building energy consumption and promote the technology and industry of energy conservation and emission reduction. The energy conservation of buildings is an urgent need to guarantee the sustainable development of China's national economy and the continuous improvement of people's living standards, and also a key field of energy conservation and emission reduction. It has been carried out all over China. The energy-saving design and construction have been comprehensively implemented to new buildings, and the energy-saving transformation to a large number of existing buildings of high energy consumption as planned, making significant contribution to save energy, protect the environment, respond to the global climate change and also promote the national, economic and social sustainable development.

The energy consumption of building envelope structures accounts for a large proportion to building energy consumption, while external walls play an important role in envelope structures. Therefore, the external wall insulation is essential to reduce building energy consumption. With the continuous improvement of energy-saving requirements for building, the external wall insulation technology is also developed considerably. Depending on the location relationship between the base wall and insulation, the external wall insulation can be divided into exterior insulation, interior insulation, sandwich insulation and self-insulation. Viewed from the history of wall insulation development at home and abroad, regardless of theoretical analysis or engineering practice verification, the external thermal insulation is the most reasonable for energy-saving external walls.

1.1.2 Development of Overseas External Thermal Insulation Technology

The external thermal insulation technology originated from Europe in the 1940s. The expanded polystyrene board (EPS board) was invented in Germany in 1950, and applied to external thermal insulation in 1957. The thin-plastered EPS board type external thermal insulation system of real engineering significance was developed successfully in 1958, and became popular in Europe in the 1960s, subject to the first weathering test.

The external thermal insulation technology was originally used to repair cracks in exterior walls of buildings damaged in the Second World War. Practical applications show that the slabs pasted on building walls effectively cover cracks of exterior walls and the composite walls have good insulation properties. Meanwhile, the external composite light insulation system of the heavy wall is the most reasonable combination of wall structures, which not only solves the problem of insulation and also reduces the corresponding wall thickness and civil construction costs. In addition, the composite wall has the optimal performance in terms of sound insulation, moisture resistance and thermal comfort, while satisfying the structural requirements.

In the 1960s, the United States introduces the external thermal insulation technology from Europe, and improved it based onits specific climatic conditions and building system characteristics. As the requirements for building energy conservation increased, the applications of external thermal insulation and decoration systems continued to increase. By the end of the 1990s, the average annual growth rate had reached 20% to 25%. The external thermal insulation technology has been widely applied from hot areas in the south to cold ones in the north of the United States, with remarkable effects. In addition to the thin-plastered EPS boards, most of the external thermal insulation systems in the United States are made of insulation material filled with the light steel or wood structure, meeting high requirements for fire resistance.

The rapid development of external thermal insulation technology occurs after theworld energy crisis in 1973. Because of the energy shortage, the market capacity of the external thermal insulation technology, driven by the vigorous promotion of governments in Europe and America, has been increasing at a rate of 15% per year. In decades of application, Europe and America have conducted a large number of experimental researches on the external thermal insulation technology, including the durability, fire safety and moisture content change of the thin-plastered external thermal insulation, condensation in cold areas, responses of systems under different impact loads, correlation between the performance test results of the laboratory and the actual performance in engineering applications, etc.

On the basis of a large number of experimental researches, European and American countries have carried out legislative work on the external thermal insulation technology, including the formulation of the compulsory certification standards for the external thermal insulation system, technical standards for materials related to this system, etc, Due to the sound standards and rigorous legislation, the 25-year durability and service life of the external thermal insulation system can be guaranteed in these countries. Actually, the application history of this system in the above-mentioned areas is far more than 25 years, and that of the earliest projects is more than 50 years. In 2000, the European Organization for Technical Approvals (EOTA) released the standard ETAG 004 "Guideline for European Technical Approval of External Thermal Insulation of Composite Systems with Rendering", which is the technical summary and specification of successful practices of the external thermal insulation technology in Europe.

Up to now, the external insulation board and thin plastering technology is stilled widely applied in Europe, mainly involving two insulation materials: flame-retardant EPS board and non-combustible rock wool board. Typically, the exterior finish is completed with paint. Take Germany as an example. The applications of thin-plastered EPS boards account for 82%, while those of thin-plastered rock wool boards for 15%. The external thermal insulation technology has a history of 70 years, but its research, application and development are conducted faster in the last 50 years. This technology is more and more mature and excellent.

1.1.3 Development of Domestic External Thermal Insulation Technology

The research on China started the building energy conservation projects relatively late starts late in China, so its the application and development of the external thermal insulation technology is much later than those of Europe, the United State and other countries. In 1986, China enacted the standard "Energy Conservation Design Standard for Civil Buildings (Heating Residential Buildings) " (JGJ 26-86) with the energy-saving rate of 30%, making marking the official start of building energy conservation in China. Previously, China's scientific research, design, construction and building material production units had carried out technical research on various kinds of external thermal insulation and tested the external thermal insulation technology. From the late 1980s to the early 1990s, a variety of external thermal insulation techniques, represented by EPS boards and composite gypsum insulation boards, were applied as main forms of external thermal insulation, due to simple production and construction, low construction costs and satisfaction of the energy-saving rate of 30%. They were primarily applied in North China. In addition, the expanded perlite and composite silicate insulation mortar also occupies some market shares. Practice shows, however, the defects of the internal thermal insulation technology in northern cold and severe cold areas are increasingly exposed and the quality of production and construction is difficult to control, resulting in engineering problems such as condensation under large indoor and outdoor temperature differences, and mildewing of interior walls. Consequently, the internal thermal insulation was gradually eliminated in the market.

In 1995, China released the design standard for building energy conservation, with the energy-saving rate of 50%. The first national conference on building energy conservation, held in 1996, summarized the previous experience and put forward the goal to focus on the promotion of the external thermal insulation. On January 1, 1998, China promulgated and implemented the *Law of the People's Republic of China on Energy Conservation*, clearly stating that "energy conservation is a long-term strategic policy for economic development of China". Since, then, China has intensified the research on and application of the external thermal insulation technology, and independently developed a variety of external thermal insulation systems, including the thin-plastered EPS board insulation system, adhesive polystyrene granule insulation system, cast-in-situ concrete composite meshed/non-meshed EPS board insulation system, and EPS wire mesh post-anchorage insulation system, meeting the requirement for the building energy conservation of 50%.

In the early 21st century, the buildings 65% in Beijing, Tianjin and other cities had successively achieved 65% of energy saving, which promoted the further development of the external thermal insulation technology. China has also independently developed some new external thermal insulation systems, including the sprayed rigid polyurethane foam insulation system and the adhesive polystyrene granule plastered board insulation system. The technical standards and atlases related to external thermal insulation have also been continuously improved and enriched to drive the development of the external thermal insulation technology and industry. A number of external thermal insulation systems have been widely applied into projects. The external thermal insulation association has been established. Relevant organizations have compiled and published monographs such as the *External Thermal Insulation Technology*, *Application Technology of External Thermal Insulation*, *100 Technical Questions & Answers of External Thermal Insulation*, *Exploration of Wall Insulation Technology*, *Construction Method of External Thermal Insulation*, *Quality Problems and Solutions of External Thermal Insulation System*, and *Construction Techniques of Building Energy Conservation Engineering*, discussing the external thermal insulation technology from the perspectives of theory and practice. At the same time, the basic experimental research has been started on the fire and weathering resistance in the industry, and corresponding technical achievements have been made. Related books such as the *Technical Research on Fire Rating Evaluation Standards of External Thermal Insulation System* have been published. With the development of external thermal insulation, new explorations are made based on the full integration of Europe and America in this field as well as the national conditions of China.

It is stipulated in the revised *Law of the People's Republic of China on Energy Conservation* promulgated on April 1, 2008 that "resource saving is one of China's basic national policies, and the energy development strategy of combining conservation and development and giving top priority to conservation shall be implemented." The new law on energy conservation further clarifies the subject of law enforcement and strengthen the legal responsibilities for energy conservation. The *Regulation on Energy Saving for Civil Buildings* formulated by the State Council was implemented on October 1, 2008, specifying that energy consumption should be reduced while the quality of the civil building functions and indoor thermal environment is guaranteed.

In the 2009 World Climate Conference in Copenhagen, the Chinese government made a commitment to cut the CO_2 emissions per unit of GDP by 40% to 45% in 2020. As a responsible power, China will, from the moment of commitment, strive to solve the major problem of coordination between sustained economic growth and energy conservation and emission reduction.

In the past decade, China, based on its national conditions, has not only learnt and introduced foreign advanced technologies, but also studied and developed the thin-plastered insulation board system, insulating decorative board system, rock wool board system, adhesive polystyrene granule system, reinforcing vertical-wire rock wool board system, inorganic insulation mortar system, vacuum insulation board system, etc., fundamentally forming a complete set of external thermal insulation system to meet the external thermal insulation requirements of the design standards of building energy conservation.

In 2013, Beijing released and implemented the energy-saving design standards for residential buildings, with the energy-saving rate of 75%. In 2015, the design standards for building energy conservation were released and implemented in Shandong, etc. China's building energy conservation has reached the advanced level of developed countries, and the development of the external thermal insulation technology in China has basically kept pace with that of developed countries. Undoubtedly, China's construction scale and building energy conservation scale are leading in the world.

1.2 Overview of China's Building Energy Conservation Standardization

1.2.1 Design Standards of Building Energy Conservation

China started the control of building energy in the 1980s and has formulated a number of standards. The *Specification for Thermotic Design of Civil Building* (JGJ 24-86) and *Energy Conservation Design Standard for Civil Buildings (Heating Residential Buildings)*, with the energy-saving rate of 30%, was promulgated in 1986; the *Design Code for Heating, Ventilation and Air-conditioning* (GBJ 19-87) in 1987; and the *Thermal Design Code for Civil Building* (GB 50176-1993) in 1993. Such standards play an important role in saving energy and improving the environment as well as economic and social benefits.

In 1995, the *Design Code for Energy Conservation of Civil Building (Heating Residential Buildings)* (JGJ 26-95) was promulgated and implemented in 1995, increasing the energy-saving rate to roughly 50%; and the *Design Standard for Energy Efficiency of Residential Buildings in Severe Code and Cold Zones* (JGJ 26-2010) in 2010, increasing the energy efficiency of residential buildings in North China to a new level and the energy-saving rate to about 65%. The *Design Standard for Energy Efficiency of Residential Buildings in Hot Summer and Cold Winter Zone* (JGJ 134-2001) was promulgated and implemented in 2001; and the *Design Standard for Energy Efficiency of Residential Buildings in Hot Summer and Warm Winter Zone* (JGJ 75-2003) in 2003, stipulating that the building energy conservation should be executed in both North and South China and the energy-saving rate in South China should be around 50%. Since 2010, the energy-saving design standards for South China have been revised with new related requirements. The *Design Standard for Energy Efficiency of Residential Buildings in Hot Summer and Cold Winter Zone* (JGJ 134-2010) and *Design Standard for Energy Efficiency of Residential Buildings in Hot Summer and Warm Winter Zone* (JGJ 75-2012) have been released successively. In 2005, China issues the *Design Standard for Energy Efficiency of Public Buildings* (GB 50189-2005), specifying that the total energy consumption of heating, ventilation, air-conditioning and lighting in one year should be reduced by 50% under the same indoor conditions, compared with those without energy-saving measures. The revision of this standard is about to be released. In addition to the above-mentioned energy-saving design standards, some provinces, municipalities and autonomous regions have also formulated local standards. Beijing issues the *Design Standard for Energy Efficiency of Residential Buildings* (DB11/891-2012) in 2012, in which the energy-saving rate of residential buildings is increased to 75% for the first time.

1.2.2 Construction Standards of Building Energy Conservation

In order to effectively implement the design

standards of building energy efficiency, China has compiled and published corresponding structural atlas, mainly including the *Building Structure of External Thermal Insulation of Outer Walls* (02J121-1, 99J121-2, 06J121-3 and 10J121), *Building Structure of Internal Thermal Insulation of Outer Walls* (03J122 and 11J122), *Building Structure of Wall Energy Saving* (06J123), *Energy-saving Renovation of Existing Buildings* (*I*) (06J908-7), *Energy-saving Structure of Public Buildings* (*severely cold and cold areas*) (06J908-1), *Energy-saving Structure of Public Buildings* (*hot summer and cold winter zone and hot summer and warm winter zone*) (06J908-2), *Roof Energy-saving Structure* (06J204), *Energy-saving Engineering Practice and Data of Building Envelope* (09J908-3), *Housing Construction Method Diagram* (*I*) - *External Thermal Insulation Method* (11CJ26/11CG13-1), etc. Provinces, municipalities and autonomous regions have compiled appropriate structural atlas of building energy conservation, such as Beijing's *External Thermal Insulation* (08BJ2-9), *Energy-saving Structure of Public Buildings* (88J2-10), *External Thermal Insulation of Class A Incombustible Material* (12BJ2-11), and *External Thermal Insulation of Buildings* (*energy-saving rate*: *75%*) (13BJ2-12). Also, China has released the related inspection and construction quality acceptance standards, such as the *Standard for Energy Efficiency Inspection of Heating Residential Buildings* (JGJ 132-2001), *Energy Efficiency Test Standard for Residential Buildings* (JGJ/T 132-2009), *Standard for Energy Efficiency Test of Public Buildings* (JGJ/T 177-2009), and *Code for Acceptance of Energy Efficient Building Construction* (GB 50411-2007). Some provinces and municipalities have compiled the local standards for building energy-saving quality acceptance, such as Beijing's *Specification for Insulation Constructional Quality Acceptance of Residential Building* (DBJ 01-97-2005), *Specification for Acceptance of Public Building Construction Quality of Energy Efficiency* (DB11 510-2007) and *Standard for On-site Testing of Energy Efficient Civil Building Engineering* (DB11/T 555-2008).

Before 2003, some quality problems aroused in external thermal insulation projects in, such as the cracking of the protective layer, hollowing and shedding of tiles, permeation of rainwater into the internal surface of the outer wall, and blowing-off of the external thermal insulation layer in serious cases. In order to standardize the technical requirements for external thermal insulation projects, guarantee the engineering quality, and achieve the advancement, safety, reliability and economical rationality, the *Technical Specification for External Thermal Insulation on Walls* (JGJ 144-2004) was compiled by research institutes and enterprises of the industry, under the leadership of the Science and Technology Development Promotion Center of the Ministry of Housing and Urban-Rural Development. The purpose of this standard is to guide the research on and application of the external thermal insulation technology in China by drawing on mature experience of advanced countries, and also to control the quality of external thermal insulation projects and promote the healthy development of the external thermal insulation industry. Five external thermal insulation systems are included in this standard, namely, the thin-plastered EPS board insulation system, adhesive polystyrene granule insulation system, EPS board and cast-in-situ concrete insulation system, EPS wire mesh and cast-in-situ concrete insulation system, and mechanically fixed EPS wire mesh insulation system. This standard stipulates the basic, structural, technical and acceptance requirements for external thermal insulation, as well as the performance, test method, design and construction requirements of the external thermal insulation system and its materials. It is one of the most essential engineering standards in the industry of external thermal insulation, with feasible and guiding effects on the existing and new external thermal insulation systems, and the escorting effects on the development of external thermal insulation. At present, its revision has been launched to add new external thermal insulation systems and focus on the fire safety of external thermal insulation. With the development of external thermal insulation, technical codes or specifications have been released successively, such as the *Technical Code for Rigid Polyurethane Foam Insulation and Waterproof Engineering* (GB 50404-2007), *Technical Specification for Thermal Insulating Systems of Inorganic Lightweight Aggregate Mortar* (JGJ 253-2011), *Technical Specification for Fire Barrier Zone of External Thermal Insulation Composite System on Walls* (JGJ 289-2012), *Technical Specification for Interior Thermal Insulation on External Walls* (JGJ/T 261-2011), and *Technical Specification for Energy Efficiency Retrofitting of Existing Residential Buildings* (JGJ/T 129-2012).

1.2.3 Standards of Building Energy Conservation Products

In order to promote the development of building energy conservation, especially the healthy development of the external thermal insulation, China has formulated system product standards, as well as supporting product standards and system test standards, and gradually improved the system of standards, as shown in Table 1-2-1.

Three product standards with the most important influence on the external wall insulation are briefly introduced below.

1.2.3.1 *External Thermal Insulation Composite System based on Expanded Polystyrene* (JG 149-2003)

The *External Thermal Insulation Composite System based on Expanded Polystyrene* (JG 149-2003) is China's first industry standard for external thermal insulation system products. It involves the non-equivalent adoption of the *Guideline for European Technical Approval of External Thermal Insulation Composite System with Rendering* (EOTA ETAG 004), *External Thermal Insulation Composite System of Expanded Polystyrene Foam Plastic and Surface Course* (NORM B6110), *Standard for External Thermal Insulation Composite System based on Expanded Polystyrene* (CEN/TC 88/WG18 N 166) and *Standard for Acceptance of Exterior Insulation and Finish System* (ICBO ES AC24). Some technical performance indicators are adjusted according to the actual conditions in China. As for the test method, the following standards are applied in a non-equivalent way: *Standard Test Method for Quick Deformation and Impact Resistance of Exterior Insulation and Finish System* (EIMA 101.86), *Test Method for Quick Deformation (impact) Resistance of Organic Coating* (ASTM D 2794-93), *Thermal Insulation Products for Building Applications-Impact Resistance of External Thermal Insulation Composite System - Specification* (PrEN 13497), *Standard Test Method for Freeze-Thaw Resistance of Exterior Insulation and Finish Systems* (EIMA101.01), *Standard Test Method for Evaluating the Tensile-Adhesion Performance of Exterior Insulation and Finish System* (ASTM E 2134-01), *Thermal Insulation Products for Building Applications-Determination of Tensile Bond Strength of Adhesive and Base Coat to Thermal Insulation* (PrEN 13494), *Standard Test Method for Determining Tensile Breaking Strength of Glass Fiber Reinforcing Mesh for Use in Class PB Exterior Insulation and Finish System after Exposure to Sodium Hydroxide Solution*, (ASTM E 2098-00), and *Thermal Insulation Products for Building Applications-Determination of Mechanical Properties of Glass Fiber Meshes* (PrEN 13496). The revised standard has been completed and issued as the national standard GB/T 29906-2013.

Standards Related to External Thermal Insulation System Products, Supporting Products and System Test Methods

Table 1-2-1

Category	Standard No.	Standard Name	Remarks
System product standard	GB/T 29906-2013	External Thermal Insulation Composite Systems based on Expanded Polystyrene	Upgraded version of JG 149-2003
	GB/T 30595-2014	External Thermal Insulation Composite Systems based on Extruded Polystyrene(XPS)	
	GB/T 30593-2014	External Wall Interior Insulation Composite Panel System	
	JG/T 158-2013	Products for External Thermal Insulation Systems based on Mineral Binder and Expanded Polystyrene Granule Plaster	Revised version of JG 158-2004
	JG/T 228-2007	Technical Requirements of External Thermal Insulation with Expanded Polystyrene Panel for In-situ Concrete	The revision has been reviewed and will be released soon
	JG/T 287-2013	Materials of External Thermal Insulation Systems based on Insulated Decorative Panel	
	JG/T 420-2013	External Thermal Insulation Composite Systems based on Rigid Polyurethane Foam	
	JG/T 469-2015	External Thermal Insulation Composite Systems based on Cellular Glass	
Single product standard	GB 26538-2011	Fired Heat Preservation Brick and Block	
	GB/T 29060-2012	Composite Thermal Insulation Brick and Block	
	GB/T 10801. 1-2002	Rigid Extruded Polystyrene Foam Board for Thermal Insulation	
	GB/T 10801. 2-2002	Rigid Extruded Polystyrene Foam Board for Thermal Insulation(XPS)	
	GB 26540-2011	EPS Board with Metal Network for Exterior Insulation and Finish Systems	
	GB/T 19686-2005	Rock Wool and Slag Wool Thermal Insulating Products for Building	
	GB/T 11835-2007	Rock Wool, Slag Wood and Products for Thermal Insulation	
	GB/T 25975-2010	Rock Wool Products for Exterior Insulation and Finish Systems(EIFS)	
	GB/T 20219-2006	Spray-applied Rigid Polyurethane Cellular Plastics	
	GB/T 21558-2008	Rigid Polyurethane Cellular Plastics for Thermal Insulation of Buildings	
	GB/T 20473-2006	Dry-mixed Thermal Insulating Composition for Buildings	
	GB/T 26000-2010	Thermal Insulating Mortar Mixed with Expanded and Vitrified Beads	
	GB/T 20974-2014	Rigid Phenolic Foam for Thermal Insulation(PF)	
	JG/T 407-2013	Self-insulation Concrete Compound Blocks	

Continued

Category	Standard No.	Standard Name	Remarks
Single product standard	JG/T 314-2012	Rigid Polyurethane Cellular Composite Insulation Board	
	JG/T 360-2012	Metal Decorative Insulation Board	
	JG/T 432-2014	Composite Insulation Board for Buildings	
	JG/T 435-2014	General Specification of Inorganic Lightweight Aggregate Board for Fireproofing and Insulation	
	JG/T 438-2014	Vacuum Insulation Panels for Buildings	
	JG/T 366-2012	Anchors for Fixing of Thermal Insulation Composite Systems	
	JG/T 229-2007	Waterproofing Flexible Putty for Exterior Thermal Insulation Systems	
	JG/T 206-2007	Environmental-friendly Silicon-acrylic Emulsion Multi-layer Coatings for External Thermal Insulation	
	JC/T 992-2006	Expanded Polystyrene Board Adhesive for Base Course Wall Thermal Insulation	
	JC/T 993-2006	Expanded Polystyrene Board Base Coat for External Thermal Insulation	
	JC/T 2084-2011	Mortar for External Thermal Composite Systems based on Extruded Polystyrene	
System test method standard	GB/T 29416-2012	Test Method for Fire-resistant Performance of External Wall Insulation Systems Applied to Building Facade	
	JG/T 429-2014	Method for Weathering Resistance Test of External Thermal Insulation	

1.2.3.2 *External Thermal Insulation Systems based on Mineral Binder and Expanded Polystyrene Granule Plaster* (JG 158-2004)

The *External Thermal Insulation Systems based on Mineral Binder and Expanded Polystyrene Granule Plaster* (JG 158-2004) is China's second industry standard of external thermal insulation products. Combining the mineral binder and expanded polystyrene granule insulation system with independent intellectual property rights based on China's national conditions, this standard involves the non-equivalent application of the *Rendering Systems for Thermal Insulation Purposes Made of Mortar Consisting of Mineral Binders and Expanded Polystyrene (EPS) as Aggregate* (Part 3 of DIN 18550) and the *Guideline for European Technical Approval of External Thermal Insulation Composite System with Rendering* (EOTA ETAG 004). Some technical performance indicators of materials are adjusted or added according to the actual situation in China. This standard has good application effects. Accordingly, solid waste such as fly ash and used polyphenylene are fully consumed, the advantages of comprehensive utilization of resources are demonstrated, and the nationwide rampant white pollution caused by used polystyrene boards is quickly eliminated by changing such boards into resources, thus promoting the development of external thermal insulation in China, and attaching great importance to China's design standard for the building energy-saving rate of 50%. Through years of application and practice, new progress has been made in the adhesive polystyrene granule insulation, not only in the fire safety and technical performance but also in the structural practice. The composite insulation technology of six-sided or five-sided cladding of polyphenyl boards with adhesive polystyrene granule has been developed, which has expanded the application scope of adhesive polystyrene granule, and met high requirements for energy conservation. The revision of the above-mentioned standard has been released and renamed as the *Products for External Thermal Insulation Systems based on Mineral Binder and Expanded Polystyrene Granule Plaster* (JG/T 158-2013), including the requirements for the external thermal insulation composite systems based on the mineral binder and EPS board.

1.2.3.3 *Technical Requirements of External Thermal Insulation with Expanded Polystyrene Panel for In-situ Concrete* (JG/T 228-2007)

The *Technical Requirements of External Thermal Insulation with Expanded Polystyrene Panel for In-situ Concrete* (JG/T 228-2007) is China's third industry standard for external insulation systems. The external thermal insulation system with EPS board and cast-in-situ concrete in external molds was independently developed by China, with its material properties, structure and construction method at the international advanced level. In this system, the insulation layer is secured by one-time pouring of the EPS boards and concrete wall and closely integrated with the wall. It can be divided into the external thermal insulation system with vertical grooved EPS boards and cast-in-situ concrete in external molds and that with wire-mesh EPS boards and cast-in-situ concrete in external molds. A fireproof and air-permeable transition layer, which is a special functional layer composed of the adhesive polystyrene granule, is added between the EPS board insulation layer and anti-cracking protective layer, thus improving the fire resistance, air permeability and weathering resistance of the insulation system, and facilitating the transition of the thermal conductivity and correction of construction errors. With the implementation of the above-mentioned standard, the external thermal insulation with EPS boards and cast-in-situ concrete in external molds is regulated effectively, and the fire safety of the external thermal insulation system is improved. Currently, this standard has been revised, including the requirements for the external thermal insulation system with cast-in-situ concrete and composite extruded polystyrene board (XPS board), as well as the structure of the fire barrier zone. It is recommended to use the reinforced vertical-wire composite rock wool board of high performance in the fire barrier zone. The revision of this standard has been reviewed in December 2014, and will be approved and issued in 2015.

1.3 Theoretical Research on External Thermal Insulation Technology at Home and Abroad

1.3.1 Basic Points of Theory of External Thermal Insulation Technology

Main basic points of theoretical research on the external thermal insulation technology:

(1) The external thermal insulation should adapt to normal deformation of the base, without cracking or hollowing.

(2) The external thermal insulation should be able to withstand its dead load, wind load and long-term outdoor climatic effects, without harmful deformation or damage.

(3) The external thermal insulation should be reliable connected with the base course wall, without falling during the earthquake.

(4) The external thermal insulation should be able to prevent flame from spreading.

(5) The external thermal insulation should be waterproof.

(6) The wall subject to composite external thermal insulation should have good insulation performance and moisture resistance.

(7) All parts of the external thermal insulation should have physical-chemical stability.

(8) The external thermal insulation should have durability and long service life.

Accordingly, the external thermal insulation is a science studying the impact of five natural destructive forces (thermal stress, water, fire, wind and earthquake) on building walls.

(1) Thermal stress: insulation practices have significant effects on the temperature change inside the structural layer. If the external thermal insulation is applied on the outer wall, the temperature of the building structure changes little; and if the internal thermal insulation, sandwich insulation and self-insulation is applied on the outer wall, the temperature of the building structure changes greatly, and the resulting thermal stress is one important factor of cracking of the building structure. Currently, attention should be paid to the damage of thermal stress arising from the unreasonable location of the wall insulation to the building structure. Human and material resources need to be invested in research on the thermal insulation system. Additionally, surface cracking and stripping arising from thermal stress of the wall surface with the unreasonable performance design of insulation materials should be resolved in the external thermal insulation engineering.

(2) Water: there are three forms of water in the natural environment. The change in three phases of water, motion and migration of various forms of water inside and outside the external thermal insulation system and even the phase change inside this system will have significant impact on the durability and function of the external thermal insulation system. The water resistance, permeability and anti-condensation of the external thermal insulation system are an important aspect of basic theoretical research of external thermal insulation.

(3) Fire: the proportion of high-efficiency organic insulation materials in the external thermal insulation system is more than 80% at present, which results in poor fire safety of the external thermal insulation system and fire during construction of the external thermal insulation. With the design standards of building energy conservation and increase of the building height, the fire safety plays an increasingly important role. Thus, the fire safety of the external thermal insulation system is a vital technical requirement for reasonable applications of the external thermal insulation. This book will discuss the method of properly viewing and reasonably implementing fire protection in the external thermal insulation, the means of technical research on the fire safety of the external thermal insulation, and the necessity to classify its fire safety.

(4) Wind: it is not uncommon that the external thermal insulation is blown off by strong winds in actual projects. In particular, the safety after destruction of the external thermal insulation system with finishing tiles is a technical problem that must be addressed.

(5) Earthquake: the earthquake usually affects the external thermal insulation system with finishing tiles. The external thermal insulation system is a non-bearing structure attached to the outer wall. That is, it does not bear the load or seismic action on the main structure. However, it is necessary to study the deformation capacity of the external thermal insulation system corresponding to the displacement of the main structure, in order to prevent this system from excessive internal stress and unacceptable deformation in the case of displacement of the main structure under large earthquake loads.

Additionally, the focuses of the industry are mentioned in this book, such as the safety of facing bricks, technical requirements for the facing brick finish, application of solid wastes in an energy-saving and environmentally-friendly way in the external thermal insulation, and energy consumption or pollution of external thermal insulation materials in the production process.

1.3.2 Progress of Theoretical Research on External Thermal Insulation Technology

Currently, China's research on external thermal insulation technology is carried out in depth: from the lack of basic research data to accumulation of related information; from direct introduction of foreign advanced techniques to research and development in conjunction with China's national conditions; and from independent research and development suitable for China's national conditions to gradual improvement of the technological system. Some gaps of basic research have been filled, such as the impact of thermal stresses of different insulation structures on the external thermal insulation system; the fire safety and weathering resistance test of thermal insulation systems; and the safety of the external thermal insulation system with the facing brick finish.

Despite the solid fundamentals, the basic experimental research on external thermal insulation is still in the exploratory stage. That is, there is a lot of work to do, or the previous results should be enriched or verified or corrected. In short, unremitted efforts should be made in the basic experimental research in order to achieve the long-term development and excellent durability of external thermal insulation. The government should invest in this area to support related researches. Enterprises should also increase the investment to seek their own and industrial sustainable development. Meanwhile, the basic theoretical research and engineering practice should be actively integrated to accelerate the development of external thermal insulation, and provide external thermal insulation products compatible with the building life, thus reducing social costs and saving social resources.

2. Theoretical Research on Temperature Field of Outer Insulated Wall

The outer insulated wall is a composite wall composed of the structural layer and insulation layer. The structural layer, i. e. outer wall body bearing forces, is known as the base course wall, in which concrete or masonry is used the most commonly at present. The insulation layer, i. e. non-bearing layer made of insulation material, is usually built with the polystyrene foam panels and insulation mortar. The base course wall and insulation layer of the composite wall should be bonded by technical means, with the outer surface protected and decorated. Therefore, additional layers such as the bonding layer, protective layer and finishing layer are needed between the base course wall and insulation layer and on the outer surface of the insulation layer.

Based on the relative positions of the insulation layer and base course wall in the composite wall, the external thermal insulation of the building is divided into four types: internal thermal insulation on the indoor side of the base course wall; external thermal insulation on the outdoor side of the base course wall; sandwich insulation between two base course walls; and self-insulation of the base course wall. The internal thermal insulation is located inside the structural layer, so the temperature and its stress of the structural layer of the wall will change greatly. The external thermal insulation is applied outside the structural layer, so the temperature of the structural layer will be relatively stable due to the insulation layer, and the corresponding temperature stress will change little, thus protecting the wall and extending the building life.

The infrared thermal images of projects show that different insulation practices have significant effects on the temperature change of the structural layer of the wall. If the external thermal insulation is applied, the temperature of the structural layer will change little. If the internal thermal insulation, sandwich insulation, integrated external and internal insulation or incomplete external thermal insulation is applied, however, the temperature of the structural layer will change greatly, leading to large temperature stress. This is one of important factors causing the cracking of the wall structure. Therefore, the location of the insulation layer affects both the thermal insulation and the temperature field inside the structural layer. The unreasonable location of the insulation layer will damage the temperature stress of the building structure, so attention needs to be paid to the location of the insulation layer in wall insulation projects.

The external environment, including solar radiation and atmospheric temperature, constantly changes in the actual application of the building. This is why the results of the temperature field distribution of the insulated wall obtained in the conventional steady-state heat transfer analysis method are greatly different from those under actual conditions. With the development of the computer technology and numerical simulation theory, numerical simulation has been increasingly applied to analyze real projects. Research results show that it is feasible and necessary to perform the real-time numerical calculation and analysis of the temperature field of the insulated wall.

The first two sections of this chapter discuss the impact of the location of the insulation layer on the temperature field (taking into account the one-dimensional heat transfer) and temperature stress of the outer wall of the building, and mainly explores the rules of change in the temperature field and temperature stress of the outer walls in different positions (eastern, western, southern and northern) over time, subject to the typical external thermal insulation, internal thermal insulation, sandwich insulation and self-insulation. The rules of the temperature field and temperature stress distribution of the outer walls with various insulation layers are obtained by means of theoretical calculation. The comparison results of the temperature field and temperature stress of the base course wall indicate that the external thermal insulation is the most beneficial to the building life.

In the third section of this chapter, the two-dimensional temperature field and temperature stress of some extension (thermal bridges) of the outer wall subject to external thermal insulation are calculated with the ANSYS software. It is concluded that the external thermal insulation must be applied on extensions; otherwise, the wall may be hollowing, cracking, shedding, etc.

2.1 Numerical Simulation of Temperature Field of Insulated Outer Wall

2.1.1 Calculation Model for Temperature Field of Insulated Outer Wall

2.1.1.1 Heat Conduction Equation

The calculation of the temperature field inside the insulated wall under complex boundary conditions such as the atmospheric temperature change, solar radiation and air convection is fundamental for the stress and durability analysis of the insulated wall. This chapter establishes a calculation model for the temperature field of a typical insulated wall under atmospheric temperature changes, involving the external thermal insulation, internal thermal insulation, sandwich insulation and self-insulation of the outer wall. Wall structures in four directions and under typical climatic conditions of each season were selected are studied. For the convenience of comparison, a calculation model is also established for the ordinary wall without insulation, assuming that the wall has a consistent and continuous multi-layer composite structure, different layers are bonded tightly and the thermal resistance between layers is ignored. The temperature (T) at any time (t) in any position (x, y or z) of the wall meets the following requirements:

$$\frac{\partial T}{\partial t}=\frac{\lambda}{c\rho}\left(\frac{\partial^2 T}{\partial x^2}+\frac{\partial^2 T}{\partial y^2}+\frac{\partial^2 T}{\partial z^2}\right) \tag{2-1-1}$$

Where, λ —thermal conductivity, kJ/(m · h · ℃);

c — specific heat of the material, kJ/(kg · ℃);

t —time, h;

ρ —material density, kg/m^3.

For the outer wall (except the door, window, extension structure and thermal bridge, of which the heat transfer process is subject to numerical simulation with the ANSYS software in Section 2.3 of this chapter) of the building, its internal temperature changes little in the length (y) and width (z) direction, i.e. $\partial T/\partial y = \partial T/\partial z \approx 0$, but drastically in the thickness direction. So, the heat conduction equation of the outer wall under normal conditions can be simplified into a one-dimensional heat conduction equation in the wall thickness direction, namely:

$$\frac{\partial T}{\partial t}=\frac{\lambda}{c\rho}\frac{\partial^2 T}{\partial x^2} \tag{2-1-2}$$

The solution of the temperature field in the thickness direction inside the wall is equivalent to that of the partial differential equation (2-1-2) under the given boundary conditions and time.

2.1.1.2 Initial and Boundary Conditions

The heat conduction equation expresses the relationship of the temperature inside an object with the time and space, but there are infinite solutions. Initial and boundary conditions must be available to determine the desired temperature field. Initial conditions refer to the instantaneous temperature distribution inside the object at the beginning, while boundary conditions refer to the rules of temperature interaction between the outer wall surface and surrounding medium (such as air or water). The initial and boundary conditions are collectively known as boundary value conditions (or definite conditions).

There are usually four types of boundary conditions related to heat conduction, only two of which are explained in this chapter.

1. First type of boundary conditions

The heat flow through the surface of the object is proportional to the difference between the surface temperature (T) and air temperature (T_a), namely:

$$-\lambda \frac{\partial T}{\partial n}=\beta(T-T_a) \qquad (2\text{-}1\text{-}3)$$

Where, β—heat transfer coefficient (heat released or absorbed by the object per unit area in unit time, corresponding to the temperature change of 1℃), kJ/(m^2 · h · ℃).

2. Second type of boundary conditions

If two solids with different properties are in good contact, the temperature and heat flow of the contact surface will be continued, with the boundary conditions as follows:

$$T_1=T_2,\quad \lambda_1 \frac{\partial T_1}{\partial n}=\lambda_2 \frac{\partial T_2}{\partial n} \qquad (2\text{-}1\text{-}4)$$

2.1.2 Solution to One-dimensional Heat Conduction Equation of Wall with External Thermal Insulation by Finite Difference Method

The finite difference method, a basic numerical method for solving the differential equation, is applied to solve the temperature field of the external thermal insulation system. Its principle is as follows: a continuous definite solution area is substituted with a grid composed of finite discrete points (grid nodes); the function of the continuous variable of the continuous definite solution area is approximated by a discrete variable function defined on the grid; the differential quotient of the original equation and definite solution conditions is approximated with a difference quotient; and the original differential equation and definite solution conditions are approximating substituted in an algebraic equation, namely a finite differential equation set, which can be solved to obtain the approximate values of discrete points of the original equation.

The insulated wall structure is divided into several layers. The corresponding geometric dimensions and material properties of each layer are entered. In order to compare the performances of two types of commonly used insulation, i. e. internal thermal insulation and external thermal insulation, the same values of material properties are applied to each layer in the modeling process. Figure 2-1-1 shows the structural diagram of functional layers of the typical wall subject to the external thermal insulation based on the adhesive polystyrene granule. The finite difference equation for solving the temperature field is established accordingly.

The nodes of the insulated wall in the thickness direction are divided into inner nodes (inside one material), inner surface codes, outer surface nodes and connection nodes of two different materials. Boundary conditions of the inner surface mainly include the indoor air heat convection, and those of the outer surface include the solar radiation, outer surface radiation, outdoor air convection, etc.

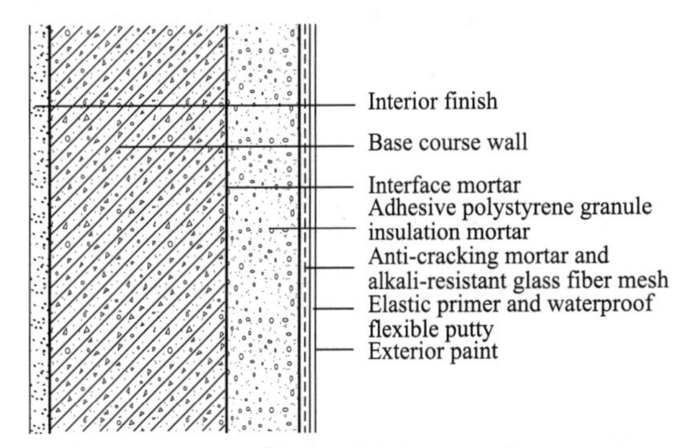

Figure 2-1-1

2.1.2.1 Heat Convection Boundary Conditions of Inner Surface of Outer Wall

Convection is a process of energy transport by mixing and flow of cold and hot fluids due to the relative displacement of parts therein. The heat transfer under the effects of convection and heat conduction between the wall surface and fluid is known as heat convection. The hot fluid density is proportional to the temperature difference between the wall (outer surface) and main fluid area (atmosphere). For the inner surface of the wall, assuming that the indoor air temperature is $T_{in}(t)$, the coefficient of heat convection between the indoor air and inner surface is β_{in}, the temperature of the inner surface is $T_1(t)$ (first node), the mutual thermal radiation between the indoor and inner surface as well as that between various layers are ignored. Correspondingly, the amount of convective heat transfer between the inner surface of the wall and the indoor air can be expressed as:

$$q_{in} = \beta_{in}[T_{in}(t) - T_1(t)] \qquad (2\text{-}1\text{-}5)$$

The detailed value of β_{in} is stipulated in the *Thermal Design Code for Civil Building* (GB 50176-1993) of China. "β_{in} = 8.7W/(m² · ℃)" (Table 2-1-1 is applied in subsequent calculations).

2.1.2.2 Boundary Conditions for Convective Heat Transfer of Outer Surface of Outer Wall

For the outer surface of the wall, similar to its inner surface, assuming that the outdoor air temperature is $T_{out}(t)$, the convective heat transfer coefficient between the outdoor air and outer surface is β_{out}, the temperature of the outer surface is $T_n(t)$ (n^{th} node), the mutual thermal radiation between the indoor and inner surface as well as that between various layers are ignored. Correspondingly, the amount of convective heat transfer between the outer surface of the wall and the outdoor air can be expressed as:

$$q_{out} = \beta_{out}[T_{out}(t) - T_n(t)] \qquad (2\text{-}1\text{-}6)$$

Also, the detailed value of β_{out} is stipulated in the *Thermal Design Code for Civil Building* (GB 50176-1993) of China. β_{out} is related to the wind speed on the outer surface of the building. See Table 2-1-1 for the value of β_{out} in subsequent calculations.

2.1.2.3 Numerical Simulation of Thermal Environment Related to Solar Radiation

For the thermal environment of a building, solar radiation is a very important external influence factor. It provides free heat in cold seasons but needs some costs to offset its interference to the room temperature in hot seasons.

The solar radiation reaching the ground is composed of direct radiation without directional change, and scattered radiation of no specific direction as a result of reflection of gas molecules, liquid or solid particles in the atmosphere. The sum of direct radiation and scattered radiation is the total solar radiation reaching the ground, hereinafter referred to as solar radiation. The solar radiation intensity is expressed by the energy of solar radiation received per unit area in unit time, and divided into direct solar radiation intensity, scattered solar radiation intensity and total solar radiation intensity (also known as solar radiation intensity).

Solar radiation is actually complex, affected by many factors. Its intensity is directly influenced by the day and night formed by the Earth's rotation, four seasons formed by the Earth's revolution, and changes in the solar incident angle and the distance between the Earth and sun in different periods and seasons, as well as the weather, atmospheric transparency, ground conditions, and properties of materials applied in the building surface. This chapter discusses the temperature field and stress field in the insulated wall structure under typical conditions, in which the number of special factors is minimized and minor influence factors are reasonably simplified and treated, to avoid the complexity of the research process and ensure the universality of final results. Specific calculations of the solar radiation intensity are presented below.

1. Direct radiation intensity

Direct radiation intensity is related to the atmospheric transparency and other factors. For one surface vertical to the sunlight on the Earth, the direct radiation intensity can be expressed as:

$$I_{DN}=I_0P^m \quad (2\text{-}1\text{-}7)$$

Where, I_{DN} is the direct solar radiation intensity; I_0 is the constant of solar radiation; m is the optical quality of the atmosphere, $m=\dfrac{1}{\sin(h_s)}$; h_s is the solar elevation; and P is the atmospheric transparency.

The solar radiation intensity I_{DH} and I_{DV} of the horizontal and vertical surfaces can be respectively expressed as:

$$I_{DH}=I_{DN}\sin(h_s) \quad (2\text{-}1\text{-}8)$$

$$I_{DV}=I_{DN}\cos(h_s)\cos\gamma \quad (2\text{-}1\text{-}9)$$

Where, γ is the angle between the projections of the wall normal and sunlight on the horizontal surface, $\gamma=A_s-A_w$; A_s and A_w are the solar azimuth (angle of the projection of sunlight on the horizontal surface and the south) and wall azimuth (angle of the projection of the wall normal on the horizontal surface and the south); and h_s is the solar elevation (angle between the ground plane and the connecting line of one point of the surface on the Earth to the sun), calculated from:

$$\sin(h_s)=\sin\phi\cdot\sin\delta+\cos\phi\cdot\cos\delta\cdot\cos\omega \quad (2\text{-}1\text{-}10)$$

Where, ϕ is the geographic latitude (angle between the equatorial plane of the Earth and the connecting line of one point to the center of the Earth); δ is the sun's declination angle (angle between the equatorial plane of the Earth and the connecting line of the center of the Earth to the sun); and ω is the hour angle, calculated from:

$$\omega=15(t-12) \quad (2\text{-}1\text{-}11)$$

Where, t is the local sun-time.

The solar azimuth A_s is calculated from:

$$\cos A_s=\frac{\sin h_s\sin\phi-\sin\delta}{\cos h_s\cos\phi} \quad (2\text{-}1\text{-}12)$$

2. Scattered radiation intensity

The scattered radiation received by the outer wall of the wall from the sky consists of the sky-scattered radiation, ground reflection and atmospheric long-wave radiation.

(1) Sky-scattered radiation

The sky-scattered radiation means that the sunlight is reflected and refracted by mist and dust in various directions in the atmosphere, making the light scattered by the entire sky. Its irradiance of the horizontal plane on sunny days is usually approximated by the Berlage equation, namely:

$$I_{SH}=0.5I_0\frac{1-P^m}{1-1.4\ln P}\sin h_s \quad (2\text{-}1\text{-}13)$$

The intensity of scattered radiation to vertical walls in different directions is:

$$I_{SV}=0.5I_{SH} \quad (2\text{-}1\text{-}14)$$

(2) Ground reflection

Part of solar radiation is reflected by the ground. The intensity of ground reflection to the vertical wall is:

$$I_{RV}=0.5\rho_s(I_{DH}+I_{SH}) \quad (2\text{-}1\text{-}15)$$

Where, ρ_s is the reflectivity of the ground to solar radiation. It is approximated to 0.2 under typical urban conditions and 0.7 under snow conditions in the city.

(3) Long-wave radiation

In addition to the direct solar radiation, the-atmosphere also absorbs the radiation reflected by the ground and wall, leading to the increase of temperature. Considering the radiation heat transfer between the atmosphere and ground/wall, this type of radiation is called long-wave radiation, calculated from:

$$q_c=C_{a-e}\left[(T_c/100)^4-(T_e/100)^4\right] \quad (2\text{-}1\text{-}16)$$

Where, q_c is the flow if radiation heat transferred between the atmosphere and ground/wall; C_{a-e} is the equivalent radiation coefficient; and T_c and T_e are the absolute temperature of the wall and the long-wave radiation temperature of the atmosphere, respectively.

3. Total solar radiation

Taking the ground radiation into account, the intensity of solar radiation received by the vertical wall is calculated from:

$$I_Z=I_{DV}+I_{SV}+I_{RV} \quad (2\text{-}1\text{-}17)$$

Where, I_{DV}, I_{SV} and I_{RV} are the direct solar radiation intensity, sky-scattered radiation intensity and ground reflection intensity of the vertical wall, respectively.

The solar radiation intensity in different regions and directions can be calculated as mentioned above. Surface materials have different capabilities in absorbing solar radiation. That is, the absorption rates vary from each other. The heat convection boundary conditions of the outer surface of the vertical wall, namely, solar radiation boundary conditions, are:

$$q_r = \alpha_s I_Z = \alpha_s (I_{DV} + I_{SV} + I_{RV}) \quad (2\text{-}1\text{-}18)$$

Where, α_s is the solar radiation absorption rate of the outer surface of the wall.

Figure 2-1-2 shows the relationship between the time and solar radiation intensity (calculated based on the above model) of eastern, western, southern and northern vertical walls in Beijing in one year. The solar radiation constants and latitudes are given in Table 2-1-1. Accordingly, the intensity of solar radiation received by walls is significantly affected by the season, direction and time. It is the highest on the southern wall in winter, spring and autumn, and the lowest on the northern wall in four seasons. In summer, the solar radiation intensity of the eastern and western walls is higher than that of the southern wall. The duration of solar radiation on each wall also varies in different seasons.

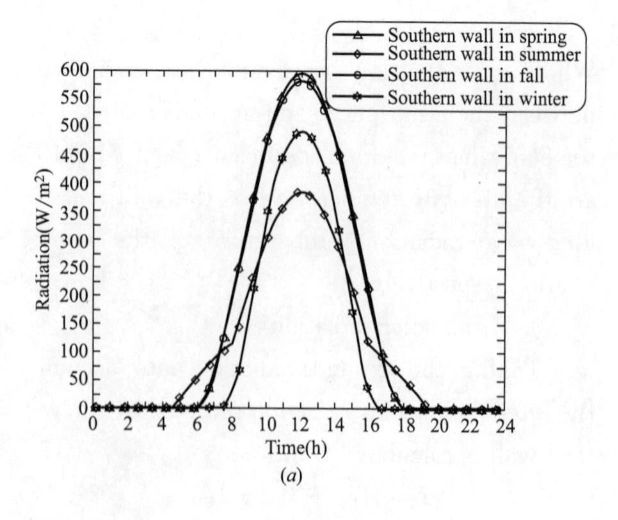

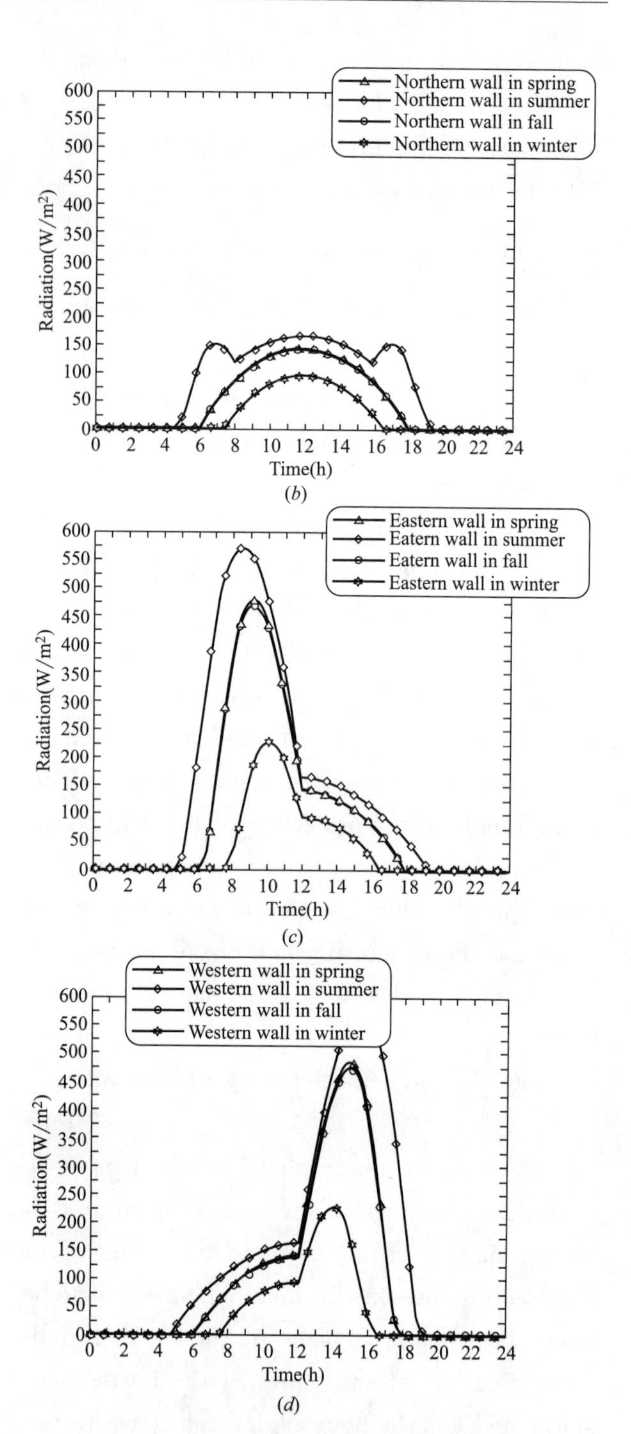

Figure 2-1-2 Solar Radiation Intensity in Different Directions and Seasons

(*a*) southern wall; (*b*) northern wall; (*c*) eastern wall; (*d*) western wall

Convection and Radiation Parameters Table 2-1-1

Season	Coefficient of Convection Heat Transfer[W/(m² · K)]		Declination Angle	Solar Constant (W/m²)
	Inner surface	Outer surface		
Spring (March)	8.7	21.0	0°	1365

Continued

Season	Coefficient of Convection Heat Transfer[W/(m² · K)]		Declination Angle	Solar Constant (W/m²)
	Inner surface	Outer surface		
Summer (June)	8.7	19.0	+23.45°	1316
Autumn (September)	8.7	21.0	0°	1340
Winter (December)	8.7	23.0	−23.45°	1392

2.1.2.4 Finite Difference Equation for Calculation of Temperature Field of Wall with External Thermal Insulation

Depending on the insulated wall structure, nodes for finite difference calculation are divided into four categories: inner nodes (inside one material), inner surface nodes (inner surface of the wall), outer nodes (outer surface of the wall), and connection nodes of two materials with different properties. Figure 2-1-3 shows the schematic diagrams of four typical categories of nodes for solution of the one-dimensional temperature field of the outer wall. The related difference equations are derived as follows.

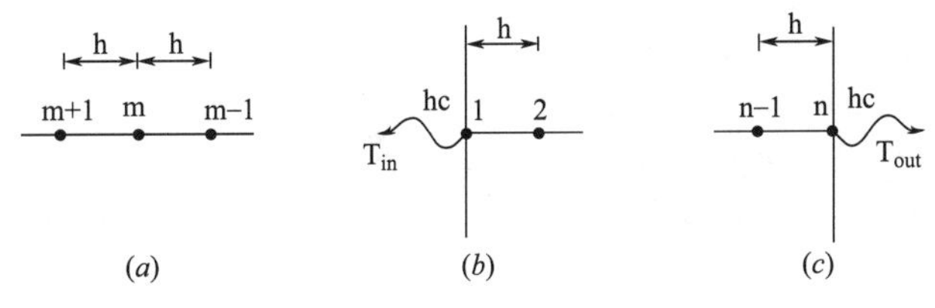

Figure 2-1-3 Schematic Diagram of Typical Nodes of Wall
(*a*) inner nodes; (*b*) inner surface nodes; (*c*) outer surface nodes

1. Inner nodes

For the inner node m (Figure 2-1-3a), the second-order differential in the location x at the time t and temperature T based on the finite difference principle can be approximately expressed as:

$$\left(\frac{\partial^2 T}{\partial x^2}\right)_{m,t} \approx \frac{1}{h^2}(T_{m+1,t} + T_{m-1,t} - 2T_{m,t}) \tag{2-1-19}$$

The rate of change in the node m at the time t and temperature T can be approximated as:

$$\frac{\partial T}{\partial t} \approx \frac{T_{m,t+\Delta t} - T_{m,t}}{\Delta t} \tag{2-1-20}$$

Substituting the equations (2-1-19) and (2-1-20) into the one-dimensional heat conduction equation (2-1-2), the temperature of the node m after the time interval (Δt) can be expressed as:

$$T_{m,t+\Delta t} = (1-2r)T_{m,t} + r(T_{m+1,t} + T_{m-1,t}) \tag{2-1-21}$$

Where, $r = \lambda \Delta t/(c\rho h^2)$.

The temperature $T_{m,\ t+\Delta t}$ of the node m at the time $t+\Delta t$ can be directly calculated by the equation (2-1-21), based on the temperature of three adjacent nodes at the time t, instead of the solution of the equation set. This is called the explicit difference method. The precondition for the stable solution to the equation (2-1-21) is $1-2r \geqslant 0$, i.e. $\Delta t \leqslant c\rho h^2/(2\lambda)$.

2. Inner surface nodes

For the inner surface nodes exposed to air (Figure 2-1-3b), assuming that the coefficient of convection heat transfer of the concrete surface is β_{in} [kJ/(m² · ℃)], the following equation can be obtained based on the energy balance principle:

$$T_{1,t+\Delta t} = (1-2r-2rB_1)T_{1,t} + 2r(T_{2,t} + B_1 T_{in,t}) \tag{2-1-22}$$

Where, $r = \lambda \Delta t/(c\rho h^2)$; $B_1 = \beta_{in} h/\lambda$; and precondition for the stable solution: $\Delta t \leqslant c\rho h^2/2(\lambda + \beta_{in} h)$.

3. Outer surface nodes

For the outer surface nodes exposed to air (Figure 2-1-3c), solar radiation needs to be added in the heat transfer of the wall surface. Assuming that the coefficient of convection heat transfer of the outer surface is β_{out} [kJ/(m² · ℃)], the total solar radiation is I_z (solar radiation energy received per unit area in unit time, kJ/(m² · h)), and the coefficient of solar radiation absorption of the wall surface is α_s, the following equation can be obtained based on the energy balance principle:

$$T_{n,t+\Delta t} = (1-2r-2rB)T_{n,t} + 2r(T_{n-1,t} + BT_{out,t}) + \frac{2\alpha_s I_z \Delta t}{c\rho h} \tag{2-1-23}$$

Where, $r = \lambda\Delta t/(c\rho h^2)$; $B_1 = \beta_{out}h/\lambda$; and precondition for the stable solution: $\Delta t \leqslant c\rho h^2/2(\lambda + \beta_{out}h)$.

4. Connection nodes of materials with different properties inside the wall

Assuming that the thermal conductivity of the material on the left side of the nth node is λ_1, the spacing of left nodes is h_1, the thermal conductivity of the material on the right side is λ_2, and the spacing of right nodes is h_2 (similar to that shown in Figure 2-1-3a), the following equation can be derived under the above-mentioned second type of boundary conditions:

$$T_{n,t+\Delta t} = \frac{\frac{\lambda_1}{h_1}(4T_{n-1,t+\Delta t} - T_{n-2,t+\Delta t}) + \frac{\lambda_2}{h_2}(4T_{n+1,t+\Delta t} - T_{n+2,t+\Delta t})}{3\left(\frac{\lambda_1}{h_1} + \frac{\lambda_2}{h_2}\right)} \tag{2-1-24}$$

The temperature of the interface node at the time ($t + \Delta t$) can be calculated by the equation (2-1-24) according to the temperature of the non-interface node at the time ($t + \Delta t$).

2.1.3 Temperature Field Calculation Results and Analysis

2.1.3.1 Wall Insulation for Calculation

Using the above calculation mode, the real-time temperature fields of the eastern, western, southern and northern walls in Beijing in four seasons, corresponding to the outdoor solar radiation and ambient temperature change, were calculated comprehensively, subject to the external thermal insulation with the adhesive polystyrene granule finish (Figure 2-1-4), internal thermal insulation (making the insulation layer on the inner side of the base course wall by adjusting functional layers), aerated concrete self-insulation (example: composite wall with 20mm cement mortar + 200mm aerated concrete + 200mm cement mortar), and sandwich insulation (example: wall subject to composite insulation with 50mm concrete slab + 50mm rock wool board + 50mm concrete slab).

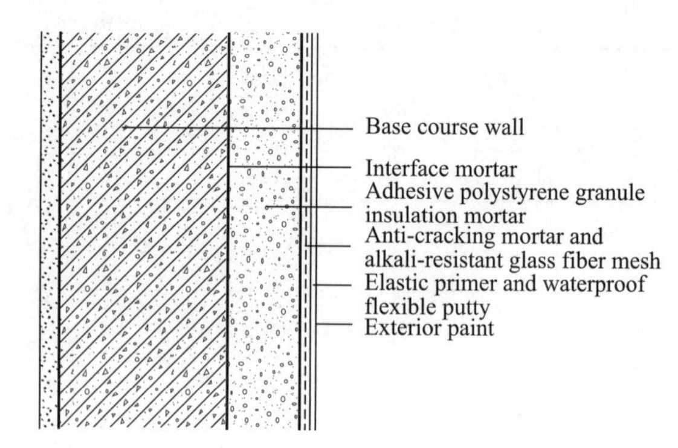

Figure 2-1-4 External Thermal Insulation with Adhesive polystyrene granule Adhesive polystyrene granule Finish

2.1.3.2 Indoor and Outdoor Air Temperature

For research conveniences, typical values of the indoor temperature were used according to its changes in four seasons. The indoor temperature was set as a constant, as shown in Table 2-1-2.

Indoor and Outdoor Temperature Parameters

Table 2-1-2

Season	Indoor Temperature (℃)	Outdoor Maximum Temperature(℃)	Outdoor Minimum Temperature (℃)
Spring (March)	23.0	25.3	4.4
Summer (June)	25.0	39.0	23.0
Autumn (September)	23.0	26.5	9.7
Winter (December)	20.0	1.5	−11.5

The outdoor temperature changes continuously at daytime and night. Affected by the weather (cloud, rain, snow, wind, etc.) and season, the above values do not change periodically. For research conveniences, the daily maximum and minimum temperature (T_{max} and T_{min}) under typical climatic conditions of each season were applied (Table 2-1-2), and the daily periodic changes in the atmospheric temperature were simulated from:

$$T_a = -\sin\left(\frac{2\pi(t_d + 2)}{24}\right)\left(\frac{T_{max} - T_{min}}{2}\right)$$

$$+\left(\frac{T_{max}+T_{min}}{2}\right) \quad (2\text{-}1\text{-}25)$$

Where, T_a is the outdoor air temperature (a function of the time and daily maximum and minimum temperature), and t_d is the time.

The typical daily maximum and minimum temperature (from the meteorological department, as shown in Table 2-1-2) in four seasons in Beijing were selected as input parameters for calculation in this study. Changes in the daily temperature are simulated by the equation (2-1-25). See Figure 2-1-5 for changes in the daily temperature (24h) in Beijing in four seasons.

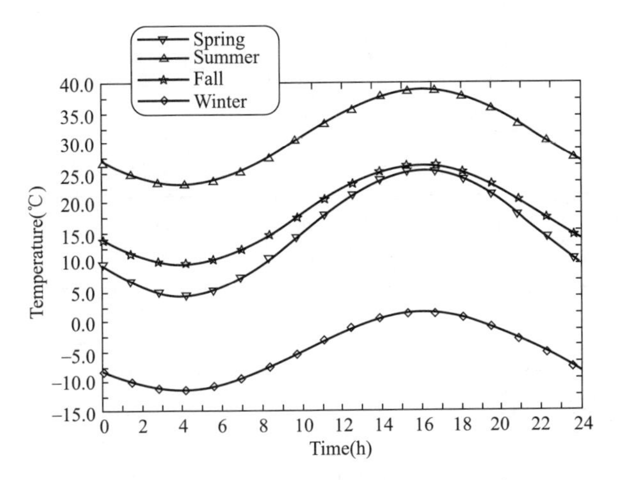

Figure 2-1-5 Simulation of Typical Daily Temperature Changes in Four Seasons in Beijing

2.1.3.3 Initial Conditions and Loading Time

The temperature field inside the outer wall at the time $t=0$ needs to be set before calculation. As boundary conditions such as the outdoor temperature and solar radiation vary in 24-hour cycles, the accurate temperature field at the specific time cannot be obtained without the calculation model. There is a certain deviation between the designated initial conditions and actual conditions, which will affect the accuracy of results calculated with the model.

If the total calculation time is set as one cycle, i.e. 24h, any deviation of initial conditions will have great impact on calculation results. Such impact can be reduced by extending the calculation time. The longer the total loading time, the smaller the impact of deviations of initial conditions on the calculation results of the last cycle is. Certainly, as the extension of total loading time may lead to the increase of calculation workload, it is impossible to extend the total loading time indefinitely. The appropriate loading time should be selected by means of calculation, with the workload and results of calculation acceptable.

Take the temperature field of the southern wall of one building in Beijing in summer as an example. The curve of wall temperature change is shown in Figure 2-1-6, corresponding to the solar radiation intensity data, material parameters of each layer, indoor temperature of 25℃ and loading time of 5 cycles (120h). Calculation results show that the temperature changes inside the wall are more and more regular over time and those of the 4th and 5th cycles are very close to each other. This means that the impact of initial conditions can be ignored from the 4th cycle. The calculated data of the 5th cycle are used as stable data in the subsequent analysis of this section.

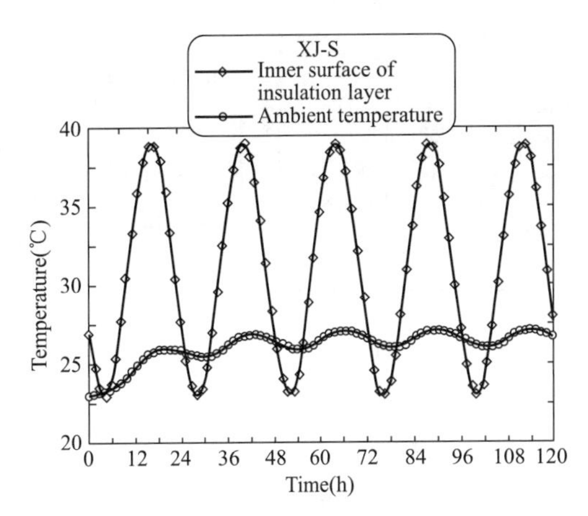

Figure 2-1-6 Changes in Inner Surface Temperature of Wall within 5 Cycles

2.1.3.4 Material Parameters

The temperature field was calculated based on the above-mentioned typical wall insulations. Refer to Table 2-1-3 and 2-1-4 for the thermal and physical parameters of related materials as well as the indoor and outdoor temperature.

Physical Parameters (I) of Materials (single insulation based on adhesive polystyrene granule) **Table 2-1-3**

Structure Type	Material Name	Thickness (mm)	Density (kg/m^3)	Specific Heat [J/(kg · K)]	Thermal Conductivity [W/(m · K)]
Paint finish for external thermal insulation of outer wall based on adhesive polystyrene granule	Inner finish	2	1300	1050	0.60
	Base course wall	200	2300	920	1.74
	Interface mortar	2	1500	1050	0.76
	Insulation mortar	60	250	1070	0.06
	Anti-cracking mortar	5	1600	1050	0.81
	Paint finish	3	1100	1050	0.50
Paint finish for internal thermal insulation of outer wall based on adhesive polystyrene granule	Inner finish	2	1300	1050	0.60
	Anti-cracking mortar	5	1600	1050	0.81
	Insulation mortar	60	250	1070	0.06
	Interface mortar	2	1500	1050	0.76
	Base course wall	200	2300	920	1.74
	Paint finish	3	1100	1050	0.50

Physical Parameters (II) of Materials (wall subject to self-insulation with aerated concrete and sandwich insulation with concrete and rock wool) **Table 2-1-4**

Structure Type	Material Name	Thickness (mm)	Density (kg/m^3)	Specific Heat [J/(kg · K)]	Thermal Conductivity [W/(m · K)]
Wall paint finish of self-insulation with aerated concrete	Inner finish	2	1300	1050	0.60
	Inner surface mortar	20	1800	1050	0.93
	Aerated concrete	200	700	1050	0.22
	Outer surface mortar	20	1800	1050	0.93
	Paint finish	3	1100	1050	0.50
Wall paint finish of sandwich insulation with concrete and rock wool	Inner finish	2	1300	1050	0.60
	Concrete slab	50	2300	920	1.74
	Rock wool board	50	150	1220	0.045
	Concrete slab	50	2300	920	1.74
	Paint finish	3	1100	1050	0.50

2.1.3.5 Calculation Results and Analysis

The temperature fields inside walls subject to typical internal thermal insulation, external thermal insulation, sandwich insulation and self-insulation were calculated with the parameters in Table 2-1-3 and 2-1-4 as model input values. Calculation results were further detailed, compared and analyzed below. Figure 2-1-7 and Figure 2-1-8 respectively show the temperature changes of typical positions of eastern, western, southern and northern walls in different directions over time in summer and winter (as the wall temperature in spring and autumn changes relatively smoothly, so related results are not given here), subject to internal and external thermal insulation with the paint finish based on the adhesive polystyrene granule.

1. Wall insulated with paint finish based on adhesive polystyrene granule

Figure 2-1-7 shows the temperature changes of each layer of the outer wall insulated with the paint finish based on the adhesive polystyrene granule in Beijing in winter, including the temperature change curves of each layer from inside to outside

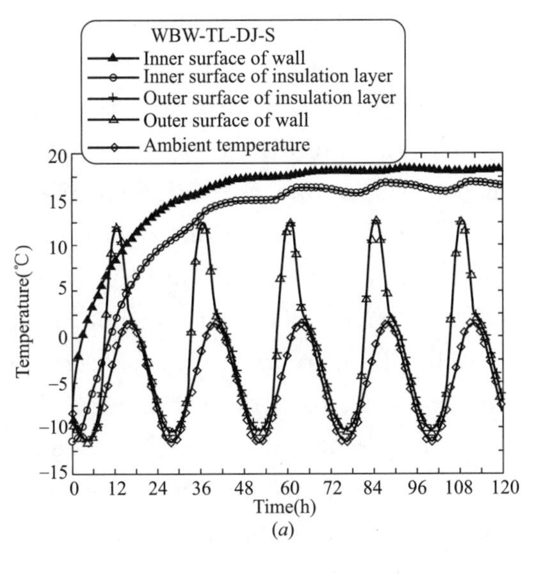

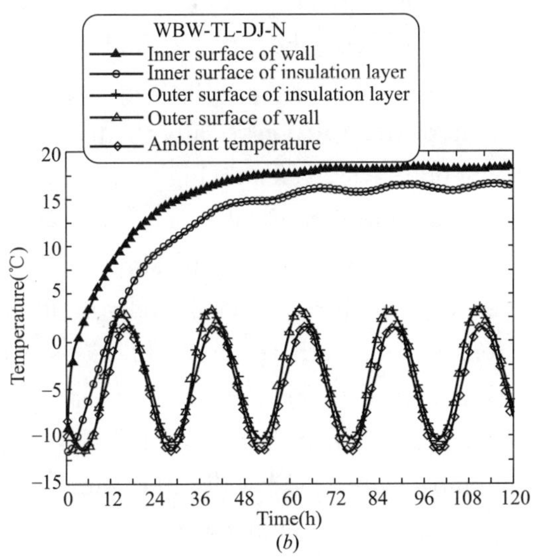

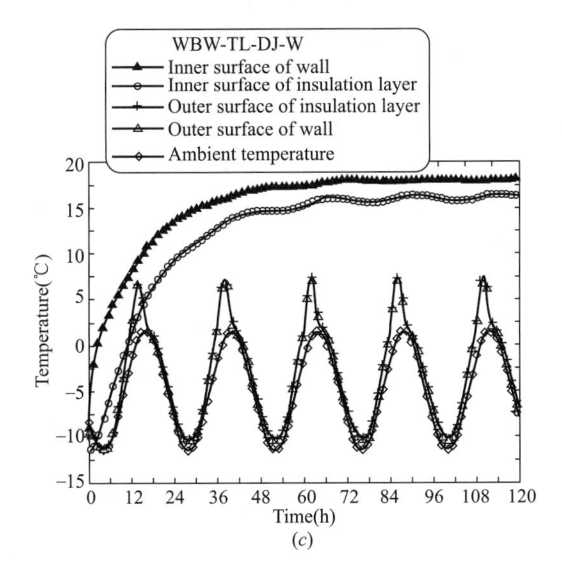

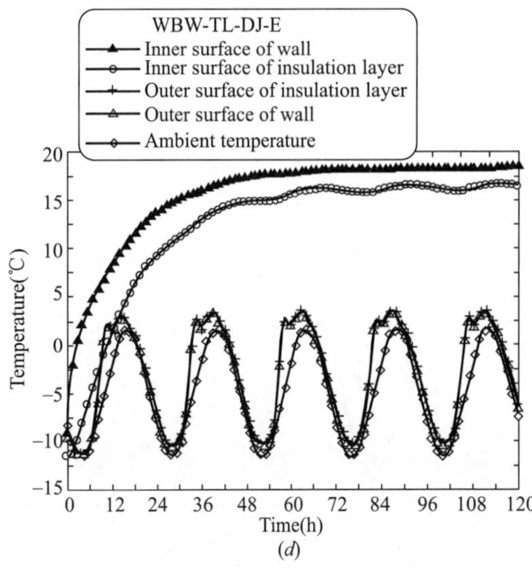

Figure 2-1-7 Temperature Changes of Different Layers of Outer Walls Insulated with Paint Finish based on Adhesive polystyrene granule over Time in Winter

(*a*) southern wall; (*b*) northern wall; (*c*) western wall; (*d*) eastern wall

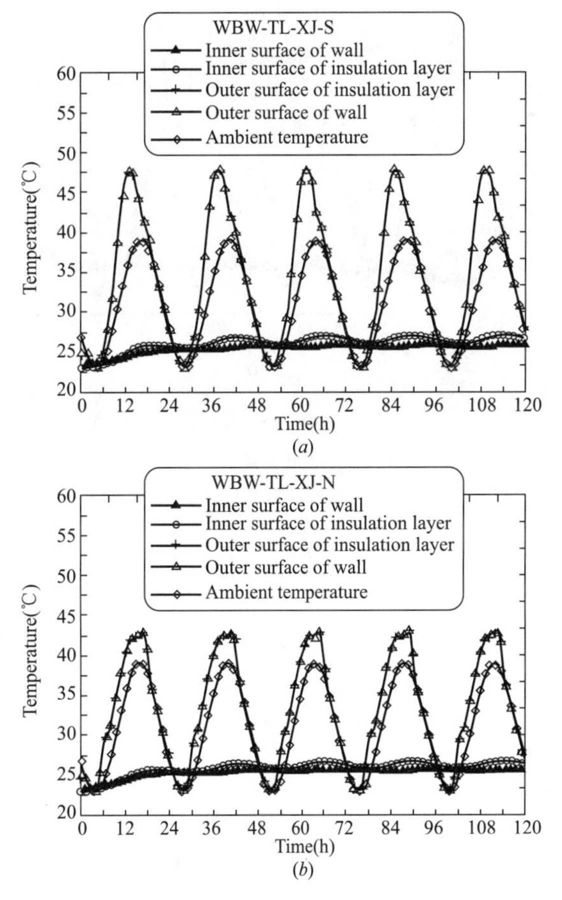

Figure 2-1-8 Temperature Changes of Different Layers of Outer Walls Insulated with Paint Finish based on Adhesive polystyrene granule over Time in Summer

(*a*) southern wall; (*b*) northern wall;

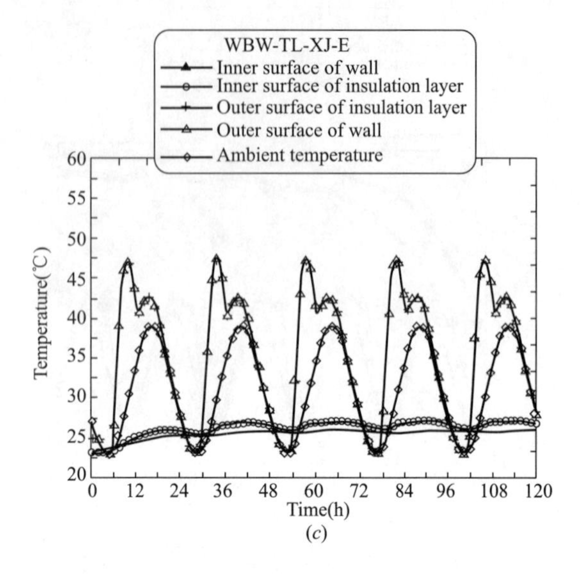

(c)

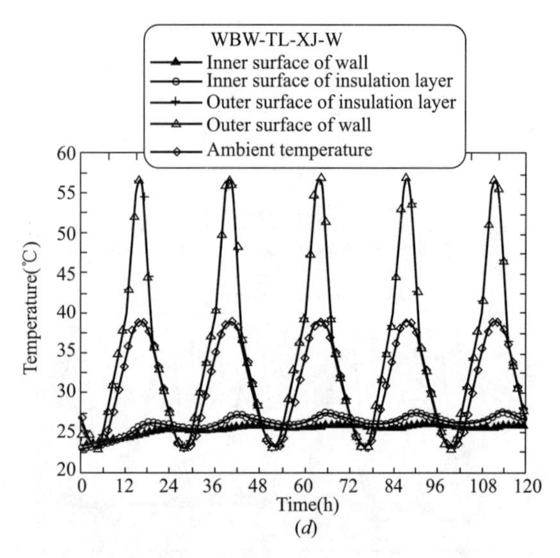

(d)

Figure 2-1-8 Temperature Changes of Different Layers of Outer Walls Insulated with Paint Finish based on Adhesive polystyrene granule over Time in Summer (Continued)

(c) eastern wall; (d) western wall

of the wall, namely, the inner surface of the wall (equivalent to the inner surface of the structural layer), inner surface of the insulation layer (equivalent to the outer surface of the structural layer), outer surface of the insulation layer, outer surface of the wall, and external environment. Results show that the model built in this study successfully simulates the impact of the outdoor temperature on the temperature field of the wall.

The temperature inside the wall varies with periodic changes in the outdoor temperature, and the magnitude of variation is related to the position inside the wall. Using the insulation material, heat transfer between the wall and external environment will decrease greatly, which obviously reduces the impact of the outdoor temperature on the indoor temperature. For example, the temperature change of the inner surface of the wall is the minimum over time, less than 3℃ on a daily basis. The closer the node is to the outer surface of the wall, the more greatly its temperature is affected by the atmospheric temperature. Therefore, the temperature of the outer surface of the wall changes the most, and the temperature of the outer surface of the outer wall changes more than the ambient temperature, depending on the wall location. Then, the intensity and duration of solar radiation have obvious effects on the temperature field of the wall, especially the part outside the insulation layer. Refer to Figure 2-1-2 for the intensity and duration of solar radiation to wall in each direction. In winter, the maximum temperature sequence of the surfaces of walls in different directions is the southern wall (12.7℃), western wall (7.5℃), eastern wall (3.4℃) and northern wall (about 3.4℃), and their minimum temperature is basically the same as the minimum ambient temperature (−11.4℃). Thus, the maximum temperature difference of the outer surfaces of walls is 24℃ at daytime and night in winter. In summer, the maximum temperature sequence of the surfaces of walls in different directions is the western wall (57℃), southern wall (48℃), eastern wall (47℃) and northern wall (43℃), and their minimum temperature is basically the same as the minimum ambient temperature (23℃). Thus, the maximum temperature difference of the outer surfaces of walls reaches 34℃ in summer.

In order to study the impact of the insulation layer location on the temperature field of the external thermal insulation system, the temperature field of the wall with the paint finish based on the adhesive polystyrene granule was calculated. Refer to Table 2-1-3 for the parameters of wall structures subject to internal thermal insulation. Except for changes in the insulation layer location, related material parameters are the same as those of external thermal insulation (Table 2-1-3). Figure 2-1-9 and Figure 2-1-10 show the temperature changes of typical parts of the eastern, western, southern and northern walls over time in summer and winter, subject to the internal thermal insulation with the paint finish based on the adhesive polystyrene granule.

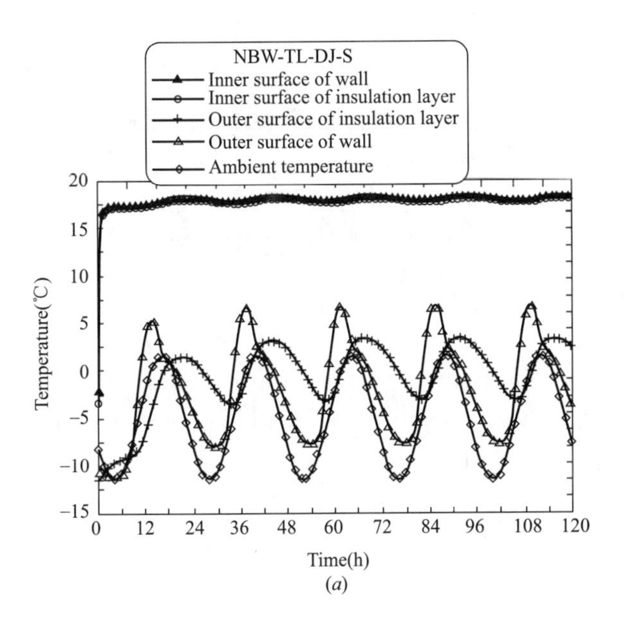

(a)

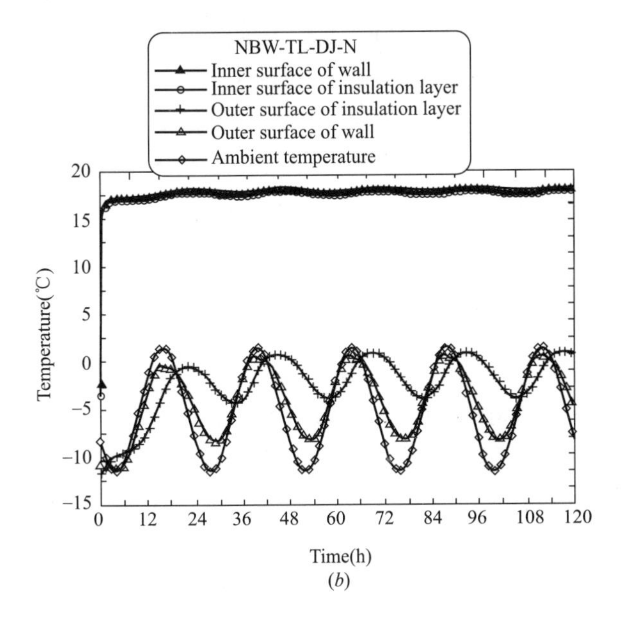

(b)

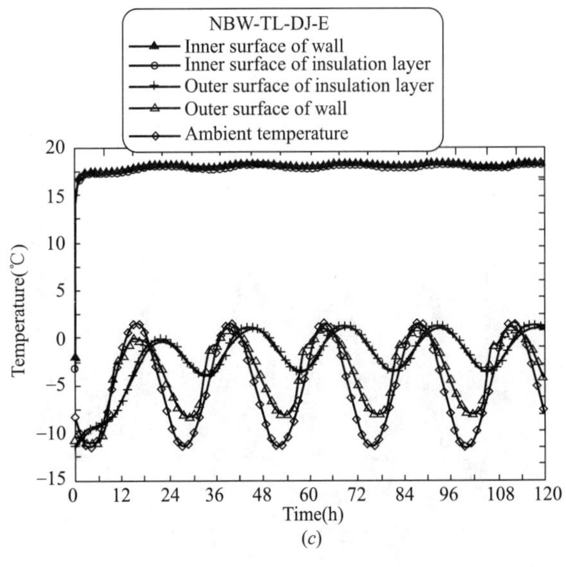

(c)

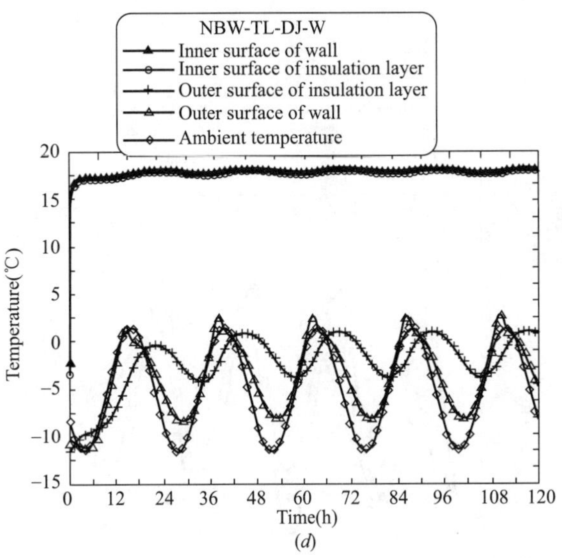

(d)

Figure 2-1-9 Temperature Changes of Different Layers of Walls subject to Internal Thermal Insulation with Paint Finished based on Adhesive polystyrene granule over Time in Winter

(*a*) southern wall; (*b*) northern wall; (*c*) eastern wall; (*d*) western wall

As shown in the figure of temperature changes of different layers of walls subject to internal thermal insulation with the paint finish based on the adhesive polystyrene granule over time:

(1) With the insulation layer, the temperature of the inner surface of the wall changes little over time, close to that of the wall subject to the external thermal insulation. That is, if the insulation layer of equivalent thickness is used, the internal thermal insulation has little difference from external thermal insulation.

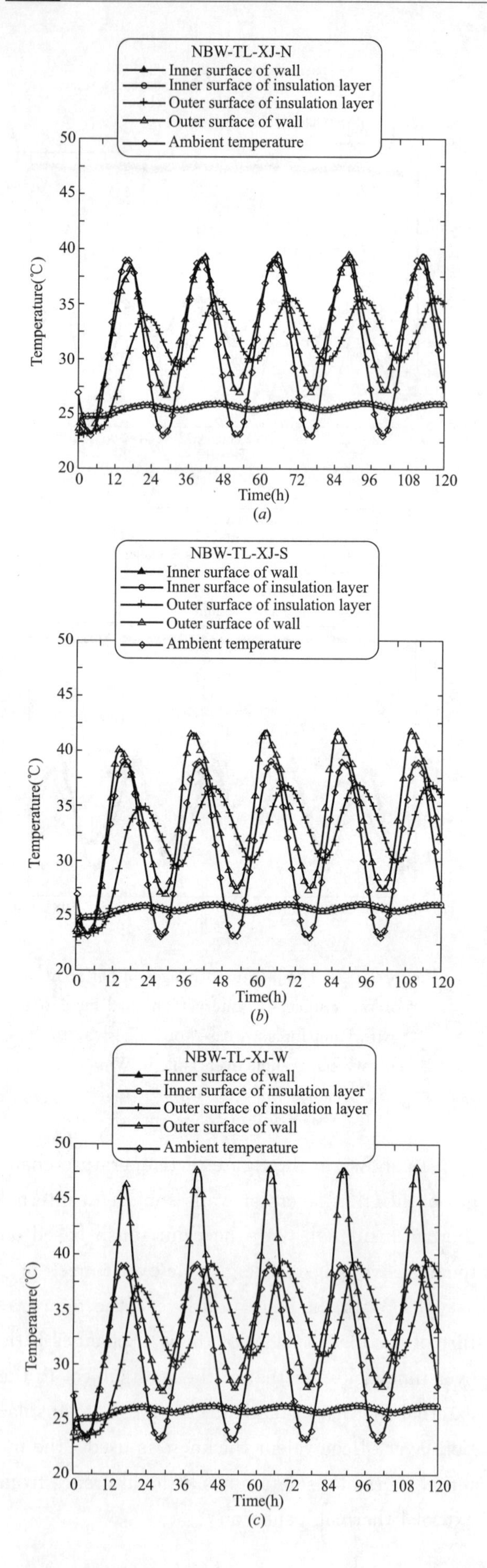

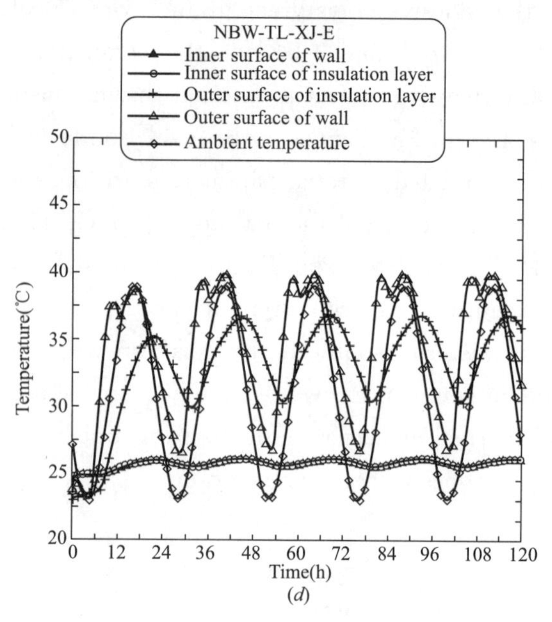

Figure 2-1-10 Temperature Changes of Different Layers of Walls subject to Internal Thermal Insulation with Paint Finished based on Adhesive polystyrene granule over Time in Summer

(*a*) northern wall; (*b*) southern wall; (*c*) western wall; (*d*) eastern wall

(2) Similar to the external thermal insulation, the closer the node is to the outer surface of the wall, the more greatly its temperature is affected by the atmospheric temperature. As the structural layer of the wall subject to internal thermal insulation is close to the outside, however, its temperature change is significantly greater than that of the structural layer of the wall subject to external thermal insulation.

(3) The maximum temperature of outer surfaces of walls subject to internal thermal insulation and in the same direction and season is lower than that of walls subject to external thermal insulation. For the outer surfaces of walls subject to the internal thermal insulation with the paint finish based on the adhesive polystyrene granule, the sequence of their maximum temperature in summer is as follows: western wall (48℃), southern wall (42℃), eastern wall (40℃) and northern wall (39℃). For the outer surfaces of walls subject to the external thermal insulation, the corresponding sequence is as follows: western

wall (57℃), southern wall (48℃), eastern wall (47℃) and northern wall (43℃). As the insulation material of large thermal resistance is close to the outer surface of the outer wall subject to external thermal insulation, heat is slowly transferred from the outer surface to inside the wall, leading to heat concentration in the outer surface of the wall. For the wall subject to internal thermal insulation, the material close to the outer surface has small thermal resistance, so heat is quickly transferred and dispersed into the wall, avoiding heat concentration and reducing the temperature of the outer wall of the wall.

(4) The temperature inside the structural layer of the wall subject to external thermal insulation can be maintained above the freezing point (16-17℃) of water under normal heating conditions, even in winter, while that of the wall subject to internal thermal insulation is mostly below 0℃ (about −2.5℃) in winter. Some walls have a positive and negative temperature cycle in 24h. From this perspective, the materials of their structural layers should be able to withstand the durability test within the freezing and melting cycle. Besides, the minimum temperature of the outer surface of the wall subject to internal thermal insulation is significantly higher than that of the wall subject to external thermal insulation. Therefore, the finish and protective mortar layer of the external thermal insulation should not change much in the case of significant temperature changes. That is, the original material design requirements need to be guaranteed under significant temperature changes.

(5) The rules of temperature changes of walls subject to internal thermal insulation are similar to those of walls subject to external thermal insulation, with specific values varying.

The impact of wall insulation (internal and external) on the temperature field is more clearly reflected by the temperature distribution at the given time and in the thickness direction of the wall. Figure 2-1-11 reveals the maximum and minimum distribution of the steadily-changing surface temperature of the western wall insulated with the paint finish based on the adhesive polystyrene granule in summer and winter. The zero point of the abscissa corresponds to the inner surface (indoor) of the wall. The distribution of the temperature of the insulated wall at any time in other seasons and in the thickness direction will fall within two boundary lines shown in the figure.

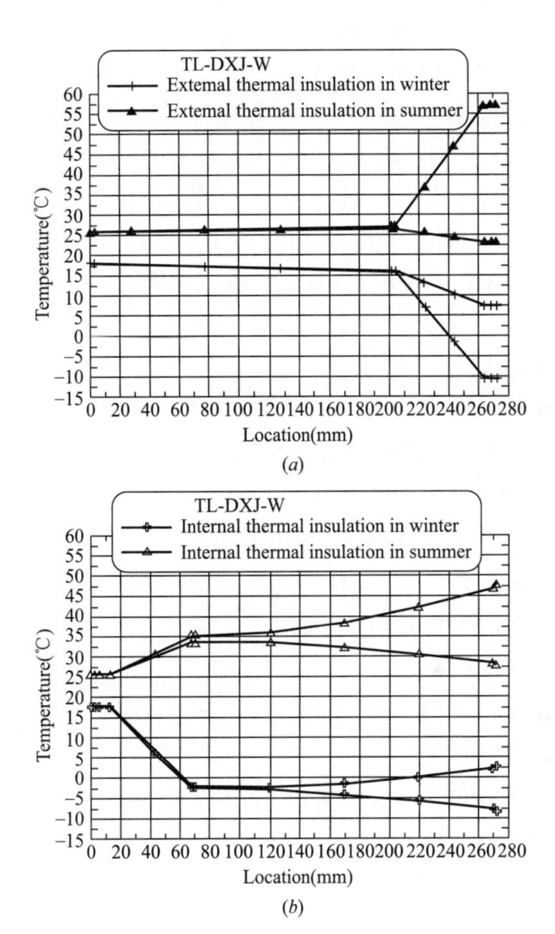

Figure 2-1-11 Maximum and Minimum Surface Temperature Distribution of Insulated Western Wall in the Thickness Direction in Winter and Summer

(*a*) external thermal insulation;

(*b*) internal thermal insulation

As shown in Figure 2-1-11:

(1) Regardless of internal or external thermal insulation, the temperature change inside the insulation layer is the most drastic. Relatively, the temperature of the external insulation layer changes more. For the western wall, the temperature of the external insulation layer is 26℃ to

57℃ in summer and −10℃ to 17℃ in winter, and that of the internal insulation layer is 25℃ to 35℃ in summer and −2.5℃ to 17℃ in winter.

(2) The temperature inside the base course wall varies significantly. The temperature of the structural layer (base course wall) changes a little in the case of external thermal insulation, only 8℃ (16℃ to 24℃), but significantly in the case of internal thermal insulation, 55℃ (−8℃ to 47℃). In this sense, the external thermal insulation is more conducive to the base course wall stability. In accordance with the follow-up calculation results, the thermal stress corresponding to the external thermal insulation is smaller.

(3) The temperature of the external insulation layer and its finish layer changes more than that of the internal insulation layer during four seasons and between the daytime and night. Thus, the external insulation layer and finish layer of the wall should have stronger resistance to the temperature deformation and thermal stress. The temperature distribution of the other walls is similar to that of the western wall, but their temperature changes are smaller than that of the western wall.

2. Temperature field of the wall without insulation layer

In order to compare the impact of the insulation layer on the temperature field of the wall under variable temperature conditions, the temperature field of the wall without the insulation layer was also calculated. The calculation results are presented in Figure 2-1-12 and 2-1-13.

As shown by the calculation results:

(1) Compared with the wall containing the insulation layer, the gap between the inner surface temperature of the wall containing no insulation layer and the indoor constant temperature is larger, and the inner surface temperature fluctuates more greatly. As more heat is transferred through the wall, more energy is needed to maintain the indoor constant temperature, leading to the increase of energy consumption.

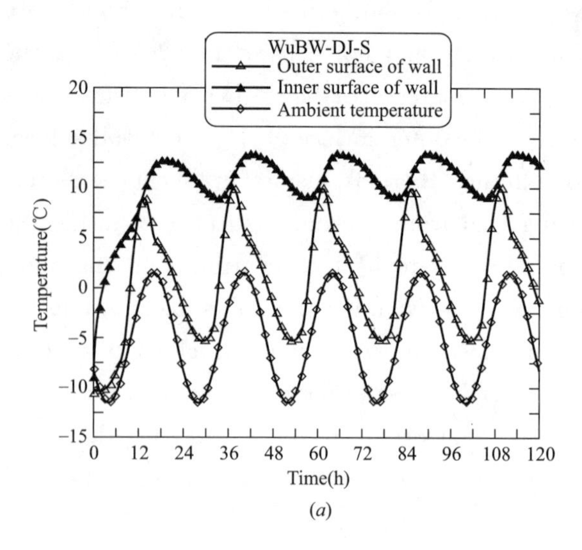

(a)

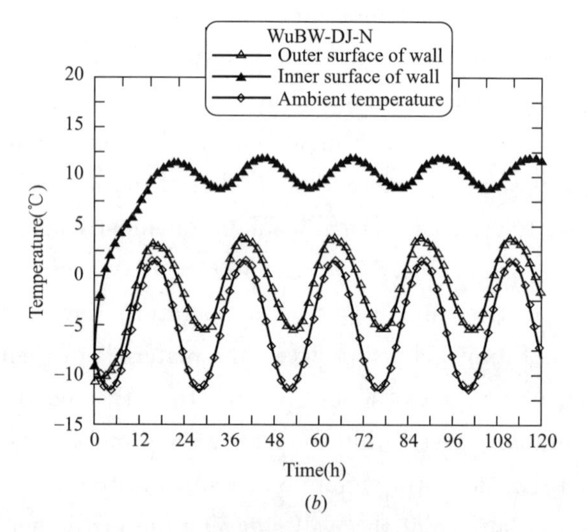

(b)

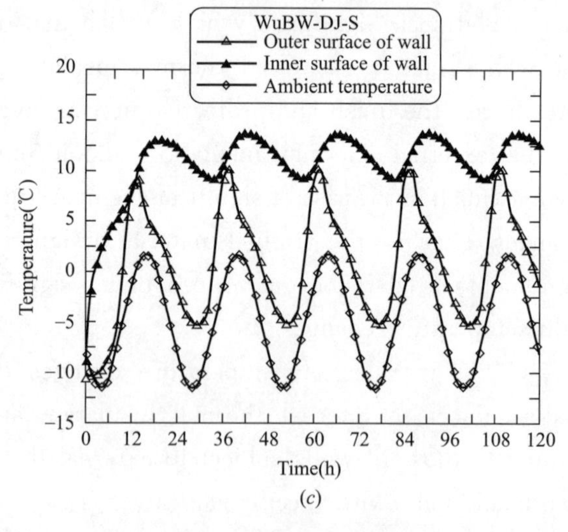

(c)

Figure 2-1-12 Temperature Change of Different Layers of Non-insulated Walls over Time in Winter

(a) southern wall; (b) northern wall; (c) eastern wall;

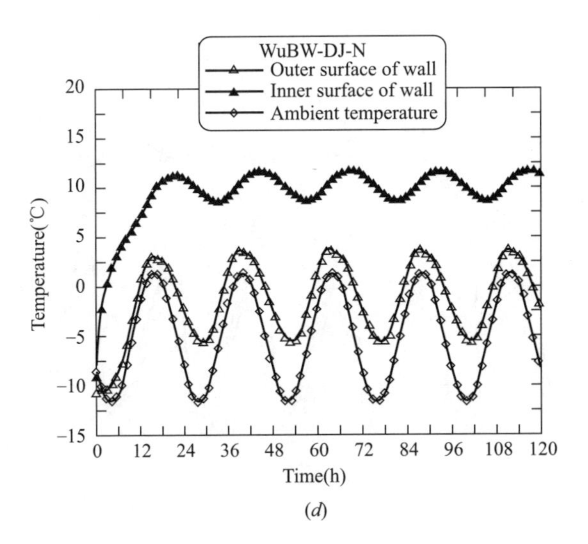

Figure 2-1-12 Temperature Change of Different Layers of Non-insulated Walls over Time in Winter (Continued)

(*d*) western wall

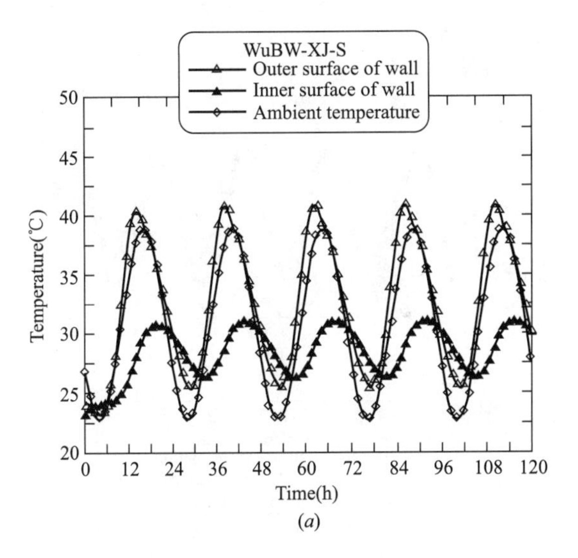

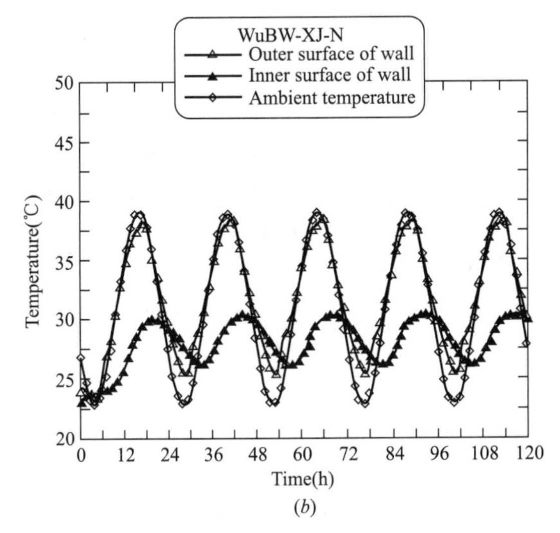

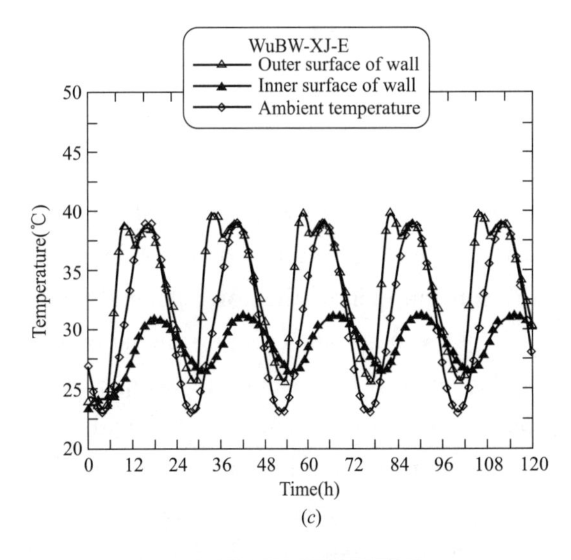

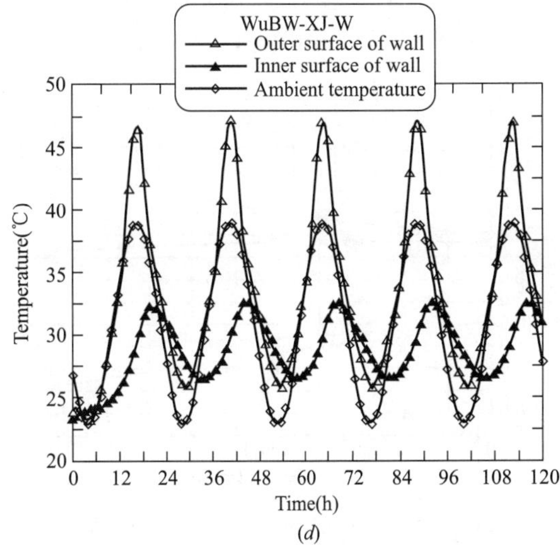

Figure 2-1-13 Temperature Change of Different Layers of Non-insulated Walls over Time in Summer

(*a*) southern wall; (*b*) northern wall;

(*c*) eastern wall; (*d*) western wall

(2) The maximum temperature of the outer surface of the wall containing the insulation layer decreases by approximately 10℃ in summer and increases by nearly 3℃ in winter (−8℃ in the presence of the insulation layer and −5℃ in the absence of the insulation layer).

Figure 2-1-14 shows the maximum and minimum distribution of the steadily-changing surface temperature of the non-insulated western wall in the thickness direction in winter and summer. The zero point of the abscissa corresponds to the inner surface (indoor) of the wall. The temperature distribution inside the base course wall obviously differs from that with the internal or external thermal insulation. Firstly, the inner surface temperature of the wall changes within a larger range, 10-30℃ in the absence of insulation (using the constant indoor temperature in four seasons) and 19-25℃ in the presence of internal or external thermal insulation. Correspondingly, the energy consumption of the non-insulated wall for heating in winter and cooling in summer is higher. Secondly, the maximum temperature of the outer surface of the non-insulated wall will decrease in summer, and its minimum temperature will increase in winter. The temperature deformation and thermal stress over the finish layer of the outer surface of the wall will drop.

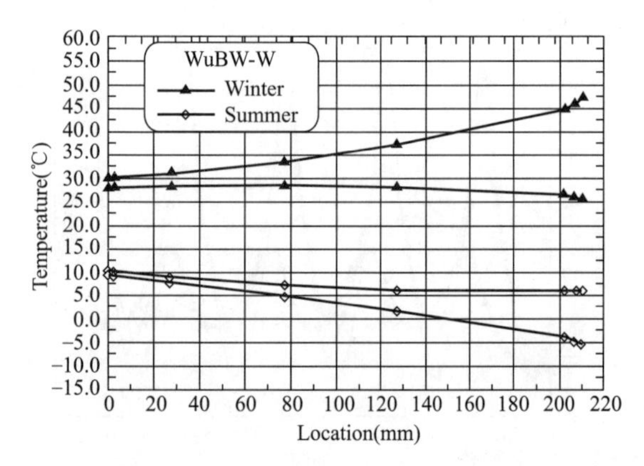

Figure 2-1-14 Maximum and Minimum Surface Temperature Distribution of Non-insulated Western Wall in the Thickness Direction in Winter and Summer

3. Temperature field of aerated concrete self-insulated wall

In order to compare the effects of different types of insulation on the temperature field of the wall, the temperature fields of the self-insulated wall (Table 2-1-3) in winter and summer were calculated. The calculation results are presented in Figure 2-1-15 and 2-1-16.

As shown by the calculation results:

(1) Compared with the wall containing the internal or external insulation layer, the gap between the inner surface temperature of the 200mm aerated concrete self-insulated wall and the indoor constant temperature is larger, and the inner surface temperature fluctuates more greatly.

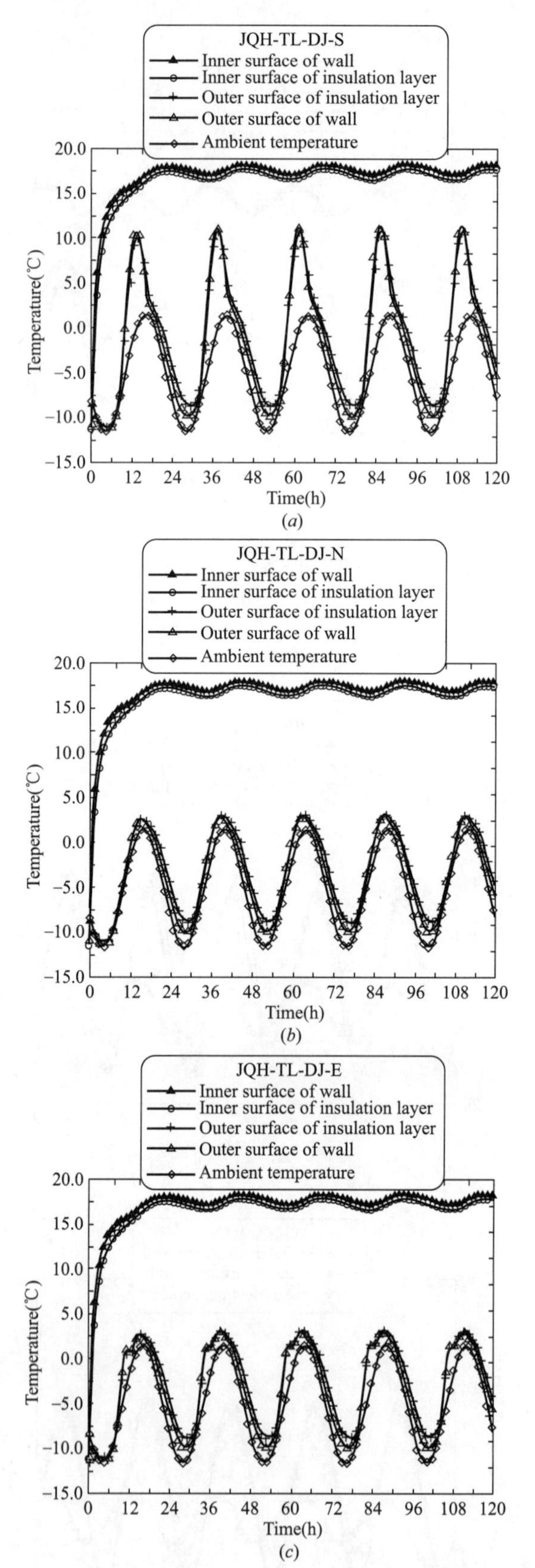

Figure 2-1-15 Temperature Changes of Different Layers of Self-insulated Walls with Aerated Concrete Paint Finish over Time in Winter

(*a*) southern wall; (*b*) northern wall; (*c*) eastern wall;

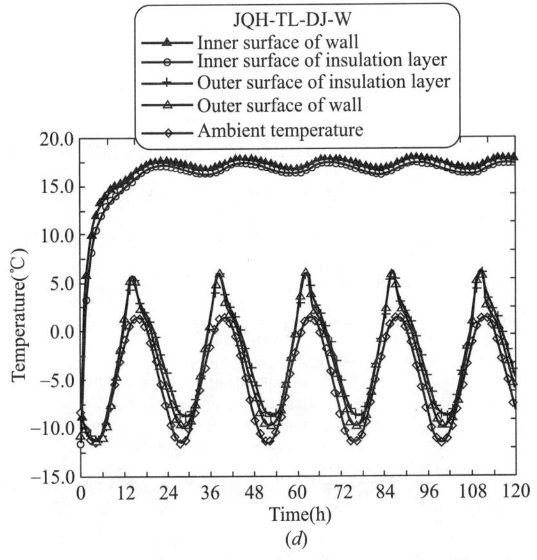

(*d*)

Figure 2-1-15 Temperature Changes of Different Layers of Self-insulated Walls with Aerated Concrete Paint Finish over Time in Winter (Continued)

(*d*) western wall

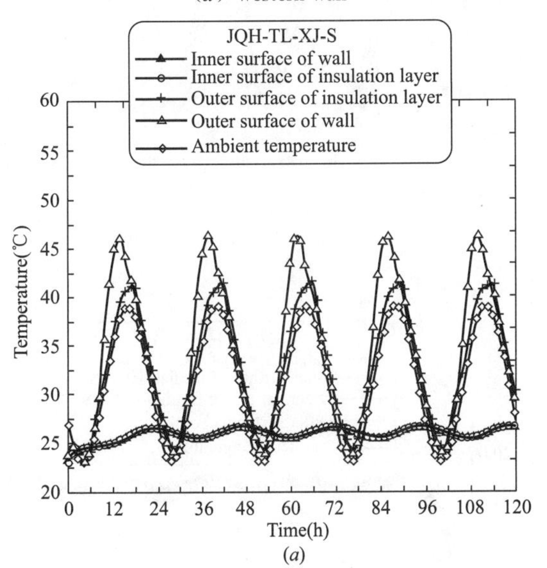

(*a*)

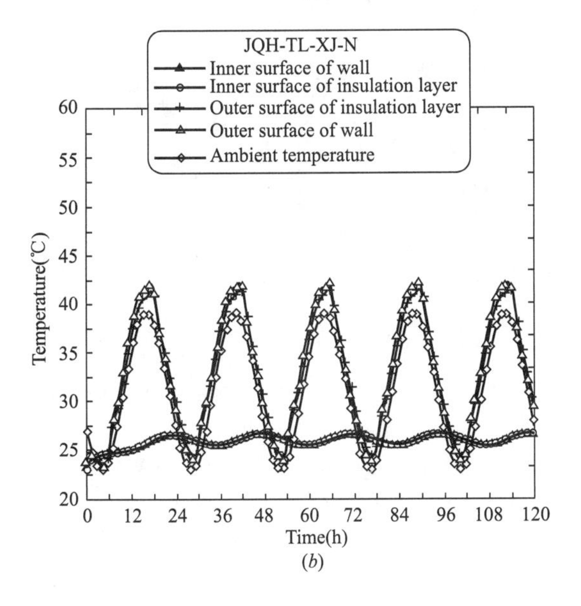

(*b*)

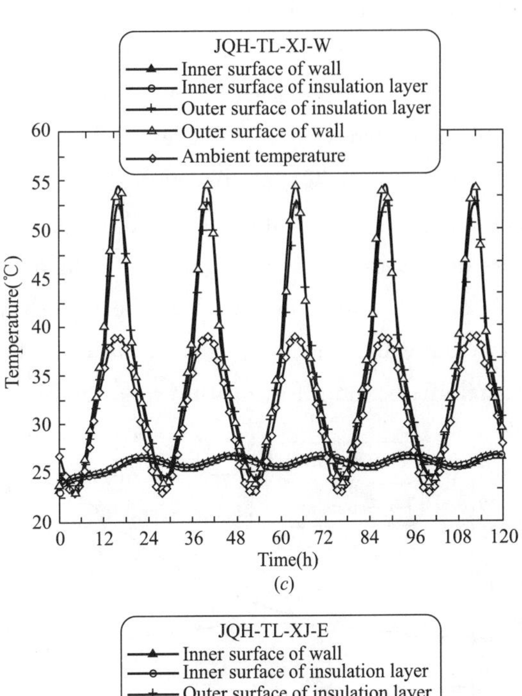

(*c*)

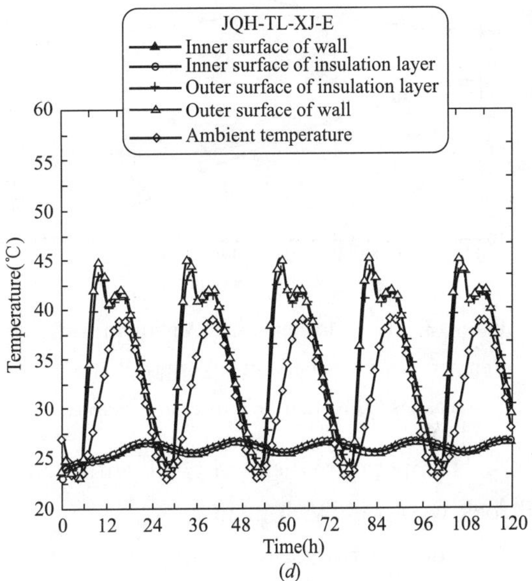

(*d*)

Figure 2-1-16 Temperature Changes of Different Layers of Self-insulated Walls with Aerated Concrete Paint Finish over Time in Summer

(*a*) southern wall; (*b*) northern wall; (*c*) western wall; (*d*) eastern wall

The insulation effect of the self-insulated wall depends on the thickness of aerated concrete.

(2) The maximum temperature of the outer surface of the self-insulated wall and its minimum temperature in winter are not significantly different from those of the wall containing the insulation layer.

Figure 2-1-17 shows the maximum and minimum distribution of the steadily-changing surface temperature of the aerated concrete self-insulated western wall in the thickness direction. The annual temperature difference of the inner surface of the self-insulated wall is 8℃, while that of the outer surface is 65℃, almost eight times of the inner surface. As a result of small deformation stress on the inner surface and large deformation stress on the outer surface, the thermal expansion of the wall is not uniform, resulting in large thermal stress and affecting the wall stability.

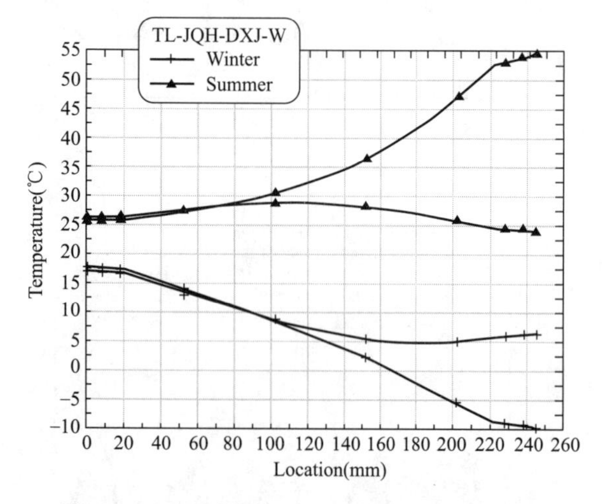

Figure 2-1-17 Maximum and Minimum Surface Temperature Distribution of Western Wall in the Thickness Direction in Winter and Summer

4. Temperature field of wall with concrete and rock wool sandwich insulation

In order to compare the impact of different insulation structures on the temperature field of the wall, the temperature fields of the wall with rock wool sandwich insulation (Table 2-1-3) were calculated. The calculation results are presented in Figure 2-1-18 and 2-1-19.

From the above calculation results, it can be seen that the expected insulation effect can be achieved by means of sandwich insulation as long as the insulation thickness is appropriate. The temperature distribution of the wall with the sandwich insulation is similar to that with the external thermal insulation.

Figure 2-1-20 shows the maximum and minimum distribution of outer surface temperature of the western wall with rock wool sandwich insulation in the thickness direction in winter and summer. The annual temperature difference of the inner wall is 8℃, while that of the outer wall is 65℃, almost eight times of the inner surface. Thus, attention should be paid to the structural stability under the thermal stress and change.

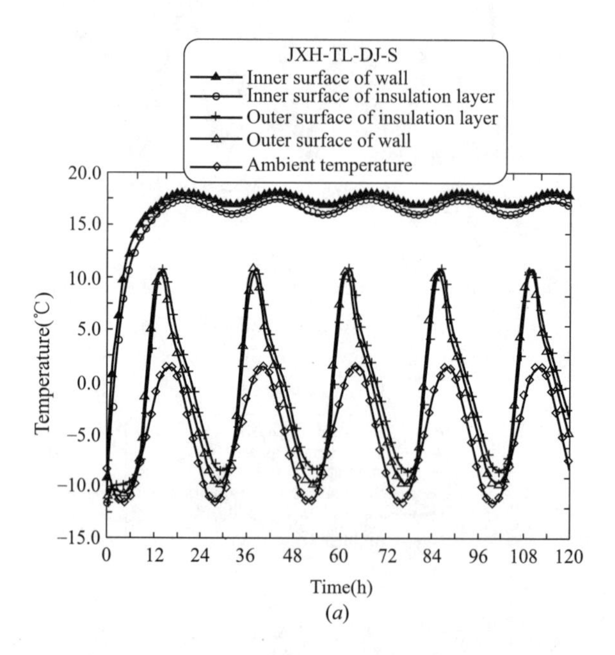

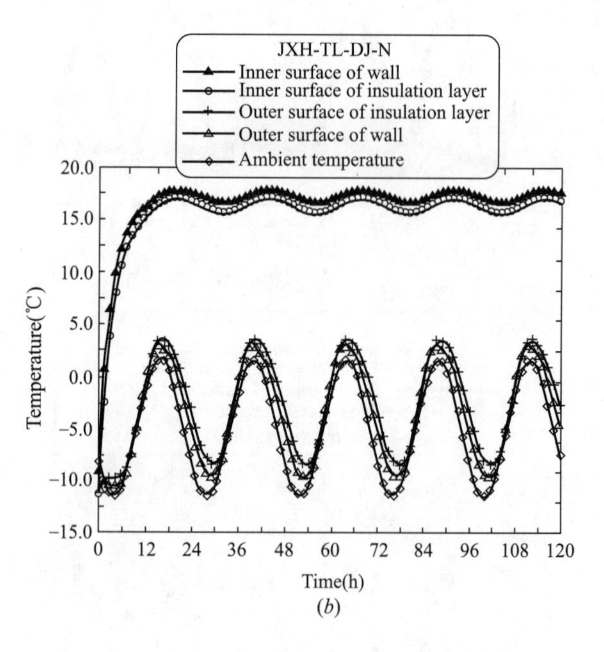

Figure 2-1-18 Temperature Changes of Different Layers of Walls with Paint Finish of Rock Wool Sandwich Insulation over Time in Winter

(*a*) southern wall; (*b*) northern wall;

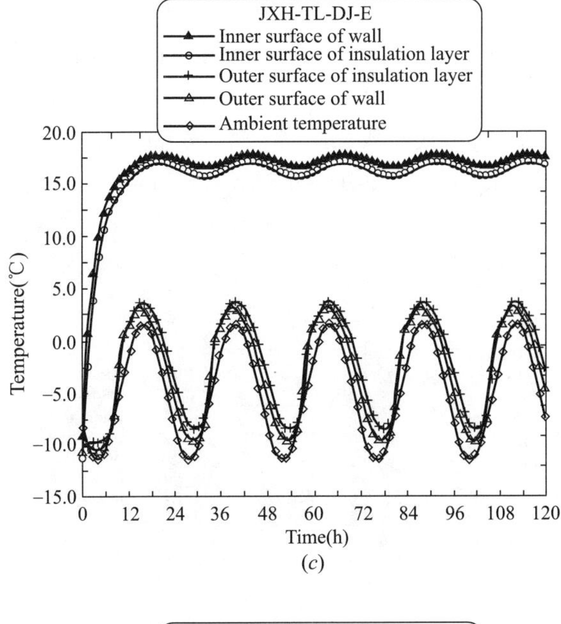

(c)

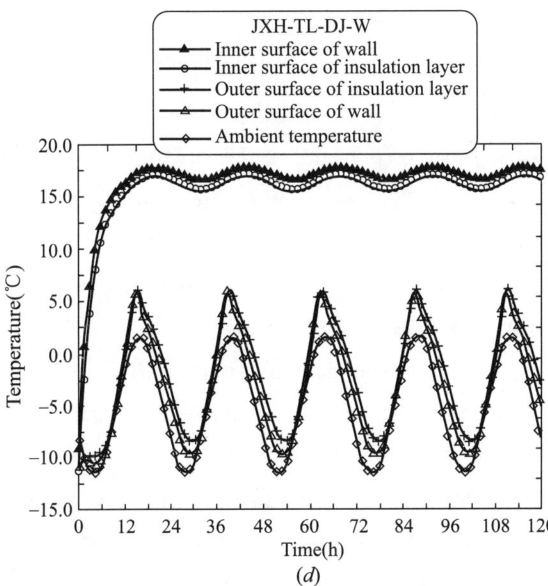

(d)

Figure 2-1-18 Temperature Changes of Different Layers of Walls with Paint Finish of Rock Wool Sandwich Insulation over Time in Winter (Continued)

(*c*) eastern wall; (*d*) western wall

5. Analysis of temperature differences of critical parts of walls with four types of insulation over time

The results of comparison of temperature changes of critical structures of walls with different types of insulation over time in Figure 2-1-11 to 2-1-20 are presented in Figure 2-1-21 to 2-1-24 (involving the outer surface of the outer walls and inner surface of the inner wall in the case of sandwich insulation).

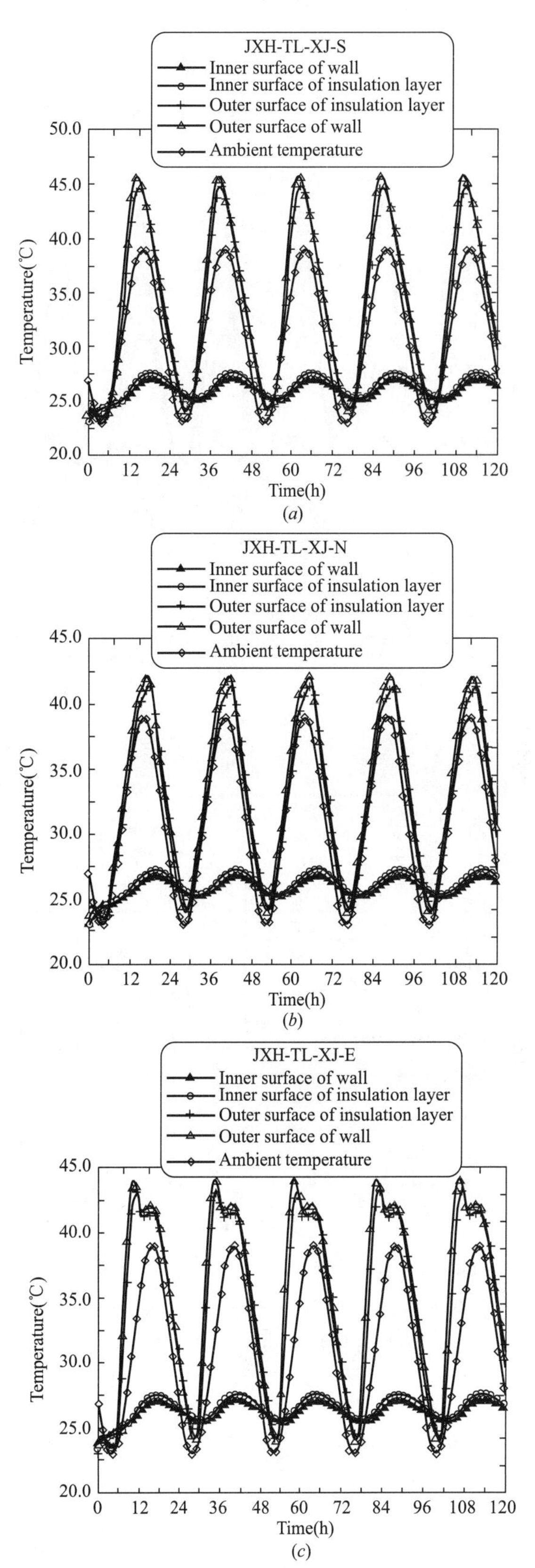

(c)

Figure 2-1-19 Temperature Changes of Different Layers of Walls with Paint Finish of Rock Wool Sandwich Insulation over Time in Summer

(*a*) southern wall; (*b*) northern wall;

(*c*) eastern wall;

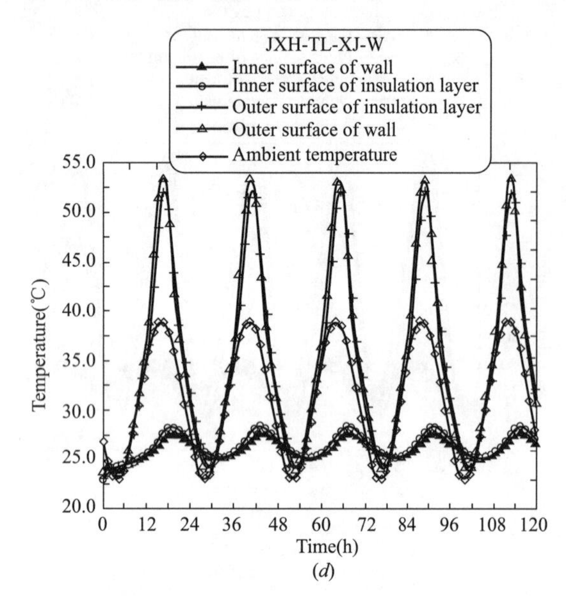

Figure 2-1-19 Temperature Changes of Different Layers of Walls with Paint Finish of Rock Wool Sandwich Insulation over Time in Summer (Continued)
(*d*) western wall

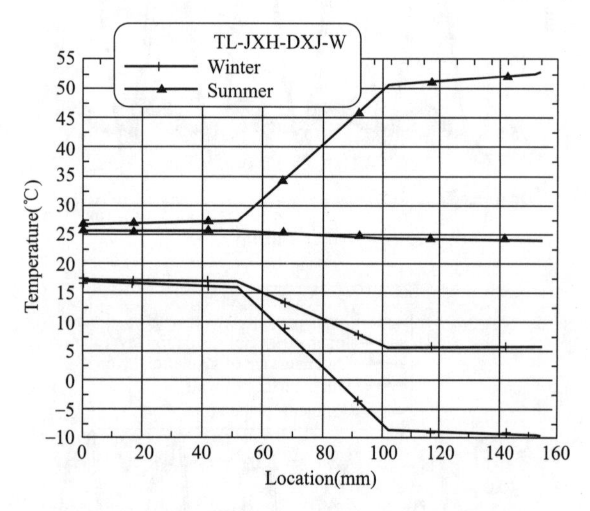

Figure 2-1-20 Maximum and Minimum Surface Temperature Distribution of Insulated Western Wall in the Thickness Direction in Winter and Summer

Refer to Figure 2-1-21 for the annual temperature differences of critical parts of walls with different types of insulation. The annual temperature difference of the inner surface of the wall subject to external thermal insulation, self-insulation or sandwich insulation is small, while that of the wall subject to internal thermal insulation is relatively large. The annual temperature difference of the outer surface of the wall subject to external thermal insulation is small, while that of the wall subject to the other three types of insulation is relatively large. Regardless of the insulation type, the temperature difference of the inner surface of the insulation layer is small while that of its outer surface is large.

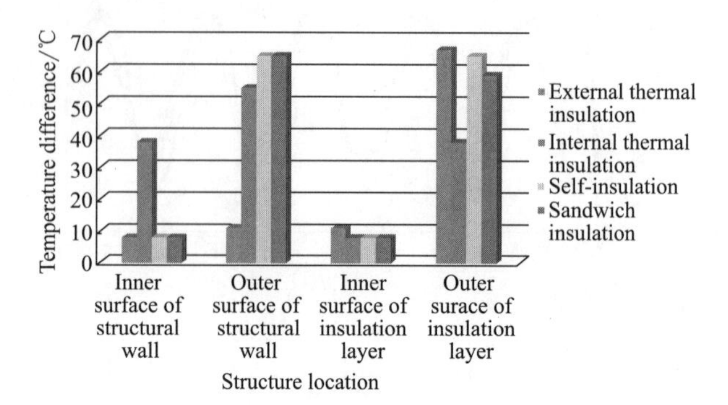

Figure 2-1-21 Annual Temperature Differences of Critical Parts of Walls with Different Types of Insulation

As shown in Figure 2-1-22, the temperature of critical parts of walls with different types of insulation varies greatly between the daytime and night in summer. The temperature differences of the outer surfaces of structural walls are relatively small between the daytime and night, less than 2℃ (minimum) in the presence of external thermal insulation and 19-30℃ in the presence of the other three types of insulation. The temperature differences of the inner surfaces of structural walls are also relatively small between the daytime and night. Due to protection by the wall, the temperature difference of the outer surface of the internal insulation layer is small, about 2℃. For the other types of insulation, the temperature difference is much great, about 25-35℃.

Figure 2-1-23 shows the temperature differences of critical parts of walls with different types of insulation between the daytime and night in winter. From Figure 2-1-22 and 2-1-23, it is indicated that the temperature differences in winter are similar to those in summer. As the difference between the outdoor and indoor temperature in winter is smaller than that in summer, however, the temperature differences between the daytime and night in winter are slightly smaller than those in summer.

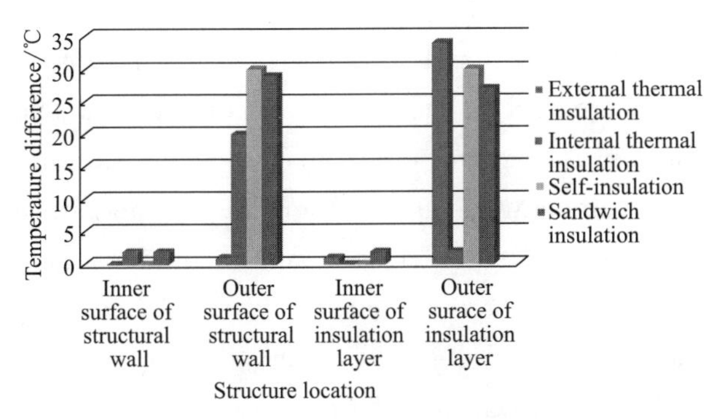

Figure 2-1-22 Temperature Differences of Critical Parts of Walls with Different Types of Insulation between Daytime and Night (Summer)

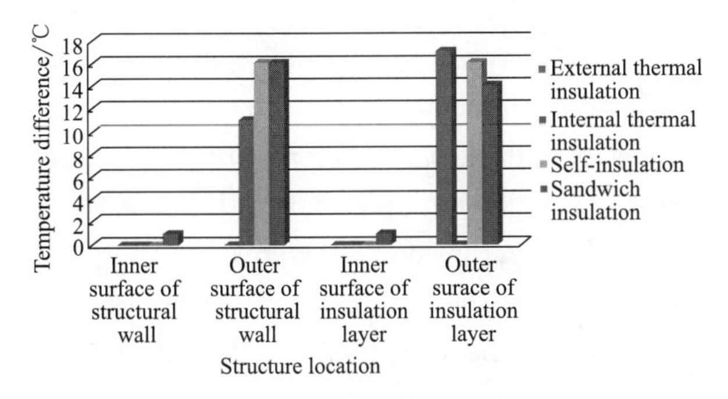

Figure 2-1-23 Temperature Differences of Critical Parts of Walls with Different Types of Insulation between Daytime and Night (Winter)

Figure 2-1-24 shows the temperature differences of the inner and outer surfaces of structural walls under the maximum and minimum temperature in different seasons. It can be seen that the temperature differences of the inner and outer surfaces of structural walls are small in the presence of external thermal insulation, but obvious in the presence of internal thermal insulation. For structural walls with self-insulation and sandwich insulation, the temperature differences of their inner and outer surfaces are remarkable in the hottest summer and coldest winter.

The location of the insulation layer affects the temperature change of the structural wall, while the resulting thermal stress affects the wall stability. Unreasonable insulation will aggravate the damage of thermal stress to the structural

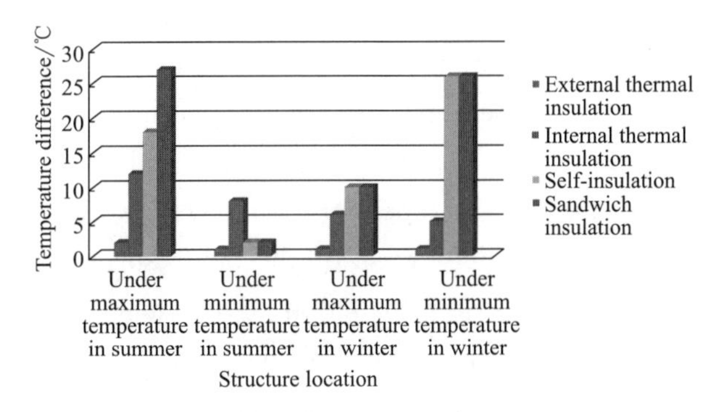

Figure 2-1-24 Temperature Differences of Inner and Outer Surfaces of Structural Walls in Different Seasons

wall and shorten its life.

Among four types of insulation, the annual temperature differences of the inner and outer surfaces of the structural wall with external thermal insulation are the smallest, 8℃ for the inner surface and 11℃ for the outer surface. The temperature difference of the inner surface of the wall with internal thermal insulation is 38℃, while that of the outer surface is 55℃, which are five times of that of the structural wall with external thermal insulation, resulting in far lower stability. In the case of self-insulation and sandwich insulation, the temperature of the inner surface of the structural wall is relatively stable, but the annual temperature difference of the inner surface is 8℃ and that of the outer surface reaches 65℃. Therefore, the structural walls with internal thermal insulation, self-insulation and sandwich insulation are in the instable state in a long time due to drastic changes in the temperature.

With the external thermal insulation, the temperature change of the inner surface of the structural wall is always the same as that of the outer surface, despite changes in the ambient temperature, and the temperature difference between the inner and outer surfaces is less than 2℃ (Figure 2-1-21). For the other types of insulation, the temperature change of the inner surface is always different from that of the outer surface, resulting in differences in the deformation speed and amount and finally affecting the life of the structural wall. Take the sandwich insulation as an example. Under the maximum temperature in summer, the temperature difference between the outer surface of the outer structural wall and the inner surface of the inner structural wall is 30℃, and the expansion stress of the outer surface is significantly greater than that of the inner surface, making the temperature difference 28℃ higher than that with external thermal insulation and affecting the structural stability.

The outer side of the insulation layer of the external thermal insulation system is subject to severe changes in a year.

The temperature changes of the plaster layer and finish layer of the wall with the external thermal insulation are greater than those with other types of insulation (meeting the same energy-saving requirements) at daytime and night and in four seasons. The annual temperature change of the exterior finish of external thermal insulation is up to 67℃ (−10℃ to 57℃).

Therefore, the plaster layer and finish layer of the wall with the external thermal insulation should comply with higher requirements for the resistance to temperature deformation and fatigue thermal stress.

2.1.4 Conclusions

Based on the calculation and analysis of the rules of changes in the temperature fields of walls with various types of insulation in the thickness direction in four seasons in Beijing, the following conclusions can be drawn.

(1) A model has been established for numerical calculation of the real-time temperature field of the outer wall of the building, taking into account the solar radiation and changes in the ambient temperature. The temperature distribution and rules of temperature changes of the wall over time can be calculated easily and quickly with this model. The establishment of this model lays a foundation for calculation of the thermal stress of each functional layer of the external thermal insulation system.

(2) The temperature distribution of the outer wall of the building is significantly affected by solar radiation, and the extent of such impact is closely related to the season and orientation. The surface of the western wall has the maximum peak temperature and the greatest temperature fluctuation in summer; the surface of the southern wall has the maximum peak temperature and the greatest temperature fluctuation in winter; the northern wall has the minimum average temperature and the smallest temperature fluctuation in four seasons; and the temperature of walls in spring and autumn falls within that in winter and summer.

(3) Regardless of internal or external thermal insulation, aerated concrete self-insulation or rock wool sandwich insulation, the temperature inside the insulation layer changes the most drastically, but that inside the external thermal insulation layer changes more. The temperature of parts outside the insulation layer changes more than that of parts inside the insulation layer.

(4) The annual temperature difference of the outer wall with the internal thermal insulation is five times of that with external thermal insulation. The annual temperature difference of the outer wall of the structure with sandwich insulation is eight times of that of the inner wall. The annual temperature difference of the outer surface of the self-insulated wall is also eight times of that of the inner surface. Therefore, the internal thermal insulation, sandwich insulation and self-insulation will reduce the stability of the structural wall and shorten its life.

(5) Currently, the design life of the concrete structure is roughly 70 years. The external thermal insulation is conducive to the stability of concrete walls. In order to achieve the 100-year life of buildings, the external thermal insulation should be vigorously promoted.

(6) The external thermal insulation is conducive to the maintenance and repair of the plaster

layer and its finish. If there is any problem in the external thermal insulation surface and even insulation layer, the functions of energy conservation and insulation can be retained by means of repair and renovation to continuously protect the structural wall.

(7) The temperature of the external thermal insulation surface changes the most severely, so its plaster layer and finish layer should meet higher requirements for resistance to the temperature deformation and fatigue thermal stress.

2.2 Thermal Stress Calculation of Insulated Wall

The rules of changes in temperature fields of walls with the internal thermal insulation, external thermal insulation, no insulation, self-insulation and sandwich insulation under the ambient temperature change and solar radiation in Beijing have been obtained by means of numerical simulation. In addition to analysis of the changes in the wall temperature, insulation effects of the insulation layer, differences in insulations and design needs of insulation structures, another purpose of temperature field calculation is to study the amount and change rules of thermal stress arising from temperature changes inside walls with different types of insulation, as well as the long-term durability problems caused by temperature changes of the insulation layer and its additional functional layers under operating conditions. This section will describe the numerical simulation of the thermal stress of each layer of walls with different types of insulation under ambient temperature changes, and analyze the potential cracking of walls as well as safety and durability problems of finishes under the thermal stress.

2.2.1 Model for Thermal Stress Calculation of Insulated Wall

2.2.1.1 Model of Wall Thermal Stress

Generally, the length and width of the outer wall of a building are much greater than its thickness (usually 15-20 times). With the ambient temperature changing, the temperature of the outer wall will change only in the thickness direction (z). In this sense, the temperature field (T) of a wall can be regarded as a function of the time (t) and wall thickness (t), namely:

$$T=f(t,z) \tag{2-2-1}$$

Given no impact of local structures with the door or window, the outer wall of a building can be deemed as a wall of infinite length and height. Thus, this study focuses on the distribution and amount of thermal stress in the thickness direction of the wall.

It is assumed that the plane size of a wall is greater than 10 times of its thickness (plane stress), the height is in the y direction, width in the x direction and z direction (Figure 2-2-1), the temperature changes only in the thickness direction at the given time, the modulus of elasticity of the material is E , the Poisson's ratio is μ , the thermal deformation coefficient is α , and the initial temperature (elastic modulus: zero) is T_0, $T_0=f(z,\ t_0)$. According to the generalized Hooke's law,

$$\begin{cases} \varepsilon_x=\dfrac{1}{E}(\sigma_x-\mu\sigma_y)+\alpha(T-T_0) \\ \varepsilon_y=\dfrac{1}{E}(\sigma_y-\mu\sigma_x)+\alpha(T-T_0) \end{cases} \tag{2-2-2}$$

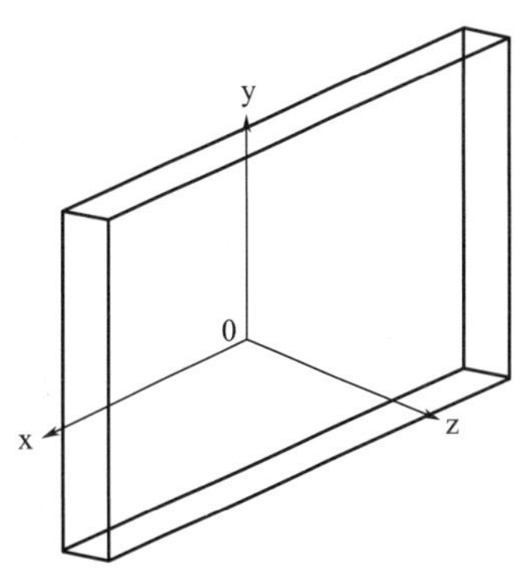

Figure 2-2-1 Schematic Diagram of Coordinates for Wall Thermal Stress Calculation

The thermal stress arising from the temperature change ($T-T_0$) was calculated based on the categories of wall constraints.

1. Temperature stress of embedded wall

The embedded wall refers to a wall that is constrained completely (in the x and y direction) and not able to turn or move up and down or right and left. Under complete constraints, $\varepsilon_x = 0$ and $\varepsilon_y = 0$. If there conditions are substituted in the equation (2-2-2), then:

$$\begin{cases} \dfrac{\sigma_x - \mu\sigma_y}{E} + \alpha(T - T_0) = 0 \\ \dfrac{\sigma_y - \mu\sigma_x}{E} + \alpha(T - T_0) = 0 \end{cases} \quad (2\text{-}2\text{-}3)$$

Solve this equation set, then:

$$\sigma_x = \sigma_y = -\frac{E\alpha(T - T_0)}{\mu - 1} \quad (2\text{-}2\text{-}4)$$

Meanings on the sign of thermal stress: if the temperature rises, the $(T - T_0)$ will be positive, and the σ_x or σ_y will be negative, indicating that the compressive stress is generated; and if the temperature drops, the $(T - T_0)$ will be negative, the σ_x or σ_y will be positive, indicating that the tensile stress is generated. The sign of thermal stress mentioned below has the same meanings.

2. Temperature stress of free wall

The free wall refers to a wall that is completely free of external constraints and can be freely deformed in all directions. The thermal stress inside the free wall is purely a kind of self-stress resulting from the uneven temperature distribution in the wall. The normal stress and normal strain in the wall is:

$$\begin{cases} \sigma_x = \sigma_y = \sigma_3 \\ \sigma_z = 0 \\ \varepsilon_x = \varepsilon_y = \varepsilon \end{cases} \quad (2\text{-}2\text{-}5)$$

If these conditions are substituted in the equation (2-2-5), then:

$$\sigma_x = \sigma_y = \sigma_3 = \frac{E[\varepsilon - \alpha(T - T_0)]}{1 - \mu} \quad (2\text{-}2\text{-}6)$$

In the case of wall deformation due to temperature changes, the strain distribution in the thickness direction of the wall should be in line with the plane assumptions. That is, ε can be expressed as:

$$\varepsilon = \alpha(A + Bz) \quad (2\text{-}2\text{-}7)$$

Where, A and B are parameters unassociated with the coordinate z. If the equation (2-2-7) is substituted into the equation (2-2-6), then:

$$\sigma_3 = \frac{E\alpha}{1 - \mu}[A + Bz - (T - T_0)] \quad (2\text{-}2\text{-}8)$$

The axial force and bending moment over any section at any time in the free wall should be zero, i. e. :

$$\int_{-d/2}^{d/2} \sigma \cdot dz = \int_{-d/2}^{d/2} \frac{E\alpha}{1 - \mu}[A + Bz - (T - T_0)]dz = 0$$

$$\int_{-d/2}^{d/2} \sigma \cdot zdz = \int_{-d/2}^{d/2} \frac{E\alpha}{1 - \mu}[A + Bz - (T - T_0)]zdz = 0 \quad (2\text{-}2\text{-}9)$$

Where, d is the thickness of the wall.

It should be noted that the origin of the section is its center. If the equation (2-2-9) is solved, then:

$$A = \frac{1}{d}\int_{-d/2}^{d/2}(T - T_0)dz = T_a$$

$$B = \frac{12}{d^3}\int_{-d/2}^{d/2}(T - T_0)zdz = \frac{12}{d^3}S = \frac{T_d}{d} \quad (2\text{-}2\text{-}10)$$

Where, $T_d = \dfrac{12S}{d^2}$; $S = \int_{-d/2}^{d/2}(T - T_0)zdz$; T_a is the average temperature change of the section (along the d); A is the average temperature change; T_d is the difference of the equivalent linear temperature change; and S is the torque of the temperature change $(T - T_0)$ along the section d.

The thermal stress of the free wall can be calculated by substituting the equation (2-2-10) into (2-2-8):

$$\begin{aligned} \sigma_3 &= \frac{E\alpha}{1 - \mu}A + \frac{E\alpha}{1 - \mu}Bz - \frac{E\alpha(T - T_0)}{1 - u} \\ &= -\left(-\frac{E\alpha}{1 - \mu}A\right) - \left(-\frac{E\alpha}{1 - \mu}Bz\right) + \left[-\frac{E\alpha(T - T_0)}{1 - \mu}\right] \\ &= -\sigma_1 - \sigma_2 + \sigma_T \end{aligned} \quad (2\text{-}2\text{-}11)$$

Where, σ_1 is the stress caused by the average temperature change; σ_2 is the stress caused by the linear temperature change; and σ_T is the thermal stress of the embedded wall (constrained completely).

The thermal stress σ of the embedded wall under the nonlinear temperature field can be obtained by simply transforming the equation (2-2-11), consisting of three stress components, i. e. :

$$\sigma_T = \left(-\frac{E\alpha}{1-\mu}A\right) + \left(-\frac{E\alpha}{1-\mu}Bz\right) + \left[-\frac{E\alpha[(T-T_0)-A-Bz]}{1-u}\right] = \sigma_1 + \sigma_2 + \sigma_3 \quad (2\text{-}2\text{-}12)$$

Where, the thermal stress σ_3 of the free wall is also known as the stress arising from the nonlinear temperature change.

The equation (2-2-12) shows the thermal stress of the completely constrained wall can be expressed as a sum of the stress (σ_1) caused by the average temperature change, stress (σ_2) caused by linear temperature change, and stress (σ_3) caused by the nonlinear temperature change. If the temperature distribution is linear, then $\sigma_3 = 0$.

Actually, the thermal stress inside the wall depends on the constraints of adjacent structures to the wall. Usually, there are four cases:

(1) Wall stretching and turning (free wall) are allowed:

$$\sigma = \sigma_3 = \sigma_T - \sigma_1 - \sigma_2 \quad (2\text{-}2\text{-}13)$$

(2) Wall turning is allowed but stretching is not:

$$\sigma = \sigma_T - \sigma_2 \quad (2\text{-}2\text{-}14)$$

(3) Wall stretching is allowed but turning is not:

$$\sigma = \sigma_T - \sigma_1 \quad (2\text{-}2\text{-}15)$$

(4) Wall stretching and turning are not allowed:

$$\sigma = \sigma_T \quad (2\text{-}2\text{-}16)$$

The thermal stress of a non-constrained wall can be determined by the above expression for the free wall. For the constrained wall, the stress distribution around the wall is different from that in the above expression. According to Saint-Venant's principle, however, the above results will apply if the distance to the wall edge exceeds the wall thickness.

2.2.1.2 Thermal Stress Model for Composite Wall

The outer wall with internal or external thermal insulation is a kind of composite wall with several functional materials in the thickness direction. Its main functional layers include the interior or exterior finish layer, structural layer, insulation layer and additional layers for insulation. The thermal stress calculation model was established for the multi-layer composite wall under the time-varying temperature field.

It is assumed that all layers are bonded properly, the temperature distribution function of each layer is $T_i = f(z_i, t)$, and $z_i = 0$ is located in the middle of each layer. According to the plane assumption, the strain of each layer is:

$$\begin{cases} \varepsilon_1(z) = \alpha_1(A_1 + B_1 z_1) \\ \varepsilon_2(z) = \alpha_2(A_2 + B_2 z_2) \\ \varepsilon_3(z) = \alpha_3(A_3 + B_3 z_3) \\ \cdots\cdots \\ \varepsilon_n(z) = \alpha_n(A_n + B_n z_n) \end{cases} \quad (2\text{-}2\text{-}17)$$

Where, the temperature distribution parameters A_i and B_i of each layer can be obtained based on its temperature field, i. e. :

$$\begin{cases} A_1 = \dfrac{1}{h_1}\displaystyle\int_{-d_1/2}^{d_1/2}[T_1(z_1) - T_{10}]dz_1 \\ B_1 = \dfrac{12}{d_1^{\,3}}\displaystyle\int_{-d_1/2}^{d_1/2}[T_1(z_2) - T_{10}]z_2 dz_2 \\ \cdots\cdots \\ A_n = \dfrac{1}{h_n}d\displaystyle\int_{-d_n/2}^{d_n/2}[T_n(z_n) - T_{n0}]dz_n \\ B_n = \dfrac{12}{h_n^{\,3}}\displaystyle\int_{-d_n/2}^{d_n/2}[T_n(z_n) - T_{n0}]z_n dz_n \end{cases} \quad (2\text{-}2\text{-}18)$$

Therefore, the thermal stress components of the ith layer of the wall can be expressed as:

$$\sigma_{iT} = -\frac{E_i\alpha_i(T_i - T_{i0})}{1-\mu} \quad (2\text{-}2\text{-}19)$$

$$\sigma_{i1} = -\frac{E_i\alpha_i}{1-\mu}A_i \quad (2\text{-}2\text{-}20)$$

$$\sigma_{i2} = -\frac{E_i \alpha_i}{1-\mu} B_i z_i \quad (2\text{-}2\text{-}21)$$

Where, σ_{iT} is the thermal stress of the ith layer of the completely constrained wall; σ_{i1} is the stress caused by the average temperature change of the ith layer; and σ_{i2} is the stress caused by the linear temperature change of the ith layer.

There are usually four cases depending on the actual constraints:

(1) Wall stretching and turning are allowed (free wall):

$$\sigma_i = \sigma_{i3} = \sigma_{iT} - \sigma_{i1} - \sigma_{i2} \quad (2\text{-}2\text{-}22)$$

(2) Wall turning is allowed but stretching is not:

$$\sigma_i = \sigma_{iT} - \sigma_{i2} \quad (2\text{-}2\text{-}23)$$

(3) Wall stretching is allowed but turning is not:

$$\sigma_i = \sigma_{iT} - \sigma_{i1} \quad (2\text{-}2\text{-}24)$$

(4) Wall stretching and turning are not allowed:

$$\sigma_i = \sigma_{iT} \quad (2\text{-}2\text{-}25)$$

Based on the constraints and forces, the composite wall with an insulation layer can be roughly divided into a structural layer mainly composed of a concrete or block base course wall, insulation layer, and protective structure (known as the additional layer) attached on the insulation layer. Most of constraints result from bonding with the base course wall, such as anchorage. The mechanical properties of the additional layer are significantly different from those of the structural layer. If the insulation layer is built as a main part, on one hand, the additional layer itself has no solid constraining effect, but is deformed along with the structural layer. On the other hand, the additional layer is normally small and the mechanical properties (such as the elastic modulus) of its materials far lower than those of concrete and other materials of the structural layer. By comparison, the structural layer is obviously softer than the additional layer. The structural layer has dominant impact on the additional layer. In turn, the additional layer only has the limited impact on the structural layer. The constraints imposed by the architectural structure on the structural layer are relatively complex, usually including mutual constraints of floor slabs, connecting beam slabs and adjacent walls. This is the basic reason why the thermal stress within an architectural structure under normal conditions cannot be quantified. In short, main constraints to the additional layer are imposed by the structural layer, and the temperature deformation difference between the structural layer and additional layer is one of main causes of thermal stress inside the additional layer.

The thermal stress of each layer under simple constraints was quantitatively calculated with the above thermal stress calculation model. At the same time, the distribution and development rules of thermal stress inside walls in different periods and seasons were qualitatively studied. This provides theoretical guidance for the structural design of the insulated wall, especially the material design and selection of the finish layer of the external thermal insulation.

2. 2. 1. 3 Material Parameters

The thermal stress was calculated with the above-mentioned typical wall insulation forms and models. The thermodynamic parameters of related materials used in the calculation are listed in Table 2-2-1 and 2-2-2.

2. 2. 2 Calculation Results and Analysis of Thermal Stress of Insulated Wall

The thermal stresses of walls with typical internal and external thermal insulation, walls with aerated concrete self-insulation and walls with concrete and rock wool sandwich insulation were calculated with the calculation results of the temperature field and the parameters (as the model input parameters) listed in Table 2-2-1 and 2-2-2. The initial temperature (T_0) of 15℃ was applied in the calculation. The true physical meaning of this parameter is the temperature corresponding to the thermal stress of zero. The specific value

Thermodynamic Parameters of Materials (single insulation based on adhesive polystyrene granule) (Ⅰ) Table 2-2-1

Structure Type	Material Name	Thickness (mm)	Density (kg/m³)	Thermal Deformation Coefficient $\times 10^{-6}$(1/K)	Elastic Modulus (GPa)
Paint finish for external thermal insulation of outer wall based on adhesive polystyrene granule	Inner finish	2	1300	10	2.00
	Base course wall	200	2300	10	20.00
	Interface mortar	2	1500	8.5	2.76
	Insulation mortar	60	200	8.5	0.0001
	Anti-cracking mortar	5	1600	8.5	1.50
	Paint finish	3	1100	8.5	2.00
Paint finish for internal thermal insulation of outer wall based on adhesive polystyrene granule	Inner finish	2	1300	10	2.00
	Anti-cracking mortar	5	1600	8.5	1.50
	Insulation mortar	60	200	8.5	0.0001
	Interface mortar	2	1500	10	0.76
	Base course wall	200	2300	10	20.00
	Paint finish	3	1100	8.5	2.00
Facing brick finish for external thermal insulation of outer wall based on adhesive polystyrene granule	Inner finish	2	1300	10	2.00
	Base course wall	200	2300	10	20.00
	Interface mortar	2	1500	8.5	2.76
	Insulation mortar	60	200	8.5	0.0001
	Anti-cracking mortar	10	1600	10	3.87
	Bonding mortar	5	1500	10	5.00
	Facing brick finish	8	2600	10	20.00
Facing brick finish for internal thermal insulation of outer wall based on adhesive polystyrene granule	Inner finish	2	1300	10	2.00
	Anti-cracking mortar	5	1600	10	3.87
	Insulation mortar	60	200	8.5	0.0001
	Interface mortar	2	1500	8.5	2.76
	Base course wall	200	2300	10	20.00
	Bonding mortar	10	1500	10	5.00
	Facing brick finish	8	2600	10	20.00

Thermodynamic Parameters of Materials (Aerated Concrete Self-insulation and Rock Wool Sandwich Insulation) (Ⅱ) Table 2-2-2

Structure Type	Material Name	Thickness (mm)	Density (kg/m³)	Thermal Deformation Coefficient $\times 10^{-6}$(1/K)	Elastic Modulus (GPa)
Wall paint finish of self-insulation with aerated concrete	Inner finish	2	1300	10	2.00
	Inner surface mortar	20	1800	10	20.00
	Aerated concrete	200	700	10	2.00
	Outer surface mortar	20	1800	10	20.00
	Paint finish	3	1100	8.5	2.00
Wall paint finish of sandwich insulation with concrete and rock wool	Inner finish	2	1300	10	2.00
	Concrete slab	50	2300	10	20.00
	Rock wool board	50	150	8.5	0.10
	Concrete slab	50	2300	10	20.00
	Paint finish	3	1100	8.5	2.00

is difficult to determine in the actual situation. For the cast-in-situ concrete or mortar, it is the initial setting (transition point from the plastic to elastic status of cement slurry) temperature of concrete or mortar, and closely associated with the construction season and time. As cracking is more likely to occur in concrete structures in hot seasons and periods, raw materials are often cooled. That is, the value of T_0 is decreased.

As the structural layer, insulation and additional layer of the insulated wall are completed in different periods, it is more difficult to determine the value of T_0 in the thermal stress calculation. For unified comparison of calculation results, the same initial temperature was selected to all layers in the calculation.

Due to the greatest changes in the temperature in winter and summer, the resulting stress inside the wall is the largest. Therefore, the thermal stresses inside walls, corresponding to the greatest changes in the temperature in winter and summer, were calculated.

2.2.2.1 Wall with External Thermal Insulation by Paint Finish based on Adhesive Polystyrene Granule

Figure 2-2-2 to 2-2-4 show the relationship between the time (in 24h) and the thermal stress of the exterior finish, the outer surface of the insulation mortar and the outer surface of the bearing base course wall of the western wall with the external thermal insulation by the paint finish based on the adhesive polystyrene granule under typical climatic conditions in winter and summer, respectively. They illustrate the thermal stress inside the wall surface in the presence of four typical constraints and under the same temperature changes. These four typical constraints are as follows: (1) wall stretching and turning are not allowed (embedded wall); (2) wall turning is allowed but stretching is not; (3) wall stretching is allowed but turning is not; and (4) wall stretching and turning are allowed (free wall).

As shown by Figure 2-2-2:

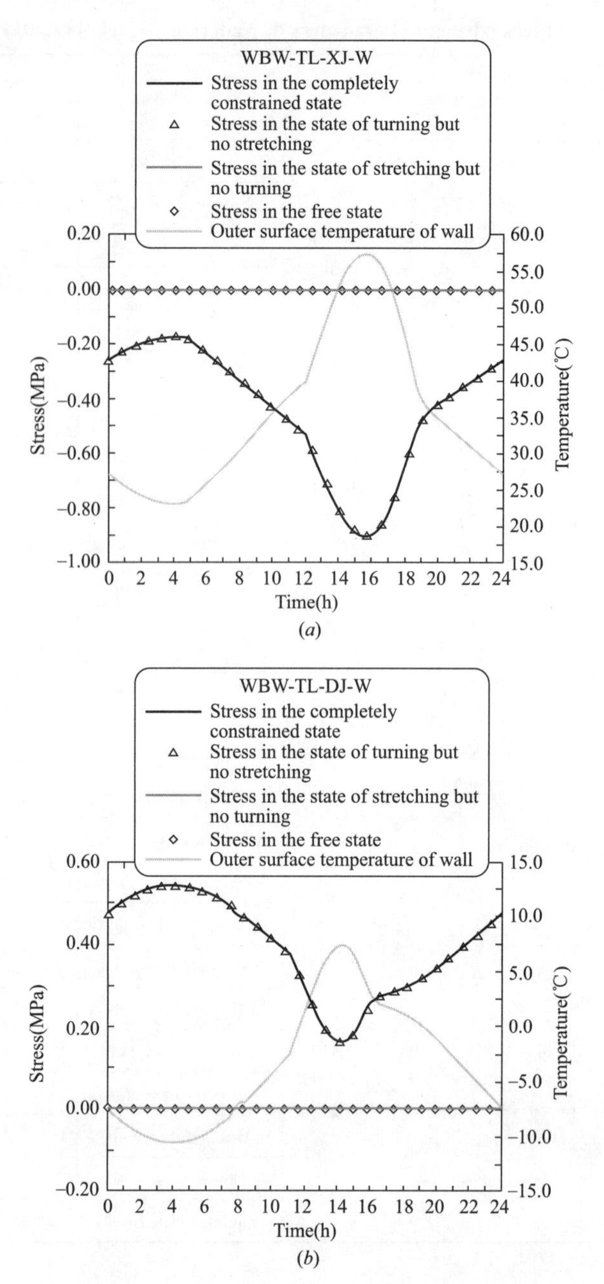

Figure 2-2-2 Changes in Stress of Outer Surface of Wall with External Thermal Insulation by Paint Finish based on Adhesive Polystyrene Granule over Time

(a) Winter; (b) Summer

(1) There is a small difference in thermal stress of the outer surface of the finish layer under the first two types of constraint, indicating that the bending stress of the outer surface of the wall is small. That is, the temperature difference between the inner and outer surfaces of the finish layer is small, and the thermal stress is mainly caused by horizontal and vertical constraints.

(2) The thermal stress of the free wall is

close to zero, indicating the small nonlinearity of temperature distribution inside the exterior finish layer (the greater the nonlinearity, the larger the free stress).

(3) The thermal stress increases and decreases at the same time with the temperature. The peak temperature corresponds to the peak stress.

(4) The wall surface is subject to tensile stress in winter and compressive stress in summer, so it is more likely to crack in winter.

(5) When the initial temperature is 15℃, the surface is subject to tensile stress (maximum value: 0.54MPa) in winter and compressive stress (maximum value: 0.9MPa) in summer. The stress of the wall surface changes from 0.54MPa to 0.16MPa in winter and from −0.90MPa to −0.17MPa in summer on a daily basis. Therefore, the stress magnitude of the wall surface is −0.90MPa to 0.54MPa in a year (winter and summer).

The actual thermal stress falls between the thermal stress under complete constraints and that in the case of free deformation, and depends on constraints (although actual constraints are difficult to determine, the calculated stress values can be compared, and their sequence is constant).

As shown by Figure 2-2-3:

(1) Similar to the finish layer, there is a small difference in thermal stress under the first two types of constraint, indicating that the bending stress inside this layer is also small. That is, the temperature difference between the inner and outer surfaces of the insulation mortar layer is small, and the thermal stress is mainly caused by horizontal and vertical axial constraints.

(2) The thermal stress of the free wall is close to zero, indicating the small nonlinearity of temperature distribution inside the anti-cracking mortar layer (the greater the nonlinearity, the larger the free stress).

(3) The thermal stress decreases with the temperature rising and increases with the temperature falling. The peak temperature corresponds to the peak stress.

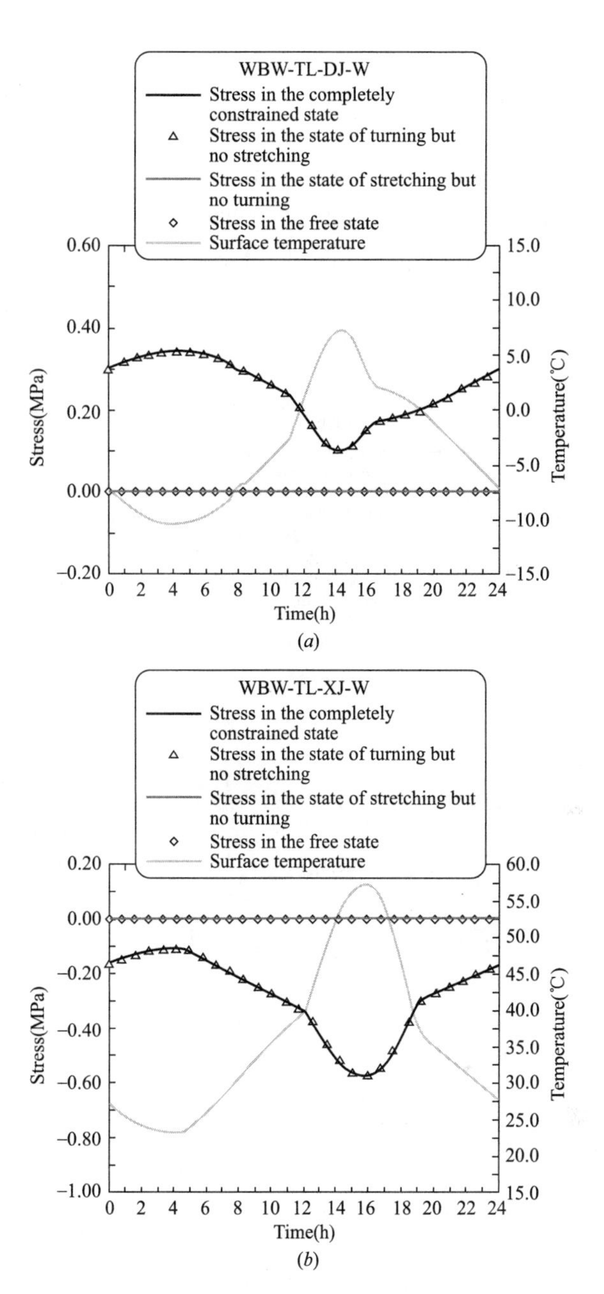

Figure 2-2-3 Stress Changes of Outer Surface of Insulation Mortar Layer of Wall with External Thermal Insulation by Paint Finish based on Adhesive Polystyrene Granule over Time

(*a*) Winter; (*b*) Summer

(4) When the initial temperature is 15℃, the surface of the insulation mortar layer is subject to tensile stress in winter and compressive stress in summer, but the magnitude of tensile and compressive stress is smaller than that of the exterior finish. The maximum tensile stress and compressive stress are 0.35MPa and −0.57MPa

in 24h, respectively.

(5) When the initial temperature is 15℃, the stress of the wall surface is from 0.10MPa to 0.35MPa in winter and from −0.57MPa to −0.11MPa in summer on a daily basis. Therefore, the stress magnitude of the inner surface is −0.57MPa to 0.35MPa in a year.

As shown by Figure 2-2-4:

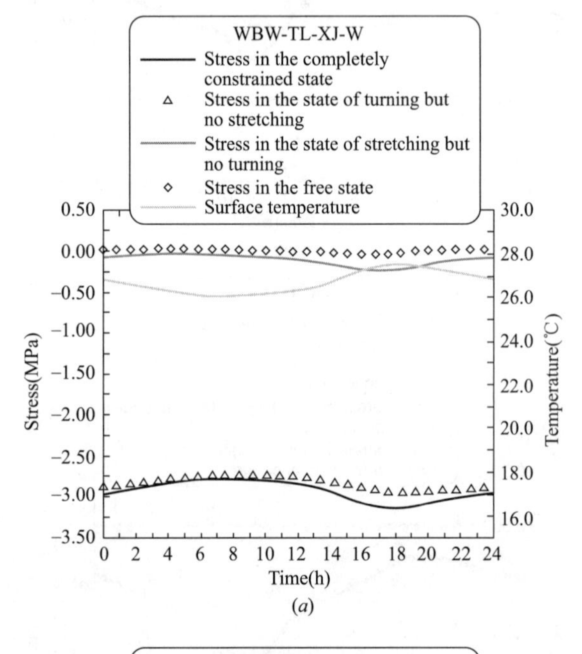

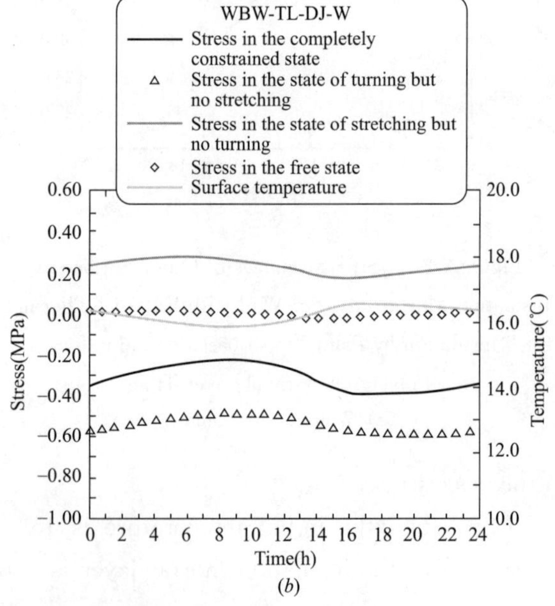

Figure 2-2-4 Stress Change of Base Course Wall Surface with External Thermal Insulation by Paint Finish based on Adhesive Polystyrene Granule over Time

(a) Winter; (b) Summer

(1) Although the temperature of the inner surface of the base course wall is not significantly different from that of the outer surface, the thermal stress of the thick base course wall wall are different under the first two types of constraint, indicating that the bending stress of the outer surface of the wall is greater than that of the exterior finish and anti-cracking mortar layer. Still, the thermal stress is mainly caused by horizontal and vertical axial constraints (involving the significant change in average temperature).

(2) The thermal stress of the free wall is close to zero, indicating the small nonlinearity of temperature distribution inside the base course wall.

(3) Similarly, the thermal stress decreases with the temperature rising and increases with the temperature falling. The peak temperature corresponds to the peak stress. As the temperature change inside the base course wall is not significant, the stress changes little.

(4) When the initial temperature is 15℃, the base course wall surface is still subject to compressive stress under complete constraints, even in winter. Therefore, the base course wall is almost unlikely to crack under temperature changes if the external thermal insulation is applied.

(5) When the initial temperature is 15℃, the stress of the base course wall is −0.39MPa to −0.22MPa in winter and the stress of the wall surface is −3.44MPa to −2.78MPa in summer on a daily basis.

The impact of material creep on the thermal stress was not taken into account in the above stress analysis. If this factor is taken into consideration, the thermal stress will decrease.

Figure 2-2-5 clearly shows the change in thermal stress of the surface of typical functional layers inside the wall with internal thermal insulation over time in 24h in winter and summer.

The distribution of thermal stress of the insulated wall in the thickness direction can reflect the rules of thermal stress distribution in each

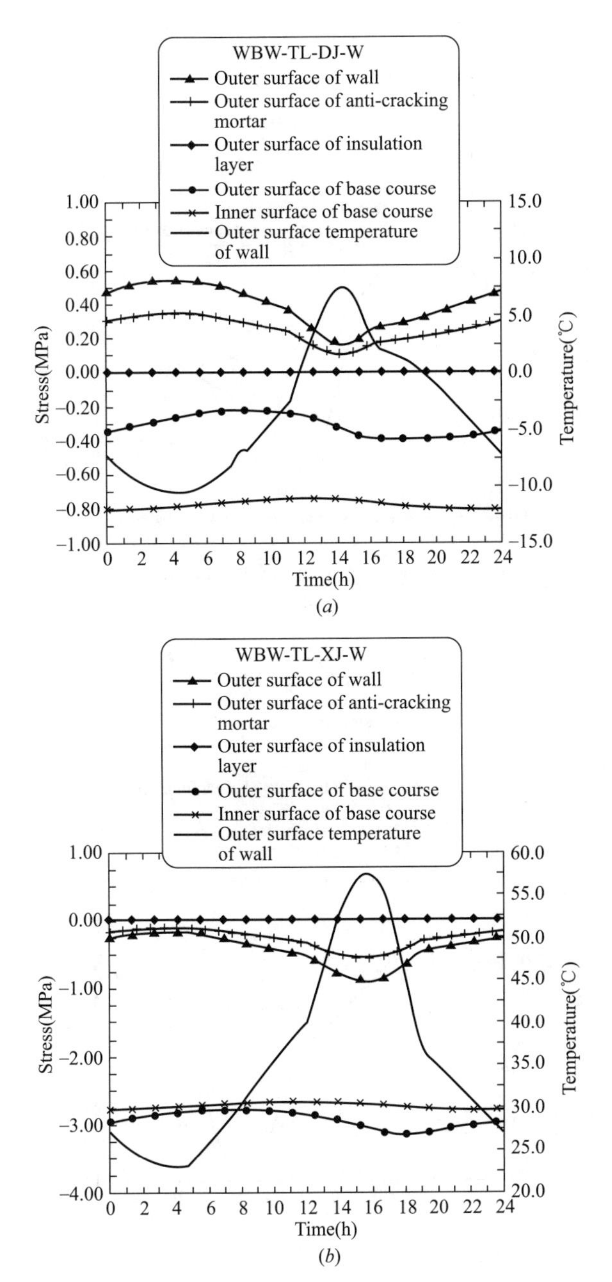

Figure 2-2-5 Changes in Surface Stress of Each Typical Functional Layer of Wall with External Thermal Insulation by Paint Finish based on Adhesive Polystyrene Granule over Time

(a) Winter; (b) Summer

layer more intuitively. Figure 2-2-6 shows the thermal stress distribution in the thickness direction of the completely constrained wall with the paint finish for external thermal insulation based on the adhesive polystyrene granule, corresponding to the minimum temperature of the outer surface in winter and its maximum temperature in summer. The temperature distribution in the thickness direction of the wall in the corresponding period is shown in Figure 2-2-5.

As can be seen from Figure 2-2-6, the thermal stress under complete constraints is distributed in a stepping form. In winter, the thermal stress gradually increases from indoors to outdoors, the base course wall is mainly subject to the compressive stress, and the insulation mortar layer and finish layer are under tensile stress; the stress inside the insulation layer is almost zero (the elastic modulus of the insulating material is only 0.1MPa); and the maximum tensile stress of the exterior finish layer is 0.54MPa. In summer, the wall is mainly subject to the compres-

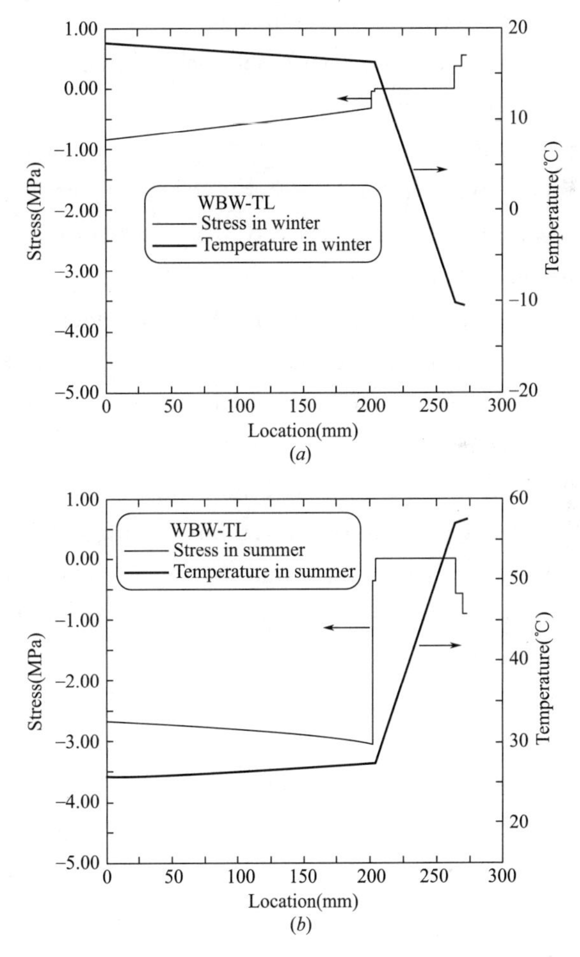

Figure 2-2-6 Thermal Stress Distribution in the Thickness Direction of Completely Constrained Wall with External Thermal Insulation by Paint Finish based on Adhesive Polystyrene Granule under Minimum and Maximum Temperature of Outer Surface in Winter and Summer

(a) Winter; (b) Summer

sive stress as the wall temperature is higher than the initial temperature (T_0); the internal thermal stress of the base course wall gradually increases from indoors to outdoors, and its maximum compressive stress is 3.44MPa; the stress inside the insulation layer is almost zero, and the maximum compressive stress of the finish layer is 0.9MPa.

2.2.2.2 Wall with External Thermal Insulation by Facing Brick Finish based on Adhesive polystyrene granule

Figure 2-2-7 to 2-2-10 show the relationship between the time (in 24h) and the thermal stress of the exterior finish, the outer surface of the insulation mortar and the outer surface of the bearing base course wall with the facing brick finish for external thermal insulation based on the adhesive polystyrene granule under typical climatic conditions in winter (southern wall) and summer (western wall), respectively. They illustrate the thermal stress inside the wall surface in the presence of four typical constraints and under the same temperature changes. These four typical constraints are as follows: (1) wall stretching and turning are not allowed (embedded wall); (2) wall turning is allowed but stretching is not; (3) wall stretching is allowed but turning is not; and (4) wall stretching and turning are allowed (free wall).

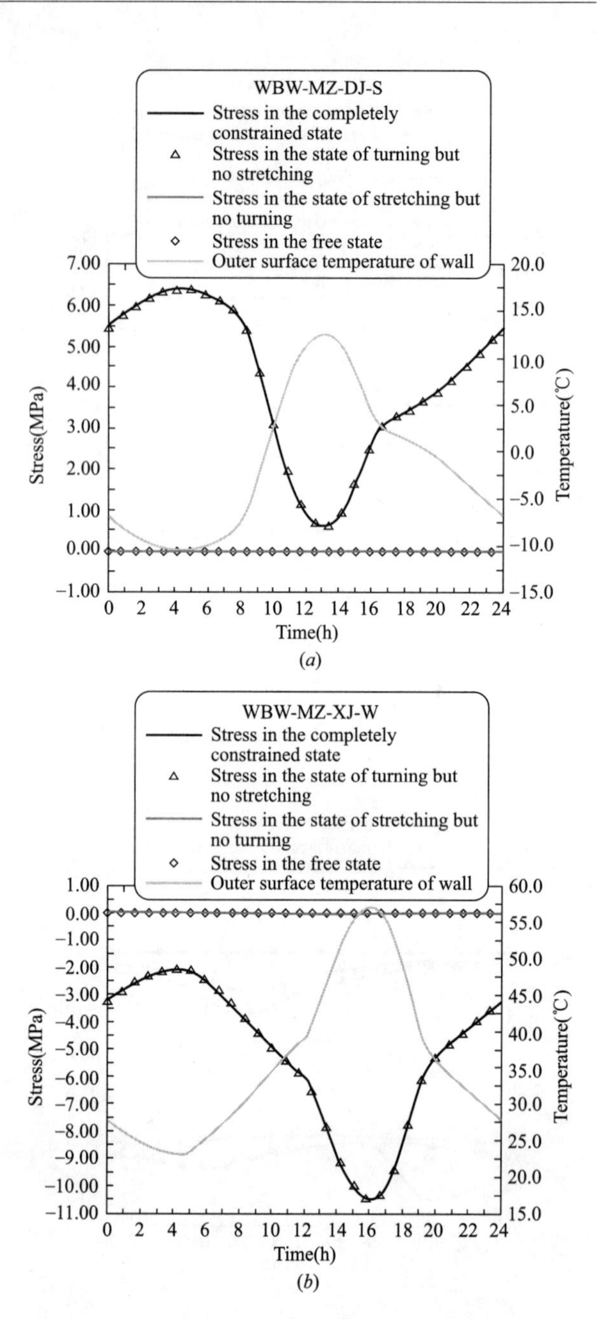

Figure 2-2-7 Changes in Stress of Outer Surface of Wall with External Thermal Insulation by Facing Brick Finish based on Adhesive Polystyrene Granule over Time

(*a*) Winter; (*b*) Summer

As shown by Figure 2-2-7:

(1) Similar to the paint finish, there is a small difference in the thermal stress of the outer surface of the finish layer under the first two types of constraint, indicating that the bending stress of the outer surface of the wall is small. That is, the temperature difference between the inner and outer surfaces of the finish layer is small, and the thermal stress is mainly caused by horizontal and vertical constraints.

(2) The thermal stress of the free wall is close to zero, indicating the small nonlinearity of temperature distribution inside the exterior finish layer (the greater the nonlinearity, the larger the free stress). This can also be verified by the temperature field calculation results.

(3) The thermal stress decreases with the temperature rising and increases with the temperature falling. The peak temperature corresponds to the peak stress.

(4) The wall surface is subject to tensile stress in winter and compressive stress in sum-

mer, so it is likely to crack in winter.

(5) When the initial temperature is 15℃, the surface is subject to tensile stress (maximum value: 6.38MPa) in winter and compressive stress (maximum value: 10.48MPa) in summer. The stress of the wall surface changes from 0.62MPa to 6.38MPa in winter and from −10.48MPa to −0.17MPa in summer on a daily basis. Therefore, the stress magnitude of the wall surface is −10.48MPa to 6.38MPa in one year (winter and summer).

The most significant difference between the facing brick finish and paint finish is that the thermal stress of the surface of the former is much greater than that of the latter, despite the almost identical changes in the surface temperature. This is mainly because that the elastic modulus of the facing brick ($E \approx 20$GPa) and insulation mortar is higher than that of the paint finish ($E \approx 2$GPa) (Table 2-2-1). If the decorative material of the surface of the insulated wall is too stiff, cracking will be inevitable. For example, if the tensile stress of ordinary mortar is made equivalent (about 6MPa) to that of tiles, cracking will be unavoidable.

Considering the impact of the facing brick size and caulking material on the overall stiffness of the finish layer, the overall stiffness of the facing brick finish should be lower than that of the facing brick. The specific calculation is as follows: using a representative unit composed of the facing bricks and caulking material (Figure 2-2-8), and assuming that the facing brick has the elastic modulus of E_1 and unit length of l_1, the caulking material has the elastic modulus of E_2 and unit length of l_2, the elastic modulus (E) of the composite unit can be expressed as the following equation according to the principle of composite materials:

$$E=\frac{(l_1+l_2)E_1E_2}{E_2l_1+E_1l_2} \tag{2-2-26}$$

It can be seen that the elastic modulus of the composite facing brick finish is a function of the elastic modulus and length of the facing brick and caulking material. Given E_1 equals 20GPa, l_1 equals 40mm, E_2 equals 5GPa and l_2 equals 5mm, then E will be 15GPa. The thermal stress of the surface, calculated with the elastic modulus of the facing brick finish equal to 15GPa, is consistent with the trend in Figure 2-2-7, but slightly decreases. When the initial temperature is 15℃, the maximum tensile stress of the surface will decrease to 4.8MPa in winter and 7.875Pa in summer. The magnitude of the stress of the wall surface is 0.45MPa to 4.8MPa in winter and −7.875MPa to −1.538MPa in summer on a daily basis, and −7.875MPa to 4.8MPa in one year (winter and summer). Although the stress magnitude decreases, the surface is still likely to crack. In winter, the surface may be easily subject to tensile fatigue shedding under cyclic tensile loads.

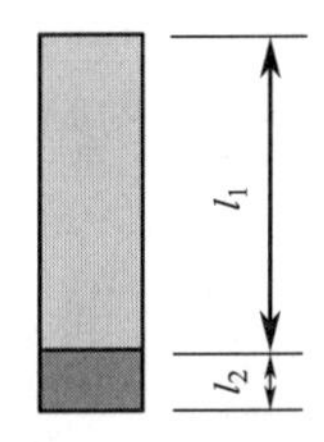

Figure 2-2-8 Schematic Diagram of Representative Unit with Facing Brick and Caulking Material

Figure 2-2-9 shows the change in the thermal stress of the outer surface of the insulation mortar layer under various constraints over time, according to the rules similar to those of the paint finish. As the elastic modulus of the insulation mortar of the facing brick finish is higher than that of the paint finish, the thermal stress of the outer surface of the insulation mortar layer slightly increases. When the initial temperature is 15℃, the magnitude of the surface stress of the insulation mortar layer is 0.10MPa to 1.04MPa in winter and −1.71MPa to −0.34MPa in summer on a daily basis, and −1.71MPa to 1.04MPa in a year. It can be concluded that the maximum tensile stress of the insulation mortar layer is close to

1MPa. Therefore, the cracking of insulation mortar is one major challenge for the wall with external thermal insulation.

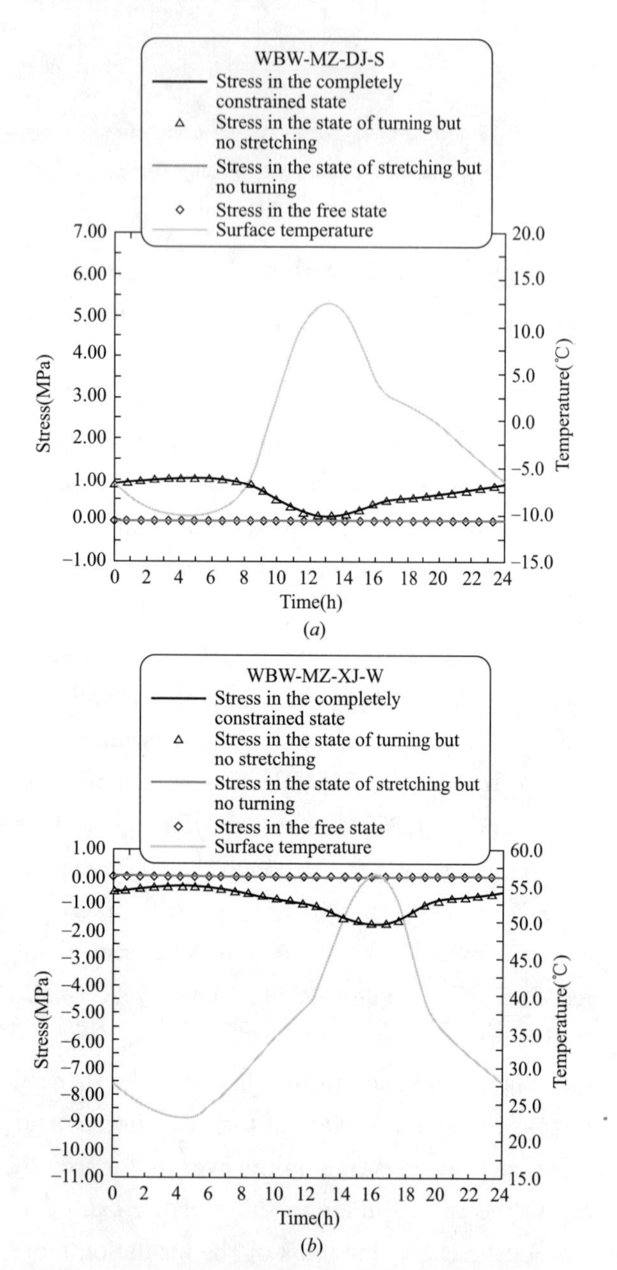

Figure 2-2-9 Changes in Surface Stress of Anti-cracking Mortar Layer of Wall with External Thermal Insulation by Facing Brick Finish based on Adhesive Polystyrene Granule over Time

(*a*) Winter; (*b*) Summer

Figure 2-2-10 shows the relationship between the time and the thermal stress of the outer surface of the base course wall with the facing brick finish for external thermal insulation based on the adhesive polystyrene granule under various con-

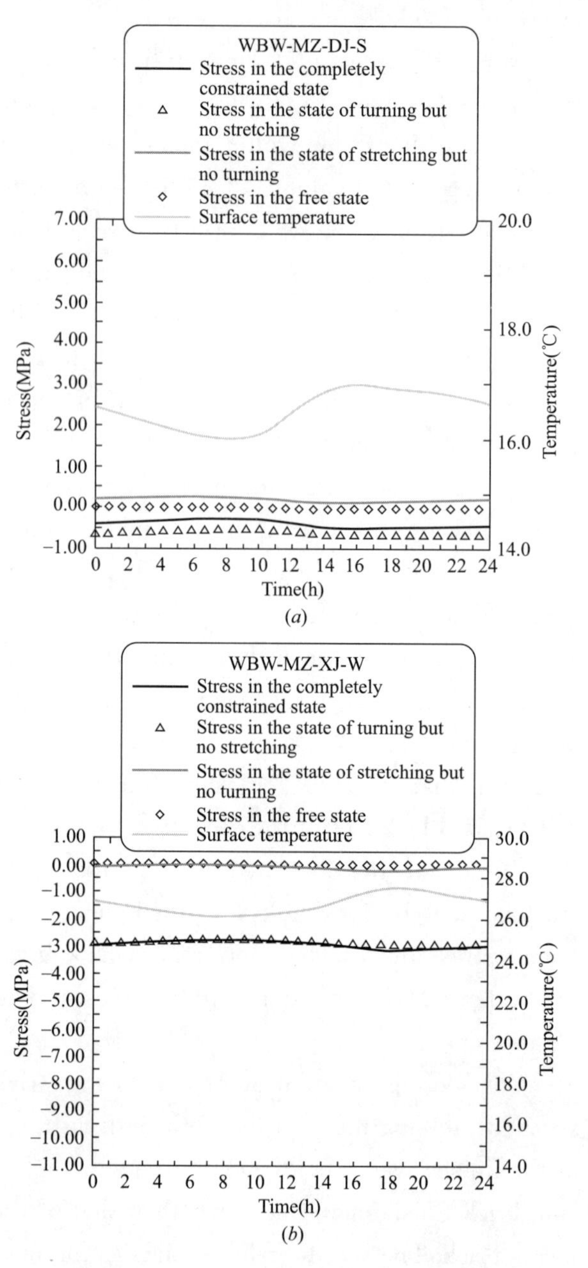

Figure 2-2-10 Changes in Surface Stress of Base Course Wall of Wall with External Thermal Insulation by Paint Finish based on Adhesive Polystyrene Granule over Time

(*a*) Winter; (*b*) Summer

straints. The results shown in this figure are similar to those of the paint finish, involving a small difference in the stress.

Figure 2-2-11 clearly shows the change in the thermal stress of typical functional layers in the insulated wall with the facing brick finish over time in 24h in winter and summer. Except that the peak stress of the surface and insulation mortar

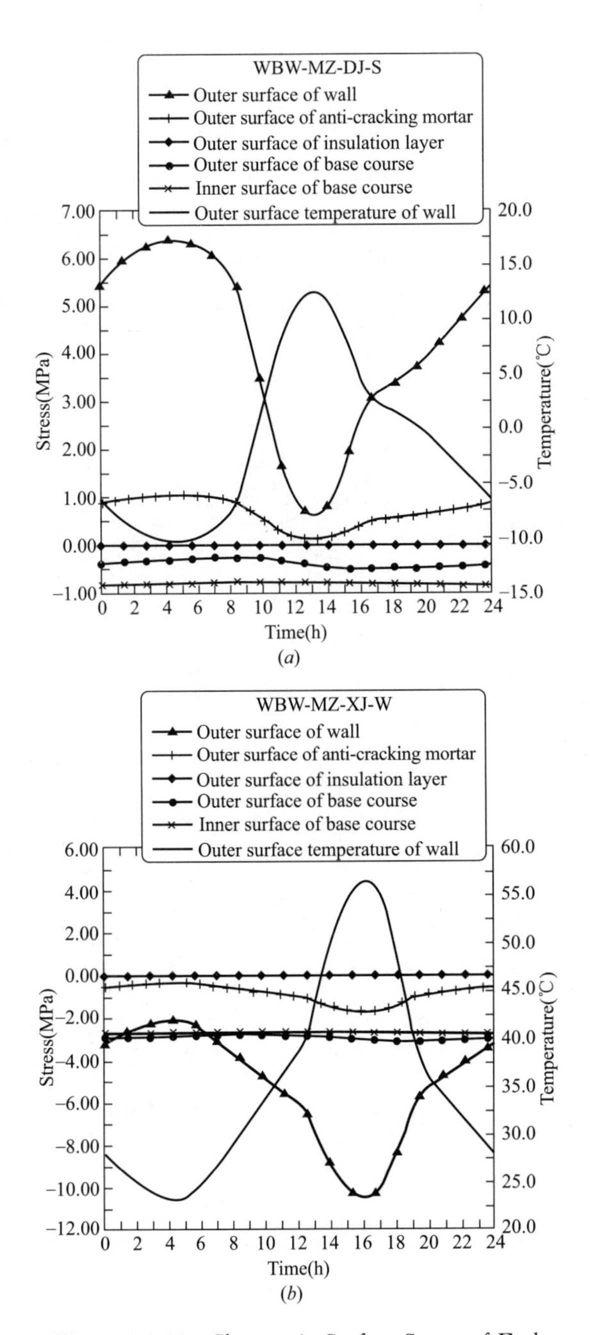

Figure 2-2-11 Changes in Surface Stress of Each Typical Functional Layer of Wall with External Thermal Insulation by Facing Brick Finish based on Adhesive Polystyrene Granule over Time

(*a*) Winter; (*b*) Summer

layer is higher than that of the paint finish, the other peak values are similar to those of the paint finish, and the rules of changes in the thermal stress are also the same.

The distribution of thermal stress of the insulated wall in the thickness direction can reflect the rules of thermal stress distribution in each layer more intuitively. Figure 2-2-12 shows the thermal stress distribution in the thickness direction of the completely constrained wall with the facing brick finish (finish layer stiffness: 20GPa) for external thermal insulation based on the adhesive polystyrene granule, corresponding to the minimum temperature of the outer surface in winter and its maximum temperature in summer. The temperature distribution in the thickness direction of the wall in the corresponding period is shown in

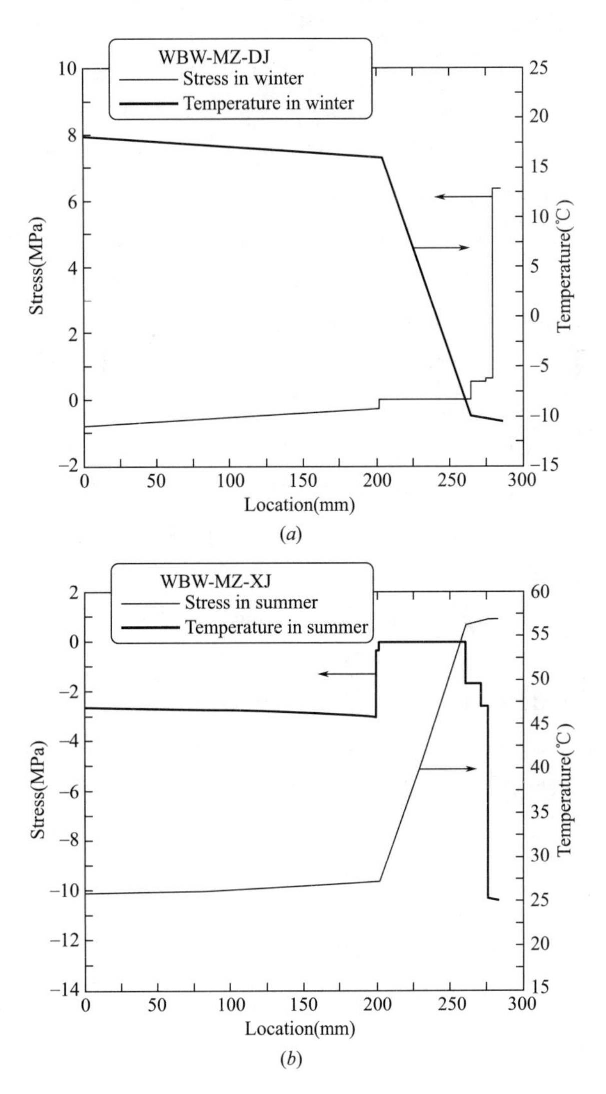

Figure 2-2-12 Thermal Stress Distribution in the Thickness Direction of Completely Constrained Wall with External Thermal Insulation by Facing Brick Finish (Surface Stiffness: 20GPa) based on Adhesive Polystyrene Granule under Minimum and Maximum Temperature of Outer Surface in Winter and Summer Respectively

(*a*) Winter; (*b*) Summer

Figure 2-2-12.

Similar to the paint finish, the thermal stress under complete constraints is distributed in a stepping form. In both winter and summer, the base course wall is subject to compressive stress, and the internal stress of the insulation layer is almost zero (elastic modulus: 0.1MPa). The insulation mortar layer and its finish layer outside the insulation layer are subject to tensile stress in winter and compressive stress in summer. The stiffness of the exterior finish has significant impact on the stress. The maximum tensile stress and compressive stress of the finish are respectively 6.38MPa and −10.48MPa when the elastic modulus of the surface is 20GPa, and 4.8MPa and −7.575MPa when the elastic modulus of the surface is 15GPa. The maximum stress inside the insulation mortar layer of the facing brick finish is far higher than that of the paint finish. The maximum tensile stress of the facing brick finish is 1.04MPa in winter, and that of the paint finish is 0.35MPa. This is mainly caused by their difference in the elastic modulus (stiffness).

2.2.2.3 Wall with Internal Thermal Insulation by Paint Finish based on Adhesive Polystyrene Granule

In order to study the effects of the location of the insulation layer on the thermal stress of the external thermal insulation system, the temperature field of the wall with internal thermal insulation by the paint finish based on the adhesive polystyrene granule was also calculated (the insulation layer and its additional layer were located in the structural layer). The dimensions and thermodynamic parameters of the structural layer are listed in Table 2-2-1. Except for changes in the location of the insulation layer, the other related material parameters were the same as those of the wall with external thermal insulation. For conveniences in comparison, the construction method of each functional layer of internal thermal insulation was the same as that of external thermal insulation (see Table 2-2-1).

Similar to the wall with corresponding external thermal insulation, the thermal stress of the wall with internal thermal insulation in 24h under different constraints was calculated first. Figure 2-2-13 to 2-2-15 show the relationship between the time (in 24h) and the thermal stress of the exterior finish, the outer surface of the bearing base

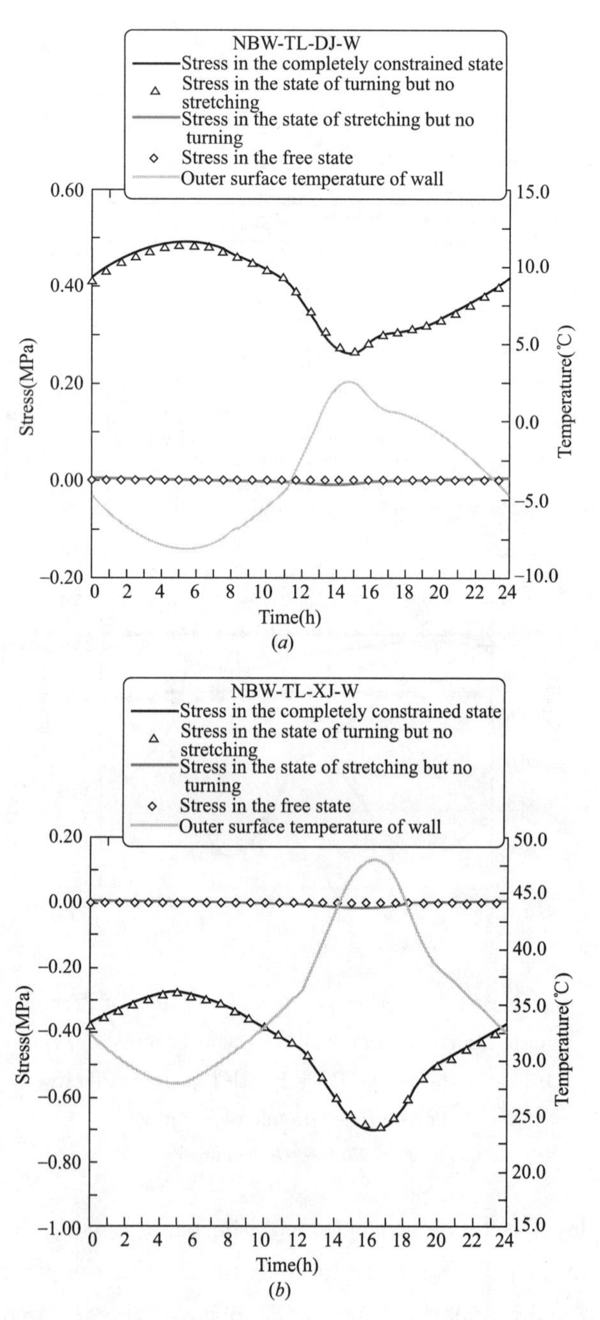

Figure 2-2-13 Changes in Stress of Outer Surface of Wall with Internal Thermal Insulation by Paint Finish based on Adhesive Polystyrene Granule over Time

(*a*) Winter; (*b*) Summer

course wall and the interior finish of the western wall with the external thermal insulation by the paint finish based on the adhesive polystyrene granule under typical climatic conditions in winter and summer, respectively. They illustrate the thermal stress inside the wall surface in the presence of four typical constraints and under the same temperature changes. These four typical constraints are as follows: (1) wall stretching and turning are not allowed (embedded wall); (2) wall turning is allowed but stretching is not; (3) wall stretching is allowed but turning is not; and (4) wall stretching and turning are allowed (free wall).

As shown by Figure 2-2-13:

(1) The peak thermal stress of the surface of the exterior finish (paint) of the wall with internal thermal insulation is slightly lower than that of the wall with corresponding external thermal insulation (as the peak temperature is slightly lower), but with no significant difference. The rules of changes in the thermal stress in the presence of internal thermal insulation are the same as those with external thermal insulation. There is a small difference in the thermal stress of the outer surface of the decorative layer under the first two types of constraint, indicating the small bending stress of the outer surface of the wall and also the small temperature difference between the inner and outer surfaces of the decorative layer. The thermal stress is mainly caused by horizontal and vertical constraints.

(2) The thermal stress of the free wall is close to zero, indicating the small nonlinearity of temperature distribution inside the exterior finish layer (the greater the nonlinearity, the larger the free stress). This can also be verified by the temperature field calculation results.

(3) The thermal stress decreases with the temperature rising and increases with the temperature falling. The peak temperature corresponds to the peak stress.

(4) The wall surface is subject to tensile stress in winter and compressive stress in summer, so it is likely to crack in winter.

(5) When the initial temperature is 15℃, the surface is subject to tensile stress in winter, max. 0.49MPa (wall with corresponding external thermal insulation: 0.54MPa), and compressive stress in summer, max. 0.7MPa (wall with corresponding external thermal insulation: 0.9MPa). The stress of the wall surface changes from 0.26MPa to 0.49MPa in winter and from −0.70MPa to −0.27MPa in summer on a daily basis. Therefore, the stress magnitude of the wall surface is −0.70MPa to 0.49MPa in a year (winter and summer).

Figure 2-2-14 shows the relationship between the time and the thermal stress of the outer surface of the base course wall under various constraints. It can be seen in this figure that the peak thermal stress under various constraints is obviously higher than that of the wall with corresponding external thermal insulation. When the initial temperature is 15℃, the internal stress of the base course wall is basically tensile stress in winter, and the surface stress of the base course wall with external thermal insulation is compressive stress, indicating that the external thermal insulation has obvious protective effects on the base course wall (structural layer). Therefore, the structural layer of the wall with external thermal insulation is less likely to crack than the base course wall with internal thermal insulation under temperature changes. The surface stress of the base course wall changes from 3.21MPa to 5.68MPa on a daily basis in winter in the presence of internal thermal insulation, and from −0.39MPa to −0.22MPa in the presence of external thermal insulation. The surface stress of the wall with internal thermal insulation is from −8.00MPa to −3.34MPa on a daily basis in summer in the case of internal thermal insulation, and from −3.44MPa to −2.78MPa in the case of external thermal insulation. The analysis results of the above calculations are verified by the ther-

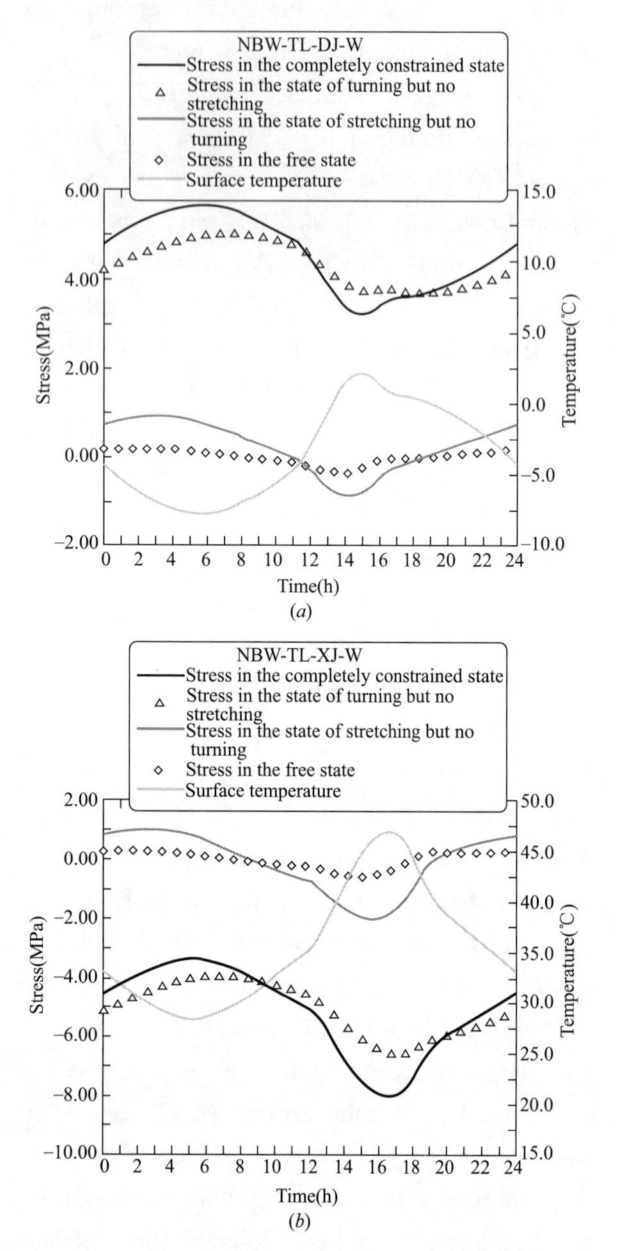

Figure 2-2-14 Changes in Stress of Base Course Wall Surface with Internal Thermal Insulation by Paint Finish based on Adhesive Polystyrene Granule over Time

(*a*) Winter; (*b*) Summer

mal stress distribution along the wall section.

In addition, as the base course wall with internal thermal insulation is thick and the temperature difference between its inner and outer surfaces is significant, the thermal stress under the first two types of constraint is obviously higher than that with external thermal insulation. This means that the bending stress of the outer surface of the wall with internal thermal insulation is higher than that of the wall with external thermal insulation. However, the thermal stress in this case is mainly caused by horizontal and vertical axial constraints (involving greater changes in the average temperature). The thermal stress of the free wall is higher than the wall with external thermal insulation, meaning that the temperature distribution inside the base course wall with internal thermal insulation is more nonlinear than that of the wall with external thermal insulation. Similarly, the thermal stress decreases with the temperature rising and increases with the temperature dropping, and the peak temperature corresponds to the peak stress. Due to insignificant temperature changes inside the base course wall, the thermal stress does not change greatly.

Figure 2-2-15 shows the changes in the surface stress of the interior finish of the wall with internal thermal insulation by the paint finish based on the adhesive polystyrene granule over time. It can be seen that, for the interior finish, the thermal stress of its surface is obviously lower than that of the exterior finish of the wall with external thermal insulation in both winter and summer. Therefore, the material performance requirements for the interior finish may be lower than those of the exterior finish. This practice has been applied in actual projects and will not be repeated here.

Figure 2-2-16 clearly shows the changes in the thermal stress of typical functional layers of the completely constrained wall with internal thermal insulation over time in winter and summer (two typical seasons). The major difference between the walls with internal and external thermal insulation lies in the stress of the base course wall.

The thermal stress distribution in the thickness direction of the insulated wall can reflect the rules of thermal stress distribution in each layer more intuitively. Figure 2-2-17 shows the thermal stress distribution along the section of the com-

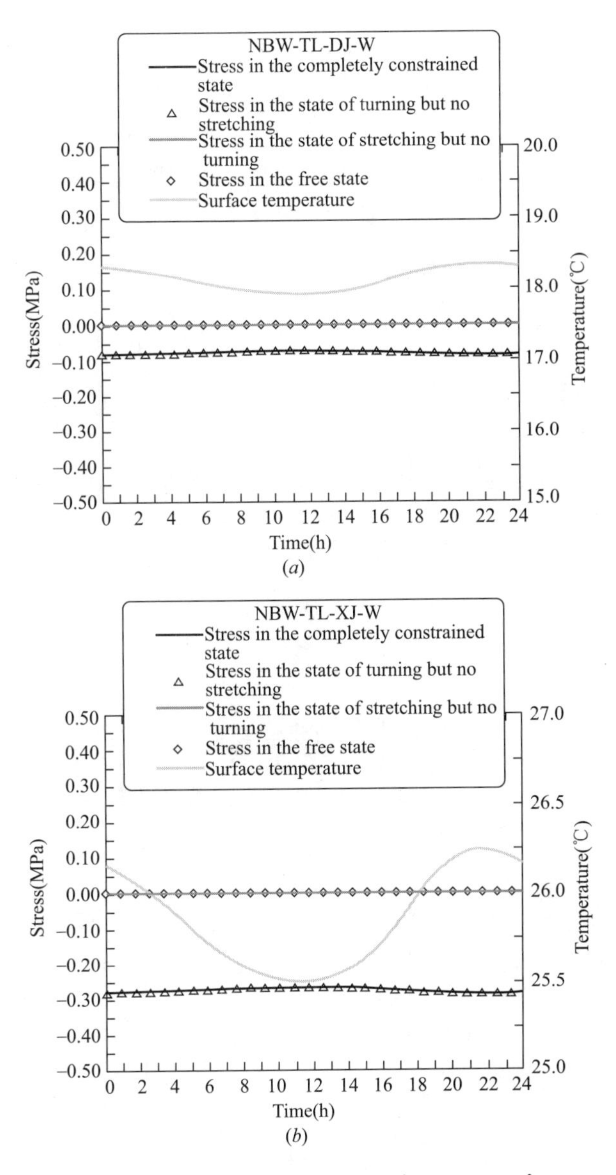

Figure 2-2-15 Changes in Surface Stress of Interior Finish of Wall with Internal Thermal Insulation by Paint Finish based on Adhesive Polystyrene Granule over Time

(*a*) Winter; (*b*) Summer

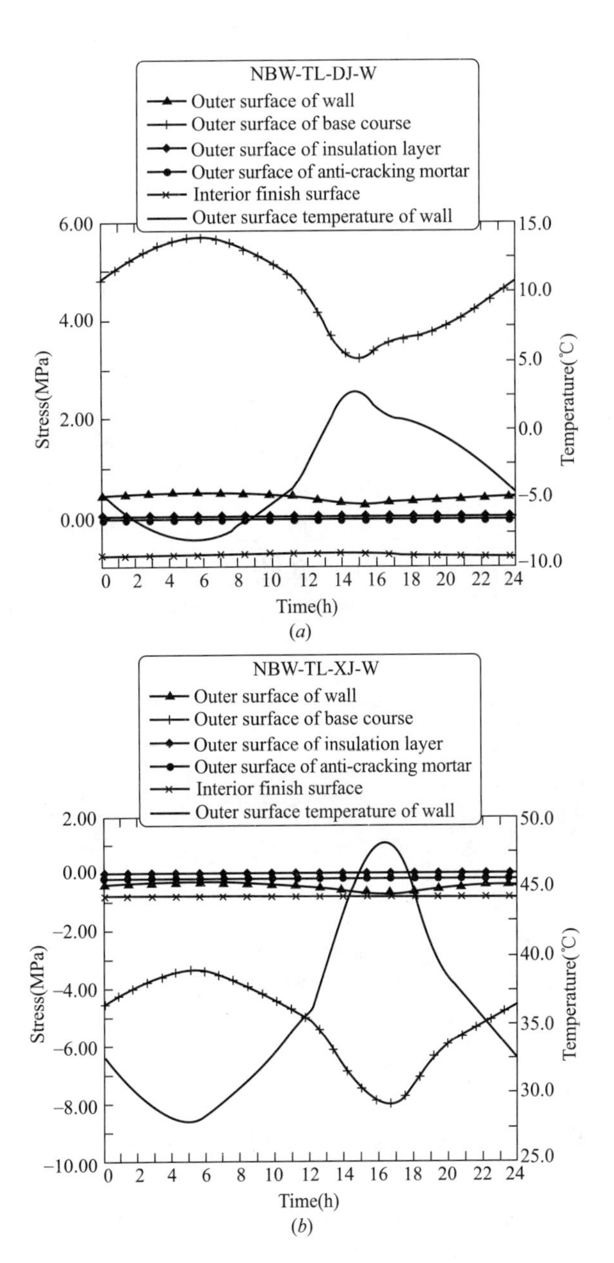

Figure 2-2-16 Changes in Surface Stress of Each Typical Functional Layer of Wall with Internal Thermal Insulation by Paint Finish based on Adhesive Polystyrene Granule over Time

(*a*) Winter; (*b*) Summer

pletely constrained wall with internal thermal insulation by the paint finish based on the adhesive polystyrene granule, corresponding to the minimum temperature of the outer surface of the wall in winter and its maximum temperature in summer. The corresponding temperature distribution in the thickness direction of the wall is also presented in this figure.

As can be seen from Figure 2-2-17, similar to the wall with other types of insulation, the thermal stress distribution of the completely constrained wall with internal thermal insulation by the paint finish based on the adhesive polystyrene granule is distributed in a stepping form in the thickness direction. In winter, the temperature gradually decreases from indoors to outdoors, and the stress is basically divided into three sections, namely, the interior finish and insulation layer, the base course wall and the exterior fin-

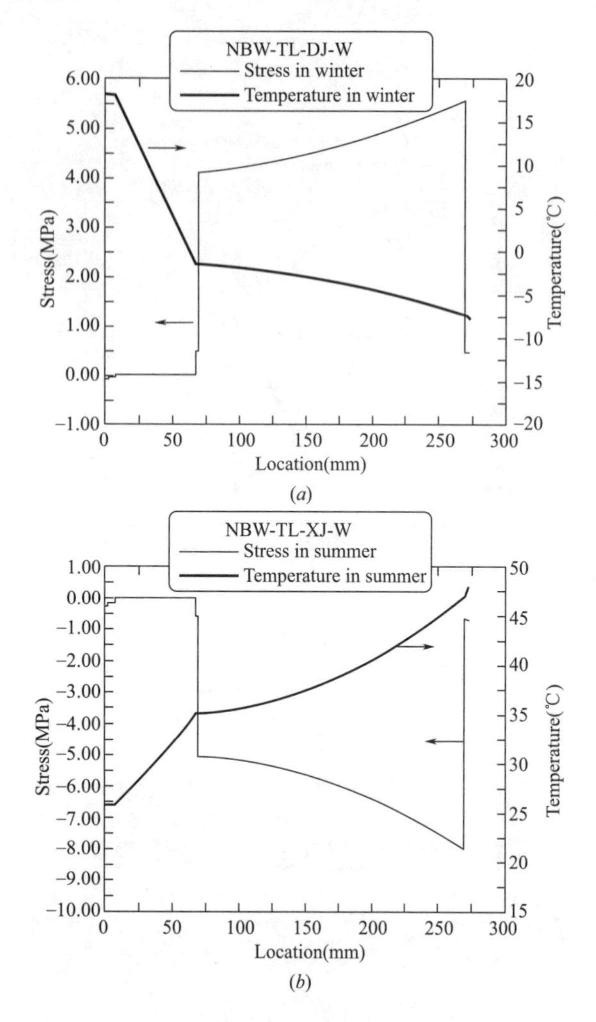

Figure 2-2-17 Thermal Stress Distribution in the Thickness Direction of Completely Constrained Wall with Internal Thermal Insulation by Paint Finish based on Adhesive polystyrene granule under Minimum and Maximum Temperature of External Surface in Winter and Summer

(*a*) winter; (*b*) summer

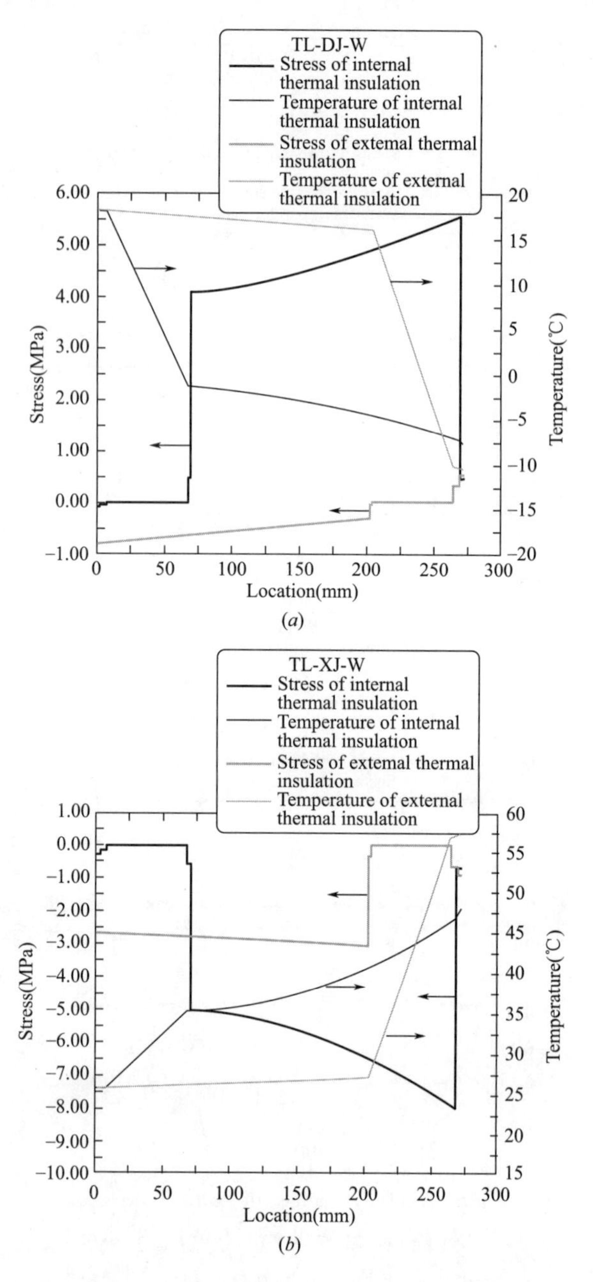

Figure 2-2-18 Thermal Stress Distribution along the Section of Completely Constrained Wall with Internal and External Thermal Insulation by Paint Finish based on Adhesive polystyrene granule under Minimum and Maximum Temperature of External Surface in Winter and Summer

(*a*) winter; (*b*) summer

ish. The base course wall is mainly subject to the tensile stress (max. 5.68MPa). The interior finish and insulation layer are subject to little stress, while the exterior finish to tensile stress (max. 0.49MPa) less than that of the base course wall. In summer, as the wall temperature is higher than the initial temperature (T_0), the temperature gradually rises from indoors to outdoors, and the wall is subject to compressive stress, which is basically divided into three sections, namely, the interior finish and insulation layer, the base course wall and the exterior finish. The entire base course wall is subject to the compressive stress (max. 8.00MPa), the interior finish and insulation layer to little stress, and the exterior finish to compressive stress (max. 0.70MPa), which is less than that of the base course wall.

To compare the impact of the internal and

external thermal insulation on thermal stress, Figure 2-2-18 shows the thermal stress distribution in the thickness direction of the completely constrained wall with internal and external thermal insulation by the paint finish based on the adhesive polystyrene granule, corresponding to the minimum and maximum temperature of the external surface in winter and summer, respectively. The temperature distribution is also illustrated in this figure. It can be seen that the thermal stress distributions in the presence of internal and external thermal insulation vary completely due to differences in the temperature distribution of the wall. The temperature range and stress of the structural layer of the wall are reduced under ambient temperature changes in the case of external thermal insulation; and the temperature variation of the structural layer is exacerbated in the presence of internal thermal insulation (compared with the wall with no insulation layer), thus increasing the thermal stress inside the base course wall. As the temperature range and corresponding thermal stress of the exterior finish are increased in the case of external thermal insulation, the material properties of the exterior finish as well as its insulation and additional layers should meet higher requirements, and their long-term durability, such as the resistance to high temperature and fatigue, are challenges during development of the external thermal insulation.

2.2.2.4 all with Internal Thermal Insulation by Face Brick Finish based on Adhesive polystyrene granule

The thermal stress of the wall with internal thermal insulation by the face brick finish based on the adhesive polystyrene granule (the insulation layer and its additional layer were located in the structural layer) was also calculated in this study. The dimensions and thermodynamic parameters of the structural layer of the wall with internal thermal insulation are listed in Table 2-2-1. Except for changes in the location of the insulation layer, the other related material parameters were the same as those of the wall with external thermal insulation. For conveniences in comparison, the construction method of each functional layer of the wall with internal thermal insulation was the same as that with external thermal insulation.

Similar to the wall with corresponding external thermal insulation, the thermal stress of the wall with internal thermal insulation in 24h under different constraints was calculated first. Figure 2-2-19 to 2-2-22 show the relationship between the time (in 24h) and the thermal stress of the exterior finish (involving two kinds of surface stiffness), the bonding mortar layer, the bearing base course wall and the inner finish of the western wall with internal thermal insulation by the face brick finish based on the adhesive polystyrene granule under typical climatic conditions in winter and summer, respectively. They illustrate the thermal stress of the wall surface in the presence of four typical constraints and under the same temperature changes. These four typical constraints are as follows: (1) wall stretching and turning are not allowed (embedded wall); (2) wall turning is allowed but stretching is not; (3) wall stretching is allowed but turning is not; and (4) wall stretching and turning are allowed (free wall).

As can be seen from Figure 2-2-19, similar to the wall with corresponding external thermal insulation, the thermal stresses of the face brick finish and paint finish for the internal thermal insulation of the wall mainly differ in that the thermal stress of the former is greatly higher than that of the paint finish (their maximum tensile stresses in winter were 5.78MPa and 0.49MPa, respectively), though their temperature changes are basically the same. This is mainly because the elastic modulus ($E \approx 20$GPa) of the face brick finish is higher than that ($E \approx 2$GPa) of the paint finish and the elastic modulus of the anti-cracking mortar is also high (Table 2-2-1). Accordingly, cracking will be inevitable if the fin-

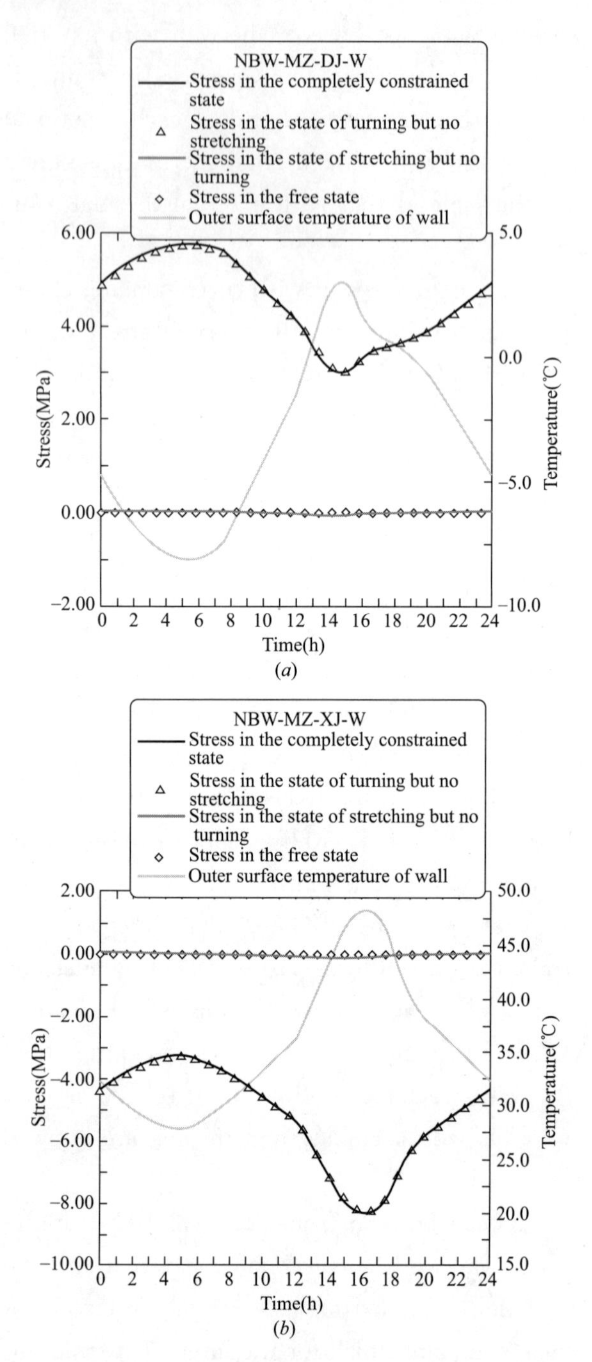

Figure 2-2-19 Changes in Stress of External Surface (stiffness: 20GPa) of Wall with Internal Thermal Insulation by Face Brick Finish based on Adhesive polystyrene granule over Time
(*a*) winter; (*b*) summer

ishing material of the insulated wall is too stiff. Considering the impact of the face brick size and caulking material, the overall stiffness of the face brick finish, the overall stiffness of the face brick finish should be lower than that of the face brick. Provided that the elastic modulus of the finish is 15GPa, the trend of the stress-time relationship is the same as that in Figure 2-2-19, but specific values are only 3/4 of the latter. Comparatively, the peak thermal stress of the surface of the exterior finish (face brick) of the wall with internal thermal insulation is slightly lower than that with external thermal insulation (as the peak temperature is slowly lower), but they are not significantly different from each other.

Figure 2-2-20 shows the changes in the thermal stress of the bonding mortar surface of the completely constrained wall with face bricks. Accordingly, the bonding mortar is still subject to high tensile stress in winter, which means that the surface is likely to fall off. Therefore, it is necessary to take appropriate measures against falling off. The rules of changes in the thermal stress of the base course wall and interior finish are the same as those of the wall with internal thermal insulation by the paint finish.

Figure 2-2-23 shows the thermal stress distribution in the thickness direction of the completely constrained wall with internal thermal insulation by the face brick finish based on the adhesive polystyrene granule, corresponding to the minimum temperature of the external surface in winter and its maximum temperature in summer. The temperature distribution in the corresponding period is also shown in this figure. The rules of temperature changes were analyzed like those with internal or external thermal insulation by the paint finish.

2.2.2.5 Aerated Concrete Self-insulated Wall

Figure 2-2-24 to 2-2-28 show the relationship between the time (in 24h) and the thermal stress of the exterior finish as well as the external surfaces of the outer ordinary cement mortar layer, aerated concrete layer and inner cement mortar layer of the wall composed of the 20mm cement mortar, 200mm aerated concrete and 20mm cement mortar under typical climatic conditions in winter and summer, respectively. They illustrate the thermal stress of the wall surface in the pres-

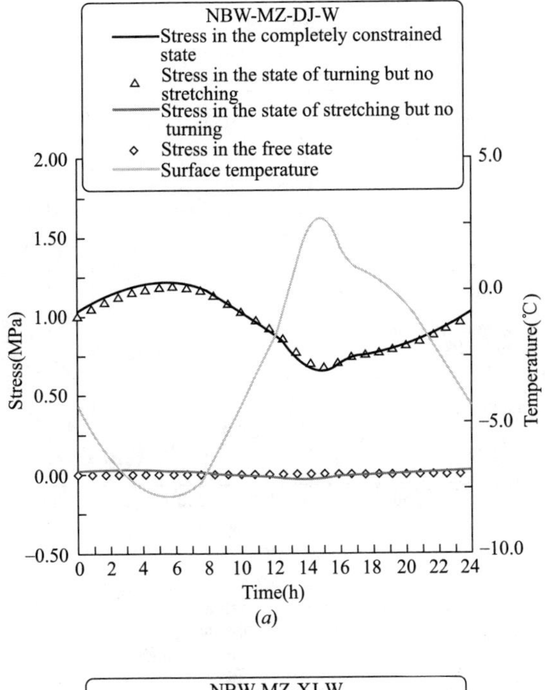

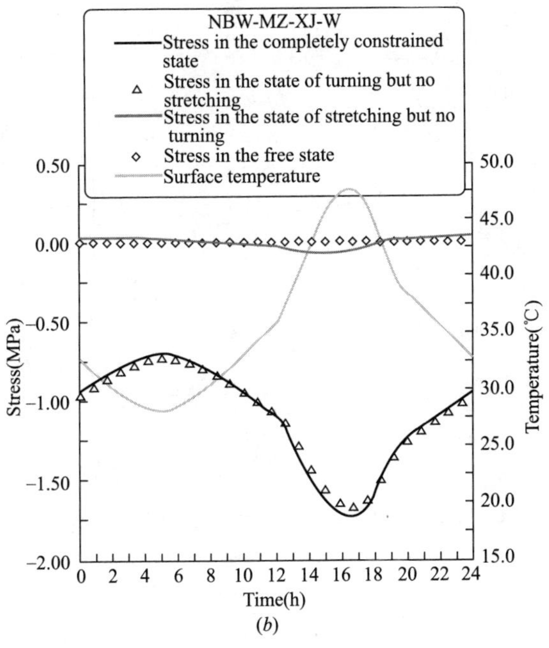

Figure 2-2-20 Changes in Surface Stress of Bonding Mortar of Wall with Internal Thermal Insulation by Face Brick Finish based on Adhesive polystyrene granule over Time

(*a*) winter; (*b*) summer

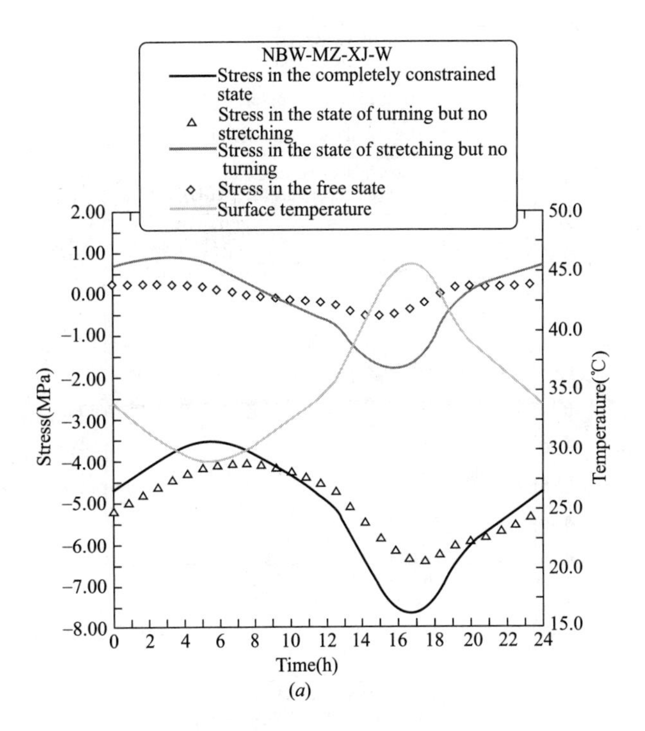

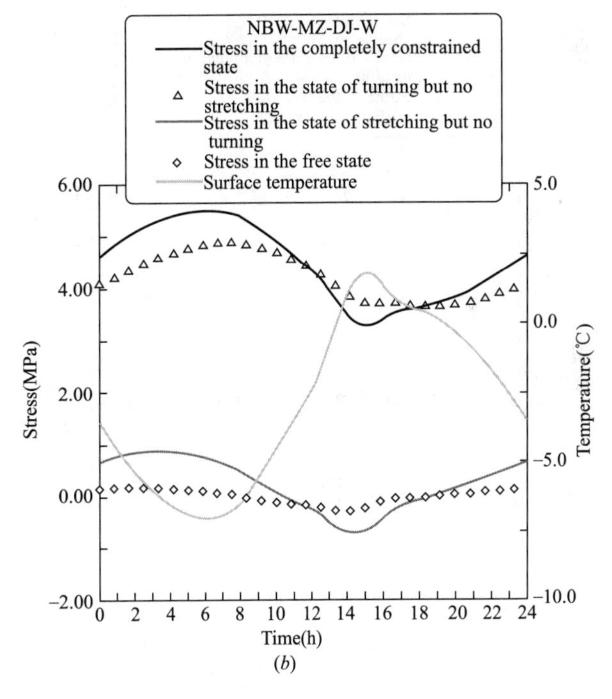

Figure 2-2-21 Changes in Surface Stress of Base Course Wall with Internal Thermal Insulation by Face Brick Finish based on Adhesive Polystyrene Granule over Time

(*a*) winter; (*b*) summer

ence of four typical constraints and under the same temperature changes. These four typical constraints are as follows: (1) wall stretching and turning are not allowed (embedded wall); (2) wall turning is allowed but stretching is not; (3) wall stretching is allowed but turning is not; and (4) wall stretching and turning are allowed (free wall).

As shown in Figure 2-2-24:

(1) The rules of changes in the thermal

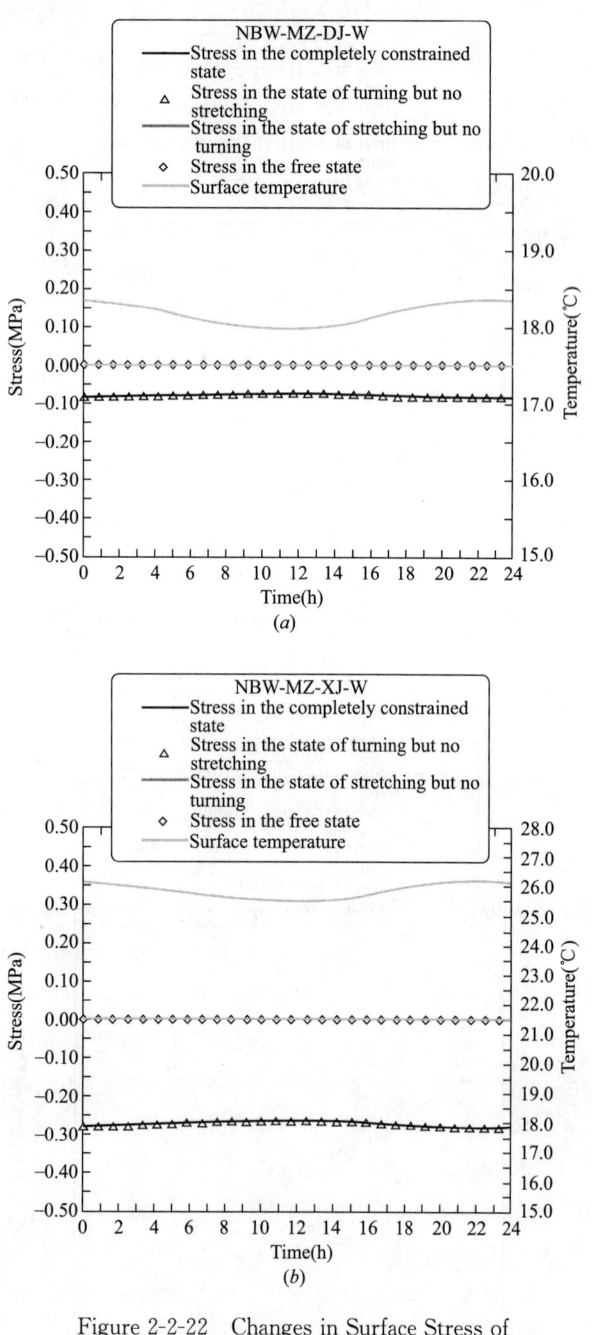

Figure 2-2-22 Changes in Surface Stress of Interior Finish of Wall with Internal Thermal Insulation by Face Brick Finish based on Adhesive polystyrene granule over Time

(*a*) winter; (*b*) summer

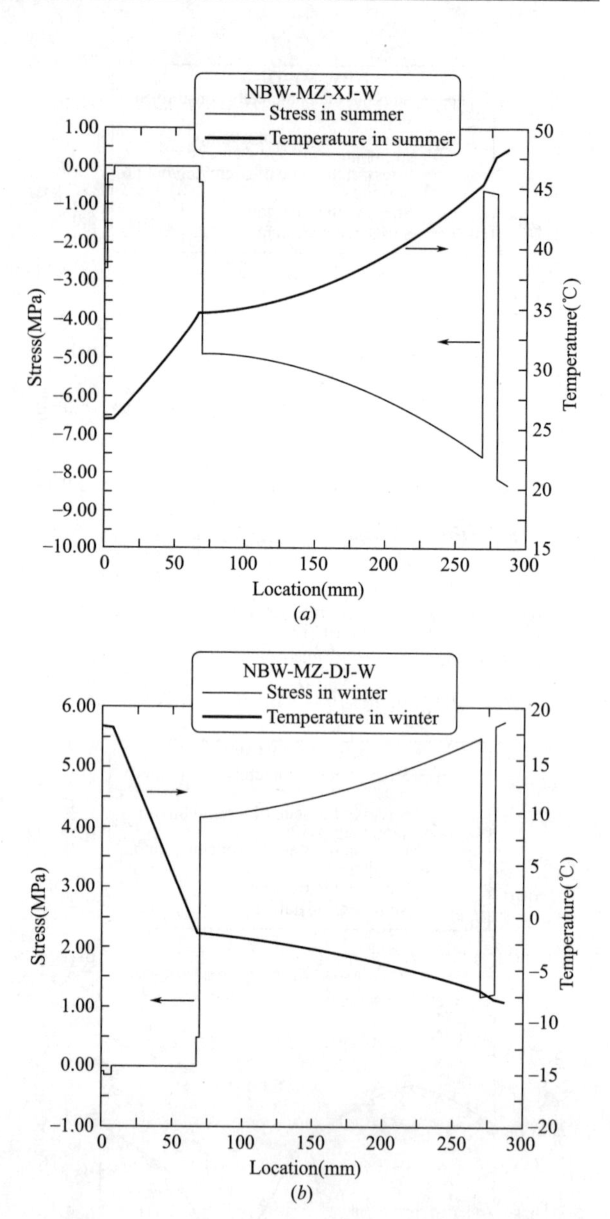

Figure 2-2-23 Thermal Stress Distribution in the Thickness Direction of Completely Constrained Wall with Internal Thermal Insulation by Face Brick Finish (Surface Stiffness: 20GPa) based on Adhesive polystyrene granule under Minimum and Maximum Temperature of External Surface in Winter and Summer

(*a*) winter; (*b*) summer

stress of the external surface of the finishing layer over time are similar to those of the exterior insulation finish based on the ordinary adhesive polystyrene granule. That is, the thermal stress vary little under the first two types of constraint, indicating the small bending stress of the external surface of the wall and also the small temperature difference between the inner and external surfaces of the finishing layer. The thermal stress is mainly caused by horizontal and vertical constraints.

(2) The thermal stress of the free wall is close to zero, indicating the small nonlinearity of temperature distribution inside the exterior finish layer.

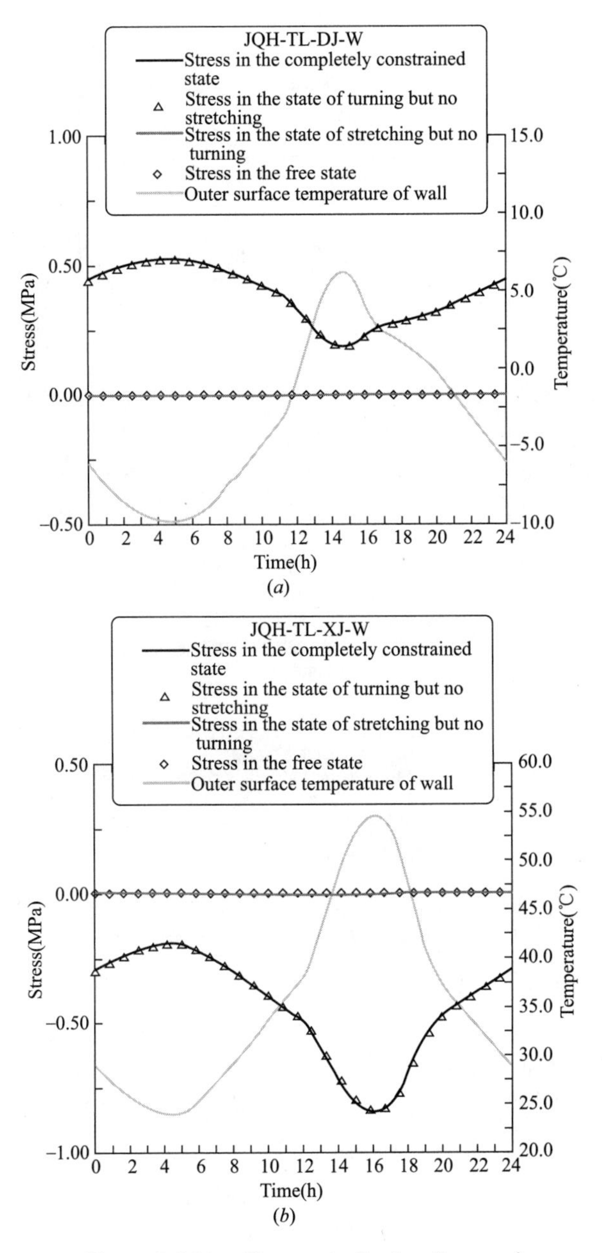

Figure 2-2-24 Changes in Surface Stress of Exterior Finish of Wall with Aerated Concrete Self-insulation by Paint Finish

(*a*) winter; (*b*) summer

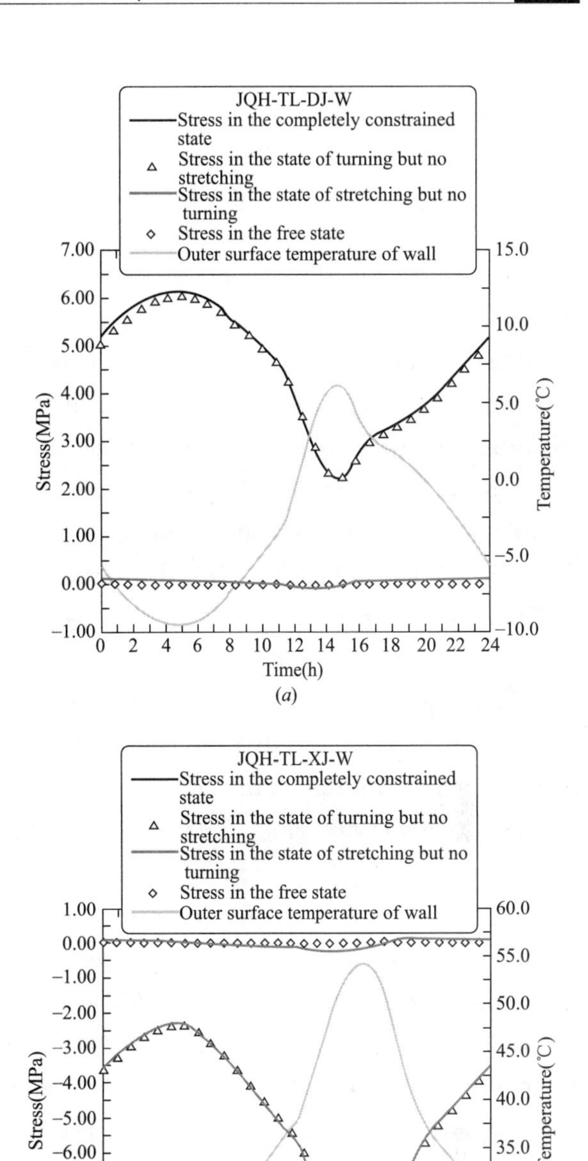

Figure 2-2-25 Changes in External Surface Stress of Outer Mortar Layer of Wall with Aerated Concrete Self-insulation by Paint Finish

(*a*) winter; (*b*) summer

(3) The thermal stress decreases with the temperature rising and increases with the temperature falling. The peak temperature corresponds to the peak stress.

(4) The wall surface is subject to tensile stress in winter and compressive stress in summer, so it is likely to crack in winter.

As shown in Figure 2-2-25:

(1) There is a small difference in the outer ordinary mortar layer under the first two types of constraint, indicating the bending stress is small inside this layer. The thermal stress is mainly caused by horizontal and vertical axial constraints.

(2) The thermal stress of the free wall is close to zero, indicating the small nonlinearity of

temperature distribution inside the mortar layer.

(3) The thermal stress decreases with the temperature rising and increases with the temperature falling. The peak temperature corresponds to the peak stress.

(4) When the initial temperature is 15℃, the surface of the outer mortar layer is subject to tensile stress in winter and compressive stress in summer, but specific values of such tensile and compressive stress are greater than those of the exterior finish. Under complete constraints, the maximum tensile stress and compressive stress are 6.14MPa and −9.80MPa in 24h, respectively. This means that there are greater risks of cracking in winter and hollowing in summer. Taking mortar shrinkage into consideration, there is almost no doubt that the wall built with ordinary cement mortar as the outer protection of aerated concrete will crack under the thermal shrinkage stress.

(5) When the initial temperature is 15℃, the surface stress of the outer mortar layer of the wall falls between 2.21MPa and 6.14MPa in winter and between −9.80MPa to −2.28MPa in summer on a daily basis. Therefore, the magnitude of stress change of the wall surface is −9.80MPa to 6.14MPa in a year.

Figure 2-2-26 shows the changes in the thermal stress of the external surface of the aerated concrete layer in winter and summer, respectively. According to this figure:

(1) The thermal stress of the external surface of the aerated concrete layer and its change over time are different from that of the base course wall insulated with ordinary adhesive polystyrene granule. There is a significant difference in the thermal stress under the first two types of constraint, indicating that the bending stress of the external surface of the wall is large. This is mainly caused by a great temperature difference between its inner and external surfaces. The thermal stress is caused by horizontal and vertical constraints, as well as bending stress arising from the temperature difference between the inner and external surfaces.

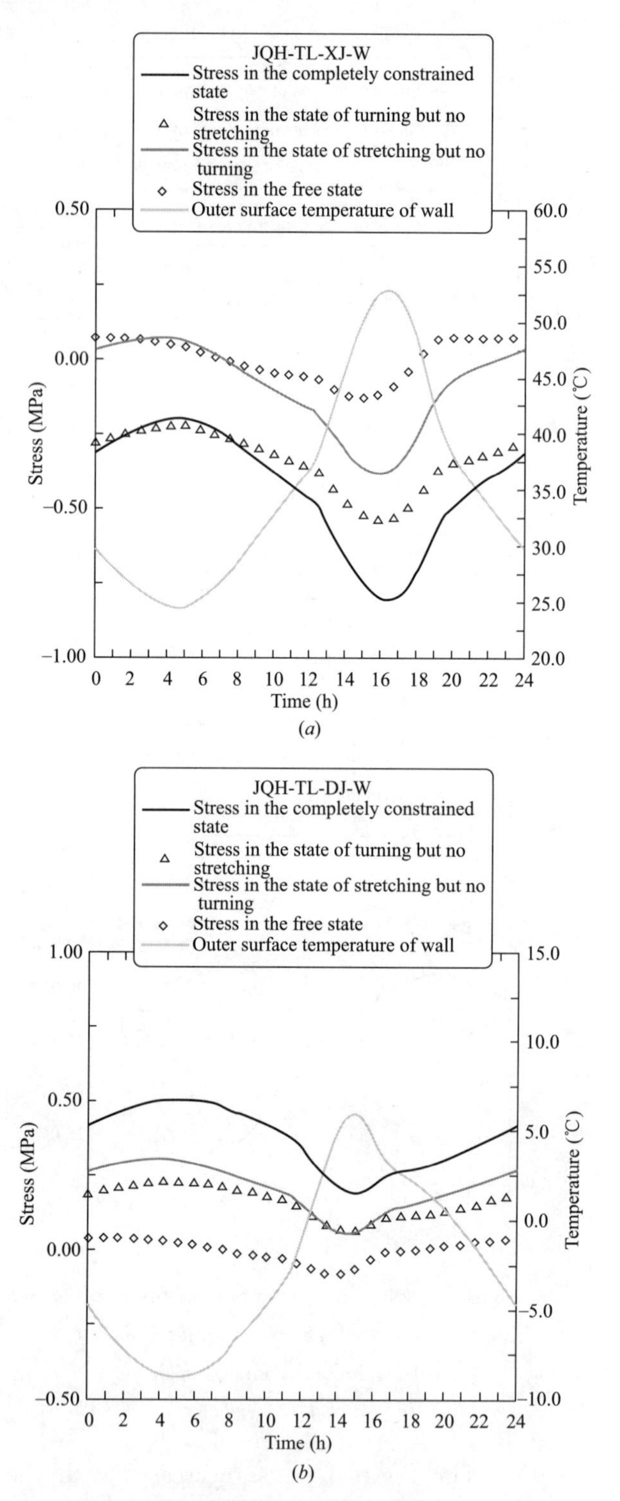

Figure 2-2-26 Changes in Stress of External Surface of Aerated Concrete Self-insulated Wall with Paint Finish

(a) winter; (b) summer

(2) The thermal stress of the free wall is close to zero, indicating the nonlinearity of tem-

perature distribution inside the aerated concrete layer.

(3) The thermal stress decreases with the temperature rising and increases with the temperature falling. The peak temperature corresponds to the peak stress. When the initial temperature is 15℃, the wall surface is subject to tensile stress in winter and compressive stress in summer.

Figure 2-2-27 shows the changes in the thermal stress of the external surface of the mortar layer over time in winter and summer, respectively. Compared with that of the external surface of the outer mortar layer, the thermal stress of the inner mortar layer is much lower due to the insulation of aerated concrete. When the initial temperature is 15℃, there is low compressive stress (with an insignificant difference in temperature) inside the mortar layer in both winter and summer, which has little impact on the protection features of the interior finish.

The distribution of thermal stress of the insulated wall in the thickness direction can reflect the rules of thermal stress distribution in each layer more intuitively. Figure 2-2-28 shows the thermal stress distribution in the thickness direction of the completely constrained wall with the paint finish for aerated concrete self-insulation, corresponding to the minimum temperature of the external surface in winter and its maximum temperature in summer. The temperature distribution in the thickness direction of the wall in the corresponding period is also shown in this figure.

As can be seen from Figure 2-2-28, the thermal stress under complete constraints is distributed in a stepping form. In winter, the temperature gradually drops from indoors to outdoors, whilst the thermal stress gradually rises. The inner mortar layer is mainly subject to compressive stress, but the aerated concrete layer to tensile stress. Compared with the aerated concrete layer and inner mortar layer, the thermal stress of the outer mortar layer increases suddenly, max. 6.14MPa. The maximum tensile stress of the ex-

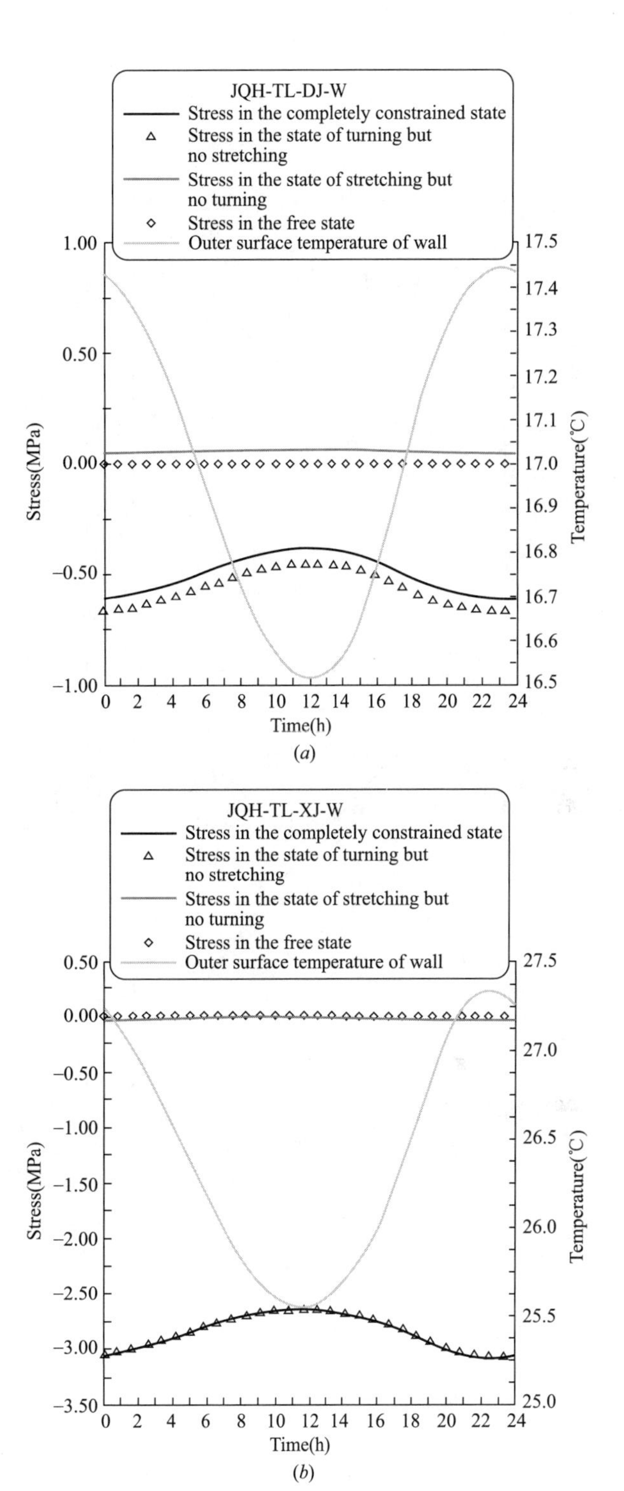

Figure 2-2-27 Changes in Stress of Inner Mortar Layer of Aerated Concrete Self-insulated Wall with Paint Finish

(*a*) winter; (*b*) summer

terior finish is 0.65MPa. In summer, the wall is mainly subject to the compressive stress as the wall temperature is higher than the initial temperature (T_0). The maximum compressive stress

(9.8MPa) occurs in the outer mortar layer.

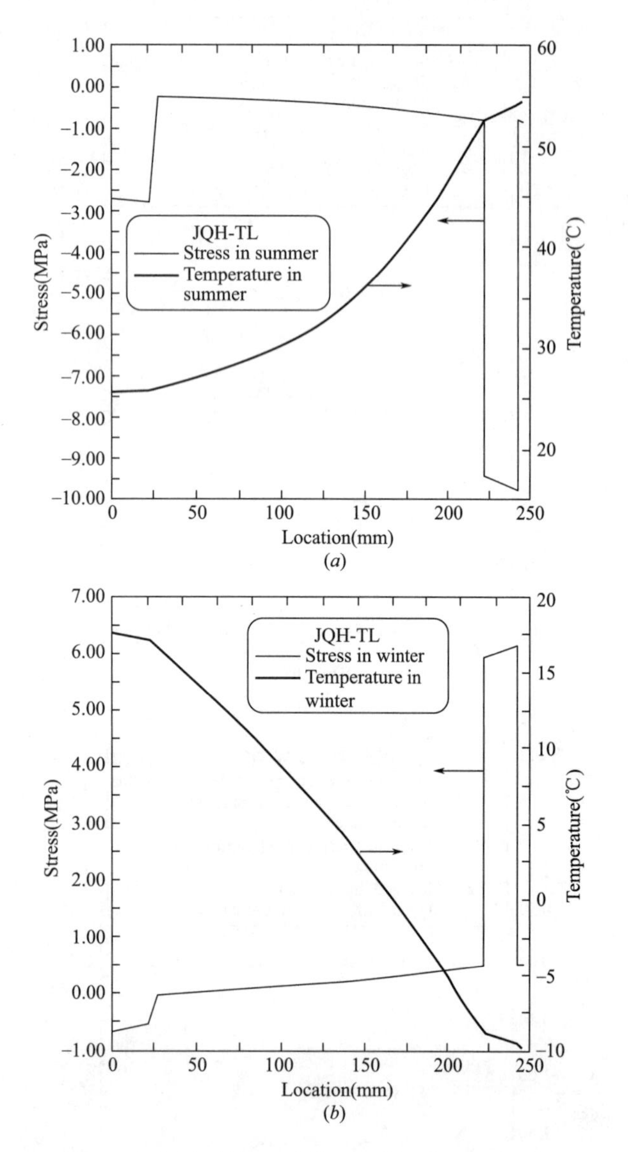

Figure 2-2-28 Thermal Stress Distribution in the Thickness Direction of Completely Constrained Wall with Aerated Concrete Self-insulation Paint Finish under Minimum and Maximum Temperature of External Surface in Winter and Summer

(*a*) winter; (*b*) summer

2.2.2.6 Wall with Concrete and Rock Wool Sandwich Insulation

Figure 2-2-29 to 2-2-31 show the relationship between the time (in 24h) and the thermal stress of the exterior finish and the external surface of the outer and inner concrete slabs of the wall with the 50mm concrete slabs, 50mm rock wool insulation boards and 50mm concrete slabs under typical climatic conditions in winter and summer, respectively. They illustrate the thermal stress of the wall surface in the presence of four typical constraints and under the same temperature changes. These four typical constraints are as follows: (1) wall stretching and turning are not allowed (embedded wall); (2) wall turning is allowed but stretching is not; (3) wall stretching is allowed but turning is not; and (4) wall stretching and turning are allowed (free wall).

According to Figure 2-2-29, as expected, the rules of changes in the thermal stress of the external surface of the finish over time are similar to those of the exterior finish of the wall insulated based on the ordinary adhesive polystyrene granule. That is, there is a small difference in the thermal stress under the first two types of constraint, indicating the small bending stress of the external surface of the wall and small temperature difference between the inner and external surfaces of the finish. The thermal stress is mainly caused by horizontal and vertical constraints. Due to differences in insulation, the peak temperature of the exterior finish of the concrete and rock wool concrete wall is a little lower than that of the wall insulated based on the ordinary adhesive polystyrene granule, resulting in differences in their specific values of thermal stress.

As shown in Figure 2-2-30:

(1) Similar to the outer mortar layer of the aerated concrete self-insulated wall, there is a small difference in the thermal stress of the outer wall of sandwich insulation under the first two types of constraint, indicating that the bending stress of the outer wall is small. The thermal stress is mainly caused by horizontal and vertical axial constraints.

(2) The thermal stress of the free wall is close to zero, indicating the small nonlinearity of temperature distribution inside the concrete slab.

(3) The thermal stress decreases with the temperature rising and increases with the temperature falling. The peak temperature corresponds to the peak stress. When the initial temperature is

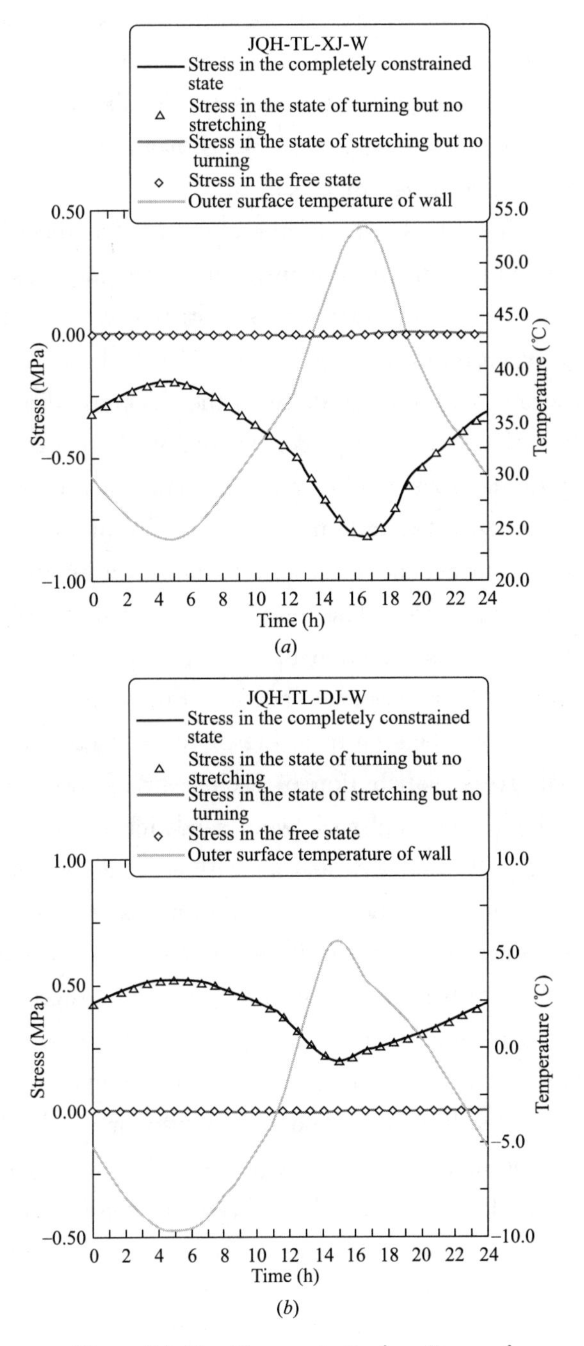

Figure 2-2-29 Changes in Surface Stress of Exterior Finish of Wall with Concrete and Rock Wool Sandwich Insulation over Time

(*a*) winter; (*b*) summer

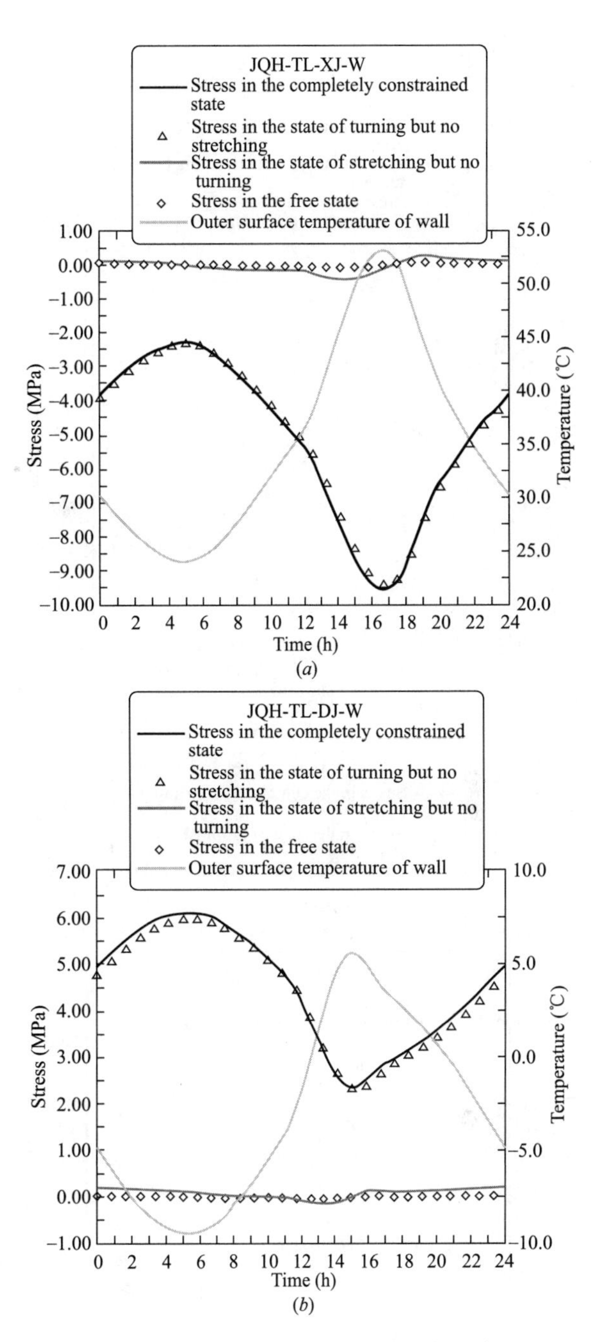

Figure 2-2-30 Changes in External Surface Stress of Outer Wall of Concrete and Rock Wool Sandwich Insulation over Time

(*a*) winter; (*b*) summer

15℃, the outer concrete slab surface is subject to tensile stress in winter and compressive stress in summer, but the values of such tensile and compressive stress are greater than those of the exterior finish. Under complete constraints, the maximum tensile stress and compressive stress are 6.12MPa and −9.55MPa in 24h, respectively. The outer concrete slabs are likely to crack in winter.

(4) When the initial temperature is 15℃, the stress of the outer concrete slab surface of the wall falls between 2.35MPa and 6.12MPa in winter and between −9.55MPa and −2.26MPa in summer. Therefore, the magnitude of change in the surface stress is −9.55MPa to 6.12MPa.

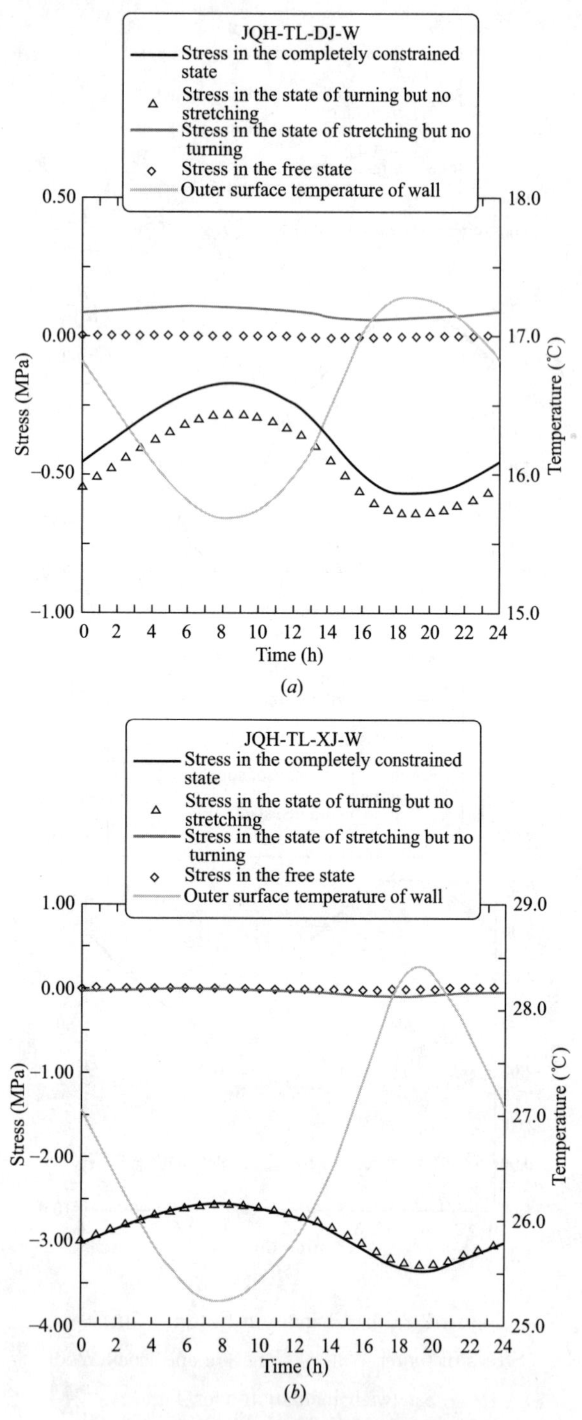

Figure 2-2-31 Changes in External Surface Stress of Inner Wall of Concrete and Rock Wool Sandwich Insulation over Time

(*a*) winter; (*b*) summer

Figure 2-2-31 shows the changes in the thermal stress of the external surface of the inner wall of sandwich insulation in winter and summer, respectively. Compared with that of the outer wall, the thermal stress of the inner wall is much lower. When the initial temperature is 15℃, the mortar layer is subject to low compressive stress (with an insignificant change in temperature) in both winter and summer, which has little impact on the wall performance.

The distribution of thermal stress of the insulated wall in the thickness direction can reflect the rules of thermal stress distribution in each layer more intuitively. Figure 2-2-32 shows the thermal stress distribution in the thickness direction of the completely constrained wall with concrete and rock wool sandwich insulation, corresponding to the minimum temperature of the external surface in winter and its maximum temperature in summer. The temperature distribution in the thickness direction of the wall in the corresponding period is also shown in Figure 2-2-32.

As can be seen from Figure 2-2-32, the thermal stress distribution of each layer of the wall with concrete and rock wool sandwich insulation is basically the same as that with aerated concrete self-insulation. They are different in that the aerated concrete layer is still subject to stress, while rock wool boards are essentially not. The thermal stress is distributed in a stepping form under complete constraints. In winter, the temperature gradually drops from indoors to outdoors, whilst the thermal stress gradually rises. The inner concrete slabs are mainly subject to compressive stress. Compared with inner concrete slabs, the thermal stress of outer concrete slabs increases suddenly, and its peak is 6.12MPa under complete constraints. In summer, the wall is mainly subject to compressive stress as the wall temperature is higher than the initial temperature (T_0). The maximum compressive stress (9.55MPa) occurs to the outer concrete slab surface.

2.2.3 Conclusions

Using the temperature field results calculated in the numerical method, the thermal stress of each functional layer of the insulated wall under four types of constraint was calculated. The fol-

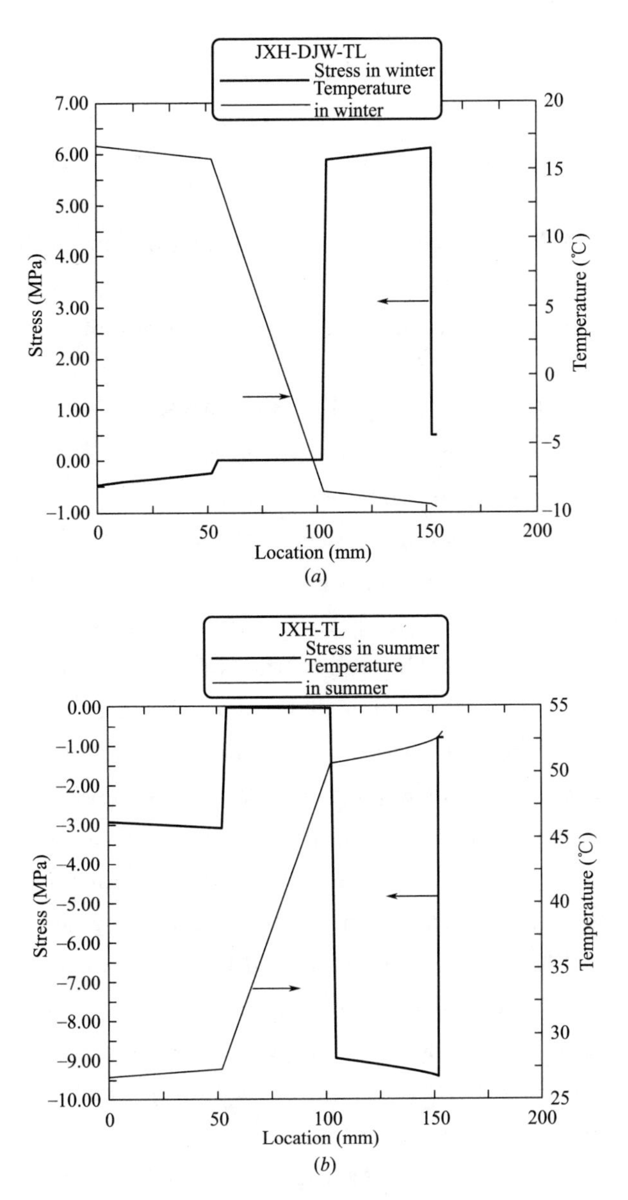

Figure 2-2-32 Thermal Stress Distribution in the Thickness Direction of Completely Constrained Wall with Concrete and Rock Wool Sandwich Insulation under Minimum and Maximum Temperature of External Surface in Winter and Summer

(*a*) winter; (*b*) summer

lowing conclusions can be reached according to the thermal stress calculation results and related analysis.

(1) Under complete constraints of the insulated wall (turning and stretching are not allowed), the thermal stress is a sum of the stress (σ_1) caused by average temperature changes, stress (σ_2) caused by linear temperature changes and stress (σ_3) caused by nonlinear temperature changes.

(2) The thermal stress inside the wall depends on the initial temperature (T_0), temperature change ($T - T_0$), elastic modulus of the material, etc. When T_0 is 15℃, the wall is mainly subject to tensile stress in winter and compressive stress in summer. Therefore, the external surface (relative to the insulation layer) of the insulated wall has risks of cracking in winter and hollowing in summer.

(3) There are insignificant temperature changes and gradients inside the structural layer of the wall with external thermal insulation, resulting in the low thermal stress, but significant temperature changes and gradients in the presence of internal thermal insulation, leading to the high thermal stress. Therefore, the external thermal insulation has good protective effects on the base course wall of the building.

(4) With the insulation layer, the temperature of each functional layer outside the insulation layer changes drastically in winter and summer, resulting in the high thermal stress. Therefore, the cracking and hollowing caused by the thermal stress of the outer layer (relative to the insulation layer) of the insulated wall should be taken into consideration in the future research and development of insulating and supporting materials.

2.3 Temperature Field and Thermal Stress Simulation of External Thermal Insulation System with ANSYS Software

2.3.1 Principle of Temperature Field and Thermal Stress Calculation with ANSYS Software

The simulation analysis of the temperature field with ANSYS software is an important means to theoretically explain the energy-saving and insulating effect of the external thermal insulation system. The ANSYS software is a kind of large-scale and general-purpose finite element analysis software integrating the analysis of the structure, fluid, electric field, magnetic field and sound field. Developed by ANSYS in the United States, one of the world's largest finite element analysis software companies, it is compatible with most of CAD software interfaces for sharing and exchange of data, such as Pro/Engineer, NASTRAN, Alogor, I-DEAS and Auto CAD. It is also one of advanced CAE tools in the modern product design. The simulation of the temperature field of the external thermal insulation system with ANSYS software is essentially a process of thermal analysis of this system. The solutions to the thermal stress of the two-dimensional and three-dimensional temperature fields of special nodes (thermal bridges) of the wall with the external thermal insulation can be obtained.

The fundamentals of thermal analysis with ANSYS software are briefly introduced below.

Thermal analysis is performed to calculate the temperature distribution and other thermal physical parameters of one system or component, such as the heat gain or loss, heat gradient, and heat flow density (heat flux). Thermodynamic analysis of structures is needed in engineering applications, including internal combustion engines, turbines and electronic components. The temperature field should be simulated during concrete pouring and welding in the civil engineering.

The thermal analysis of ANSYS software is a heat balance equation based on the energy conservation principle, in which the temperature of each node is calculated by the finite element method and other thermal physical parameters are derived. Three types of heat transfer are involved in the thermal analysis of ANSYS software, namely, heat conduction, heat convection and heat radiation. In addition, the phase change, internal heat source and thermal contact resistance can be analyzed. The thermal analysis function of ANSYS software is generally included in five modules, i. e. ANSYS/Multiphysics, ANSYS/Mechanical, ANSYS/Thermal, ANSYS/FLOTRAN, and ANSYS/ED, among which the ANSYS/FLOTRAN excludes the thermal analysis of phase change.

2. 3. 1. 1 Temperature Field Simulation with ANSYS Software

1. Classification of thermal analysis of ANSYS software

The thermal analysis in the ANSYS software program is divided into two types.

(1) Steady-state heat transfer

The temperature field of the system does not change over time. This mode applies to the analysis of impact of steady thermal load on the system or component. Generally, the steady-state thermal analysis is conducted before the transient-state thermal analysis, to determine the initial temperature field. By means of the steady-state thermal analysis, the parameters such as the temperature, heat gradient, heat flow rate and heat flow density, arising from the steady thermal load, can be calculated with finite elements.

If the net heat flow rate of the system is zero, that is, the heat flowing in to the system and generated by the system equals to the heat flowing out of the system ($q_{流入} + q_{生成} - q_{流出} = 0$), the system will be regarded in the steady thermal state. The temperature of any node does not change over time in the steady-state thermal analysis. The corresponding energy balance equation is:

$$[K]\{T\} = \{Q\} \tag{2-3-1}$$

Where, $[K]$ is a conduction matrix, involving the thermal conductivity, convection coefficient, radiation coefficient and form factor; $\{T\}$ is a node temperature vector; and $\{Q\}$ is a vector of node heat flow rate, involving heat generation.

The ANSYS software is used for generating [K], {T} and {Q} with the geometric parameters of the model, thermal performance parameters of the material, and boundary conditions applied.

(2) Transient-state heat transfer

Transient-state heat transfer is a process of system heating or cooling, in which the temperature, heat flow rate, thermal boundary conditions and internal energy of the system change significantly over time. According to the energy conservation principle, the transient-state heat balance can be expressed as (in a matrix form):

$$[C]\{T^{\&}\}+[K]\{T\}=\{Q\} \quad (2\text{-}3\text{-}2)$$

Where, [K] is a conduction matrix, involving the thermal conductivity, convection coefficient, radiation coefficient and form factor; [C] is a specific heat matrix, taking into account the increase of internal energy of the system; {T} is a node temperature vector; $\{T^{\&}\}$ is derivative of the temperature to time; and $\{Q\}$ is a vector of node heat flow rate, involving heat generation.

The transient-statc thermal analysis is conducted to calculate the temperature field and other thermal parameters of one system over time. In engineering applications, it is generally applied to calculate the temperature field as a thermal load for stress analysis.

The basic steps of transient-state thermal analysis are similar to those of steady-state thermal analysis. Their major difference is that the load in the transient-state thermal analysis changes over times. In order to express the load changing over time, the load-time curve must first be divided into load steps. The load and time of each load step must be defined. At the same time, the load step must be of gradual change or stepping type.

2. Boundary and initial conditions for thermal analysis of ANSYS software

The boundary or initial conditions for thermal analysis of ANSYS software can be divided into seven types, namely, the temperature, heat flow rate, heat flow density, convection, radiation, thermal insulation and heat generation.

The temperature is applied as a DOF (degree of freedom) constraint on the boundary with the known temperature. The heat flow rate, as a concentrated load of the node, must be used on the linear element model. The positive heat flow rate means that the heat flows into the node. That is, the element is provided with heat. The negative heat flow rate means that the element outputs heat. The convection is applied as a surface load on the external surface or surface effect element of a solid to calculate the heat exchange between the solid and fluid. The heat flow density refers to the heat flow rate per unit area, and is applied as a surface load on the external surface or surface effect element of a solid. The positive heat flow density means that the heat flows into the element. The heat generation rate is applied as a body load on an element. The heat generated by chemical reaction or current can be simulation.

3. Thermal analysis modeling of ANSYS software

Thc finite element model is built according to the engineering conditions to be analyzed. The type, options and real constant of the element as well as the thermal parameters of the material are defined or set in the pre-processing program of ANSYS software. For steady-state heat transfer, only the thermal conductivity needs to be defined under normal conditions. This parameter may be constant or changing with the temperature. Finally, a geometric model is created and finite element meshes are divided.

4. Common elements of thermal analysis of ANSYS software

Thermal analysis involves about 40 types of element.

(1) Thermal analysis elements

There are 14 types of element dedicated to thermal analysis, as shown in Table 2-3-1.

Thermal Analysis Elements Table 2-3-1

Element Type	ANSYS Element	Description
Linear element	LINK31	2-node heat radiation element
	LINK32	2D 2-node heat conduction element
	LINK33	3D 2-node heat conduction element
	LINK34	2-node heat convection element

Continued

Element Type	ANSYS Element	Description
2D solid	PLANE35	6-node triangular element
	PLANE55	4-node quadrilateral element
	PLANE75	4-node axisymmetric element
	PLANE77	8-node quadrilateral element
	PLANE78	8-node axisymmetric element
3D solid	SOLID70	8-node hexahedron element
	SOLID87	10-node tetrahedron element
	SOLID90	20-node hexahedron element
Shell	SHELL57	4-node
Point	MASS71	Mass element

(2) Surface effect element

The surface effect element is formed by nodes of a solid surface and covered on the solid surface. Therefore, the number of nodes will not be increased. There are four types of surface effect element for thermal analysis of ANSYS software, namely SURF19 (2D), SURF151 (2D), SURF22 (3D), and SURF152 (3D). With the surface effect element, the surface load can be flexibly defined for thermal analysis of ANSYS software.

If used at the same time, one of the heat flow density and heat convection must be applied on the solid surface, while the other on the surface effect element.

2.3.1.2 Thermal Stress Simulation of ANSYS Software

The wall will expand or contract with its temperature changing. In the event of differences in expansion or contraction of different parts or constraints to expansion and contraction, the thermal stress will be generated to the wall. Therefore, attention should be paid to the changes in the temperature field of the structure, but also the wall stress distribution arising from temperature changes. Engineering disasters caused by cracks in the system under the thermal stress often occur in the external thermal insulation system. The thermal stress distribution of the external thermal insulation system can be simulated with ANSYS software, to illustrate the failure mechanism of thermal stress and thus avoid the thermal stress via engineering measures or effectively release the thermal stress. This is also of practical significance for the healthy development of the entire external thermal insulation industry.

1. Classification of thermal stress analysis with ANSYS software

There are three optional methods of thermal stress analysis in ANSYS software, depending on specific conditions.

(1) Direct method

This method is applicable when the node temperature is known. In the thermal stress analysis of the wall, the node temperature will be directly applied as a body load on the node, according to BF, BFE or BFK commands.

(2) Indirect method

This method is applicable when the node temperature is unknown. The thermal analysis should be performed first. Then the node temperature obtained should be applied as a body load on the node involved in the thermal stress analysis of the wall.

(3) Heat-structure coupling method

Taking the heat-wall coupling effect into account, a coupling element with the temperature and displacement freedom is used to simultaneously obtain the results of thermal analysis and thermal stress distribution of the wall.

The thermal stress of the external thermal insulation system is analyzed in the indirect method in this book.

2. Corresponding elements of thermal analysis and structural analysis

In order to calculate the thermal stress after thermal analysis, the thermal analysis elements and structural elements need to be converted. Refer to Table 2-3-2 for correspondences between thermal analysis elements and structural elements.

Correspondences between Thermal Analysis Elements and Structural Elements Table 2-3-2

Thermal Analysis Element	Structural Element
LINK32	LINK1
LINK33	LINK8
PLANE35	PLANE2
PLANE55	PLANE42
SHELL57	SHELL63
PLANE67	PLANE42
LINK68	LINK8
SOLID70	SOILD45
MASS71	MASS21
PLANE75	PLANE25
PLANE77	PLANE82
PLANE78	PLANE83
SOLID87	SOLID92
SOILD90	SOLID95
SHELL157	SHELL63

2.3.2 Example of Temperature Field and Thermal Stress Calculation

The simulation and analysis of the temperature field and thermal stress of the external thermal insulation system with ANSYS software are illustrated by the following calculation example.

Compared with the internal thermal insulation system, the external thermal insulation system should meet higher material performance and design requirements as it is located on the surface of the base course wall. Cracks are usually found in nodes between the thermal insulation system and base course wall in the engineering practice as a result of improper design and treatment of local nodes of the external thermal insulation system. For example, the thermal insulation of the internal surface is often neglected in the thermal insulation design of the outer wall of the parapet. Figure 2-3-1 (*a*) shows the cracks in the absence of thermal insulation measures on the internal surface of the parapet with the external thermal insulation. In addition, the thermal insulation of overhanging parts are often neglected in the design, such as the balcony, awning, balcony railings close to the outer wall, shelf for the outdoor unit of the air-conditioner, pilaster, bay window, decorative line, balcony partition close to the outer wall, eave gutter, as well as inner, outer and coping sides of the parapet, which will result in cracks. Figure 2-3-1 (*b*) shows the crack of the connection between the overhanging part and main wall. These cracks will not only have a bad sensory and psychological impact on the user, but also seriously affect the stability and service life of the external insulation system and shorten the service life of the base course wall. This section simulates the temperature field and thermal stress of the wall with ANSYS software, and an-

(*a*)

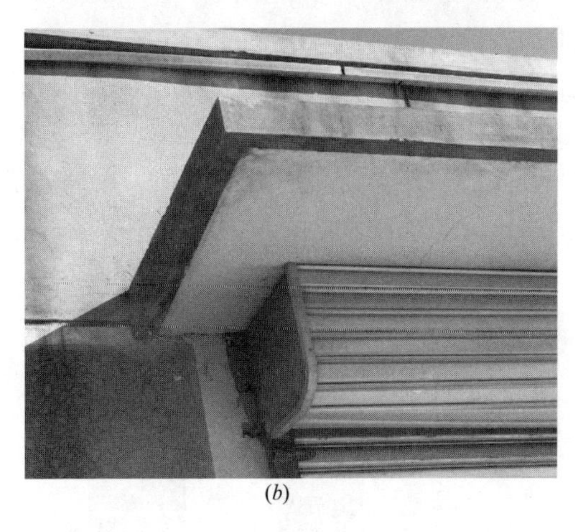

(*b*)

Figure 2-3-1 Photos of Cracks on Non-insulated Internal Surface and Overhanging Parts of Parapet
(*a*) Crack of Non-insulated Internal Surface of External Thermal Insulation of Parapet;
(*b*) Crack of Connection between Overhanging Part and Main Wall

alyzes the causes and mechanisms of wall cracking, thus providing a scientific and theoretical basis for the design of local nodes of the external thermal insulation system and disposal of related engineering disasters.

2.3.2.1 Calculation Model

It is assumed that one overhanging structure (awning, as shown in Figure 2-3-2) is built, the external thermal insulation system (see Table 2-3-3 for its parameters) based on the adhesive polystyrene granule is applied on the outer side of the base course wall, and the awning is directly connected with the base course wall but not insulated. For conveniences, the temperature field and thermal stress of the wall with the external thermal insulation but no overhanging structure was used as a comparison object (Figure 2-3-2).

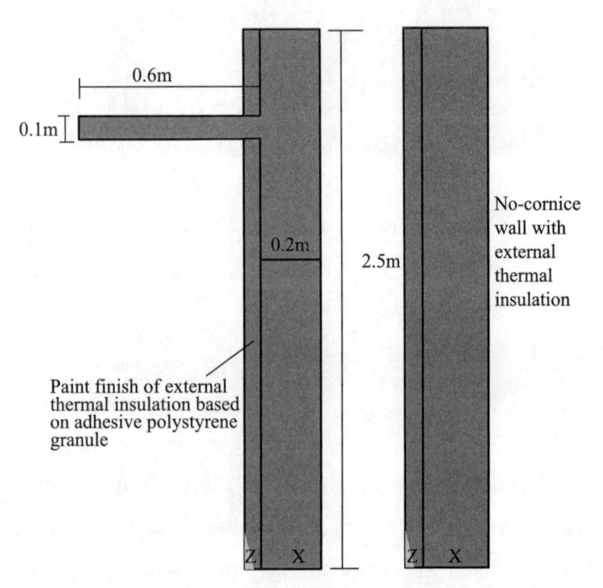

Figure 2-3-2 Basic Dimensions of Overhanging Structure (awning) and Wall

The average outdoor ambient temperature in summer and winter in Beijing was used as an additional temperature load of the structure model. The temperature curve is shown in Figure 2-3-3. Solar radiation was not taken into consideration here.

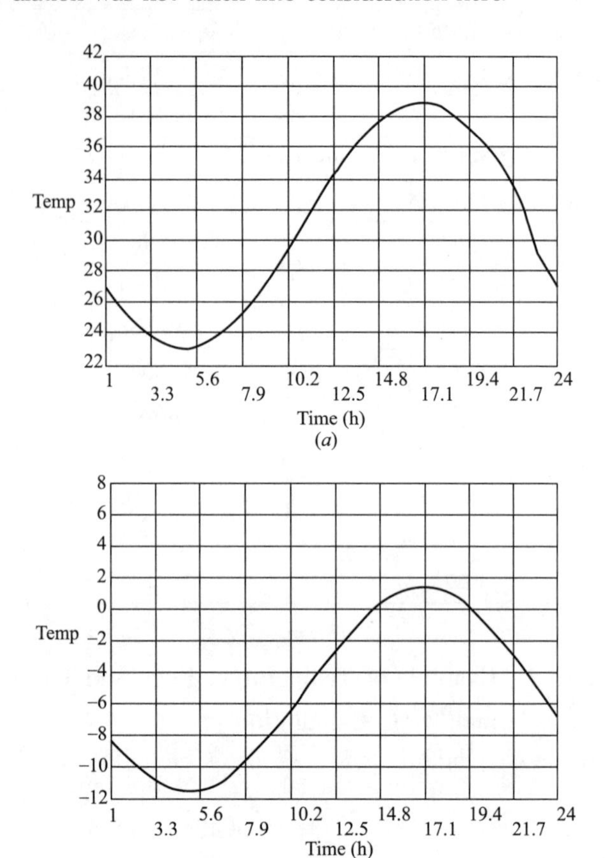

Figure 2-3-3 Temperature Curves of Beijing in Summer and Winter

(*a*) summer; (*b*) winter

It is assumed that the indoor temperature is25℃ in summer and 20℃ in winter; convective heat transfer is formed between the inner wall surface and indoor air, of which the coefficient is 8.7W/(m^2 • ℃); and the initial temperature of the model is 15℃.

Thermophysical Parameters of Materials **Table 2-3-3**

Material Name	Thickness (mm)	Density (kg/m^3)	Specific Heat [J/(kg • K)]	Thermal Conductivity [W/(m • k)]	Thermal Expansion Coefficient 10^{-6}(1/K)	Elastic Modulus (GPa)
Interior finish	2	1300	1050	0.60	10	2.00
Base course wall	200	2300	920	1.74	10	20.00
Interface mortar	2	1500	1050	0.76	8.5	2.76
Insulation mortar	50	200	1070	0.07	8.5	0.0001
Anti-cracking mortar	5	1600	1050	0.81	8.5	1.50
Paint finish	3	1100	1050	0.50	8.5	2.00

The ANSYS engineering software was used in the simulation calculation. The temperature field was calculated first. Then the temperature field obtained was applied as an external load on the structure for thermal stress calculation.

2. 3. 2. 2 Analysis of Temperature Field Calculation Results

Figure 2-3-4 shows the results of simulation calculation (only the results of the summer are given here). The temperature distribution is described according to the temperature field distribution cloud chart and temperature distribution curve diagram.

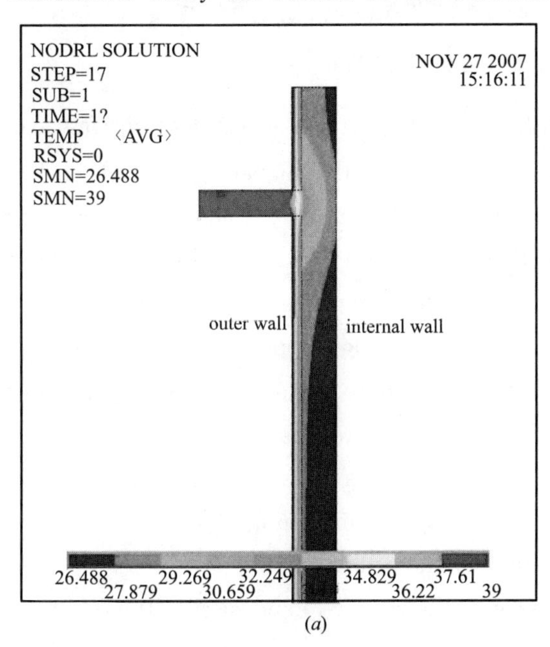

(*a*)

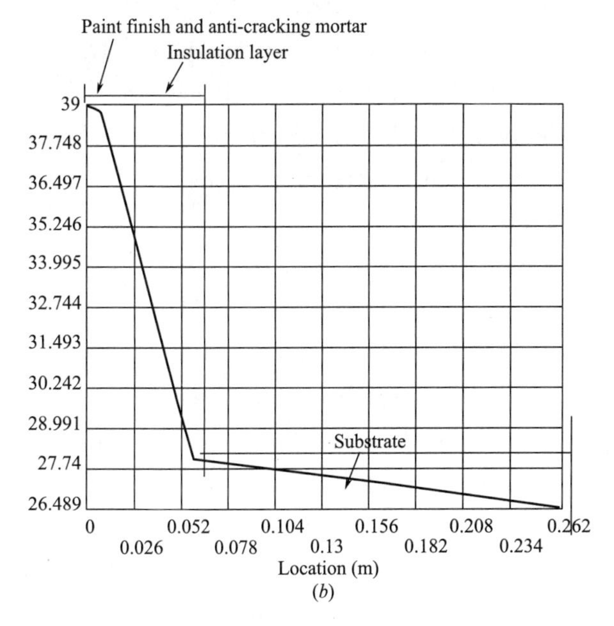

(*b*)

Figure 2-3-4 Simulation Results of Temperature Field in Summer (outdoor temperature: 39℃)

(*a*) temperature field distribution diagram;

(*b*) curve of temperature distribution in the thickness direction of base course wall

According to the temperature field distribution cloud chart in Figure 2-3-4 (*a*), the awning with no insulation layer has obvious "thermal bridge" effects, and a large amount of thermal energy is transferred through the wall. The magnitude of temperature change of the internal surface of the outer wall near the awning reaches 9-10℃ in one day. According to the results of simulation calculation, the magnitude of temperature change of the internal surface of the outer wall with the insulation layer on the awning is about 1℃ in one day. This "thermal bridge" is very harmful to the external thermal insulation system and wall, which will often result in condensation of the external thermal insulation system in winter and not be conducive to the insulation effects of the whole system. The infrared test results of actual projects show that such neglected parts are obvious thermal bridges, as shown in Figure 2-3-5. Compared with those insulated, the temperature of these parts is significantly affected by the environment, and the resulting thermal stress will cause cracks of connection between these parts and the main wall. At the same time, the thermal bridges have adverse effects on the overall energy conservation. The infrared test results are basically consistent with the numerical simulation results.

2. 3. 2. 3 Analysis of Temperature Stress Calculation Results

This section illustrates the thermal stress in the thickness direction of the wall with the awning and ordinary wall under the outdoor temperature of 39℃ in summer (the thermal stress in winter is not given here).

According to the thermal stress simulation results in Figure 2-3-6, if the overhanging structure (such as the awning) of the wall is not subject to external thermal insulation, the stress of

Figure 2-3-5 Thermal Bridge of Overhanging Part in Infrared Test Results

its paint finish in the vertical direction of the wall surface is significantly higher than (even 100 times) that of the ordinary wall. Meanwhile, the paint finishes of these two types of wall are subject to high compressive stress along the wall surface (wall with the awning: approximately 0.425MPa; ordinary wall: approximately 0.404MPa) under the outdoor temperature of 39℃. It can be concluded that the compressive stress of the paint finish of the wall with the awning is 0.02MPa higher than that of the ordinary wall.

Due to the thermal expansion of the overhanging structure, the stress of the paint finish of the connection between the overhanging structure and external thermal insulation system changes suddenly in the vertical direction of the wall surface, and the maximum tensile stress is 0.13MPa. If the tensile strength between the paint finish and anti-cracking mortar is not enough, the paint finish is easy to fall off. At the same time, the stress of the paint finish of the connection between the overhanging structure and external thermal insulation system changes suddenly in the vertical direction of the wall surface, max. 0.45MPa. The paint finish of this connection is easier than other parts to expand, which will result in cracks and engineering disasters.

The thermal shear stresses of the ordinary wall and the wall with awning, which are lower than other types of stress, can be basically ignored.

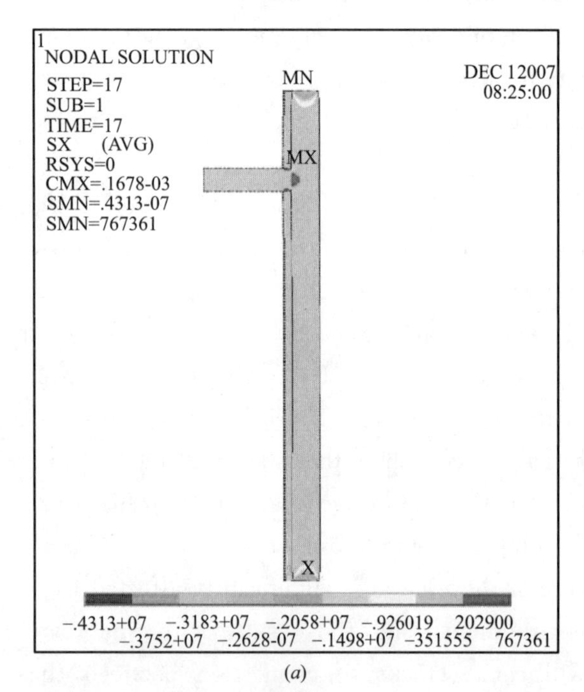

(a)

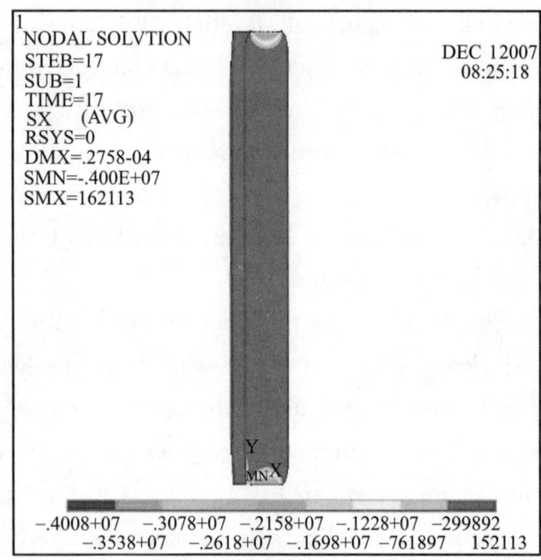

(b)

Figure 2-3-6 Simulation Results of Thermal Stress in Summer (39℃)

(a) wall with awning; (b) ordinary wall

Two temperature fields that are different significantly will be formed by the awning and base course wall due to the outdoor temperature effects. The awning with no external thermal insulation is greatly affected by the outdoor temperature. The temperature field of the base course wall is basically stable due to the external thermal insulation system. In this case, the thermal stress generated by the awning will be harmful to the base course wall. The thermal stress of the awning and base course wall connection is analyzed below. According to the comparative analysis results, when the temperature is 39℃ in summer, the stress of the connection between the overhanging structure and wall changes suddenly along the wall surface (about 3.6MPa) and in the vertical direction (about 0.95MPa) of the wall surface. When the temperature is -11.5℃ in winter, the stress suddenly changes into about 1.2MPa along the wall surface and 0.4MPa in the vertical direction of the wall surface, obviously increasing the stress burden near the contact point. The service life of structures in these areas will be greatly shortened in a long time. In addition, cracks in these areas are often neglected.

2.3.2.4 Thermal Deformation

As shown by the deformation results in Figure 2-3-7, the external thermal insulation system near the awning is squeezed due to significant thermal expansion of the non-insulated awning in summer. According to the above thermal stress analysis, the exterior finish of the paint layer may be pressed off. In winter, the external thermal insulation system near the awning is "dragged" to crack due to significant contraction of the awning. In the event of improper treatment of the connection, the external thermal insulation system may be directly separated from the awning, resulting in cracks. If such engineering disasters occur, rain and snow may seep into the insulation layer, which has severe impact on the insulation effect and durability of the external thermal insulation system.

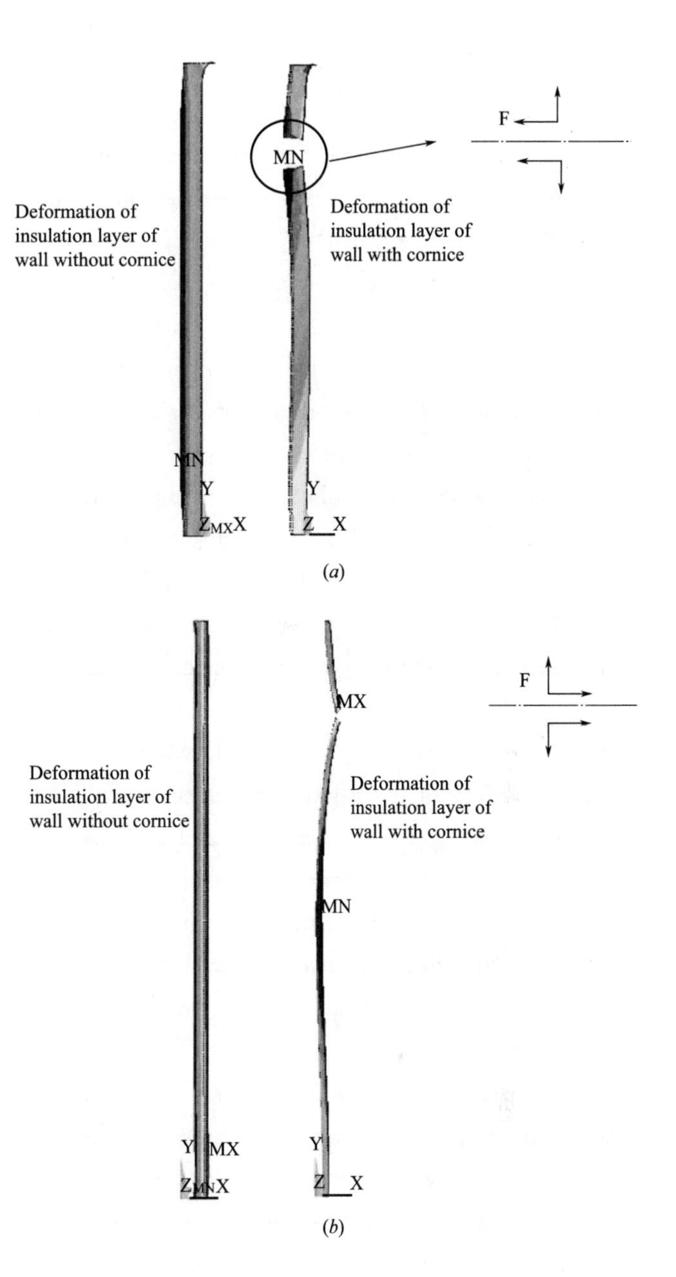

Figure 2-3-7 Simulation of Insulation Layer Deformation in Winter and Summer

(*a*) insulation layer deformation in summer (39℃);

(*b*) insulation layer deformation in winter (−11.5℃)

2.3.3 Conclusions

Through the numerical simulation and analysis of the temperature field and thermal stress of the non-insulated awning of the wall with the external thermal insulation, it can be concluded that no insulation of overhanging structures has a number of adverse effects, which are mainly reflected in the following aspects.

(1) The non-insulted overhanging structure will have the obvious "thermal bridge" effect on the overall wall and reduce the insulation effects. At the same time, when the temperature of the overhanging structure (corresponding surface of the interior finish) is below the dew point in winter, vapor tends to condense on its surface, thus forming dews.

(2) The thermal expansion and contraction of the non-insulated overhanging structure will lead to the sudden change of the stress in the connection between the insulation layer and overhanging structure. The connection of the insulation layer will be subject to high compressive stress along the wall surface as a result of thermal expansion in winter, and high tensile stress along the wall surface because of contraction in winter. These stresses may cause hollowing, cracking and falling of the related insulation layer.

(3) The thermal expansion and contraction of the non-insulated overhanging structure will lead to the stress concentration in the connection of the overhanging structure and wall, which will damage the structural layer of the wall and affect its service life.

In accordance with the analysis results, external thermal insulation measures should be taken to the overhanging structure, which will help to stabilize the temperature and deformation of the overhanging structure and prevent the common quality problem of cracking caused by the overhanging structure to the external thermal insulation system and base course wall.

2.4 Summary

As shown by the results of numerical simulation of changes in the temperature field and thermal stress of walls with the external thermal insulation, internal thermal insulation, sandwich insulation and self-insulation with the time and azimuth, the wall with the external thermal insulation has many advantages, mainly including the following items.

(1) The main structure of the wall with the external thermal insulation has a small difference of temperature in one year, season or day, thus greatly decreasing the fluctuation of its thermal stress, effectively protecting the main structure, improving the durability of the wall structure and prolonging the building life. In this sense, the external thermal insulation is the most reasonable method to save resources and energy among the above-mentioned four types of insulation.

(2) The thermal capacity of the base course wall is generally much larger than that of the insulation layer, so the external thermal insulation is beneficial to the thermal stability of a room. If not supplied evenly in the heating period, a large amount of heat will be stored in the base course wall, which will prevent the temperature of the internal surface of the outer wall from dropping sharply, thus avoiding the quick drop of the room temperature. Due to the shielding effect of the external thermal insulation system, the temperature of the base course wall rises slowly under high outdoor temperature in summer, which will make the indoor temperature stable and help to improve the comfort. In short, if the external thermal insulation is applied to the wall, the room will be warm in winter and cool in summer, and the energy consumption of heating and air-conditioning will be reduced.

(3) With the external thermal insulation, the heat loss of the thermal bridge will be decreased, and local condensation will be avoided in case that the temperature of the internal surface of the thermal bridge is too low in winter. The problem of serious heat losses of thermal bridges cannot be solved by other types of insulation.

3. Study on Waterproofing Performance and Air Permeability of Insulated Outer Wall

Thermal insulation measures are widely taken in cold and severe cold areas, to improve the insulation performance and air-tightness of the envelope, reduce the heat losses of heat transfer and cold vapor permeation and achieve the energy-saving effects in the energy efficiency standards of buildings. If the air-tightness of the envelope is improved, however, moisture transfer may be affected, and a lot of domestic vapor indoors cannot be easily discharged to outdoors through the envelope. If too much water is absorbed or accumulated in the insulation material of the envelope, condensate, dew (on the internal surface), mildew and bacteria may be caused inside the envelope in winter, and even the envelope may be subject to frost heaving and damage, which will shorten the service life of buildings.

This chapter first analyzes the problems of heat and moisture transfer of outer walls with the internal thermal insulation, sandwich insulation and self-insulation in the severe cold and cold areas, and proposes that the external thermal insulation should be vigorously promoted. This chapter also systematically introduces the water-absorbing capacity, hydrophobicity, waterproofing performance and vapor permeability of the external thermal insulation material, and analyzes the impact of vapor condensation on the service life of the external thermal insulation system. In addition to the great impact of climatic conditions such as the temperature, sunlight and air humidity, the external thermal insulation system is affected by the repeated effects of moisture and heat. The thermal insulation and energy conservation of the outer wall are also influenced by its external or internal moisture, moisture of building materials or condensate on the internal surface of the outer wall. Major factors affecting the bonding performance of the external thermal insulation are mainly analyzed. It is recommended to adopt the external thermal insulation process, combining bonding (main method) and anchorage of organic polymer cement mortar and mechanical anchors, to improve the overall performance of the external thermal insulation system. Finally, the following opinions and appropriate recommendations on the waterproofing performance and vapor permeability of the external thermal insulation system are put forward: (1) improve the hydrophobicity of materials (to avoid and reduce liquid water); (2) add a waterproof barrier structure, i. e. polymer elastic primer layer (to prevent liquid water); and (3) add a vapor transfer and dispersion structure, i. e. moisture dispersion layer (to smoothly discharge gaseous water), so that the external thermal insulation system has good functions of moisture discharge and waterproofing.

Waterproofing and moisture discharge of the external thermal insulation system must not be ignored, as they have significant impact on the energy efficiency of buildings, durability of rooms and comfort of the indoor environment.

(1) Impact on the energy efficiency of buildings

1) The permeation of rainwater into the external thermal insulation system will reduce the effective thermal resistance of the insulation layer; 2) the intrusion of vapor will lead to damping, condensation or freeze-thaw damage inside the external thermal insulation system and affect the insulation performance; and 3) heating equipment is needed for conformity of the indoor temperature to design requirements, which will increase the energy consumption of buildings.

(2) Impact on the durability of buildings

If rainwater infiltrates into the external surface of the building, ice will be generated in the eroded building material, which will generate the frost heaving stress and damage the external surface of the building in freeze-thaw cycles. In particular, the freezing and thawing of the face brick system will destruct the bonding strength of materials, reduce the adhesion of each layer and weathering resistance of the entire system, and also accelerate the ageing of organic insulation materials and degradation of inorganic materials, thereby affecting the durability of buildings.

(3) Impact on the comfort of buildings

The destructive effect of vapor is mainly caused by its transfer driven by the differential pressure between the inside and outside of the envelope. In the event of the unreasonable structure and material of the external thermal insulation system, the dispersion of vapor will be affected, resulting in condensation on the inner side of the wall in winter, damping of the insulation layer, and even condensate flow. In this case, interior decorative materials and furniture will be affected with damp and deformed, and large-scale dark spots, mildew and mould will be generated on the internal surface of the outer wall. Such mildew and bacteria will change into contaminants in the humid environment in a long time, exerting adverse effects on the indoor air quality. The vapor flow will badly affect the indoor thermal environment and reduce the living comfort.

To sum up, the energy conservation, durability and comfort of buildings are closely associated with the waterproofing performance and vapor permeability of their external thermal insulation systems. Accordingly, the study and discussion of the impact of natural destructive forces on the thermal and structural performance of the wall with the external thermal insulation system as well as the service life of the building will necessarily involve the rules of water motion in this system, and the analysis of damage caused by water to this system. The external thermal insulation system must be permeable and resistance to water and condensate, so that this system has the same service life as the building and meets the national standards for the energy conservation of buildings.

3.1 Basic Principle of Moisture Transfer

Most of building materials are porous, and the building wall structure itself is a typical porous medium. The porous medium means that, for any phase, the other phases are dispersed in the space with the heterogeneous substance. It is extremely complicated to analyze and calculate the simultaneous movement of heat and moisture inside the porous material.

The porosity (ϕ) of the porous medium is calculated from the equation (3-1-1):

$$\phi=\frac{v_p}{v} \tag{3-1-1}$$

Where, ϕ — volume porosity, %;

v_p —pore volume, m^3;

v —total volume of the porous medium, m^3.

The volume weight (ρ_p) of the porous medium is calculated from the equation (3-1-2):

$$\rho_p=\frac{M}{V} \tag{3-1-2}$$

Where, ρ_p —mass per unit volume of the porous medium, kg/m^3;

M —mass of the porous medium, kg.

The relationship between the volume weight and density is given in the equations (3-1-3) and (3-1-4):

$$M=\rho_p V \tag{3-1-3}$$

$$M=\rho V(1-\phi) \tag{3-1-4}$$

Where, ρ —density of the porous medium, kg/m^3.

Differences in the water-absorbing capacity of materials are mainly caused by those in the porosity, pore size and surface tension; and differences in the vapor permeability of materials result from those in the porosity and pore size.

(1) Surface tension

According to the water absorption, materials are divided into hydrophilic, hydrophobic and water-saturating materials.

1) There is particularly strong adhesion between the molecules of the hydrophilic material and water. Many building materials are hydrophilic to different degrees. The contact angle (θ) between the surfaces of the hydrophilic material and water is less than $\pi/2$. When the tension of the liquid-solid contact interface is less than that of the solid-gas contact interface (i. e. solid surface tension), the solid will be wetted. Water can be absorbed into the hydrophilic material through capillaries.

2) The molecules of the hydrophobic material repel those of water. When the contact angle (θ) between the surfaces of the hydrophobic material and water is more than $\pi/2$, the solid will not be wetted. That is, when the tension of the liquid-solid contact interface is more than the solid surface intension, the solid will not be wetted. Generally, water seepage into capillaries of the hydrophobic material is hindered, thus reducing the water absorption.

3) The contact angle (θ) between the surfaces of the water-saturating material and water is $\pi/2$, so the water-saturating material is a boundary of hydrophilic and hydrophobic materials.

If a dry hydrophilic material is kept in the air, water molecules in the air will be absorbed by the molecules on the pore surfaces, forming a monomolecular coating of water molecules. Due to the thermal motion of molecules, however, internal water molecules will jump out of this coating and external water molecules continuously enter this coating. Similarly, if a material of high moisture content is kept in the dry air, an opposite process will occur. This is what is called water absorption and desorption. The number of water molecules flowing into and out of the coating is related to the concentration of water molecules in the air. A dynamic equilibrium will be formed with the motion of water molecules in the air and material in a sufficiently long period, this is the mass exchange between the outside and material. Meantime, moisture transfer will occur because of the partial pressure of vapor, water concentration, temperature difference (non-uniformity), and vapor permeation.

(2) Pore size

If the pore size of the material exposed to water is relatively large, water will be absorbed fast, but saturation will be achieved quickly due to the restriction of the absorption height. If there are a lot of thin capillary pores in the material exposed to water, water will be absorbed slowly, but saturation cannot be achieved easily due to the great height of absorption under capillary effects. After this material is separated from water, water will still rise along capillary pores. The pore size of the material has little impact on the vapor permeability.

When there is any pressure difference (partial or total pressure of vapor), humidity difference (moisture content of the material) and temperature difference inside the material, moisture of the material will be transferred. The moisture contained in the material may exist in three states, i. e. gas (vapor), liquid (liquid water), and solid (ice), among which the water and its vapor are the most common. Only two types of water can be transferred inside the material, i. e. vapor diffusion (also known as the vapor permeation) and liquid water permeation through capillaries.

When the moisture content of the material is lower than the maximum degree of water absorption, the moisture in the material is still a kind of absorbed water. In this case, the absorbed water will be first vaporized and then diffused in the gaseous form in the direction of decreasing partial pressure of vapor or flow of the heat. When the moisture content of the material is higher than the maximum degree of water absorption, there will be free water inside the material, which will be transferred along capillaries from the part of high moisture content to that of low moisture content. Main mechanisms of moisture transfer include the capillary action, liquid diffusion, and molecular diffusion. Additionally, the air pressure difference, water pressure difference, water gravity and temperature difference will lead to moisture transfer. There are many theories associated with moisture transfer. This section only introduces the laws of moisture transfer of liquid and gaseous water.

3.1.1 Liquid Water Flow in Porous Material

Darcy's law was originally applied to describe the liquid water flow in a porous medium. When water flows vertically and isothermally through a loose sand column at a low rate, the volume flow rate (Q) is found to be proportional to the head difference (ΔH_d) and the cross-sectional area (A) of the column, and inversely proportional to the height of the sand column, i. e. :

$$Q=\frac{K_c A \Delta H_d}{\Delta S} \qquad (3\text{-}1\text{-}5)$$

Where, $H_d = Z + P/\rho g$;

K_c —proportional constant, namely, hydraulic conductivity, m/d;

Z —height, m;

P —pressure, Pa;

ρ —fluid density, kg/m^3.

The Darcy velocity is defined as:

$$V=\frac{Q}{A} \qquad (3\text{-}1\text{-}6)$$

Then:

$$V=-K_c\frac{d}{ds}\left(z+\frac{p}{\rho g}\right) \qquad (3\text{-}1\text{-}7)$$

Where, the negative sign indicates that the flow direction is opposite to the hydraulic gradient.

Later studies show that the hydraulic conductivity is proportional to the fluid density but inversely proportional to the fluid viscosity (μ). If the intrinsic permeability of the porous medium is defined as $K=\mu K_c/\rho g$, then the equation (3-1-7) can be changed into:

$$V=-\frac{k}{\mu}\frac{d}{ds}(p+\rho g z) \qquad (3\text{-}1\text{-}8)$$

For the water absorption and desorption of the wall surface of the building, only the transfer of vapor in the porous medium is taken into account, and the equation (3-1-8) can be changed into:

$$V_v=-\frac{k}{\mu}\frac{dp}{dx} \qquad (3\text{-}1\text{-}9)$$

Where, V_v —velocity of vapor, m/s;

p —pressure of vapor, Pa;

x —distance to the internal surface in the vertical direction of the wall, m.

3.1.2 Vapor Transfer in Porous Material

The natural humidity of a material is the result of heat and moisture balance between the material and surroundings. It varies at the same time with the humidity of the surrounding air. Under the temperature difference between the material and surrounding environment and the partial pressure of vapor, vapor will flow into and out of the porous material through its surface, forming a dynamic equilibrium on this surface. The transfer of heat and vapor may lead to changes in the temperature inside the material and partial pressure of vapor. Vapor will be transferred, driven by the temperature gradient as well as the gradient of partial pressure of vapor. The transfer of vapor through the porous material is treated according to the Fick's law. That is, the vapor flow (W_v) per unit area of the wall surface with the permeability ε and thickness l is a function of the vapor pressure difference ($P_{vi}-P_{vo}$), i. e. :

$$W_v=\varepsilon\frac{p_{vi}-p_{vo}}{l} \qquad (3\text{-}1\text{-}10)$$

This equation cannot properly reflect the transfer of vapor in the material. Instead, it only gives a calculation method as a whole, and greatly varies from the actual process of transfer. The limit state of complete monomolecular coating will be achieved under a certain relative humidity. Correspondingly, the relationship between the amount (W_{L1}) of water absorbed by the material and the surface area of pores is as follows:

$$W_{L1}=KS \quad (3\text{-}1\text{-}11)$$

Where, S —surface area of pores, m^2;

K —proportional coefficient, equal to the monomolecular coating weight per unit area.

Where,

$$K=\left(\frac{18}{N}\right)^{\frac{1}{8}} \text{g/cm}^2 \quad (3\text{-}1\text{-}12)$$

Where, N —Avogadro constant, 6.023×10^{22}/(g · mol).

In order to explain the characteristics of water absorption and desorption of the material, vapor infiltrates based on its partial pressure gradient corresponding to the equilibrium humidity limit. Direct permeation of vapor will not occur until the partial pressure on two sides of the building wall is extremely low. The permeation process can be expressed as:

$$q_m=-\varepsilon\frac{de}{dx} \quad (3\text{-}1\text{-}13)$$

Where, q_m —specific flow of vapor, kg/(m^2 · h);

$\frac{de}{dx}$ —partial pressure gradient of vapor.

The permeability (ε) is generally determined by some tests.

The moisture in the porous material exists in the liquid state under normal conditions. When the relative humidity is less than 95%, however, the moisture in the material can be regarded diffused in the vapor form.

3.2 Moisture Protection of Building Wall

The humidity inside and outside a building is high in four seasons in the areas of hot summer and cold winter as well as those of hot summer and warm winter in China, especially in summer with abundant rain. Typically, rainwater is discharged through appropriate roofs, the base is elevated and natural ventilation measures are taken, to greatly reduce the impact of high humidity. Currently, the dew point temperature method and vapor pressure curve/partial pressure curve crossing method are applied in the moisture-proof design of domestic buildings under construction.

3.2.1 Vapor Permeation

When the moisture contents in the indoor and outdoor air are not equal to each other, that is, there is a difference in the partial pressure of vapor between both sides of a building wall, vapor molecules will be diffused from the side of high partial pressure to the other side of low partial pressure through the building wall. This is known as vapor permeation. It is a transfer process of vapor molecules, referred to as the mass or moisture transfer.

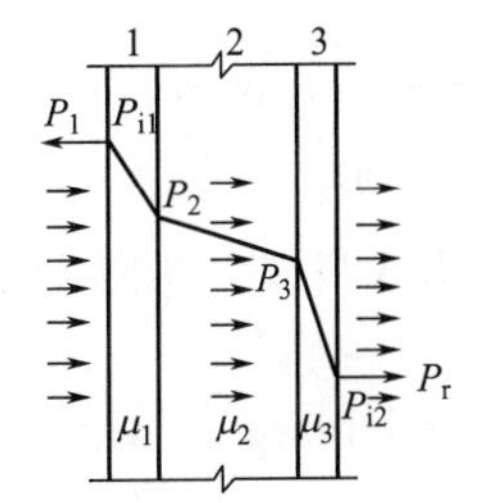

Figure 3-2-1 Vapor Transfer of Envelope

The moisture transfer of the building wall involves vapor infiltration arising from the partial pressure difference and vapor transfer caused by the temperature difference (heat transfer). The transfer of saturated vapor and liquid water also occurs in the condensation zone. Hence it is very complicated to conduct the calculation of moisture transfer. At present, the moisture of the building wall is calculated by means of rough analysis, namely, the pure infiltration of vapor in the steady state (or the quasi-steady state-average of a certain period, such as the winter or summer, equivalent to the mean value theorem for integrals; it is of no practical significance to study the instantaneous value). In the calculation, the partial pressure of indoor and outdoor vapor was regarded as a constant, the mutual impact of heat and moisture transfer (heat and mass transfer) was ignored, and the transfer of liquid water in the building wall was not taken into consideration. The calculation of pure vapor infiltration in the steady state is similar to that of the steady-state heat transfer. That is, the amount of vapor infiltration per unit area of the building wall in unit time is proportional to the partial pressure difference of indoor and outdoor vapor and inversely proportional to the resistance in the infiltration process under the steady conditions, calculated from (Figure 3-2-1):

$$\omega = \frac{1}{H_o}(P_i - P_e) \qquad (3\text{-}2\text{-}1)$$

Where, ω —vapor permeability, $g/(m^2 \cdot h)$;

H_o —total resistance of vapor permeation of the building wall, $(m^2 \cdot h \cdot Pa)/g$;

P_i —partial pressure of vapor of internal air, Pa;

P_e —partial pressure of vapor of external air, Pa.

The total resistance of vapor permeation of the building wall is determined by:

$$H_o = H_1 + H_2 + H_3 + \cdots H_n + \Lambda = \frac{d_1}{\mu_1} + \frac{d_2}{\mu_2} + \frac{d_3}{\mu_3} + \cdots + \frac{d_n}{\mu_n} + \Lambda \qquad (3\text{-}2\text{-}2)$$

Where, d_1, d_2, d_3, $\cdots$, d_n —thickness of each structural layer from the first layer of the inner side of the outer wall, m;

μ_1, μ_2, μ_3, $\cdots$, μ_n —vapor permeability coefficient of each structural layer from the first layer of the inner side of the outer wall, $g/(m \cdot h \cdot Pa)$;

H_1, H_2, H_3, $\cdots$, H_n —vapor permeation resistance of each structural layer from the first layer of the inner side of the outer wall, $(m^2 \cdot h \cdot Pa)/g$.

The vapor permeability coefficient refers to the amount of vapor passing through one $1m^2$ area of a 1m thick object under the pressure difference of 1Pa on both sides. The vapor permeability (μ) of the material is associated with its compaction and pore structure. The greater the porosity and pore percentage, the higher the vapor permeability. The glass and metal are not permeable for vapor. It should also be noted that the vapor permeability coefficient is also related to the temperature and relative humidity, but the average (under the normal temperature and humidity) is applied in calculations of thermal engineering of buildings.

The vapor permeation resistance refers to the time needed for 1g vapor to pass through one $1m^2$ area of a building wall or material layer under the partial pressure difference of 1Pa on both sides, in $(m^2 \cdot h \cdot Pa)/g$. Compared with the vapor permeation resistance of the structural layer, the resistance (Λ) moisture transfer between the inner and external surfaces (external air convection layer) of the building wall is insignificant, so it can be neglected in the calculation of the total resistance of vapor permeation. Thus, the partial pressure of vapor in the inner and external surfaces of the building wall can be approximated as P_i and P_e. The partial pressure of vapor in the internal surface of any layer of the building wall can be calculated from (similar to the internal temperature):

$$P_m = P_i - \frac{\sum_{j-1}^{m-1} H_j}{H_o}(P_i - P_e) \qquad (3\text{-}2\text{-}3)$$

$$m = 2, 3, 4, \cdots, n$$

Where, $\sum_{j-1}^{m-1} H_j$ is the sum of vapor permeation resistance from the first to $(m-1)^{th}$ layer, calculated from the inner side.

The saturated vapor pressure $P_{s,m}$ ($m=1, 2, \cdots, n+1$, in Pa) of the internal surface of any layer of the outer wall was calculated. The saturated vapor pressure of the external surface of the outer wall was calculated in the case of $m=n+1$. Under the certain atmospheric pressure (the standard atmospheric pressure was used), the saturated vapor pressure corresponds to the temperature. The temperature t_m (unit: ℃) of the internal surface of any layer of the outer wall was calculated from the following equation (material sequence: from indoors to outdoors) after consulting the table of saturated vapor pressure under the standard atmospheric pressure but different temperature conditions.

$$t_m = t_i - \frac{R_i + \sum_{j=0}^{m-1} R_j}{R_0}(t_i - t_e) \qquad (3\text{-}2\text{-}4)$$

$$m = 1, 2, 3, \cdots, n+1$$

Where, R_0—heat transfer resistance of the outer wall, $m^2 \cdot K/W$;

R_i—heat transfer resistance of the internal surface of the outer wall, $m^2 \cdot K/W$;

$\sum_{j=0}^{m-1} R_j$—sum of the heat transfer resistance of the first to $(m-1)^{th}$ layer from the indoor side.

The heat transfer resistance of the outer wall is calculated from the equation (3-2-6):

$$R_0 = R_i + R_1 + R_2 + \cdots + R_n + R_e$$
$$= R_i + \frac{d_1}{\lambda_1} + \frac{d_2}{\lambda_2} + \cdots + \frac{d_n}{\lambda_n} + R_e \qquad (3\text{-}2\text{-}5)$$

Where, R_e—heat transfer resistance of the external surface of the outer wall, $m^2 \cdot K/W$;

d_n—thickness of one layer, m;

λ_n—thermal conductivity of the material of one layer, $W/(m \cdot K)$;

R_n—heat transfer resistance of the material of one layer, $m^2 \cdot K/W$.

The thermal conductivity is only associated with the material itself, and has the following physical meaning: heat transferred through one $1m^2$ area of one 1m thick material under the temperature difference of 1℃ of both sides.

3.2.2 Air Temperature and Humidity

3.2.2.1 Relative Humidity and Dew Point Temperature

Vapor that can be accommodated in the air under the specific pressure and temperature has a saturation value. Once vapor is saturated, vapor will condense into liquid water. The partial pressure of vapor corresponding to the saturated moisture content is called the partial pressure of saturated vapor, which changes with the air temperature. Figure 3-2-2 shows the relationship between the air temperature and partial pressure of saturated vapor under the normal pressure.

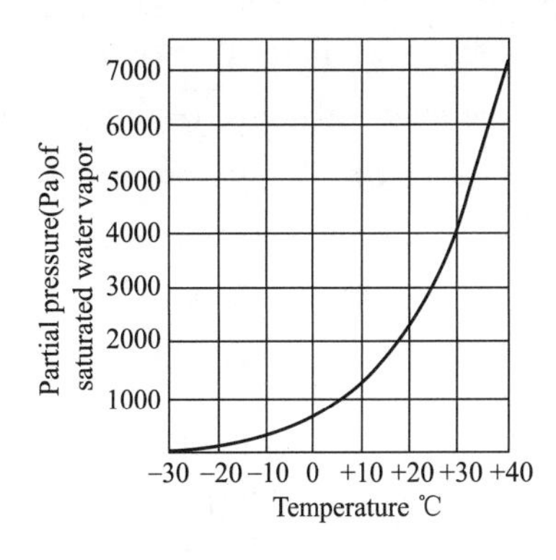

Figure 3-2-2 Relationship between Partial Pressure of Saturated Vapor and Temperature

As mentioned above, the relative humidity (φ) of air is the ratio of the actual partial pressure (P_i) of vapor to that (P_E) of saturated vapor under the same pressure, i.e. $\varphi = (P_i/P_E) \times 100\%$. As shown in Figure 3-2-2, the partial pressure of saturated vapor changes with the air temperature. Therefore, when the actual moisture content of air remains unchanged, that is the actual partial pressure (P_i) of vapor is not changed, the relative humidity will gradually rises with the air temperature falling; and when the relative humidity reaches 100% but the temperature continues to fall, vapor in the air will be condensed. The temperature corresponding to the relative humidity of 100% and saturated state of air is called "dew point temperature", usually expressed as t_d .

3.2.2.2 Wet Bulb Temperature and Air Temperature and Humidity Diagram

The relative humidity of indoor air can be measured with a wet bulb thermometer by wrapping its lower end with a gauze containing water (Figure 3-2-3). As the gauze is wet, the partial pressure of the surrounding vapor is higher than that in the air, so vapor in the gauze will be diffused around and the vaporization heat will be absorbed, thus reducing the temperature of the gauze to be lower than the ambient temperature. Then some heat will be transferred from the surrounding air to gauze. When the vaporization heat generated by the gauze is balanced with the heat transferred from the air to gauze, the temperature of the wet bulb thermometer will not drop, and the corresponding reading of this thermometer is the wet bulb temperature (t_W). Because the evaporation rate of moisture in the gauze is directly associated with the dryness of

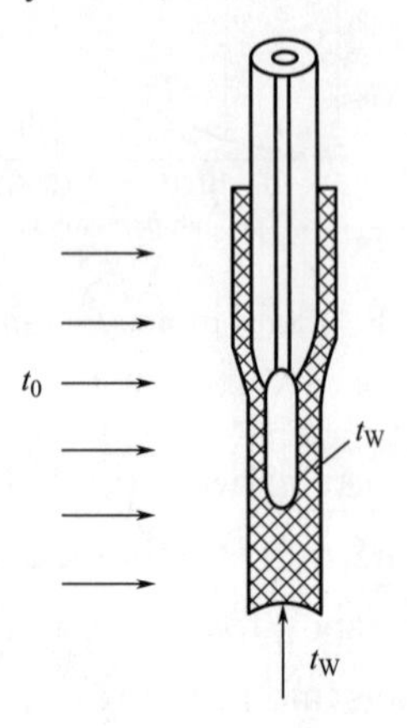

Figure 3-2-3 Wet Bulb Temperature

surrounding air, the relative humidity of air and partial pressure of vapor can be roughly obtained from the air temperature and humidity chart (Figure 3-2-4) after measurement of the dry and wet bulb temperature.

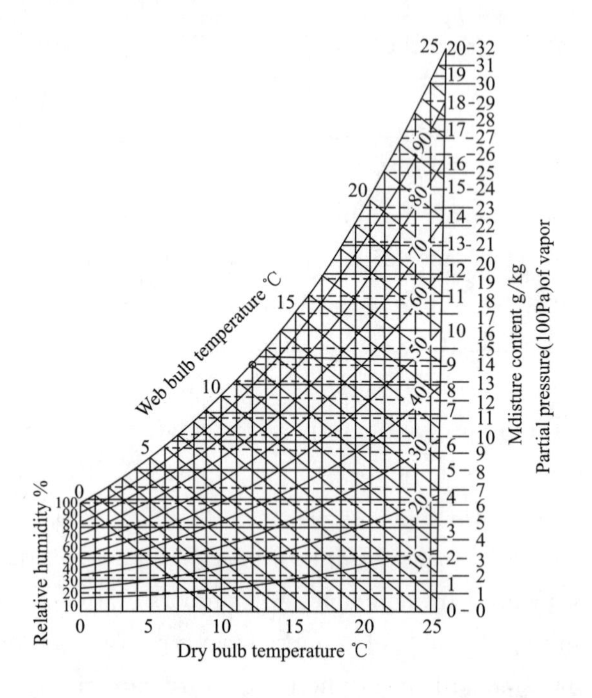

Figure 3-2-4 Air Temperature and Humidity Chart

The air temperature and humidity chart is a tool diagram drawn according to the physical properties of wet air, representing the relationship of the air temperature (dry bulb temperature), wet bulb temperature, relative temperature and partial pressure of vapor under the standard atmospheric pressure.

3.2.2.3 Indoor Air Humidity

With indoor and outdoor air convection, the moisture content of outdoor air directly affects the indoor air humidity. With the indoor temperature rising in a room heated in winter, the partial pressure of saturated vapor of indoor air will be higher than that of outdoor air. Vapor is generated in activities of indoor equipment and personnel, thus increasing the indoor air humidity and making the actual partial pressure of indoor vapor higher than that of outdoor vapor. Due to the significant temperature difference between the indoor and outdoor air in winter, however, there is a great difference in the partial pressure of saturated vapor, and thus the relative humidity of indoor air is usually low. In general, the more frequent the air convection, the lower the relative humidity of indoor air. Even additional humidification measures should be taken to meet the normal comfort requirements.

3.2.3 Inspection of Internal Condensation and Condensate Amount

The internal condensation has great invisible hazards to the building wall. It should be analyzed in the structural scheme at the beginning of design, so that measures can be taken to eliminate or control the impact of condensation.

3.2.3.1 Condensation Judgment

(1) Determine the partial pressure (P_i and P_e) of vapor in accordance with the temperature and relative humidity (t and φ) of indoor and outdoor air. Then calculate the partial pressure of vapor of each layer of the building wall by the equation (3-2-3), and draw the " P " distribution line. For a heated room, use the average temperature and average relative humidity of outdoor air in the local heating period as outdoor air calculation parameters in the design (as the whole heating period instead of the specific time is studied).

(2) According to the temperature (t_i and t_e) of indoor and outdoor air, determine the temperature of each layer, and draw the corresponding distribution line of the partial pressure " P_s " of saturated vapor.

(3) Judge whether condensation occurs based on the intersection between the " P " and " P_s " lines. In the case of no intersection [Figure 3-2-5*a*], there is no internal condensation; otherwise, there is internal condensation [Figure 3- 2-5*b*].

As mentioned above, the internal condensation usually occurs in a composite building wall. If the material layer of low vapor permeation resistance is laid first and then that of high vapor permeation resistance in the direction of vapor permeation, there will be higher vapor permeation resistance in the intersection of these two layers, which will lead to condensation. Typically, this interface where condensation occurs the most easily and seriously is called the "condensation interface" of the building wall. As shown in Figure 3-2-6, the condensation interface is often the intersection between the dense material and the external surface of the insulation material.

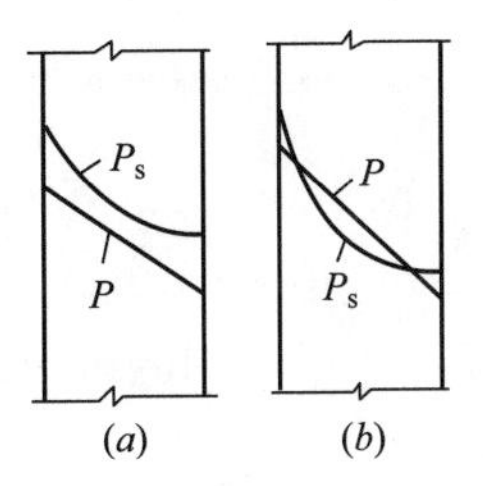

Figure 3-2-5 Judgment of Internal Condensation of Envelope

(*a*) without internal condensation; (*b*) with internal condensation

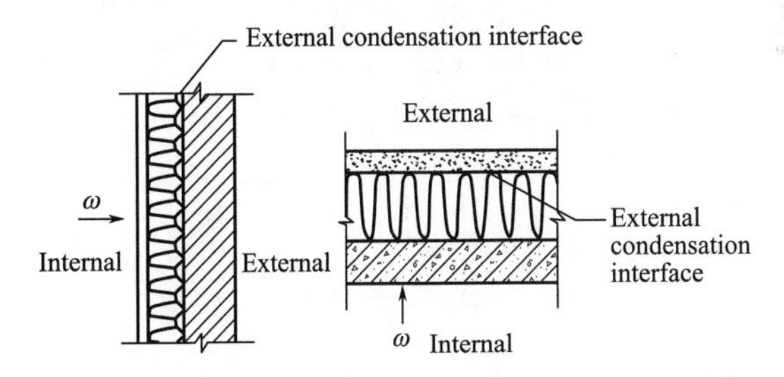

Figure 3-2-6 Condensation Interface

3.2.3.2 Condensation Intensity Calculation

Obviously, vapor of the condensation interface is saturated in the event of internal condensation, and the partial pressure of saturated vapor is $P_{s,c}$. Provided the permeability ω_1 of vapor from the air of high partial pressure to the condensation interface, and the permeability ω_2 from the condensation interface to the air of low partial pressure, their difference is the condensation intensity ω_c of this interface (Figure 3-2-7), i.e.:

$$\omega_c = \omega_1 - \omega_2 = \frac{P_A - P_{s,c}}{H_{o,i}} - \frac{P_{s,c} - P_B}{H_{o,e}} \tag{3-2-6}$$

Where, ω_c — condensation interface of the interface, $g/(m^2 \cdot h)$;

ω_1 and ω_2 —permeability of vapor on both sides

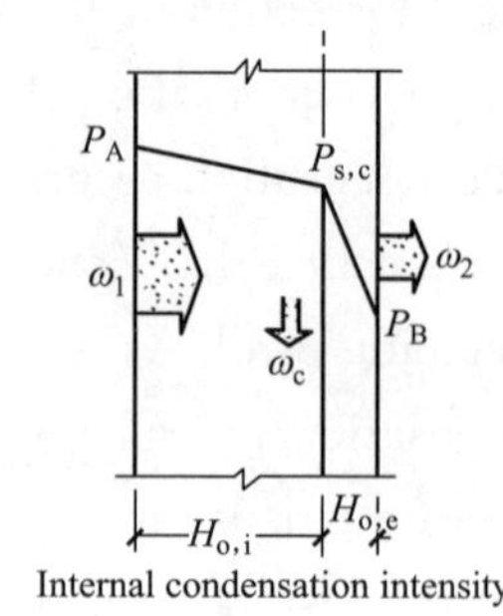

Figure 3-2-7 Determination of Internal Condensation Intensity

of the interface, $g/(m^2 \cdot h)$;

P_A —partial pressure of vapor of air on the side of high partial pressure, Pa;

P_B —partial pressure of vapor of air on the side of low partial pressure, Pa;

$P_{s,c}$ —partial pressure of saturated vapor at the condensation interface, Pa;

$H_{o,i}$ —permeation resistance of vapor flowing into the condensation interface, $m^2 \cdot h \cdot Pa/g$;

$H_{o,e}$ —permeation resistance of vapor flowing out of the condensation interface, $m^2 \cdot h \cdot Pa/g$.

3.2.3.3 Estimation of Cumulative Condensation during Heating Period

Generally, vapor permeation and condensation occur slowly in building walls. When the partial pressure of indoor vapor is close to that of outdoor vapor along with climatic changes in the late period of heating, vapor will not permeate, but condensate in the building wall will be gradually diffused to indoors and outdoors in other seasons. If the vapor condensation in the building wall has little impact on the insulation material in the heating period, a small amount of condensate is allowed.

The total amount of condensation during the heating period is calculated from:

$$\omega_{c,o} = 24\omega_c Z_h \tag{3-2-7}$$

Where, $\omega_{c,o}$ —total amount of condensation per square meter of the building wall during the heating period, g;

ω_c —condensation intensity at the interface, $g/(m^2 \cdot h)$;

Z_h —number of heating days, d.

The humidity increment (by weight) of the insulation material during the heating period is calculated from:

$$\Delta\omega = \frac{24\omega_c Z_h}{1000 d_i \rho_i} \times 100\% \tag{3-2-8}$$

Where, $\Delta\omega$ —the moisture increment by weight, %;

d_i —thickness of the insulation material, m;

ρ_i —density of the insulation material, kg/m^3.

It should be noted that rough estimation is performed above. The mechanism of liquid-phase moisture transfer within the condensation range must be taken into account to obtain accurate results.

The vapor permeation resistance needed under normal moisture conditions should be guaranteed in the building wall. A small amount of condensate is allowed in the wall of ordinary heated rooms, as it will evaporate out of the structure in the warm season. In order to ensure the structural durability, however, the humidity increase arising from internal condensation of the insulation material of the building wall must not exceed the limit during the heating period. Table 3-2-1 lists the allowable humidity increments of some insulation materials.

Allowable Humidity Increments (by weight) of Insulation Materials during Heating Period Table 3-2-1

Insulation Material	[$\Delta\omega$](%)
Porous concrete(foamed concrete and aerated concrete), ρ_o =500-700kg/m^3	4
Cement expanded perlite, cement expanded vermiculite, etc., ρ_o =300-500kg/m^3	6
Asphalt expanded perlite, cement expanded vermiculite, etc., ρ_o =300-400kg/m^3	7
Cement fiber board	5

Continued

Insulation Material	[$\Delta\omega$](%)
Mineral wool, rock wool, mineral wool and their products(board or felt)	3
Polyethylene foam	15
Slag and slag filler	2

3.2.4 Internal Surface Condensation and Preventive Measures of Building Wall

The temperature of the internal surface of a building wall is often lower than the indoor air temperature in winter. When the internal surface temperature is lower than the dew point temperature of indoor air, vapor in the air will condense on the internal surface. Therefore, condensation on the internal surface mainly depends on whether its temperature is lower than the dew point temperature. Figure 3-2-8 shows the process of surface condensation of the building wall.

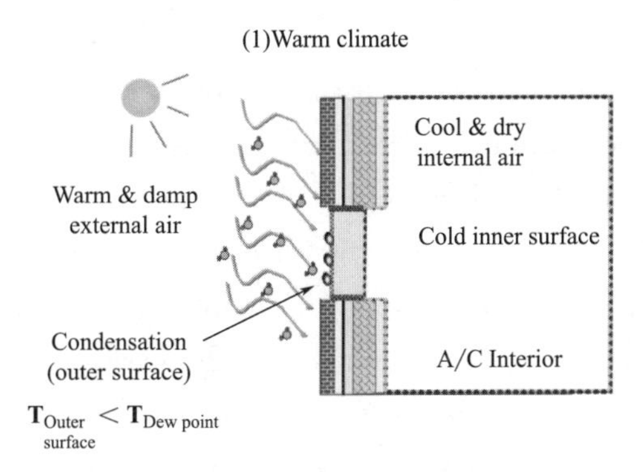

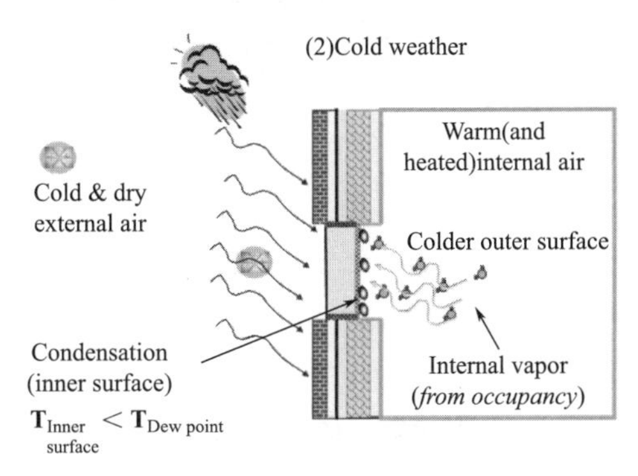

Figure 3-2-8 Schematic Diagram of Surface Condensation of Building Wall

It is one basic requirement in the thermal engineering design of the building to prevent condensation on the internal surfaces of the wall and roof. Specific preventive and control measures are as follows.

(1) The outer wall of the building should have sufficient insulation capacity, and its heat transfer resistance should be at least more than the specified minimum value. In addition, attention should be paid to condensation at cold (thermal) bridges.

(2) If the indoor air humidity is too high, ventilation measures should be taken to reduce the humidity.

(3) Preferably, the internal surface of the wall of an ordinary room should be made of hygroscopic material, so that a small amount of condensate generated in the low-temperature period in one day can be absorbed by the internal surface of the structure, and return into the indoor air in the case of high indoor temperature but low relative humidity, which is called the "breathing" effect.

(4) When the indoor humidity is high and condensation on the internal surface can be avoided, such as the public bath, the internal surface should be made of smooth and non-absorbent material and water guide facilities should be added to drain condensate.

(5) For the light wall made of "single material" or with composite structure as well as the wall made of light material and with internal thermal insulation, the minimum heat transfer resistance calculated should also be verified according to the provisions of Table 4.1.2 of the *Thermal Design Code for Civil Building* (GB 50176-1993). The additional value needs to be taken into account in the minimum heat transfer resistance of the light outer wall. The outer walls of energy-efficient buildings, almost all of which meet the additional value requirements, must be checked for conformity.

3.2.5 Impact of Insulation Layer Location on Vapor Permeation of Wall

The amount of rain that a wall receives depends on the inclination of the wall, instead of its thermal characteristics. On clear nights, however, the amount of dew is decisive to the heat storage and transfer of the outer wall. There is also long wave radiation between the surface of the external thermal insulation system and its surroundings. Since no sunlight is available at night and the atmospheric reflection is much less than the radiation of the external surface of the wall on sunny days (cloudless), the heat of the external surface of the composite wall will be absorbed by the atmosphere. In case that heat is not supplied from indoors and the internal surface of the outer wall has low capacity of heat storage, the outer wall will be over cold, and the internal surface temperature of the outer wall will be lower than the dew point temperature of indoor air, which will result in condensation.

Energy-efficient buildings in the northern cold and severe cold areas are designed conservatively, so their indoor temperature is often higher than that of conventional brick-concrete buildings in winter. Due to high air-tightness, the absolute humidity of indoor air of energy-efficient buildings is higher than that of conventional brick-concrete buildings. Moreover, energy-efficient buildings are equipped with energy-saving windows and doors of small gaps, thus improving the air-tightness. However, there is hardly any ventilation window (or hole) in the wall or window or appropriate facilities for ventilation, air exchange and moisture elimination. In this sense, the moisture in the indoor air cannot be properly drained, resulting in high humidity of indoor air and increasing the moisture transfer load of the wall. As most of insulation materials have a large coefficient but small resistance of vapor permeation, the location of the insulation layer is particularly important for moisture transfer of walls.

The condensation of walls with four types of insulation in severe cold areas (taking Harbin as an example) of China during the heating period was analyzed in the checking method. Results show that only the external thermal insulation structure is rational for heat and moisture transfer. The humidity increments corresponding to the internal thermal insulation, sandwich insulation and self-insulation were calculated with the allowable increments of insulation materials in the heating period in accordance with the *Thermal Design Code for Civil Building* (GB 50176-1993). It can be concluded that the internal thermal insulation cannot meet the thermal engineering design requirements of civil buildings. Although the humidity increments of walls with the sandwich insulation and self-insulation are less than the allowable values under normal conditions, the sandwich insulation and self-insulation may be subject to frost heaving in the case of condensation (below 0℃).

3.2.5.1 Condensation Analysis of Outer Wall with External Thermal Insulation

1. Structure and material parameters

Analysis was performed with Harbin (indoor temperature in the heating period: 18℃; indoor relative humidity: 60%; average outdoor temperature: −10℃; average outdoor relative humidity: 66%; number of heating days: 176) as an example. It is assumed that the basic structure of the outer wall is composed of the interior finish +200mm reinforced concrete wall+10mm EPS board and 60mm EPS board bonded with mortar +4mm surface mortar (containing one layer of alkali-resistant glass fiber mesh) + waterproof flexible putty+paint. Refer to Table 3-2-2 for the material of each layer as well as its parameters and calculation results.

2. Calculation results

(1) Partial pressure of vapor of indoor and outdoor air

$t_i = 18℃$: $P_{s,i} = 2062.5\text{Pa}$

$P_1 = 2062.5\text{Pa} \times 60\% = 1237.5\text{Pa}$

$t_e = -10℃$: $P_{s,e} = 260.0\text{Pa}$

$P_7 = 260.0\text{Pa} \times 66\% = 171.6\text{Pa}$

Material Properties of Paint Finish by Point and Frame Bondedand Thin-plastered EPS Board **Table 3-2-2**

Structural Form	Material Name/ Serial Number	Thickness d (m)	Thermal Conductivity λ [W/(m·K)]	Vapor Permeability Coefficient μ [g/(m·h·Pa)10^{-4}]	Thermal Resistance $R=\frac{d}{\lambda}$ (m^2·K)/W	Vapor Permeation Resistance $H=\frac{d}{\mu}$ (m^2·h·Pa)/g
External thermal insulation by paint finish with point and frame bonded and thin-plastered EPS board	Interior finish / 1	0.002	0.60	0.1	0.0033	200.00
	Reinforced concrete / 2	0.200	1.74	0.158	0.1149	12658.23
	EPS bonding mortar / 3	0.010	0.76	0.21	0.0132	476.19
	Molded EPS board / 4	0.080	0.041	0.162	1.95	4938.27
	Surface mortar / 5	0.004	0.81	0.21	0.0049	190.48
	Paint finish / 6	0.002	0.50	0.1	0.0040	200.00

Note: Refer to the Thermal Design Code for Civil Building (GB 50176-1993) for the above data.

(2) Partial pressure of vapor in the internal surface of each layer of the outer wall

$$P_2 = 1237.5 - \frac{200}{18663.17}(1237.5 - 171.6)$$
$$= 1226.08(\text{Pa})$$

$$P_3 = 1237.5 - \frac{12858.23}{18663.17}(1237.5 - 171.6)$$
$$= 503.13(\text{Pa})$$

$$P_4 = 1237.5 - \frac{13334.42}{18663.17}(1237.5 - 171.6)$$
$$= 475.94(\text{Pa})$$

$$P_5 = 1237.5 - \frac{18272.69}{18663.17}(1237.5 - 171.6)$$
$$= 193.90(\text{Pa})$$

$$P_6 = 1237.5 - \frac{18463.17}{18663.17}(1237.5 - 171.6)$$
$$= 183.02(\text{Pa})$$

(3) Pressure of saturated vapor in the internal surface of each layer of the outer wall

$$t_1 = 18 - \frac{0.1100}{1.7525}(18 - (-10))$$
$$= 16.625℃ \qquad p_{s,1} = 1890.92(\text{Pa})$$

$$t_2 = 18 - \frac{0.1133}{2.2416}(18 - (-10))$$
$$= 16.58℃ \qquad p_{s,2} = 1885.92(\text{Pa})$$

$$t_3 = 18 - \frac{0.2283}{2.2416}(18 - (-10))$$
$$= 15.15℃ \qquad p_{s,3} = 1721.05(\text{Pa})$$

$$t_4 = 18 - \frac{0.2414}{2.2416}(18 - (-10))$$
$$= 14.98℃ \qquad p_{s,4} = 1702.21(\text{Pa})$$

$$t_5 = 18 - \frac{2.1926}{1.7525}(18 - (-10))$$
$$= -9.39℃ \qquad p_{s,5} = 273.60(\text{Pa})$$

$$t_6 = 18 - \frac{2.1975}{2.2416}(18 - (-10))$$
$$= -9.45℃ \qquad p_{s,6} = 272.64(\text{Pa})$$

$$t_7 = 18 - \frac{2.20159}{2.2416}(18 - (-10))$$
$$= -9.5℃ \qquad p_{s,7} = 271.99(\text{Pa})$$

(4) Distribution diagram of partial pressure of vapor and saturated vapor of the outer wall (Figure 3-2-9)

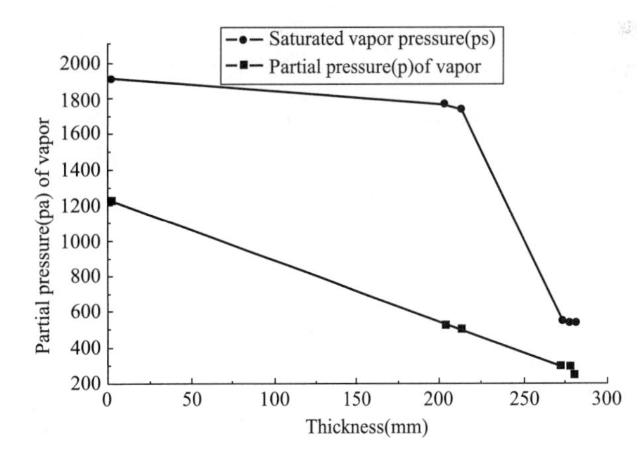

Figure 3-2-9 Distribution of Partial Pressure of Vapor of Outer Wall with External Thermal Insulation

3. Result analysis

When the partial pressure of saturated vapor in any part of the thickness direction is higher than that of vapor in the outer wall, condensation will not occur in the insulation structure.

3.2.5.2 Condensation Analysis of Outer Wall with Internal Thermal Insulation

1. Structure and material parameters

The outer wall with internal thermal insulation is composed of the interior finish + 25mm surface mortar (containing one layer of alkali-resistant glass fiber mesh) + 80mm EPS board + 10mm EPS board bonding mortar + 200mm reinforced concrete wall + waterproof flexible putty + paint. Refer to Table 3-2-3 for the material of each layer and its parameters.

2. Calculation results

See Figure 3-2-10.

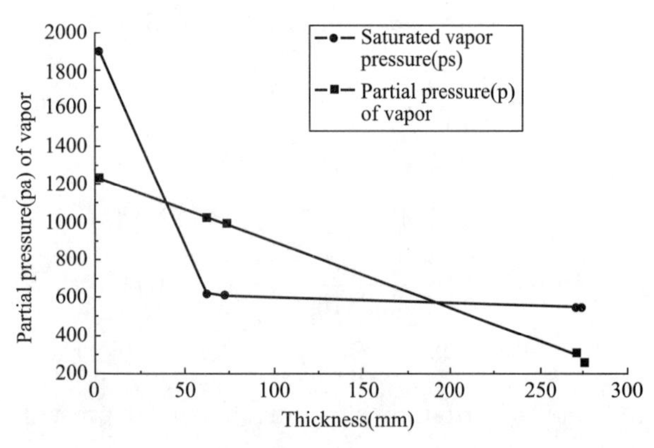

Figure 3-2-10 Distribution of Partial Pressure of Vapor of Outer Wall with Internal Thermal Insulation

3. Result analysis

When the pressure of saturated vapor in the EPS board is lower than the partial pressure of vapor, condensation will occur to the external surface of the insulation layer of the insulation structure. The humidity increment arising from internal condensation of the insulation material of the envelope should be less than the allowable value, which is 15% for the polystyrene foam in the design code. Calculated from the equation (3-2-8), the humidity increment (Δw) of the EPS board is 35.8% (assuming that the dry density ρ of the insulation material is 20kg/m^3 for the EPS board and 500kg/m^3 for the aerated concrete), which is far greater than the design value.

3.2.5.3 Condensation Analysis of Outer Wall with Sandwich Insulation

1. Structure and material parameters

The outer wall is composed of the interior finish + 160mm reinforced concrete wall + 80mm EPS board + 40mm reinforced concrete wall + waterproof flexible putty + paint. Refer to Table 3-2-4 for the material of each layer and its parameters.

Material Properties **Table 3-2-3**

Structural Form	Material Name/ Serial Number	Thickness d (m)	Thermal Conductivity λ [W/(m·K)]	Vapor Permeability Coefficient μ [g/(m·h·Pa)×10^{-4}]
Internal thermal insulation	Interior finish/1	0.002	0.60	0.1
	Surface mortar/2	0.025	0.81	0.21
	EPS board/3	0.080	0.041	0.162
	EPS board bonding mortar/4	0.010	0.76	0.21
	Reinforced concrete/5	0.200	1.74	0.158
	Paint finish/6	0.002	0.50	0.1

Material Properties **Table 3-2-4**

Structural Form	Material Name/ Serial Number	Thickness d (m)	Thermal Conductivity λ [W/(m·K)]	Vapor Permeability Coefficient μ [g/(m·h·Pa)×10^{-4}]
Sandwich insulation	Interior finish/1	0.002	0.60	0.1
	Reinforced concrete/2	0.160	1.74	0.158
	EPS board/3	0.080	0.041	0.162
	Reinforced concrete/4	0.040	1.74	0.158
	Paint finish/5	0.002	0.50	0.1

2. Calculation results

See Figure 3-2-11.

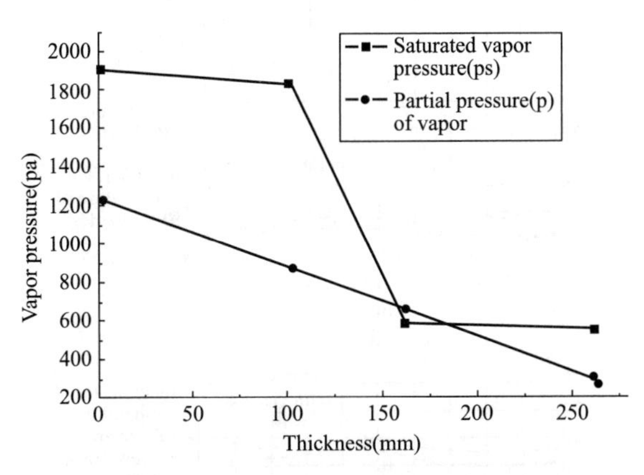

Figure 3-2-11 Distribution of Partial Pressure of Vapor of Outer Wall with Sandwich Insulation

3. Result analysis

When the pressure of saturated vapor in the EPS board is lower than the partial pressure of vapor, condensation will occur to the external surface of the insulation layer of the insulation structure. The humidity increment (Δw) is 6.1%, which is less than the specified design value.

3.2.5.4 Condensation Analysis of Outer Wall with Self-insulation

1. Structure and material parameters

The outer wall is composed of the interior finish + 10mm cement mortar + 400mm aerated concrete + 20mm cement mortar + paint. Refer to Table 3-2-5 for the material of each layer and its parameters.

2. Calculation results

See Figure 3-2-12.

3. Result analysis

When the pressure of saturated vapor in the aerated concrete is lower than the partial pressure of vapor, condensation will occur near the external

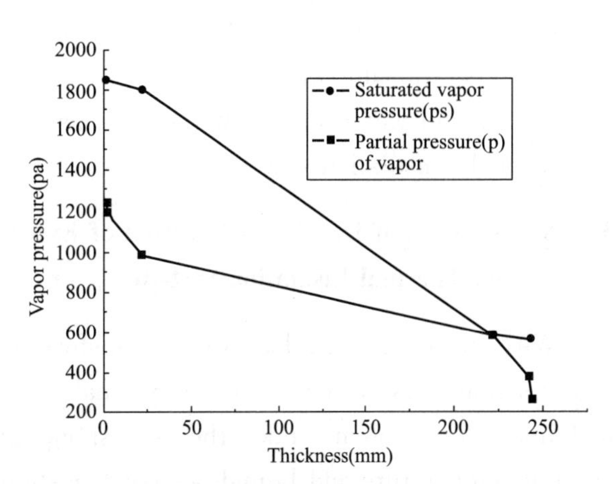

Figure 3-2-12 Distribution of Partial Pressure of Vapor of Outer Wall with Self-insulation

surface of the aerated concrete. The humidity increment (Δw) is 0.23%, which is less than the specified design value. If the temperature is often below 0℃, more serious frost heaving with ice will occur.

3.2.5.5 Comparative Analysis of Condensation Results of Four Insulation Structures

Condensation of outer walls with the internal thermal insulation, sandwich insulation and self-insulation was calculated, followed by special treatment. In accordance with the provisions of the Thermal Design Code for Civil Building (GB 50176-1993), the vapor barrier should be added on the inner side of the condensation interface or the vapor separation capacity of the existing vapor barrier should be improved. However, the vapor barrier has adverse effects because of reverse transfer throughout the year, and hence some buildings are equipped with vapor barriers on both sides. This measure must be taken with care, as liquid water generated by condensation or brought in cannot evaporate easily, which will cause greater hazards.

Material Properties **Table 3-2-5**

Structural Form	Material Name/ Serial Number	Thickness d (m)	Thermal Conductivity λ [W/(m·K)]	Vapor Permeability Coefficient μ [g/(m·h·Pa)$\times10^{-4}$]
Self-insulation	Interior finish/1	0.002	0.60	0.1
	Cement mortar/2	0.160	0.81	0.21
	Aerated concrete/3	0.080	0.19	0.998
	Cement mortar/4	0.040	0.81	0.21
	Paint finish/5	0.002	0.50	0.1

Only the external thermal insulation of the outer wall is rational to prevent the insulation system from liquid water and discharge gaseous water out of the insulation system.

3.2.6 Analysis of Dew Point Location of External Thermal Insulation System

With indoor and outdoor vapor transferred, condensation does not occur in the external thermal insulation system. Under the corresponding outdoor temperature and humidity, condensation may occur on the external surface of the insulation layer.

Beijing was taken as an example. Given the indoor temperature of 20℃ and humidity of 60% in winter and summer, the condensing temperature is about 12℃. The outdoor temperature changes of one day in summer and winter were used as the temperature input load of numerical simulation. Calculation results are shown in Figure 3-2-13 and 3-2-14.

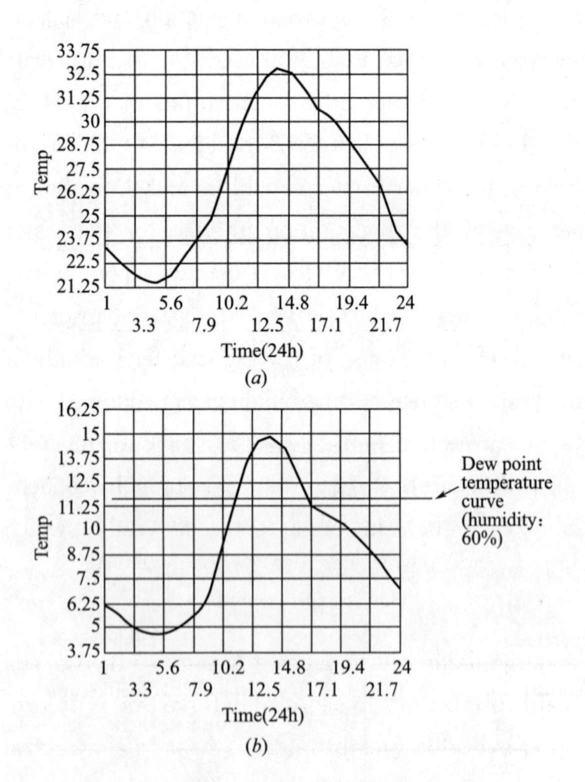

Figure 3-2-13 Temperature Change Curve of Internal Surface of Non-insulated Wall with Ambient Temperature (in one day)

(*a*) summer; (*b*) winter

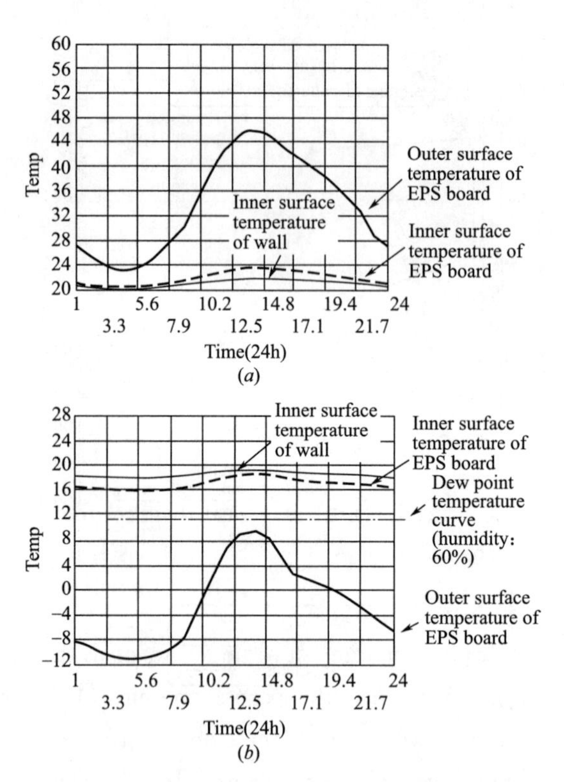

Figure 3-2-14 Temperature Change Curve of Each Surface of Wall subject to External Thermal Insulation with Ambient Temperature (in one day)

(*a*) summer; (*b*) winter

As shown in Figure 3-2-13, the temperature of the internal surface of the non-insulated wall is lower than that of the external surface in summer. If any, condensation only occurs to the internal surface. The temperature of the internal surface of the wall is much higher than the condensing temperature of air. Under the above assumptions in summer, condensation does not occur to the internal surface. In winter, condensation occurs to the internal surface roughly from 4: 00 p. m. to next 11: 00 a. m., and the temperature of the internal surface is lower than the condensing temperature, so condensation occurs to the wall.

As shown in Figure 3-2-14, when the external thermal insulation is applied, the temperature of the internal surface is much higher than the condensing temperature of air in summer, so condensation does not occur. In winter, condensation does not occur in the insulation layer as its internal temperature is higher than the condensing temperature, but occurs to the anti-cracking mortar layer of the external surface of the EPS board in the case of high humidity of air as the temperature of the external surface of the EPS board is lower than the condensing temperature. However, the thin anti-cracking mortar layer is able to absorb a small amount of liquid water. In the presence of liquid water, the anti-cracking mortar will be wetted, which will reduce its strength and adhesion. Under dry conditions, the anti-cracking mortar layer may become hollow or shed off as a result of deformation.

In the engineering case shown in Figure 3-2-15, the anti-cracking layer of the external thermal insulation system in winter is subject to freeze-thaw damage as a result of condensation in winter, and wetting in other seasons, resulting in the decline of adhesion and separation of the anti-cracking mortar of the surface from insulation boards. The regular damage to the surface has distinctive characteristics. Therefore, the water and freeze-thaw resistance of anti-cracking mortar should be taken into account in formulation of related standards, and the structure should be able to absorb condensate to prevent liquid water in the anti-cracking protective layer.

Figure 3-2-15 Shedding of Anti-cracking Protective Layer Caused by Condensation

3.3 Waterproofing Performance and Air Permeability of External Thermal Insulation System

The external thermal insulation system composed of an insulation layer and protective layer (including a decoration system) is a non-bearing structure that is attached to the surface of the structural wall and perpendicular to the ground. The external thermal insulation system should be able to withstand the intrusion of external liquid water from the perspective of the waterproofing performance, and not affect the vapor transfer under the partial pressure difference between indoor and outdoor vapor in terms of the vapor permeability.

3.3.1 Kuenzel's Outer Wall Protection Theory

In 1968, Dr. Kuenzel, a German physicist, first proposed the technical specifications of outer wall waterproofing and plastering (Kuenzel's outer wall protection thoery), which has been widely quoted in European standards for building materials:

$$W \leqslant 0.5 \qquad (3\text{-}3\text{-}1)$$

$$S_d \leqslant 2 \qquad (3\text{-}3\text{-}2)$$

$$W \cdot S_d \leqslant 0.2 \qquad (3\text{-}3\text{-}3)$$

Where, W —water absorption rate, in $kg/(m^2 \cdot h^{0.5})$, for evaluating the rate of water absorption of the material;

S_d —equivalent thickness of the static air layer, in m, for assessing the drying capacity (air permeability) of the material.

The equation (3-3-1) applies to the water absorption capacity of the material, and the equation (3-3-2) to its vapor permeability. The vapor permeability is also described by other indicators such as the wet vapor flow density and vapor permeability coefficient, which can be converted into each other. The equation (3-3-3) combining the requirements of the above two indicators represents a unified and coordinated indicator (Figure 3-3-1).

3.3.2 Water Absorption Capacity

The water absorption capacity of a material is generally expressed as a water absorption coefficient, which of one material is a constant. Prior to saturation, the amount of water absorbed by one material is proportional to the square root of the time. Once saturation is achieved, the curve of the amount of water absorbed by this material will become a straight line parallel to the time axis (Figure 3-3-2).

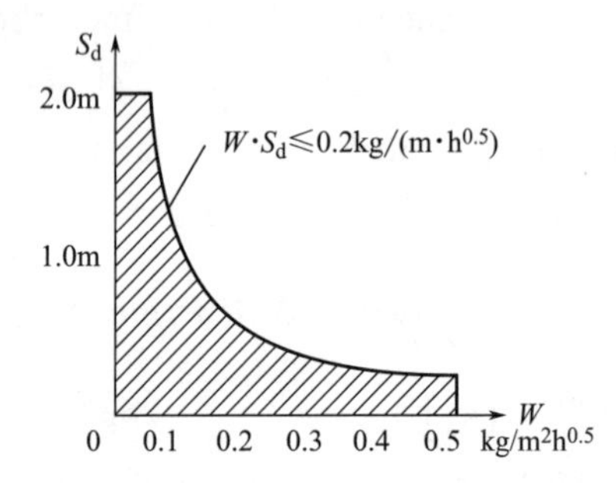

Figure 3-3-1 Relationship between Water Absorption Capacity and Vapor Permeability

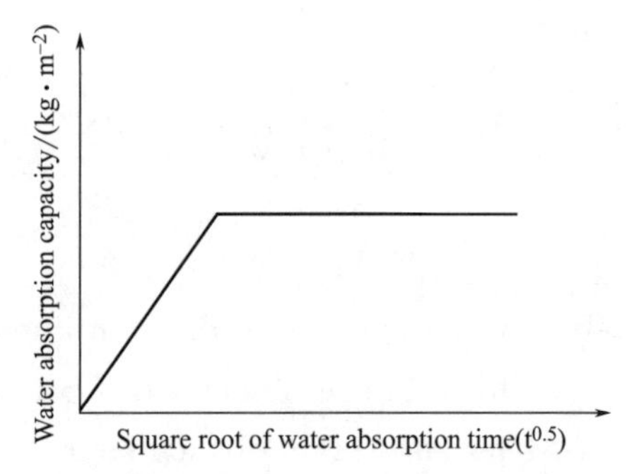

Figure 3-3-2 Relationship between Water Absorption Capacity and Time of Material

3.3.3 Hydrophobicity

If liquid is dropped on the surface of a solid base course wall, an angle (θ) will be formed at the edge of the solid-liquid contact interface, which is often referred to as the "contact angle". As shown in Figure 3-3-3, the angle θ can be applied to measure the degree of wetting of the solid base course wall by liquid, or the hydrophobicity of this base course wall.

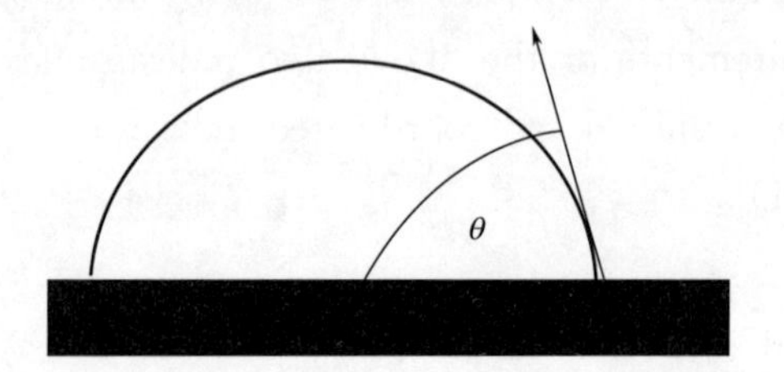

Figure 3-3-3 Contact Angle in Wetting of Substrate Surface

Specific test steps are as follows: keeping mortar on a flat table, making 2-3 drops of water through a burette onto the surface of the mortar block, and determining the hydrophobicity of this surface according to the static contact angle of water drops on the mortar surface in 1min. The criteria are listed in Table 3-3-1. Of course, the dynamic contact angle needs to be considered when the hydrophobicity is measured. It is not discussed here as complex factors are involved. In nature, the contact angle of a water drop on the surface of a plant such as the lotus leaf is larger than 130°. That is, when the contact angle of a water drop on abase course wall is large, this base course wall can be regarded with the "lotus effect". When a hydrophobic material is selected, adequate consideration should be given to relevant provisions on the material indicators of the external thermal insulation system, and the hydrophobicity should be controlled at a reasonable ratio, in order to meet the bonding requirements of various layers and guarantee the system safety.

Relationship between Surface Hydrophobicity and Initial Contact Angle Table 3-3-1

Hydrophobicity	Contact Angle θ	Base course wall Wetting Effect
Grade Ⅰ	$\theta \leqslant 30°$	Complete wetting
Grade Ⅱ	$30° < \theta \leqslant 90°$	Obvious wetting
Grade Ⅲ	$90° < \theta \leqslant 110°$	Slight wetting
Grade Ⅳ	$110° < \theta \leqslant 130°$	Good hydrophobicity
Grade Ⅴ	$\theta > 130°$	Excellent hydrophobicity

3.3.4 Vapor Permeability

The vapor permeability of a material is often expressed as its equivalent static air layer thickness or wet density.

$$S_d = \mu \cdot s \quad (3\text{-}3\text{-}4)$$

Where, S_d —equivalent static air layer thickness, m;

μ —diffusion resistance coefficient (air: $\mu = 1$);

s —coating thickness, m.

The indicators involved in the theory have significant discreteness during the experiment, so a large amount of experimental data is needed as a support. Dr. Kuenzel later mentioned in the article "Techniques for Outer Wall Plastering" that the practical application experience is a major factor in terms of the water absorption capacity and vapor permeability of a material. The test method with the vapor permeability tested as the equivalent air layer thickness is only applied in the European standard EN12866. The wet density of vapor is tested and substituted to obtain the equivalent air layer thickness in other standards. Although the absolute value is not quite accurate, it is of practical significance to compare the relative values of two materials via one test method.

3.3.5 Waterproofing Performance and Vapor Permeability

Refer to Table 3-3-2 for the excerpts related to the water absorption capacity and vapor permeability in external thermal insulation system standards, construction specifications and other standards at home and abroad.

Water Absorption Capacity and Vapor Permeability of Material

Table 3-3-2

Standard	Water Absorption Capacity	Vapor Permeability	Remarks
External Thermal Insulation Composite System based on Expanded Polystyrene (JG 149-2003)	With the 5mm protective layer, the water absorption capacity after immersion in water for 24h should be 500g/m² [equivalent to 0.1kg(m² · h^0.5)] or less.	The wet density of vapor of the protective layer and finish should be 0.95g/(m² · h)(equivalent to 1.2m) or more.	The water absorption capacity is tested according to the standard JG 149-2003; and the wet density of vapor is tested in the hydraulic method specified in the *Test Methods for Water Vapor Transmission of Building Materials* (GB/T 17146-1997).
External Thermal Insulating Rendering Systems Made of Mortar with Mineral Binder and Using Expanded Polystyrene Granule as Aggregate (JG 158-2004)	The water absorption capacity after immersion in water for 1h should be 1000g/m² [equivalent to 1kg(m² · h^0.5)] or less.	The wet density of vapor should be 0.85g/(m² · h)(equivalent to 1.2m) or more.	—
Technical Specification for External Thermal Insulation on Walls (JGJ 144-2004)	With the plaster layer only or all protective layers, the water absorption capacity after immersion in water for 1h should be 1000g/m² [equivalent to 1kg(m² · h^0.5)] or less.	The wet density of vapor should meet the design requirements.	The water absorption capacity is tested according to the standard JGJ 144-2004; and the wet density of vapor is tested in the desiccant method specified in the *Test Methods for Water Vapor Transmission of Building Materials* (GB/T 17146-1997).
External Thermal Insulation Composite Systems based on Expanded Polystyrene (EN 13499:2003)	Protective layer: ≤ 0.5kg(m² · h^0.5)	Protective layer and finish: ≥ 20.4g/(m² · h)(equivalent to 1m)	The water absorption capacity is tested according to the standard EN 1602-3: 1998 and the vapor permeability to the standard EN ISO 7783-2:1999.
Guideline for European Technical Approval of External Thermal Insulation Composite System with Rendering (ETAG 004: 2000)	The water absorption capacity after immersion in water for 24h should be 0.5g/m² [equivalent to 0.1kg(m² · h^0.5)] or less.	Protective layer and finish: ≤ 2m (foamed plastic insulation materials) and 1m (mineral insulation materials)	The water absorption capacity is tested according to the *Thermal Insulating Products for Building Applications-Determination of Short-term Water Absorption by Partial Immersion* (EN1609: 1997).

It is stipulated in the European Standard EN 1062-1: 2002 "Paints and Varnishes-Coating Materials and Coating Systems for Exterior Masonry and Concrete-Part 1: Classification" that the water absorption capacity should be determined in accordance with the standard EN 1062-3: 1998 "Paints and Varnishes-Coating Materials and Coating Systems for Exterior Masonry and Concrete-Part 3: Determination and Classification of Liquid-water Transmission Rate (Permeability)". Refer to Table 3-3-3 for coating classification during selection.

Classification based on Water Absorption Capacity

Table 3-3-3

Category	W_0	W_1	W_2	W_3
Water absorption capacity	—	High	Medium	Low
Requirement/ kg/(m² · h^0.5)	No requirement	>0.5	$0.5 \geq W_2 > 0.1$	≤0.1

It is stipulated in the European Standard EN 1062-1: 2002 "Paints and Varnishes-Coating Materials and Coating Systems for Exterior Masonry and Concrete-Part 1: Classification" that the vapor permeability should be determined in accordance with the standard EN ISO 7783-2: 1999 "Paints and Varnishes-Coating Materials and Coating Systems for Exterior Masonry and Concrete-Part 2: Determination and Classification of Water Vapor Transmission Rate (Permeability)". Refer to Table 3-3-4 for classification during selection.

Classification based on Vapor Permeability

Table 3-3-4

Category	V_0	V_1	V_2	V_3
Vapor permeability	—	High	Medium	Low
g/(m² · d)	No requirement	>150	$15 \leq V^2 > 150$	<15
Requirement / M (resistance-equivalent to the static air layer thickness)	—	<0.14	$0.14 \leq V^2 < 1.4$	≥1.4

The following equation can be obtained with the medium permeability in the standard EN ISO 7783-2: 1999 "Paints and Varnishes-Coating Materials and Coating Systems for Exterior Masonry and Concrete-Part 2: Determination and Classification of Water Vapor Transmission Rate (Permeability)":

$$1/V_{EIFS} = 1/V_{砂浆} + 1/V_{涂料} \quad (3\text{-}3\text{-}5)$$

A new equation can be obtained based on $S_d = 20.357/V$:

$$S_{d,EIFS} = S_{d,砂浆} + S_{d,涂料} \quad (3\text{-}3\text{-}6)$$

The S_d value of the external thermal insulation system can be easily calculated with the S_d value of mortar measured by the external thermal insulation manufacturer and the S_d value of paint measured by the paint manufacturer. The smaller the S_d value, the better the effect and the higher the vapor permeability of paint will be.

The test method stipulated in the industry standard "Determination and Classification of Water-vapor Transmission Rate (Permeability) for Exterior Wall Coatings" (JG/T 309-2011) is different from that in the standard "Test Methods for Water Vapor Transmission of Building Materials" (GB/T 17146-1997). In this test method, the saturated solution of ammonium dihydrogen phosphate is used instead of water, and the relative humidity is 93%, different from the relative humidity (50%) of the laboratory, thus generating the vapor flow. The result can be expressed as the vapor transmission rate V in g/(m² · d) or the vapor permeation resistance S_d (equivalent air layer thickness of diffusion) in m. their conversion relationship is $S_d = 20.357/V$. The representation of the vapor permeability of the paint by S_d is more scientific and intuitive, because the thicker the coating, the greater the S_d value and the poorer the vapor permeability. Therefore, the S_d value is a test indicator of paints of the outer wall in European standards.

In addition, the curing methods of specimens are also different. The test method stipulated in the *Test Methods for Water Vapor Transmission of Building Materials* (GB/T 17146-1997) applies to all building materials, but the paint drying process under outdoor conditions vary from that in the laboratory. Specimen ageing is added in the industry standard "Determination and Classification of Water Vapor Transmission Rate of Exterior Wall Coating" (draft for comments). If exposed outdoors, the outer wall will be subject to rain erosion and high-temperature impact. That is, soluble components of the coating will be washed away by rain, and the coating will be softened under high temperature, making the coating denser and reducing the vapor permeability. Specimens are not subject to ageing in the *External Thermal Insulation Composite System based on Expanded Polystyrene* (JG 149-2003), so the measured data will be smaller than their actual values.

3.3.6 Design Principles of Waterproofing Performance and Vapor Permeability of External Thermal Insulation System

(1) By analyzing the influence factors of water and vapor transfer, it is not difficult to conclude that the external thermal insulation system must not be made of insulation material with completely closed pores or very low porosity.

(2) The finish or protective layer exposed directly to water should be made of hydrophobic material with high capillary porosity. Water drops (lotus effect) will be formed on the wall surface exposed to water, thus achieving the waterproofing effect.

Foreign studies have shown that the insulation layer and protective surface of a building will not have good protection functions until their vapor permeability and water absorption capacity are appropriate. Normally, the water absorption capacity of a material is expressed as the water absorption coefficient abroad, namely:

$$K = W/S \cdot \sqrt{t} \qquad (3\text{-}3\text{-}7)$$

Where, K —water absorption coefficient, $kg/m^2 \cdot \sqrt{h}$;

W —amount of absorbed water, kg;

S —water absorption area, m^2;

t —water absorption time, h.

The vapor permeability coefficient μ is often used to represent the vapor permeability in the external thermal insulation of buildings. The higher the material porosity, the higher the vapor permeability will be. The vapor permeability coefficient μ of static air is $6.08 \times 10^{-4} g/(m \cdot s \cdot Pa)$. The greater the μ value, the higher the vapor permeability will be.

From the perspective of surface protection, the water absorption capacity should be minimized while the vapor permeability should be maximized. The surface of the ideal external thermal insulation system should not have the water absorption capacity or vapor diffusion resistance, though this is impossible. The typical requirement abroad for the water absorption capacity and vapor permeability is as follows: $K \leqslant 0.5 kg/m^2 \cdot h^{0.5}$.

The above data apply to the water absorption of mortar used in ordinary buildings. The water absorption capacity of concrete cannot meet such requirements. Table 3-3-5 lists the water absorption coefficients of materials of outer walls. It can be seen that the freeze-thaw resistance is not possible and cracking is inevitable in the absence of water repellency.

Water Absorption Coefficients of Some Outer Wall Materials of Buildings Table 3-3-5

Material Name	Water Absorption Coefficient($kg/m^2 \cdot h^{0.5}$)
Cement mortar	2.0~4.0
Concrete	1.1~1.8
Solid clay brick	2.9~3.5
Porous clay brick	8.3~8.9
Aerated concrete block	4.4~4.7

3.4 Waterproof Barrier and Vapor Transfer and Diffusion Structure of External Thermal Insulation System

The energy efficiency, durability and comfort of a building is closely related to the waterproofing performance and vapor permeability of its insulation system, so the insulation system must have a rational structure to ensure the waterproofing and vapor permeability and prevent the damage caused by phase changes of condensate.

3.4.1 Polymer Elastic Primer

The polymer emulsion elastic primer contains a large amount of silicone resin, in which the emulsion of dense paint and small diameter is used. The silicone resin can be applied on the surface to form a monomolecular hydrophobic structure of strong resistance to large molecules of liquid water. In this case, "water drops" can be formed by rainwater but the external surface will not become wet. At the same time, the external surface has good waterproofing performance, vapor permeability and penetrability. Nano-scale particles will penetrate in through capillary pores of the base course wall, and a thin layer of silicone resin network will be formed on the walls of capillary pores but not make capillary pores blocked, hence preventing the seepage of external water, discharging the internal vapor, and avoiding the poor drainage or condensation of the wall and moisture increase in the insulation layer.

If water in the primer of the base course wall surface evaporates, a continuous closed coating will be formed, which will reduce dissipation of water vapor in the base course wall, prevent the newly applied cement mortar from cracking as a result of excessive water losses and achieving the effects of the cement mortar curing solution. The vapor permeability of the coating is inversely proportional to its thickness, so the polymer elastic primer should be thin and even. If the thickness of the primer is controlled with construction techniques, the functions of the cement mortar curing coating will be realized and internal water vapor will be slowly released. As the primer has good permeability, the elastic coating can be closely fitted with the cement mortar layer to form an anti-cracking waterproof barrier for the insulation layer and provide a good interface basis for the finish layer. In this sense, the polymer elastic primer has dual effects of film-forming and permeation sealing, and also the capability of preventing cracking and reinforcement mesh corrosion.

(1) Cracking prevention

The polymer elastic primer has the capability of flexibility deformability, i. e. elasticity. If there are minor cracks in the anti-cracking mortar leveling layer, the coating will be subject to tensile deformation but no cracking, and the coating integrity and barrier functions will be retained. Accordingly, the elastic primer can be applied on the surface of the anti-cracking mortar leveling layer to prevent partial cracks therein from direct exposure to the air, eliminate erosion caused by water and other corrosive substances, and effectively protect the wall structure with external thermal insulation.

(2) Prevention of reinforcement mesh corrosion

The polymer elastic primer is able to prevent the corrosion of the reinforcement mesh of the external thermal insulation system by water, with the specific principle as follows: calcium and magnesium ions in the cement base are dissolved in water seeping into the cement mortar, forming alkaline solution, and the breaking strength of the alkali-resistant glass fiber mesh drops greatly in a humid and high-alkalinity environment. For the external thermal insulation system with the face brick finish, corrosive gases (such as CO_2, SO_2 and SO_3) are dissolved in water, so the skeleton effect of the hot-dip galvanized welded wire mesh damaged locally will fail as a result of rusting. Meanwhile, the surface of the finish layer is alkalized due to calcium hydroxide crystallized in the process of transfer and evaporation of aqueous solution. In the event of large-scale alkalization, the acid-base balance of internal mortar will change, which will aggravate the rusting of the internal wire mesh and reduce the cracking resistance of the surface.

3.4.1.1 Basic Principle of Waterproofing and Vapor Permeation of Coating

Basic principle of the polymer emulsion elastic primer: its micropore are smaller than water drops, and the size of the former is only about one 20, 000th of a water drop (0.05mL). This means that external water cannot infiltrate into the coating (diameter of water molecules: 4×10^{-10}m). The coating is not absolutely non-permeable. Instead, the normal diameter of water drops in nature is far greater than that of film pores, and the water pressure for seeping cannot be achieved easily. Therefore, the coating is always dry in bad weather. Water molecules are always aggregated into water drops in the presence of surface tension. Under low pressure, the surface tension is balanced with the pressure of water permeation into the coating, thus preventing seepage. If the pressure increases to a certain value, however, this balance will be broken and water drops will be broken into smaller ones.

Basic principle of the vapor permeability of the polymer emulsion elasticprimer: each micropore of the coating is 700 times of vapor molecules, so vapor can easily evaporate through the micropore of the coating. Due to the pores of the coating, the gas phases inside and outside the coating are interconnected. The pore size of the coating determines the vapor permeability. That is, the vapor permeability will be high in the case of large pores. As the "smaller volume unit" allowing vapor permeation is enlarged, however, water drops are easier to break, which will reduce the waterproofing performance. On the contrary, if there are small pores, the vapor permeability will be low, but the waterproofing performance will be enhanced (Figure 3-4-1).

With the vapor permeability coefficient basically unchanged, the polymer emulsion elastic primer can be applied on the insulating protective

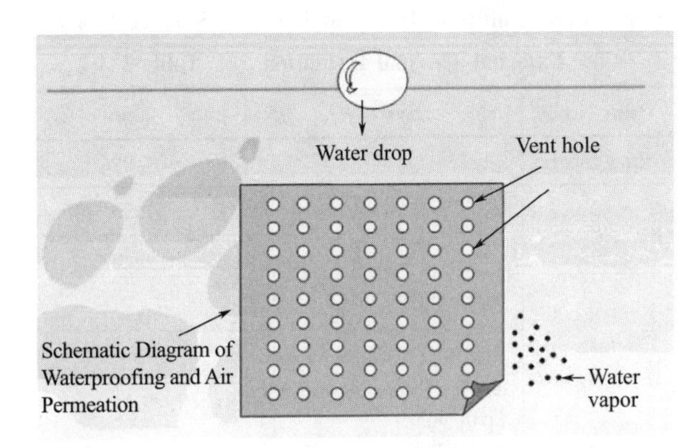

Figure 3-4-1 Schematic Diagram of Waterproofing and Vapor Permeation of Polymer Emulsion Elastic Base Coating

layer to effectively and greatly reduce the water absorption coefficient of the surface of the finish layer. Table 3-4-1 shows the contrast test data.

Contrast Test of Water Absorption Capacity and Vapor Permeability of Polymer Emulsion Elastic Primer Table 3-4-1

Item	Unit	Specimen with Polymer Emulsion Elastic Primer	Contrast Specimen
Water absorption coefficientK	$kg/m^2 \cdot \sqrt{h}$	0.12	1.11
Vapor permeability coefficient μ	g/(m·s·Pa)	9.89×10^{-9}	10.72×10^{-9}

The above data shows that when the polymer elastic primer (thickness: about100μm) is applied on the insulating protective layer, the water absorption coefficient of the surface will be greatly reduced, but the vapor permeability coefficient of the material will basically remain unchanged and the heat transfer coefficient will be guaranteed, thus improving the freeze-thaw resistance, durability and cracking resistance of the external thermal insulation system, and also meeting its vapor permeability requirements.

3.4.1.2 Factors Affecting Vapor Permeability of Coating

The vapor permeability of a coating is mainly affected by the coating thickness, PVC, ageing treatment, solvent-based paint and water-based paint (Table 3-4-2).

Vapor Permeability of Different Paints in Surface Mortar for External Thermal Insulation Table 3-4-2

Paint No.	System	S_d Value(m)
Surface mortar	—	0.22
B1	+primer(PVC45%)	0.50
	+primer+elastic middle coating(PVC38%)	1.32
	+primer+elastic middle coating + elastic topcoat (PVC29%)	2.11
B3	+primer(PVC45%)	0.39
	+ primer + middle coating(PVC85%)	0.96
	+ primer + middle coating+topcoat(varnish)	1.63
D3	+primer(PVC29%)	0.74
	+ primer + middle coating(PVC70%)	1.80
	+ primer + middle coating+topcoat	3.09
E1	+primer(oily)	0.88
	+ primer + middle coating	1.30
	+ primer + middle coating+topcoat	2.06

The above data shows:

(1) The coating thickness is inversely proportional to the vapor permeability. The thicker the coating, the greater the S_d value and the poorer the vapor permeability will be.

(2) In the case of the same thickness, the higher the PVC content of a coating, the better the vapor permeability.

(3) Ageing is one important factor affecting the vapor permeability. It has little impact on the high-PVC paint but significant impact on the low-PVC paint. The vapor permeability will change greatly after ageing (the requirements for specimen ageing are added in the Determination and Classification of Water Vapor Transmission Rate of Exterior Wall Coating: drying specimens for 14d, keeping them in a 50℃ oven for 24h, then making them immersed in water for 24h, and repeating the above cycle three times).

(4) The paint of poor vapor permeability is easy to foam, crack and fall off. For example, hollowing of the elastic paint is mainly caused by its poor vapor permeability.

3.4.2 Moisture Dispersion Structure

The ideal structure should be built based on the vapor permeability of each material. If the vapor permeability gradually rises from indoors to outdoors, a passage for smooth transfer of water vapor will be formed, which will prevent condensate in the wall and insulation finish. Analyzed from the perspective of drying, this is also conducive to moisture drainage after evaporation. In terms of the water absorption rate, the thermal insulation system will be prevented from liquid water. Compared with the surface material, the water absorption rate of the internal material may be lower. In order to solve the problem of vapor condensation in the thermal insulation system, the vapor permeability of each material should be appropriate. That is, the vapor permeability of the material close to the external surface should be higher. Nevertheless, the finish material should satisfy more rigorous requirements for the waterproofing performance to prevent liquid water.

The condensate inside the external thermal insulation system (caused by the cracking of the exterior finish, fatal damage to the elastic primer or inflow of rainwater) needs to be efficiently discharged to prevent this system from freezing and thawing, so this system must be equipped with a vapor permeation and transfer structure, namely, moisture dispersion structure. The moisture dispersion structure was developed for the external thermal insulation system (especially the external thermal insulation system based on the XPS board) of non-smooth moisture drainage as a result of poor permeability. The external thermal insulation system based on the pasted XPS board and the mineral bonder and EPS granule has excellent effects of moisture transfer and conditioning and its moisture transfer will be adjusted automatically, thereby enhancing its "breathability" (Figure 3-4-2 and 3-4-3).

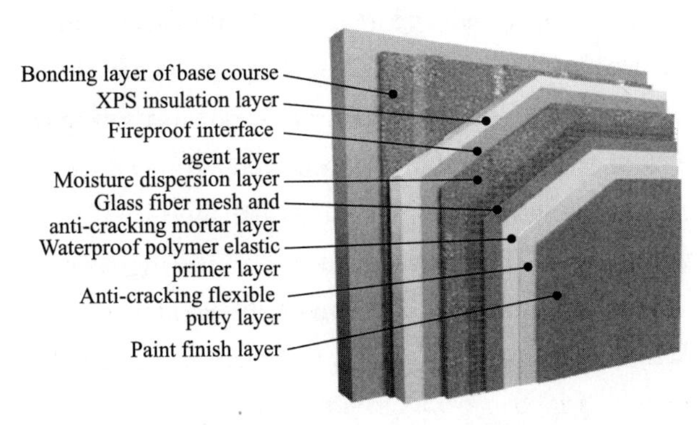

Figure 3-4-2 Basic Structure of External Thermal Insulation System based on Pasted XPS Board, Adhesive polystyrene granule

Despite of its excellent insulating and watcr-proofing functions, the XPS board (matte) has defects such as high strength, large deformation stress, low vapor permeability and poor adhesion, which have caused many engineering quality problems. It has a special molecular structure made with the particular forming process, resulting in poor vapor permeability and adhesion. Accordingly, appropriate measures are taken in the external thermal insulation system. That is, a vapor transfer and diffusion structure (i. e. leveling layer with the adhesive polystyrene granule) is added on the surface of the insulation layer, to improve the breathability of this system. The following recommendations are also put forward.

(1) Two vent holes are drilled in the XPS board (Figure 3-4-4) and filled with the adhesive polystyrene granule mortar. As the vapor permeability of the adhesive polystyrene granule mortar is greatly higher than that of the XPS board [test results show that the vapor permeability coefficient of the adhesive polystyrene granule mortar is 20.4ng/(m · s · Pa), about 10 times of that of the XPS board], the vapor permeability of the XPS board treated in the above method can be improved. Also, the bonding between the XPS board and base course wall can be enhanced if vent holes are filled with the adhesive polystyrene granule mortar.

(2) The 10mm wide joints are reserved between XPS boards (Figure 3-4-5) and filled with the adhesive polystyrene granule mortar. As the adhesive polystyrene granule mortar has excellent performance in water absorption, conditioning and transfer, this can further enhance the vapor permeability of the XPS board system, and prevent the XPS board surface from condensation, especially frost heaving in the cold and severe cold areas. In addition, six surfaces of the XPS board are wrapped with the hypoelastic adhesive polystyrene granule mortar, which improves the bonding performance. Because of compaction of board gaps and good integrity of the insulation layer, the stress arising from the thermal deformation of boards can be dispersed, absorbed

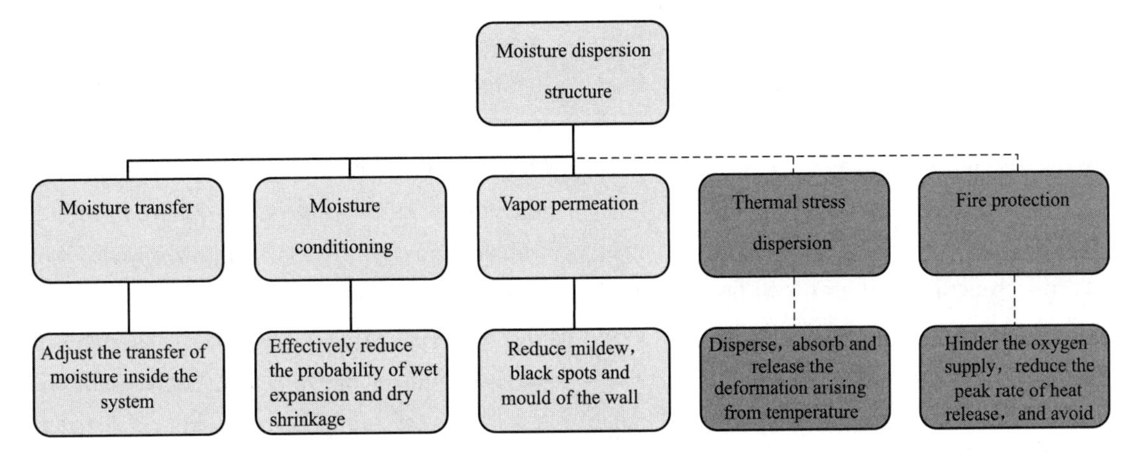

Figure 3-4-3 Schematic Diagram of Moisture Dispersion Structure

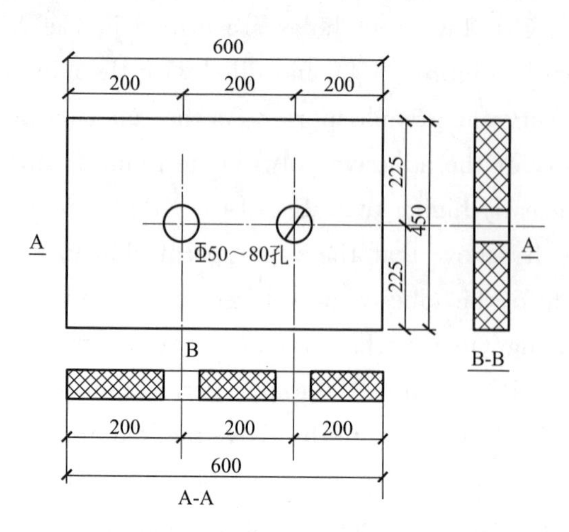

Figure 3-4-4 XPS Board Hole Design

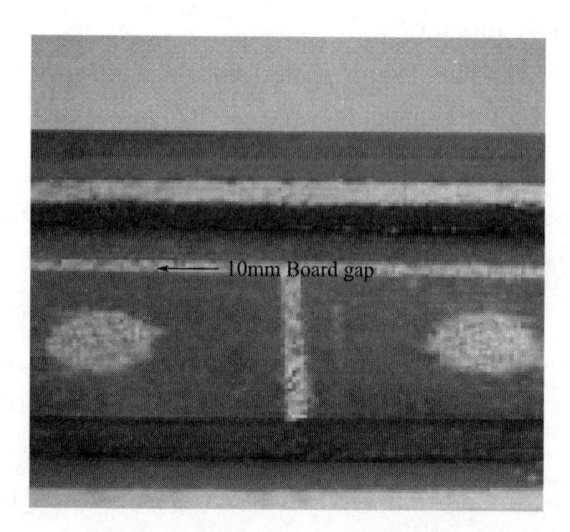

Figure 3-4-5 XPS BoardGap Design

or restricted by the adhesive polystyrene granule mortar, which will reduce the shrinkage deformation of XPS boards, improve their cracking resistance and effectively prevent board gaps from cracking.

(3) A layer of adhesive polystyrene granule mortar is plastered on the surface of the XPS board, to form a structure of moisture dispersion and vapor permeation structure. This structure is able to absorb a small amount of condensate produced because of the condensation or vapor permeability difference, so there will be no liquid water in the external thermal insulation system. Water vapor therein is discharged out of this system. Condensation occurs on the internal surface of the anti-cracking layer if the outdoor temperature is low. A small amount of condensate produced can be absorbed by and dispersed in the adhesive polystyrene granule layer with excellent performance in water absorption, conditioning and transfer, and then diffused in the gaseous form due to good vapor permeability of the adhesive polystyrene granule. This can prevent the destructive force generated by changes in three phases of the accumulated liquid water, and improve the bonding and permeability of the external thermal insulation system, thus guaranteeing the stability and safety of the external thermal insulation works of the outer wall.

3.5 Summary

The results of analysis of basic characteristics of moisture and heat in the building wall and discussion of moisture and heat transfer principles reveal the motion law of water in the porous material. According to the theoretical calculations of four insulation structures by crossing of partial pressure curves of water vapor, condensation occurs to the internal thermal insulation, self-insulation and sandwich insulation during vapor transfer from indoors to outdoors, but does not occur to the external thermal insulation. Therefore, the external thermal insulation is the most reasonable among four types of insulation.

There are condensation conditions in the anti-cracking layer of the external thermal insulation, with the outdoor temperature and humidity changing. As the thin anti-cracking layer is not able to completely absorb the condensate, there will be liquid water in this layer. Therefore, the moisture dispersion structure should be added by plastering the adhesive polystyrene granule mortar on the external surface of the insulation board to absorb the condensate. Due to the vapor permeability of the material, the destruction caused by changes in three phases of water can be avoided.

The damage of water to the external thermal insulation system is a result of one or more phenomena. In the presence of water (necessary condition), destructive forces will be produced. In addition to the hydrophobicity of the material itself, a waterproof protective layer of polymer elastic primer is needed on the anti-cracking mortar surface, to stop liquid water and discharge gaseous water. Provided that the vapor permeability coefficient meets the standards, this will greatly reduce the amount of water absorbed by the insulation system and material, effectively avoid or decrease the damage of frost heaving forces to external thermal insulation systems in cold, severe cold and humid areas, and prevent the damage to polymer mortar adhesion under wet alkaline conditions in a long term.

The external thermal insulation system must have the functions of cracking prevention, waterproofing and moisture discharge, to prevent the inflow of liquid water and facilitate the discharge of gaseous water.

4. Study on Weathering Resistance of External Thermal Insulation System

4.1 Test Introduction

With the continuous development of building energy efficiency, especially the outer wall insulation, people not only pay attention to the insulation performance of the external thermal insulation system, but also attach more importance to the safety and durability of this system. The large-scale weathering resistance test is an essential test to inspect and evaluate the overall performance of the external thermal insulation system, especially its durability.

In actual applications, the thermal insulation system is exposed to the sunlight, rain, cold and heat, and undergoes long-term repeated changes in the temperature, so its thermal stress, especially that on the protective layer on the surface, will change constantly. Heat is blocked by the insulation material and accumulated in the external thermal insulation system, making the temperature of the protective layer up to 70℃ in summer. In the case of sudden rainstorm after consecutive sunny days in summer, the change in surface temperature may reach 50℃. Drastic changes in the temperature will result in uneven deformation of each layer of the external thermal insulation system, thus generating thermal stress inside this system. When the thermal stress exceeds the limit, the external thermal insulation system will crack, and its service life will be shortened.

The weathering resistance test stipulated in theTechnical Specification for External Thermal Insulation on Walls (JGJ 144-2004) comprises the simulation of sudden rainstorm after intense sunlight in summer and repeated effect of annual temperature differences in winter and summer, and the accelerated weathering test of the large external thermal insulation system. In accordance with the above standard, following 80 heating (70℃)-water spraying (15℃) cycles and 5 heating (50℃) -freezing (−20℃) cycles, the finish layer must not be foamed or fall off, the protective layer must not become hollow or fall off, and water seepage and cracking must not occur. The system involving the unreasonable structure and nonconformity mass or compatibility of the material, which cannot withstand the above test, must not be applied in actual projects. Instead, it needs to be improved with appropriate measures based on the experience, until the test results of weathering resistance meet the requirements. The practice has proven that the large-scale weathering resistance test is closely correlated with the actual project. In order to ensure the durability of the external thermal insulation system, the large-scale weathering resistance test must be performed before this system is widely applied.

4.1.1 Test Purpose

The purpose of this test is to scientifically and systematically study external thermal insulation systems that are different in the structure and material composition. By means of collecting and analyzing the data of temperature and humidity changes, making the records on cracking and hollowing, and conducting numerical simulation of the temperature field, the regular changes in the weathering resistance of the external thermal insulation system were explored, and the forced generated were analyzed, to study the reasonable structure and material meeting the requirements of weathering resistance.

4.1.2 Test Equipment

Two temperature control boxes were applied to test the weathering resistance of four external thermal insulation systems at the same time (Figure 4-1-1 and 4-1-2).

Figure 4-1-1 Outer Box of Weathering Resistance Tester

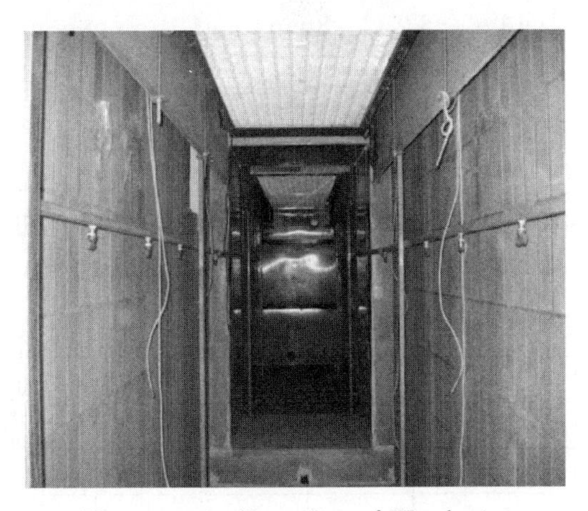

Figure 4-1-2 Inner Box of Weathering Resistance Tester

This test method applies to the comparison of advantages and disadvantages of different external thermal insulation systems and materials under the same ambient temperature. Both the test method and equipment were first applied in China, but the test results are more persuasive.

4.1.3 Test Method

(1) Refer to the Technical Specification for External Thermal Insulation on Walls (JGJ 144-2004) for this test method. Specific test steps are as follows:

1) 80 heating-water spraying cycles: 6h respectively.

(i) 3h temperature rise: heat the specimen surfaces to 70℃ and keep the surface temperature of 70±5℃ (duration of temperature rise: 1h).

(ii) 1h spraying of water: spray 1.0~1.5L (m^2 · min) water of 15±5℃ onto the specimen surface.

(iii) Keep the specimens static for 2h.

2) Status conditioning: at least 48h.

3) 5 heating-freezing cycles: 24h respectively.

(i) 8h temperature rise: heat the specimen surfaces to 50℃ and keep the surface temperature of 50±5℃ (duration of temperature rise: 1h).

(ii) 16h temperature drop: cool the specimen surfaces to −20℃ and keep the surface temperature of −20±5℃ (duration of temperature drop: 2h).

(2) Conduct observation, recording and inspection in line with the following requirements:

After every four high temperature-water spraying cycles and one heating-freezing cycle, observe whether the specimens are cracking or hollow or fall off, and make related records.

Conduct the status conditioning for 7d after the above test. In accordance with the prevailing industry standard "Test Standard for Bonding Strength of Building Face Brick" (JGJ 110), check the tensile and bonding strength between the plaster layer and insulation layer, and cut joints to the surface of the insulation layer.

4.2 Numerical Simulation of Temperature Field of Weathering-resistant Wall

The study on the weathering resistance must be supported with a large amount of data, among which the temperature of each part of the wall with the external thermal insulation at each moment is crucial. However, it is impossible to record all data in the test stage. The temperature field of the wall with the external thermal insulation in the entire test process can be calculated by means of numerical simulation. The numerical simulation of the temperature rise and drop rate of the wall involved in the weathering resistance test is helpful to the study on the temperature cracks of the protective layer, so numerical simulation will be a powerful tool to summarize the law of weathering resistance of the external thermal insulation system.

This section focuses on the method and process of numerical simulation of the temperature field.

4.2.1 Simulation of Weathering Resistance Test Environment

In accordance with the *Technical Specification for External Thermal Insulation on Walls* (JGJ 144-2004), two environments of the external thermal insulation system of the wall were simulated, namely, heat-rain environment and heat-cold environment.

(1) Heat-rain environment: 80 heating-water spraying cycles, 6h respectively, including 1h temperature rise and 1h constant temperature; 1h spraying of water (11～17℃); and 2h static treatment (air temperature inside the box: 26℃). By fitting with the measured data, the air temperature $T_w(t)$ inside the box and the atmospheric temperature $T_s(t)$ near the finish layer can be separately expressed as:

$$T_w(t)=\begin{cases}\dfrac{T_1-T_0}{t_1-t_0}(t-t_0)+T_0, t_0\leqslant t\leqslant t_1,\\ T_1,\ t_1<t\leqslant t_2,\\ \dfrac{T_6-T_1}{t_{31}-t_2}(t-t_2)+T_1, t_2<t\leqslant t_{31},\\ \dfrac{T_5-T_6}{t_4-t_{31}}(t-t_{31})+T_6,\ t_{31}<t\leqslant t_4,\\ \dfrac{T_6-T_5}{t_5-t_4}(t-t_4)+T_5, t_4<t\leqslant t_5,\\ T_6,\ t_5<t\leqslant t_6\end{cases} \tag{4-2-1}$$

$$T_s(t)=\begin{cases}\dfrac{T_1-T_0}{t_1-t_0}(t-t_0)+T_0, t_0\leqslant t\leqslant t_1,\\ T_1,\ t_1<t\leqslant t_2,\\ \dfrac{T_3-T_2}{t_3-t_2}(t-t_2)+T_2, t_2<t\leqslant t_3,\\ \dfrac{T_4-T_3}{t_4-t_3}(t-t_3)+T_3, t_3<t\leqslant t_4,\\ \dfrac{T_6-T_5}{t_5-t_4}(t-t_4)+T_5, t_4<t\leqslant t_5,\\ T_6,\ t_5<t\leqslant t_6\end{cases} \tag{4-2-2}$$

Where, $t_0=0$h, $t_1=1$h, $t_2=3$h, $t_3=3.3$h, $t_{31}=3.7$h, $t_4=4$h, $t_5=5$h, $t_6=6$h (water spraying time: t_2-t_4); and $T_0=26$℃, $T_1=75$℃, $T_2=11$℃, $T_3=15$℃, $T_4=17$℃, $T_5=21$℃, $T_6=26$℃.

(2) Heat-cold environment: 5 heating-freezing cycles, 24h respectively, including 1h of temperature rise (−20℃ to 48℃) +7h of constant temperature (48℃) =8h in total; 0.6h of temperature drop (48℃ to 0℃, linear) +1.4h of temperature drop (0℃ to −20℃, parabolic) = 3h in total; and 13h of constant temperature (−20℃). Based on the measured data, the ambient temperature inside the box under the heat-cold conditions is fitted as:

$$T(t)=\begin{cases}\dfrac{T_1-T_0}{t_1-t_0}(t-t_0)+T_0,\ t_0\leqslant t\leqslant t_1,\\ T_1, t_1<t\leqslant t_2,\\ \dfrac{T_2-T_1}{t_3-t_2}(t-t_2)+T_1, t_2<t\leqslant t_3,\\ \dfrac{T_2-T_3}{(t_3-t_4)^2}(t-t_4)^2+T_3,\ t_3<t\leqslant t_4,\\ T_3, t_4<t\leqslant t_5\end{cases} \tag{4-2-3}$$

Where, $t_0=0$h, $t_1=1$h, $t_2=8$h, $t_3=8.6$h, $t_4=11$h, $t_5=24$h (node temperature and indoor temperature: 26℃); and $T_0=-20$℃, $T_1=48$℃, $T_2=0$ ℃, $T_3=-20$℃.

4.2.2 Simulation Calculation Results and Analysis

(1) Wall types and related parameters

Using the above model, the real-time temperature fields and stresses of the composite adhesive polystyrene granule paint finish system based on the EPS board and that based on the sprayed rigid polyurethane foam in heat-cold and heat-rain environments were calculated, respectively. The dimensions and thermodynamic parameters of each layer in the calculation process are listed in Table 4-2-1.

(2) Temperature field results and analysis

Figure 4-2-1 and Figure 4-2-2 show the steadily changing temperature of each layer in the presence of the external thermal insulation by the composite adhesive polystyrene granule paint finish based on the EPS board and that based on the sprayed rigid polyurethane foam over time in one period (24h and 6h, respectively) under heat-cold and heat-rain conditions.

It can be seen from the above figures that the temperature of each layer inside the wall varies with the periodic changes in the outdoor temperature, and the magnitude of change is related to the location of each functional layer. The longer the distance to the external surface of the wall, the less the impact of the ambient temperature will be. If the insulation material is applied properly, heat transfer between the wall and external environment can be reduced effectively, the internal surface of the wall will not be significantly affected by the changes in ambient temperature, and the temperature change of nodes close to the external surface of the wall will be greatly affected. The results show that the temperature of the internal surface of the wall changes little over time. The daily temperature change of two insulation systems is less than 3℃ in the heat-cold environment and 1℃ in the heat-rain environment. The temperature of the external surface of the wall changes the most over time. The daily temperature of the EPS board insulation system is

Thermodynamic Parameters of Materials **Table 4-2-1**

Structure Type	Material Name	Thickness (mm)	Density (kg/m^3)	Specific Heat [J/(kg·K)]	Thermal Conductivity [W/(m·K)]	Thermal Deformation Coefficient [1/(m·K)]	Elastic Modulus GPa	Poisson's Ratio
Composite adhesive polystyrene granule paint finish based on EPS board	Base course wall	140	2500	882	1.37	10	25.5	0.2
	Interface mortar	1	1500	1050	0.76	8.5	2	0.2
	EPS board	60	20	1380	0.041	60	0.0091	0.4
	Insulation mortar	10	200	1070	0.07	30	0.27	0.35
	Anti-cracking mortar	4	1600	1050	0.93	12	5.18	0.28
	Paint	3	1100	1050	0.5	12	2	0.2
Composite adhesive polystyrene granule paint finish based on sprayed rigid polyurethane foam	Base course wall	140	2500	882	1.37	10	25.5	0.2
	Polyurethane	40	500	1380	0.0265	60	0.026	0.4
	Insulation mortar	30	200	1070	0.07	30	0.27	0.35
	Anti-cracking mortar	4	1600	1050	0.93	12	5.18	0.28
	Paint	3	1100	1050	0.5	12	2	0.2

from −21℃ to 52℃ in the heat-cold environment and from 26℃ to 71℃ in the heat-rain environment; and that of the polyurethane insulation system is from −17℃ to 46℃ in the heat-cold environment and from 38℃ to 67℃ in the heat-rain environment.

The external surface temperature of the wall with the EPS board insulation system is closer to that of the insulation layer (EPS board), compared with the polyurethane insulation system. It can be seen in Figure 4-2-1 and 4-2-2 that the surface temperature curve of the outer wall with the EPS board insulation system is closer to the external surface temperature curve of the insulation layer than that with the polyurethane insulation system. This is because the 30mm adhesive polystyrene granule leveling layer of low thermal conductivity [0.07W/(m · K)] is applied on the surface of the insulation layer (sprayed rigid polyurethane foam) of the polyurethane insulation system, making part of heat outside the adhesive polystyrene granule layer. The 10mm adhesive polystyrene granule leveling layer, which is thinner than that of the polyurethane insulation sys-

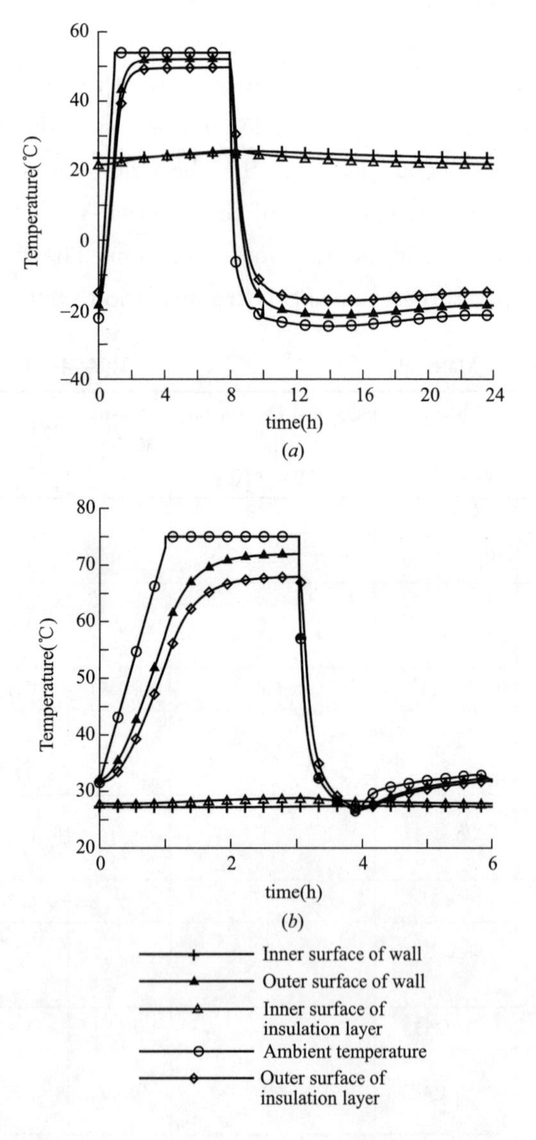

Figure 4-2-1 Temperature Change of Each Layer of Composite Adhesive polystyrene granule Paint Finish System based on EPS Board over Time

(a) heat-cold cycle; (b) heat-rain cycle

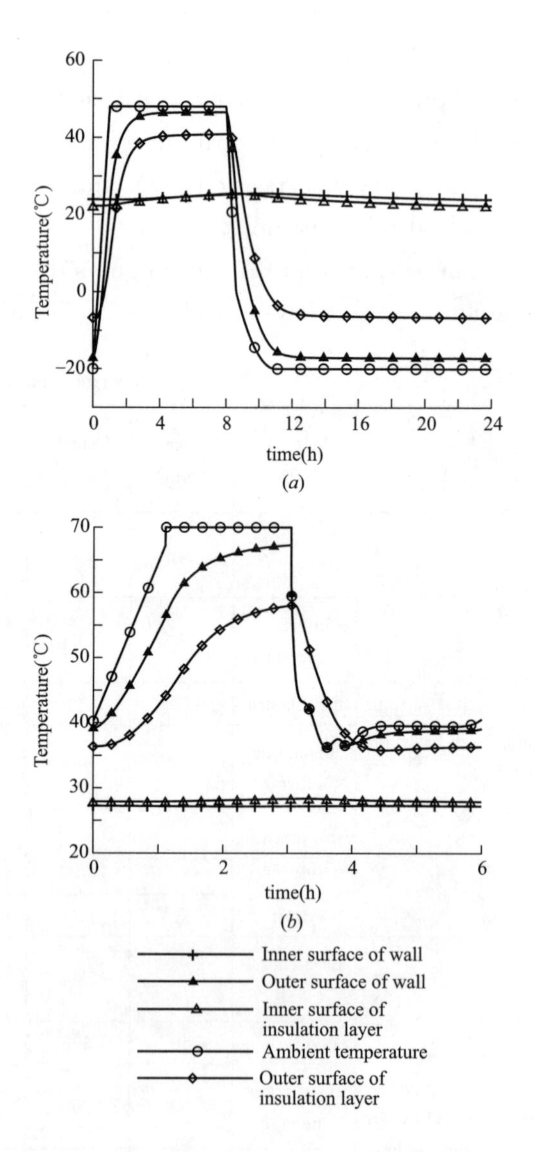

Figure 4-2-2 Temperature Change of Each Layer of Composite Adhesive polystyrene granule Paint Finish System based on Polyurethane Board over Time

(a) heat-cold cycle; (b) heat-rain cycle

tem, is applied in the EPS board insulation system. Therefore, there is a relatively great difference between the polyurethane surface temperature and the external surface temperature of the wall.

In order to verify the correctness of the temperature field calculated by the finite difference method, the theoretical calculations of the surface temperature of walls with these two insulation systems were compared with the values measured in the weathering resistance test in the heat-cold and heat-rain environments. Figure 4-2-3 and Figure 4-2-4 show the comparison between the measured values and theoretical calculations of the surface temperature of walls with these two insulation systems in the heat-rain environment. It can be concluded that the test results are fitted well with calculations under heat-rain conditions. This proves that the correct temperature field calculation model was used.

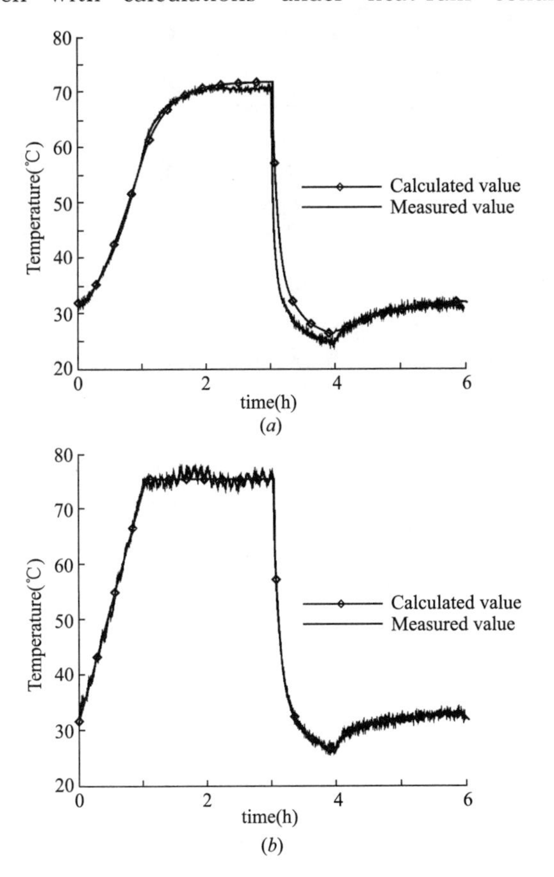

Figure 4-2-3 Comparison between Calculated and Measured Values of Composite Adhesive polystyrene granule Paint Finish System based on EPS Board under Heat-rain Conditions

(*a*) wall surface temperature; (*b*) ambient temperature

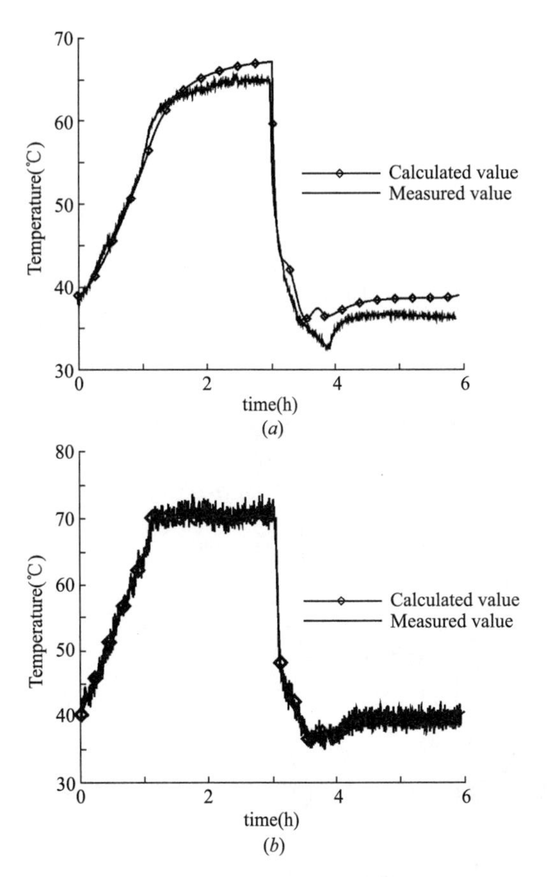

Figure 4-2-4 Comparison between Calculated and Measured Values of Composite Adhesive polystyrene granule Paint Finish System based on Polyurethane under Heat-rain Conditions

(*a*) wall surface temperature; (*b*) ambient temperature

Figure 4-2-5 and Figure 4-2-6 show the internal temperature distribution along the sections of walls with the composite adhesive polystyrene granule paint finish system based on the EPS board and that based on the sprayed rigid polyurethane foam, corresponding to the maximum and minimum ambient temperature under heat-cold and heat-rain conditions.

The following conclusions can be drawn from Figure 4-2-5 and 4-2-6:

1) The temperature inside the base course wall with the external thermal insulation system is close to the indoor temperature and changes little in heat-cold or heat-rain cycles.

2) The temperature of the insulation layer of

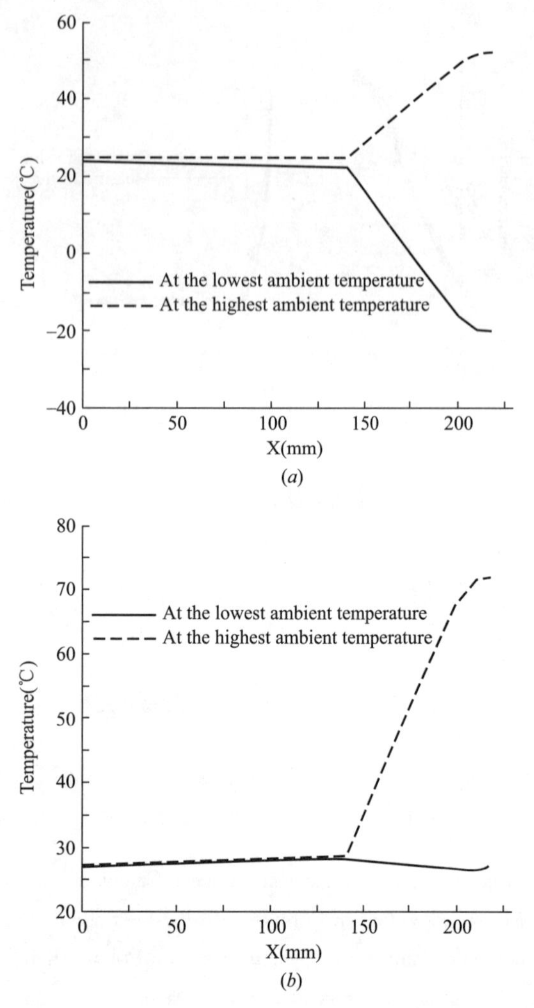

Figure 4-2-5 Temperature Field Distribution along the Section of Wall with Composite Adhesive polystyrene granule Paint Finish System based on EPS Board

(*a*) heat-cold cycle; (*b*) heat-rain cycle

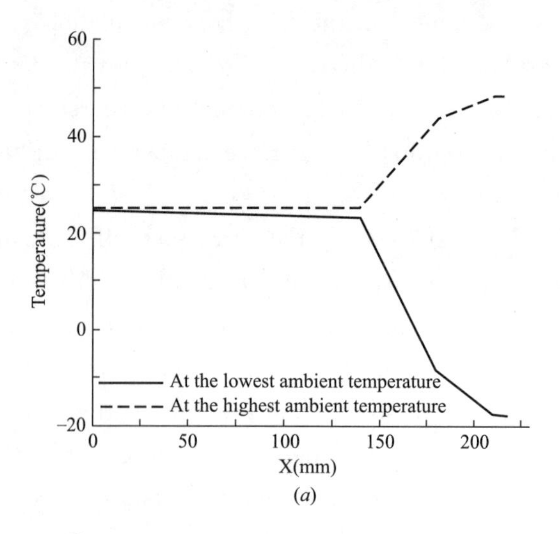

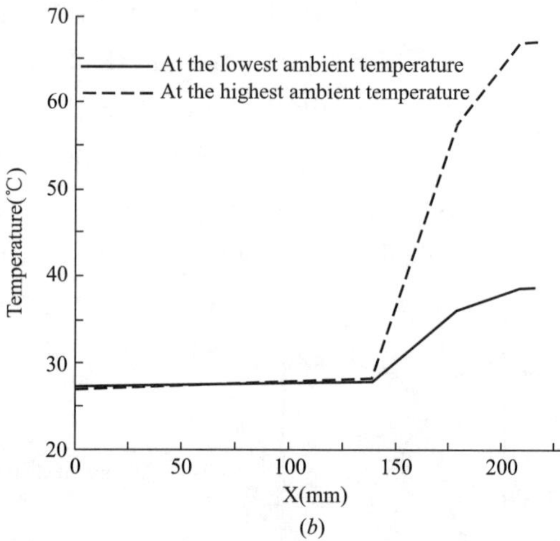

Figure 4-2-6 Temperature Field Distribution along the Section of Wall with Composite Adhesive polystyrene granule Paint Finish System based on Polyurethane

(*a*) heat-cold cycle; (*b*) heat-rain cycle

these two insulation systems changes the most violently along sections. The maximum and minimum temperature differences between the inner and external surfaces of the insulation layer of the EPS board insulation system are 38.4℃ and 39℃ respectively in heat-cold and heat-rain cycles; and those of the polyurethane insulation system are 31.6℃ and 29.4℃ respectively in heat-cold and heat-rain cycles.

3) With the insulation layer in the external thermal insulation system, the temperature of the external surface of the protective layer changes the most over time in heat-cold and heat-rain cycles. The temperature of the external surface of the protective layer of the EPS board insulation system is from −20℃ to 52℃ and from 27℃ to 72℃ respectively in heat-cold and heat-rain cycles; and that of the polyurethane insulation system is from −18℃ to 49℃ and from 39℃ to 68℃ respectively in heat-cold and heat-rain cycles.

(3) Stress field results and analysis

The stress field of the EPS board and polyurethane insulation systems were calculated, with the temperature field calculated above in heat-cold and heat-rain cycles and the thermal stress calculation parameters in Table 4-2-1 as model input values.

Figure 4-2-7 to 4-2-10 show the relationship between the time and thermal stress of the external surface of the anti-cracking mortar layer and

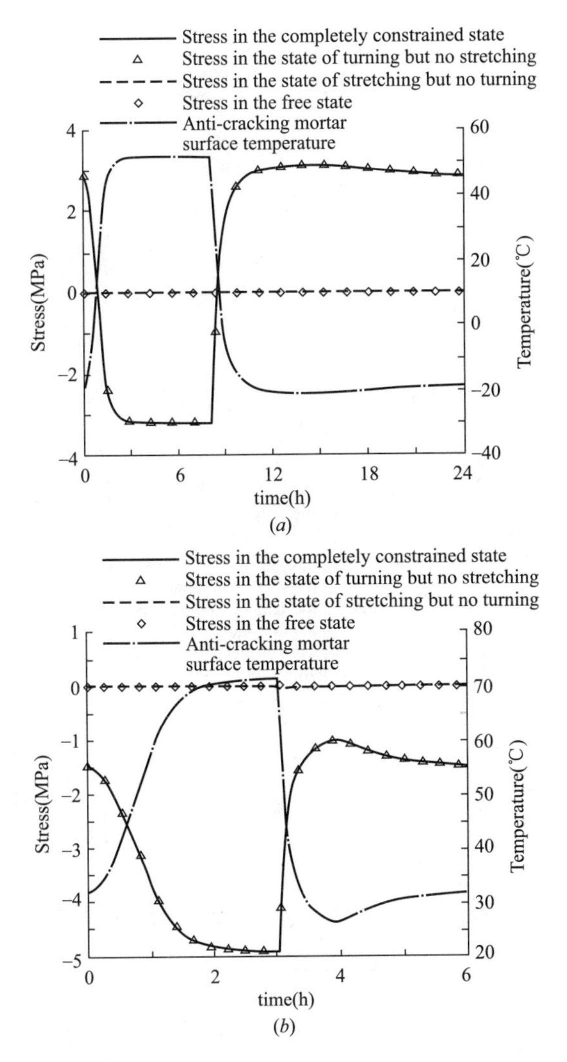

Figure 4-2-7 Change in Stress Field of Mortar Layer of Composite Adhesive polystyrene granule Paint Finish System based on EPS Board over Time under Different Constraints

(*a*) heat-cold cycle; (*b*) heat-rain cycle

the internal surface of the base course wall under typical heat-cold and heat-rain conditions, with the composite adhesive polystyrene granule paint finish system based on the EPS board and that based on the sprayed rigid polyurethane foam. They illustrate the thermal stress of the wall in the presence of four typical constraints and under the same temperature changes. These four typical constraints are as follows: (1) wall stretching and turning are not allowed (embedded wall); (2) wall turning is allowed but stretching is not; (3) wall stretching is allowed but turning is not; and (4) wall stretching and turning are allowed (free wall).

The following conclusions can be drawn from Figure 4-2-7 and 4-2-8:

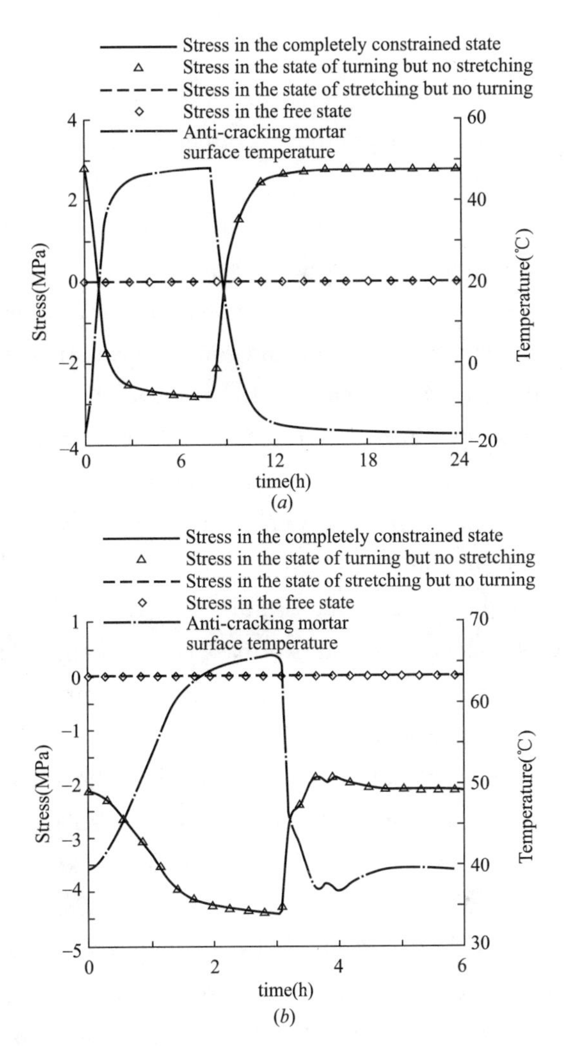

Figure 4-2-8 Change in Stress Field of Mortar Layer of Composite Adhesive polystyrene granule Paint Finish System based on Polyurethane Board over Time under Different Constraints

(*a*) heat-cold cycle; (*b*) heat-rain cycle

1) There is a small difference in the thermal stress of the external surface of the anti-cracking mortar layer under the first two types of constraint, indicating that the small bending stress of the external surface of the anti-cracking mortar layer under heat-cold and heat-rain conditions as well as the small temperature difference between the inner and external surfaces of this layer. The thermal stress is mainly caused by horizontal and vertical constraints.

2) The thermal stress of the free wall is close to zero, indicating the small nonlinearity of temperature distribution inside the anti-cracking mortar layer under the hot-cold and hot-rain conditions.

3) The thermal stress of the mortar layer changes with the ambient temperature. It decreases with the temperature rising and increases with the temperature dropping. The peak temperature corresponds to the peak stress.

4) The anti-cracking mortar layer is alternately subject to tensile and compressive stress in heat-cold cycles. That is, the compressive stress is generated in this mortar layer at high temperature and tensile stress at low temperature.

5) The anti-cracking mortar layer is only subject to compressive stress in heat-rain cycles.

Figure 4-2-9 and Figure 4-2-10 show the stress calculation results of base course walls with these two insulation systems. The following conclusions can be drawn:

1) Although there is a small temperature difference between its inner and external surfaces, the thermal stress of the thick base course wall varies under the first two types of constraint. This means that the bending stress of the internal surface of the wall is greater than that of the anti-cracking mortar layer. Most of the thermal stress of the base course wall is caused by axial constraints.

2) The stress is still close to zero in the fully free state, so the temperature change inside the base course wall is not very linear, and a large amount of heat is kept outside the insulation layer.

3) The thermal stress of the base course wall changes with the ambient temperature. It decreases with the temperature rising and increases with the temperature dropping. The peak temperature corresponds to the peak stress.

4) When the reference temperature is 15℃, the internal surface of the base course wall is always subject to compressive stress in heat-cold and heat-rain cycles, so the base course wall is unlikely to crack.

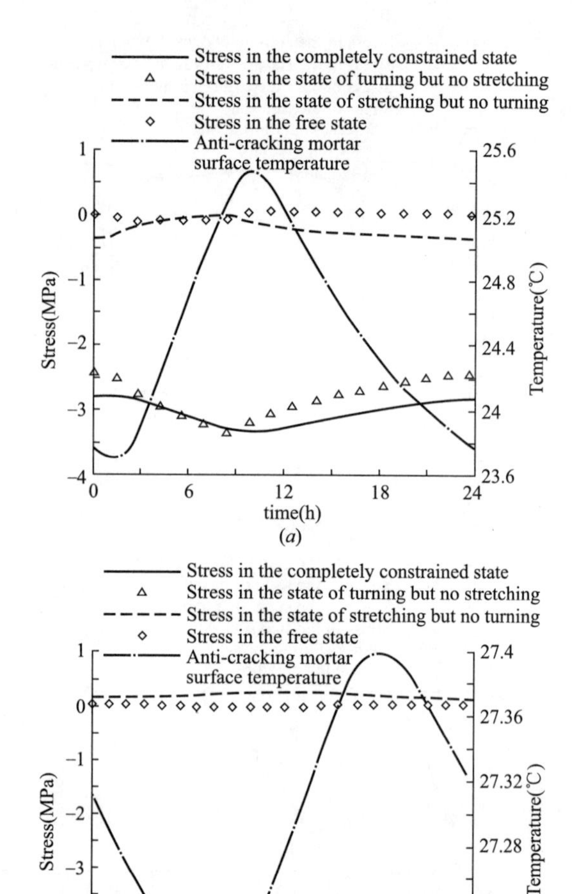

Figure 4-2-9 Change in Stress Field of Base course wall of Composite Adhesive polystyrene granule Paint Finish System based on EPS Board over Time under Different Constraints

(*a*) heat-cold cycle; (*b*) heat-rain cycle

Figure 4-2-11 and Figure 4-2-12 respectively show the thermal stress distribution along the sections of the completely constrained wall with the composite adhesive polystyrene granule paint finish system based on the EPS board and sprayed rigid polyurethane foam, corresponding to the maximum temperature gradient under heat-cold and heat-rain conditions. The corresponding temperature field distribution in the thickness direction of the wall is also illustrated in these figures.

As can be seen from Figure 4-2-11 and 4-2-12, the thermal stress under complete constraints

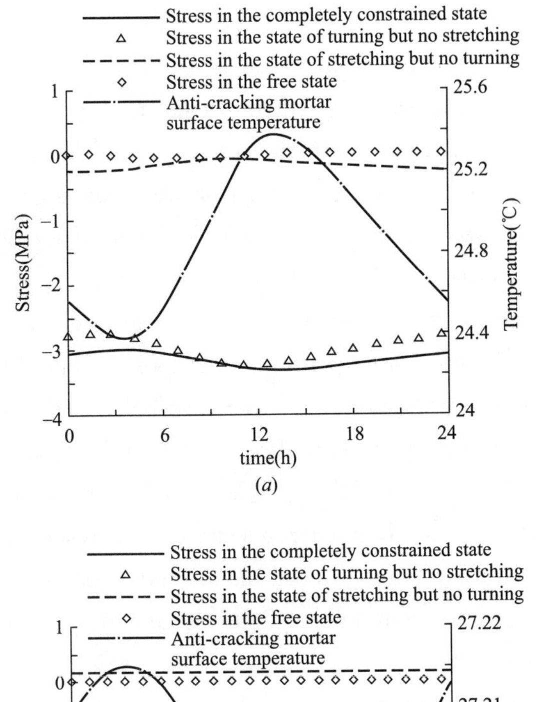

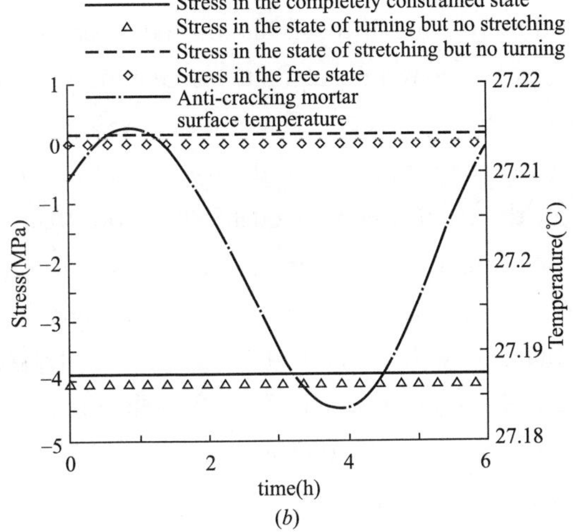

Figure 4-2-10 Change in Stress Field of Base course wall of Composite Adhesive polystyrene granule Paint Finish System based on Polyurethane Board over Time under Different Constraints

(a) heat-cold cycle; (b) heat-rain cycle

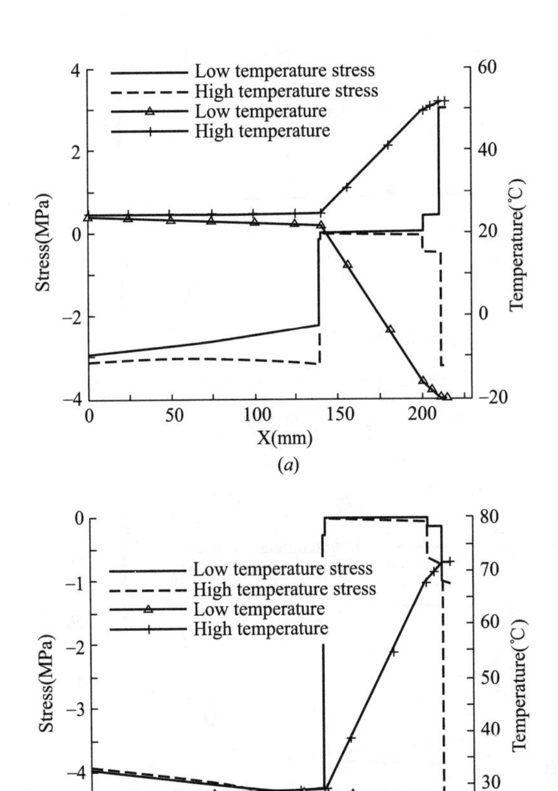

Figure 4-2-11 Stress Field Distribution along the Section of Wall with Composite Adhesive polystyrene granule Paint Finish System based on EPS Board

(a) heat-cold cycle; (b) heat-rain cycle

is distributed in a stepping form. In the low-temperature period of the heat-cold cycle, the thermal stress gradually increases from indoors to outdoors, the base course wall is mainly subject to compressive stress, the protective layer outside the insulation layer is mainly under tensile stress, and the temperature inside the insulation layer changes the most violently. Due to its small elastic modulus, the stress of the insulation layer is close to zero. In the heat-rain cycle and the high-temperature period of the heat-cold cycle, compressive stress occurs to each functional layer of the wall as the temperature is higher than the reference temperature (15℃).

It should be noted that the relaxation effect of the material creepage on thermal stress was not taken into consideration in the above stress analysis. The thermal stress will decrease if this factor is involved.

(4) Comparison with conventional finite element calculations

In order to verify the correctness of the internal thermal stress of the wall based on the generalized Hooke's law, the stress field under the maximum temperature gradient inside the wall was calculated by building the three-dimensional insulated wall model in the conventional finite element method and applying the temperature field in the form of body load on nodes, and also compared

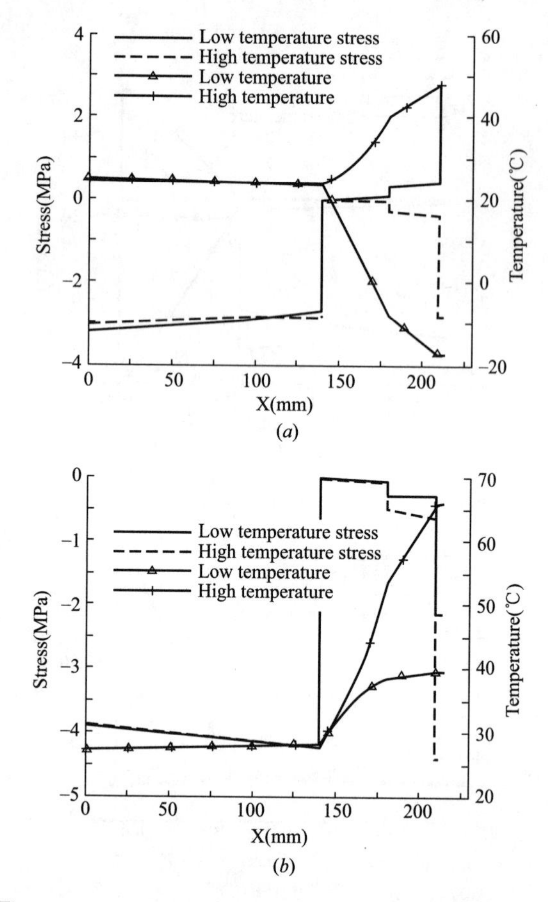

Figure 4-2-12 Stress Field Distribution along the Section of Wall with Composite Adhesive polystyrene granule Paint Finish System based on Polyurethane Board

(*a*) heat-cold cycle; (*b*) heat-rain cycle

with the value calculated above. Table 4-2-2 shows the comparison of the thermal stress at the interface of each functional layer under the maximum temperature gradient inside the completely constrained wall with the composite adhesive polystyrene granule paint finish system based on the EPS board and sprayed rigid polyurethane foam respectively in heat-cold and heat-rain cycles. The thermal stress values of each layer were compared, calculated by two methods in the low-temperature period of the heat-cold cycle and high-temperature period of the heat-rain cycle.

It can be seen that, under complete constraints, there is a small difference between the results obtained by the simplified calculation and those calculated with the conventional finite-element three-dimensional calculation model. The time needed in the modeling and solving process of the conventional finite-element three-dimensional model is far more than that in the simplified calculation. Therefore, the stress field of the insulated wall can be accurately and quickly obtained by the simplified calculation, and the accuracy of results is close to that of results obtained by the conventional finite-element three-dimensional model.

Comparison of Stress Results of Simplified Calculation and Calculation by Conventional Finite Element Method

Table 4-2-2

Insulation System	Location	Low-temperature Period of Heat-cold Cycle		High-temperature Period of Heat-rain Cycle	
		ANSYS calculation result(MPa)	Current calculation result(MPa)	ANSYS calculation result(MPa)	Current calculation result(MPa)
EPS board insulation system	Internal surface of wall	−2.97	−2.92	−3.92	−3.90
	External surface of insulation	0.16	0.16	−0.26	−0.26
	External surface of mortar	3.07	3.00	−5.00	−4.89
	External surface of wall	1.06	1.04	−1.74	−1.70
Polyurethane insulation system	Internal surface of wall	−3.05	−3.17	−3.90	−3.88
	External surface of insulation	0.08	0.06	−0.13	−0.09
	External surface of mortar	2.80	2.79	−4.58	−4.39
	External surface of wall	0.97	0.97	−1.59	−1.53

The interaction between different layers is ignored in the simplified calculation, while each layer is regarded completely bonded in the conventional finite element calculation. Results show that the thermal stress values of each layer, calculated by these two methods are close to each other. This indicates that the stress of each layer is mainly caused by boundary constraints under complete embedding conditions.

4.2.3 Conclusions

The temperature field and stress field of each functional layer of the EPS board and polyurethane insulation systems under heat-cold and heat-rain conditions were obtained by means of the large-scale weathering resistance test and numerical simulation. The following conclusions can be drawn from the test phenomena and calculation results.

(1) The temperature field measured in the weathering resistance test agrees well with that calculated by means of numerical simulation. The temperature of the outer protective layer changes the most in the weathering resistance test. The daily temperature of the EPS board insulation system is from −21℃ to 52℃ in under heat-cold conditions and from 26℃ to 71℃ under heat-rain conditions, and that of the polyurethane insulation system is from −17℃ to 46℃ under heat-cold conditions and from 38℃ to 67℃ under heat-rain conditions.

(2) The anti-cracking mortar layer is alternately subject to tensile and compressive stress in the heat-cold cycle, but only to the compressive stress in the heat-rain cycle. When the reference temperature is 15℃, the stress of the anti-cracking mortar layer of the EPS board insulation system is from −3.18MPa to 3.12MPa and from −4.88MPa to −0.99MPa respectively in the heat-cold and heat-rain cycle, and that of the polyurethane insulation system is from −2.85MPa to 2.79MPa and from −4.41MPa to −1.8MPa respectively in the heat-cold and heat-rain cycle.

(3) Under complete constraints, there is a small difference between the results of the simplified calculation and those obtained with the conventional finite-element three-dimensional model. By means of the simplified calculation, the thermal stress of each functional layer of the insulation system can be obtained more easily and quickly.

(4) If the adhesive polystyrene granule mortar is applied rationally, the cracks in the protective layer of the insulated wall can be reduced, which will improve the weathering resistance of the insulated wall.

4.3 Test Case Analysis

4.3.1 Overview

The study involving 12 large-scale weathering resistance tests and 48 external thermal insulation systems in total is a project with the most testing workload, insulation materials and insulation systems at present in the field of the weathering resistance test of the external thermal insulation system. The insulation materials here include the adhesive polystyrene granule, EPS board, XPS board, polyurethane, rock wool, and reinforced vertical-fiber rock wool composite board; the construction process and structure here include the casting-in-situ, on-site spraying, pasting, point and frame bonding, thin plastering and thick plastering; the finish layer involves the paint, face brick and finish mortar; and the majority of mainstream insulation materials and structures on the market are covered.

Refer to Table 4-3-1 for the construction methods of 48 external thermal insulation systems.

Refer to the sections 4.3.2, 4.3.3 and 4.3.4 for analysis reports of some weathering resistance tests.

Summary of System Construction Method of Large-scale Weathering Resistance Test Table 4-3-1

S/N	Bonding Layer	Insulation Layer	Leveling Layer	Finish Layer	Remarks
1. 1	15mm adhesive polystyrene granule	60mm EPS board	10mm adhesive polystyrene granule	Paint	—
1. 2	15mm adhesive polystyrene granule	65mm EPS board	—	Face brick	—
1. 3	—	50mm insulation mortar	—	Face brick	—
1. 4	5mm bonding mortar	70mm EPS board	—	Paint	—
2. 1	15mm adhesive polystyrene granule	40mm EPS board	10mm adhesive polystyrene granule	Paint	—
2. 2	15mm adhesive polystyrene granule	65mm EPS board	—	Paint	With two holes and trapezoid-shaped slot in the EPS board
2. 3	15mm adhesive polystyrene granule	65mm EPS board	—	Paint	Flat EPS board with no joint
2. 4	5mm bonding mortar	70mm EPS board	—	Paint	—
3. 1	—	60mm meshed EPS board	10mm adhesive polystyrene granule	Face brick	—
3. 2	—	60mm meshed EPS board	—	Face brick	—
3. 3	—	60mm non-meshed EPS board	10mm adhesive polystyrene granule	Paint	—
3. 4	—	60mm non-meshed EPS board	—	Paint	—
4. 1	15mm adhesive polystyrene granule	50mm XPS board	10mm adhesive polystyrene granule	Paint	—
4. 2	15mm adhesive polystyrene granule	50mm XPS board	—	Paint	No peeling but with two holes
4. 3	5mm bonding mortar	60mm XPS board	—	Paint	—
4. 4	5mm bonding mortar	60mm XPS board	—	Paint	—
5. 1	15mm adhesive polystyrene granule	60mm EPS board	10mm adhesive polystyrene granule	Face brick	—
5. 2	15mm adhesive polystyrene granule	60mm EPS board	—	Face brick	—
5. 3	15mm adhesive polystyrene granule	60mm XPS board	—	Face brick	—
5. 4	5mm bonding mortar	60mm EPS board	—	Face brick	—
6. 1	—	40mm sprayed rigid polyurethane foam	30mm adhesive polystyrene granule	Paint	—

Continued

S/N	Bonding Layer	Insulation Layer	Leveling Layer	Finish Layer	Remarks
6.2	—	40mm sprayed rigid polyurethane foam	10mm adhesive polystyrene granule	Paint	—
6.3	—	40mm sprayed rigid polyurethane foam	—	Paint	The polyurethane surface is leveled
6.4	—	40mm sprayed rigid polyurethane foam	—	Paint	The polyurethane surface is not leveled
7.1	15mm adhesive polystyrene granule	75mm EPS board	30mm adhesive polystyrene granule	Paint	—
7.2	15mm adhesive polystyrene granule	85mm EPS board	10mm adhesive polystyrene granule	Paint	—
7.3	15mm adhesive polystyrene granule	90mm EPS board	—	Paint	—
7.4	5mm bonding mortar	100mm EPS board	—	Paint	Rock wool type fire barrier
8.1	15mm adhesive polystyrene granule	60mm XPS board	10mm adhesive polystyrene granule	Paint	—
8.2	15mm adhesive polystyrene granule	60mm XPS board	10mm adhesive polystyrene granule	Paint	—
8.3	15mm adhesive polystyrene granule	65mm XPS board	—	Paint	—
8.4	5mm bonding mortar	70mm XPS board	—	Paint	Rock wool type fire barrier
9.1	5mm bonding mortar	100mm rock wool board	—	Paint	—
9.2	5mm bonding mortar	100mm rock wool board	20mm adhesive polystyrene granule	Paint	Mainly by anchorage
9.3	5mm bonding mortar	100mm rock wool board	20mm adhesive polystyrene granule	Paint	Mainly by anchorage
9.4	15mm adhesive polystyrene granule	60mm XPS board	10mm adhesive polystyrene granule	Paint	—
10.1	15mm adhesive polystyrene granule	100mm reinforced vertical-fiber rock wool board	—	Paint	—
10.2	15mm adhesive polystyrene granule	100mm reinforced vertical-fiber rock wool board	—	Face brick	To the anti-cracking layer
10.3	5mm bonding mortar	100mm reinforced vertical-fiber rock wool board	—	Paint	—
10.4	5mm bonding mortar	100mm reinforced vertical-fiber rock wool board	—	Face brick	To the anti-cracking layer
11.1	15mm adhesive polystyrene granule	60mm XPS board	10mm adhesive polystyrene granule	Paint	—
11.2	15mm adhesive polystyrene granule	60mm XPS board	—	Paint	—
11.3	15mm adhesive polystyrene granule	60mm XPS board	30mm adhesive polystyrene granule	Paint	—

Continued

S/N	Bonding Layer	Insulation Layer	Leveling Layer	Finish Layer	Remarks
11.4	5mm bonding mortar	60mm XPS board	—	Paint	—
12.1	5mm bonding mortar	40mm polyurethane composite insulation board	10mm inorganic insulation mortar	Paint	—
12.2	5mm bonding mortar	40mm polyurethane composite insulation board	—	Paint	Double-layer alkali-resistant glass fiber mesh
12.3	5mm bonding mortar	40mm polyurethane composite insulation board	10mm adhesive polystyrene granule	Paint	—
12.4	5mm bonding mortar	40mm polyurethane composite insulation board	—	Paint	Single-layer alkali-resistant glass fiber mesh

Note: The base course wall is a C20 concrete wall. The interface between the insulation material and the mineral binder/EPS granule is treated with an interface agent. The anti-cracking layer is composed of the 4mm anti-cracking mortar, alkali-resistant glass fiber mesh and high-elasticity primer if the paint finish is applied, and the 10mm anti-cracking mortar and hot-dip galvanized welded wire mesh if the face brick finish is applied.

4.3.2 Analysis Report of Weathering Resistance Test of Polyurethane Insulation System

4.3.2.1 Test Purpose

The sprayed polyurethane was used as the main insulation material and contrast material in the 6th weathering resistance test, in order to determine the impact of the polyurethane surface leveling, the adhesive polystyrene granule mortar leveling layer and the leveling layer thickness on the weathering resistance of the external thermal insulation system based on the polyurethane sprayed on the site.

The polyurethane composite insulation board was used as the main insulation material and contrast material in the 12th weathering resistance test, in order to determine the impact of the use and type of the insulation mortar leveling layer on the weathering resistance of the external thermal insulation system based on the polyurethane composite insulation board.

4.3.2.2 System Structure and Material Selection

Refer to Table 4-3-2 and 4-3-3 for the system structure and material selection.

Structure and Material Selection of External Thermal Insulation System based on Sprayed Rigid Polyurethane Foam

Table 4-3-2

System	Structure						
	Base course wall	Interface layer	Bonding layer	Insulation layer	Leveling layer	Anti-cracking layer	Finish layer
Paint finish system with sprayed rigid polyurethane foam + 30mm adhesive polystyrene granule mortar leveling layer	Concrete wall	Moisture-proof polyurethane primer	None	40mm polyurethane	30mm adhesive polystyrene granule	Dry-mixed anti-cracking mortar + alkali-resistant glass fiber mesh + elastic primer	Waterproof flexible putty + paint
Paint finish system with sprayed rigid polyurethane foam+10mm adhesive polystyrene granule mortar leveling layer	Ditto	Ditto	None	Ditto	10mm adhesive polystyrene granule	Ditto	Ditto

Continued

System	Structure						
	Base course wall	Interface layer	Bonding layer	Insulation layer	Leveling layer	Anti-cracking layer	Finish layer
Paint finish system (leveled) based on the sprayed rigid polyurethane foam	Ditto	Ditto	None	Ditto	None	Ditto	Ditto
Paint finish system (not leveled) based on the sprayed rigid polyurethane foam	Ditto	Ditto	None	Ditto	None	Ditto	Ditto

Note: The fireproof polyurethane interface agent was applied on the polyurethane surface layer.

Structure and Material Selection of External Thermal Insulation System based on Polyurethane Composite Insulation Board

Table 4-3-3

System	Structure						
	Base course wall	Interface layer	Bonding layer	Insulation layer	Leveling layer	Anti-cracking layer	Finish layer
External thermal insulation system based on polyurethane composite insulation board and inorganic insulation mortar plaster	Concrete wall	None	Adhesive of polyurethane composite insulation board	40mm polyurethane composite insulation board	10mm inorganic insulation mortar	Dry-mixed surface mortar / double-component surface mortar + alkali-resistant glass fiber mesh + elastic primer	Waterproof flexible putty + paint
External thermal insulation system based on polyurethane composite insulation board and double-layer glass fiber mesh	Ditto	None	Ditto	Ditto	None	Dry-mixed surface mortar / double-component surface mortar + double-layer alkali-resistant glass fiber mesh + elastic primer	Ditto
External thermal insulation system based on polyurethane composite insulation board as well as adhesive polystyrene granule plaster	Ditto	None	Ditto	Ditto	10mm adhesive polystyrene granule paste	Dry-mixed surface mortar / double-component surface mortar + alkali-resistant glass fiber mesh + elastic primer	Ditto
External thermal insulation system based on polyurethane composite insulation board and thin plaster	Ditto	None	Ditto	Ditto	None	Dry-mixed surface mortar / double-component surface mortar + alkali-resistant glass fiber mesh + elastic primer	Ditto

Note: The wall was divided into two parts in this weathering resistance test. The double-component surface mortar was applied on one half with windows, and the dry-mixed plaster mortar on the other half without windows.

4.3.2.3 Weathering Resistance Test Record and Analysis

1. Record and analysis of the 6th weathering resistance test

(1) Cracking and hollowing record

In this weathering resistance test, the system without the adhesive polystyrene granule layer was cracking but the system with the composite adhesive polystyrene granule layer was not in the curing stage. The system with the 10mm adhesive polystyrene granule leveling layer was also cracking, and the system without the EPS granule layer was cracking more seriously, but the system with the 30mm EPS granule leveling layer was basically not cracking. Refer to Table 4-3-4 for details.

Cracking and Hollowing Record of Weathering Resistance Test of External Thermal Insulation System based on Sprayed Rigid Polyurethane Foam

Table 4-3-4

Category	Before the Test	During the Test	After the Test
Insulation system with sprayed rigid polyurethane foam + 30mm adhesive polystyrene granule mortar leveling layer	No cracking or hollowing	No cracking Total number of cracks: 0; no hollowing	None
Insulation system with sprayed rigid polyurethane foam + 10mm adhesive polystyrene granule mortar leveling layer	No cracking or hollowing	Cracks appeared from the 8th heat-rain cycle, and expanded and increased later. Total number of cracks: 8; no hollowing	No crack expansion
Sprayed rigid polyurethane foam system (leveled)	Cracks appeared in the curing stage. Number of cracks: 15	New cracks appeared from the 2nd heat-rain cycle, and expanded and increased later. Total number of cracks: 20; no hollowing	No crack expansion
Sprayed rigid polyurethane foam system (not leveled)	Cracks appeared in the curing stage. Number of cracks: 20	New cracks appeared from the 2nd heat-rain cycle, and expanded and increased later. Total number of cracks: 31; no hollowing	No crack expansion

(2) Temperature curve record and analysis

Figure 4-3-1 shows the changes in the temperature of the external surface of the weathering-resistant wall and the air (water) temperature inside the box over time, recorded by the software of the weathering resistance tester in one steady period of the heat-rain cycle test of four sprayed rigid polyurethane foam systems.

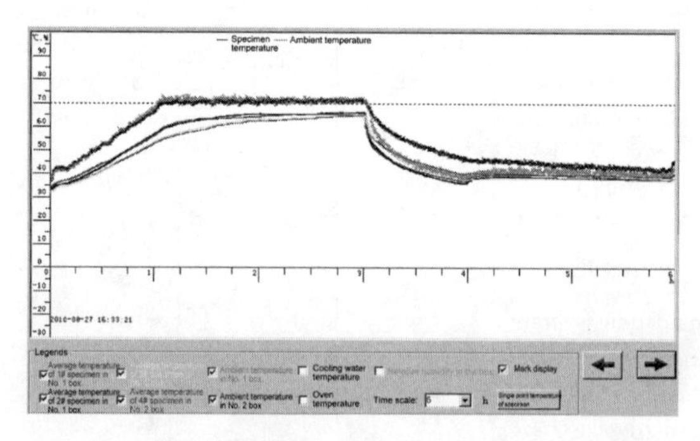

Figure 4-3-1 Temperature Recorded by Tester in One Period of Heat-rain Cycle of Sprayed Rigid Polyurethane Foam System

Figure 4-3-2 shows the curves of the air temperature inside the box and the external surface of the exterior finish, obtained by means of numerical simulation in the same period of the heat-rain cycle of four sprayed polyurethane systems. Comparison results show that the numerical simulation data is basically consistent with the actual data collected.

Figure 4-3-3 shows the numerical simulation results of the temperature change rates of anti-cracking layers of four sprayed rigid polyurethane foam systems.

Table 4-3-5 lists the numerical simulation results

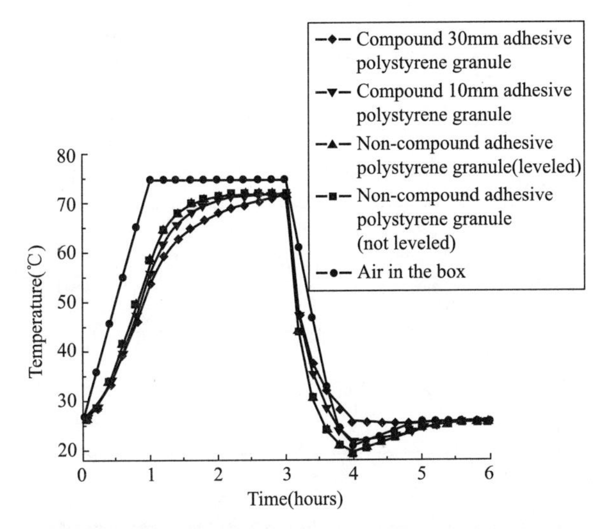

Figure 4-3-2 Temperature of External Surface of Each Wall in One Period of Heat-rain Cycle

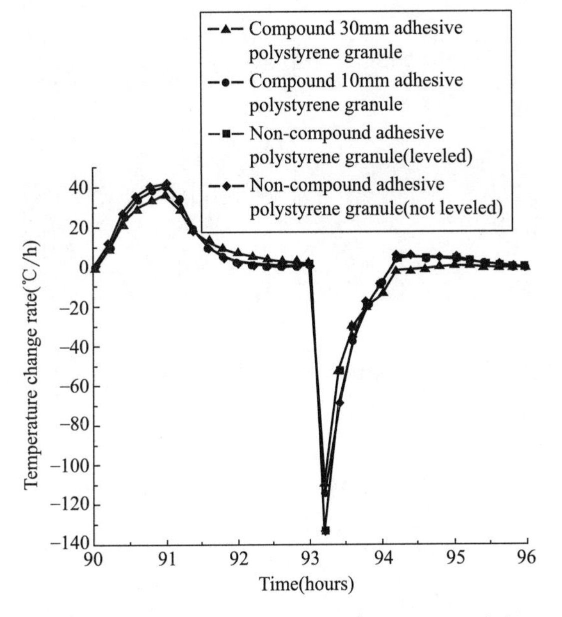

Figure 4-3-3 Temperature Rise/Drop Rate of the Middle Part of Anti-cracking Mortar in the 16th Heat-rain Cycle

of temperature drop rates of anti-cracking layers of four sprayed rigid polyurethane foam systems in the water-spraying stage of the heat-rain cycle.

It can be seen in Figure 4-3-1, Figure 4-3-2 and Table 4-3-4 that the temperature of the composite 30mm adhesive polystyrene granule system rises (drops) the most slowly, followed by the composite 10mm adhesive polystyrene granule system and the system without the composite adhesive polystyrene granule system in sequence. The temperature crack on the surface of the external thermal insulation system is associated with the temperature of the finish and also the change rate of the surface temperature. As shown in Figure 4-3-3 and Table 4-3-5, the thicker the adhesive polystyrene granule layer outside the polyurethane, the lower the temperature change rate of the anti-cracking mortar layer is. This is conducive to the reduction of thermal deformation of the anti-cracking mortar layer and prevention of surface cracking. In combination with the temperature crack analysis of four XPS board systems, if the ordinary organic insulation material of low thermal conductivity is used, the surface temperature may change too fast and temperature cracks may appear. If the adhesive polystyrene granule transition layer with the thermal conductivity between those of the mortar and organic insulation material is applied outside the original insulation layer, the change in surface temperature will slow down. As the thermal expansion coefficient of the adhesive polystyrene granule is between those of the organic insulation material and surface mortar, the difference in deformation rates of adjacent materials will be reduced.

Numerical Simulation Results of Temperature Drop Rates of Anti-cracking Layers of Sprayed Rigid Polyurethane Foam Systems in the Water-spraying Stage of Heat-rain Cycle **Table 4-3-5**

Item	System			
	Composite 30mm adhesive polystyrene granule system	Composite 10mm adhesive polystyrene granule system	Polyurethane system(leveled)	Polyurethane system(not leveled)
Temperature change rate(℃/h) after 0.1h spraying of water	−129.6	−131.7	−151.8	−151.8

Continued

Item	System			
	Composite 30mm adhesive polystyrene granule system	Composite 10mm adhesive polystyrene granule system	Polyurethane system(leveled)	Polyurethane system(not leveled)
Temperature change rate(℃/h)after 0.2h spraying of water	−88.6	−96.0	−114.0	−114.0
Temperature change rate(℃/h)after 0.3h spraying of water	−60.3	−70.9	−80.0	−80.0

In the sprayed rigid polyurethane foam system, the surface and polyurethane are constrained by each other because of their differences in the thermal expansion and contraction, which will lead to the constraint stress. The sprayed rigid polyurethane foam, as a whole different from the insulation block, produces stronger constraints to the surface, thereby generating higher thermal stress. Cracking may occur when the thermal stress exceeds the material strength of the surface. If the sprayed polyurethane is compounded with an appropriate layer of adhesive polystyrene granule, which has the thermal expansion coefficient between those of the polyurethane and cement mortar, non-continuous physical properties and strong capabilities in deformation absorption, stress concentration can be avoided and the surface can be deformed freely, thereby reducing the thermal stress of the surface and avoiding temperature cracks.

Figure 4-3-4 After the Weathering Resistance Test of Wall with the System of Sprayed Rigid Polyurethane Foam+30mm Adhesive polystyrene granule Mortar Leveling Layer

(3) Test results

(i) System with the sprayed rigid polyurethane foam+30mm adhesive polystyrene granule mortar leveling layer Refer to Figure 4-3-4 for the wall after the weathering resistance test. This system was not cracking before the weathering resistance test, and the test results comply with the standards. Refer to Table 4-3-6 for the pull-out test results, which also meet the standards.

Pull-out Test Results of Wall Table 4-3-6

Measurement Point No.	Pull Strength (MPa)	Average (MPa)	Cutting Location	Breaking Location
1	0.110	0.111	To the polyurethane	Adhesive polystyrene granule
2	0.122			
3	0.102			
4	0.110			

The weathering resistance test results of the sprayed rigid polyurethane foam system show that, compared with the system without the EPS granule and the system with a thin layer of EPS granule, the system with an appropriate adhesive polystyrene granule leveling layer has significant advantages in the weathering resistance. This proves that the adhesive polystyrene granule transition layer is an essential surface leveling material for the polyurethane system.

(ii) System with the sprayed rigid polyurethane foam+10mm adhesive polystyrene granule mortar leveling layer Refer to Figure 4-3-5 for the wall after the weathering resistance test. This system was not cracking in the curing stage, but eight cracks appeared after the weathering resistance test. Cracking of this system is less than that of No. 1 and 2 walls, but the weathering resistance of this system does not meet the standards.

Figure 4-3-5 After the Weathering Resistance Test of Wall with the System of Sprayed Rigid Polyurethane Foam+10mm Adhesive polystyrene granule Mortar Leveling Layer

(iii) Sprayed rigid polyurethane foam system (leveled)

Refer to Figure 4-3-6 for the wall states before and after the weathering resistance test. A large number of cracks appeared on the entire wall surface before this test, and expanded and increased after this test. The weathering resistance of this system does not conform to the standards.

(iv) Sprayed rigid polyurethane foam system (not leveled)

Refer to Figure 4-3-7 for the wall states before and after the weathering resistance test. Like No. 1 system, a large number of cracks appeared on the entire wall surface before this test, and expanded and increased after this test. The weathering resistance of this system does not conform to the standards.

(4) Analysis of test results

The anti-cracking mortar on the wall surface

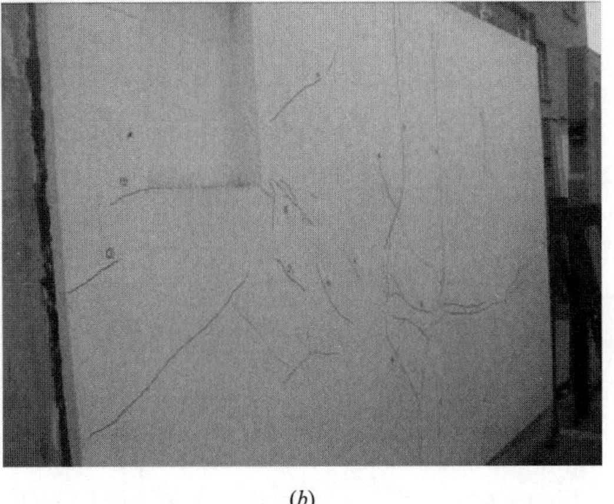

(*a*)

(*b*)

Figure 4-3-6 States of Wall with Sprayed Rigid Polyurethane Foam System (leveled) before and after Weathering Resistance Test

(*a*) curing stage; (*b*) after the weathering resistance test

was cracking in this test. Analysis results show that main reasons are as follows.

(i) The long shrinkage period and great shrinkage of polyurethane and also the constraints of the surface result in the stress concentration and further cracking of the anti-cracking mortar.

(ii) There is a significant difference between the deformation rates of the polyurethane and anti-cracking mortar. When the anti-cracking mortar is directly applied on the polyurethane surface, the thermal stress generated is far higher than the strength of the anti-cracking mortar itself, which will lead to serious cracking of the wall surface.

(iii) The performance indicators of the anti-cracking mortar itself cannot meet the corresponding performance requirements of the polyurethane system.

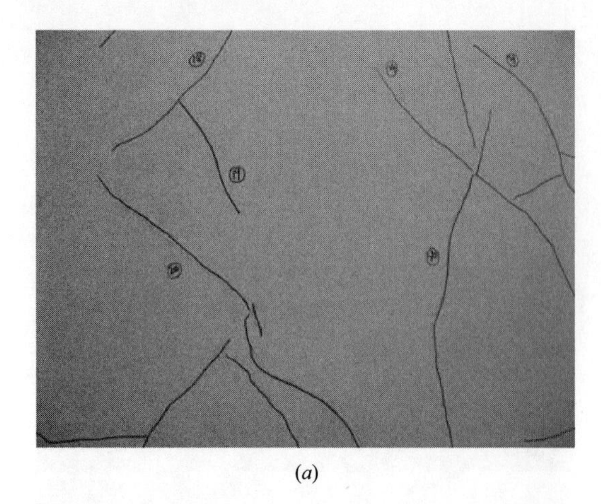

(*a*)

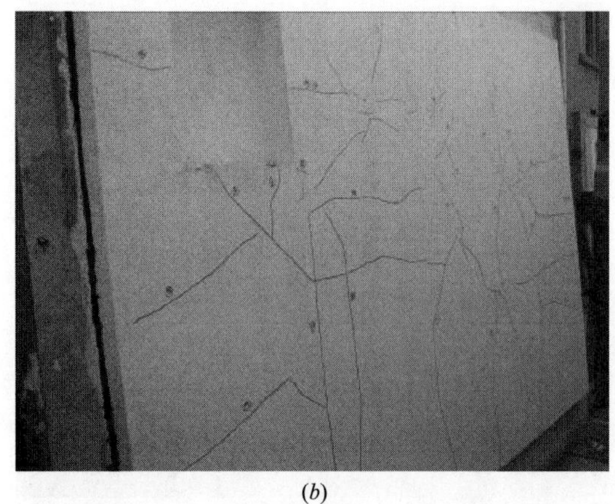

(*b*)

Figure 4-3-7 States of Wall with Sprayed Rigid Polyurethane Foam System (not leveled) before and after Weathering Resistance Test

(*a*) curing stage; (*b*) after the weathering resistance test

If an appropriate layer of adhesive polystyrene granule is compounded, the thermal stress can be dispersed well, thus reducing the cracking of anti-cracking mortar on the surface. This effect will be gradually enhanced with the thickness of adhesive polystyrene granule increasing. Accordingly, an appropriate adhesive polystyrene granule leveling layer has a strong absorption and dispersion effect on the stress generated as a result of the polyurethane shrinkage and surface temperature difference.

With the ambient temperature changing, the temperature of the insulation layer and protective layer of the adhesive polystyrene granule composite system varies less slowly than that of the thin plaster system without the adhesive polystyrene granule layer. This means that the adhesive polystyrene granule layer has the effects of deformation absorption and flexible layer-by-layer change, which can reduce the deformation rate difference of adjacent materials, thus greatly decreasing the shrinkage and thermal stress of boards and eliminating the cracking arising from non-coordinated deformation.

Cracking in the curing stage is mainly caused by polyurethane deformation after spraying. It is recommended, before surface construction, to keep the sprayed polyurethane paint finish system static until the polyurethane deformation is fully stabilized. If the adhesive polystyrene granule layer is used, the polyurethane surface can be leveled, and a transition can be formed between the polyurethane and anti-cracking mortar, which will reduce or eliminate cracking.

2. Record and analysis of the 12th weathering resistance test

The double-component surface mortar was applied on the left half of the wall surface and the dry surface mortar on the right half in the 12th weathering resistance test. There was no cracking or hollowing in the curing stage.

(1) Cracking and hollowing record

Refer to Table 4-3-7 for cracking and hollowing of the external thermal insulation system based on the polyurethane composite insulation board in the weathering resistance test.

Cracking and Hollowing Record of Weathering Resistance Test of External Thermal Insulation System based on Polyurethane Composite Insulation Board **Table 4-3-7**

Category	Before the Test	During the Test	After the Test
External thermal insulation system with polyurethane composite insulation board and inorganic insulation mortar plaster	No cracking or hollowing	Cracks appeared from the 4th heat-rain cycle, and expanded and increased later. Total number of cracks: 16; no hollowing	No crack expansion

Continued

Category	Before the Test	During the Test	After the Test
External thermal insulation system with polyurethane composite insulation board and double-layer glass fiber mesh	No cracking or hollowing	Cracks appeared from the 4th heat-rain cycle, and expanded and increased later. Total number of cracks: 12; no hollowing	No crack expansion
External thermal insulation system with polyurethane composite insulation board and adhesive polystyrene granule plaster	No cracking or hollowing	Cracks appeared from the 5th heat-rain cycle, and expanded and increased later. Total number of cracks: 10; no hollowing	No crack expansion
External thermal insulation system with polyurethane composite insulation board and thin plaster	No cracking or hollowing	Cracks appeared from the 5th heat-rain cycle. Total number of cracks: 15; no hollowing	No crack expansion

(2) Test results and analysis

Refer to Figure 4-3-8 to 4-3-11 for wall cracking after the weathering resistance test.

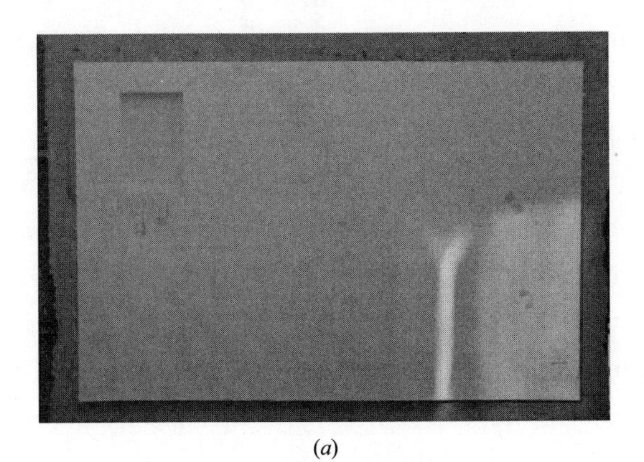

(*a*)

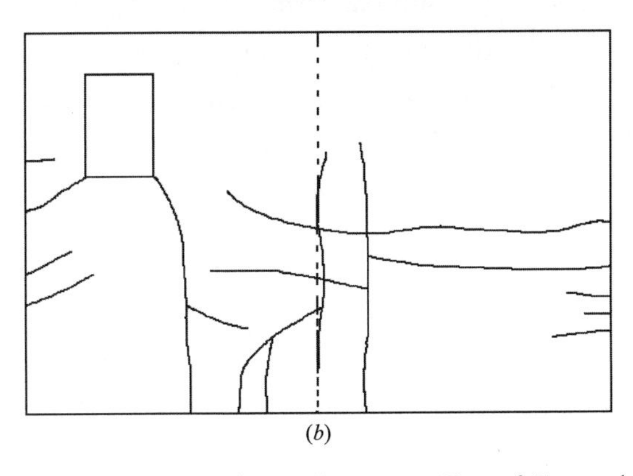

(*b*)

Figure 4-3-8 Weathering Resistance Test of External Thermal Insulation System with Polyurethane Composite Insulation Board and Inorganic Insulation Mortar Plaster

(*a*) test result; (*b*) cracking diagram

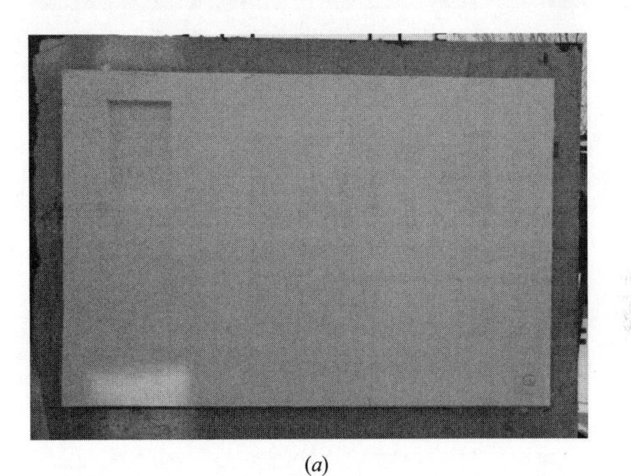

(*a*)

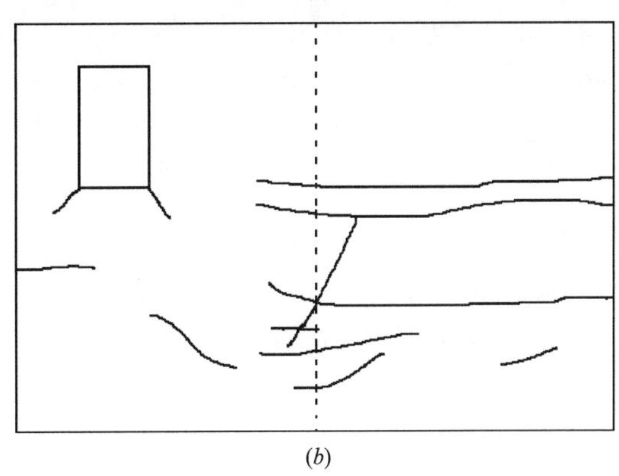

(*b*)

Figure 4-3-9 Weathering Resistance Test of External Thermal Insulation System with Polyurethane Composite Insulation Board and Double-layer Glass Fiber Mesh

(*a*) test result; (*b*) cracking diagram

As shown in Figure 4-3-8 and 4-3-9, there was no cracking or hollowing in the curing stage and cracking occurred from the 4th or 5th heat-rain cycle in this test. The external thermal insulation system with the polyurethane composite insulation board and inorganic insulation mortar plaster was cracking the most seriously, followed by the system with the polyurethane composite insulation board and thin plaster and then the system with the polyurethane composite insulation board and double-layer glass fiber mesh. The system with the 10mm adhesive polystyrene granule leveling layer was also cracking in the test, but there were a few cracks and no through cracks.

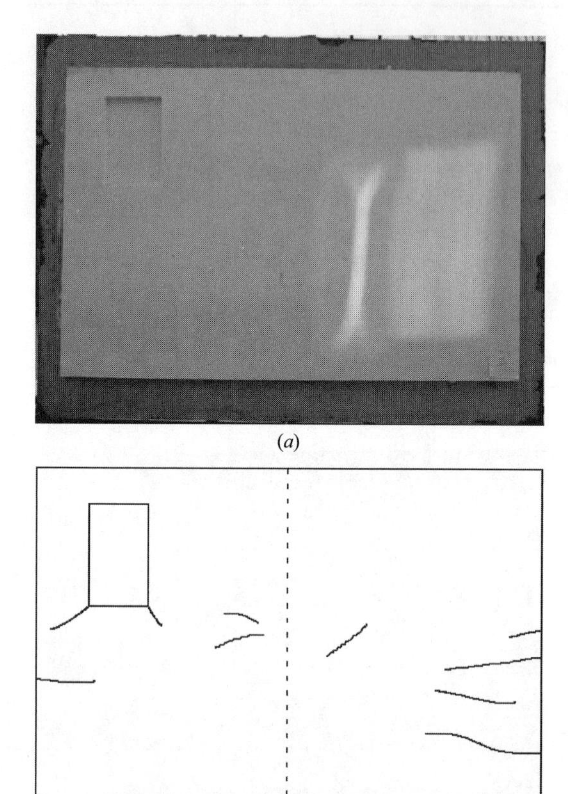

Figure 4-3-10 Weathering Resistance Test of External Thermal Insulation System with Polyurethane Composite Insulation Board and Adhesive polystyrene granule Plaster

(*a*) test result; (*b*) cracking diagram

The polyurethane composite insulation board has good insulation performance, but high strength, significant dimensional instability and large thermal deformation. It is risky to apply the method of point-and-frame bonding and thin plastering in the polyurethane composite insulation board system. Although cracking may be alleviated with the double-layer glass fiber mesh and reinforcing bars, there are still a large number of cracks in this system. When the inorganic insulation mortar (a kind of rigid material with high strength) is applied as a transition layer, the thermal stress can be reduced to some extent, but the strain generated in this system cannot be absorbed promptly and the stress cannot be released, which will easily result in cracking.

If the adhesive polystyrene granule is used as a transition layer, the surface deformation will be reduced greatly, and part of the stress will be released. Compared with the other systems, the external thermal insulation system with the adhesive polystyrene granule has higher resistance to cracking. In this test, however, the thickness of adhesive polystyrene granule was only 10mm, so cracks are not completely avoided. With reference to the results of the 6th weathering resistance test, the thickness of the adhesive polystyrene granule should be increased to fully avoid cracking.

From the perspective of overall cracking, the cracks on the side with double-component surface mortar are fewer than those on the side with dry-mixed surface mortar, and there are also fewer through cracks on the side with double-component surface mortar. The double-component surface mortar should be prepared with emulsion instead of mineral binder. The proportions of the liquid and mineral binder should be fixed. As additional water is not needed, the quality of the double-component surface mortar is easy to control. Meanwhile, the emulsion has higher flexibility than the mineral binder, so the surface mortar has higher flexibility and cracking resistance. Accordingly, the double-component surface mortar has higher cracking resistance than the dry-mixed surface mortar in the external thermal insulation system.

Figure 4-3-8 shows the pull strength of the external thermal insulation system after the weathering resistance test. Comparatively, the strong adhesion between the mortar and polyurethane composite insulation board is destructive to the latter. The strength of the adhesive polystyrene granule and the inorganic insulation mortar is not destructive to the polyurethane composite insulation board. The adhesive polystyrene granule layer is bonded well with the polyurethane composite insulation board, but the former may be damaged. The inorganic insulation mortar has high strength but is weakly bonded with the polyurethane composite insulation board, so the in-

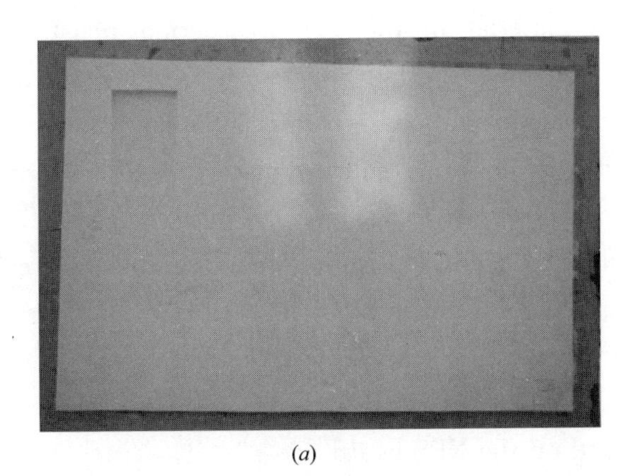

(*a*)

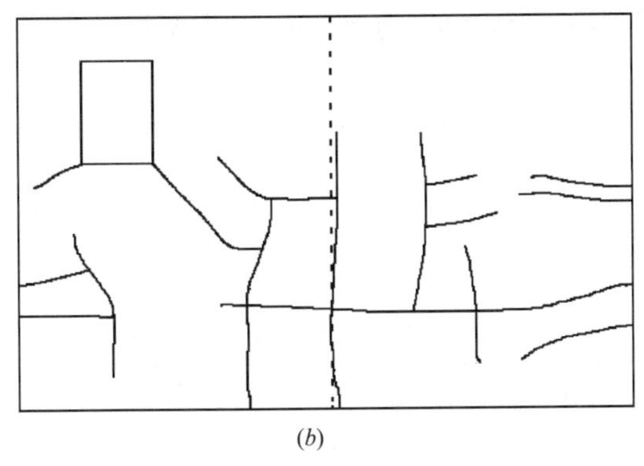

(*b*)

Figure 4-3-11 Weathering Resistance Test of External Thermal Insulation System with Polyurethane Composite Insulation Board and Thin Plaster

(*a*) test result; (*b*) cracking diagram

terface between them may be damaged.

Pull Strength and State of External Thermal Insulation System after Weathering Resistance Test Table 4-3-8

Photo of pull strength of external thermal insulation system based on the polyurethane composite insulation board and the inorganic insulation mortar plaster/Average pull strength: 0.05MPa	Photo of pull strength of external thermal insulation system based on the polyurethane composite insulation board and the double-layer glass fiber mesh/Average pull strength: 0.12MPa

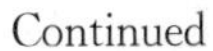

Continued

	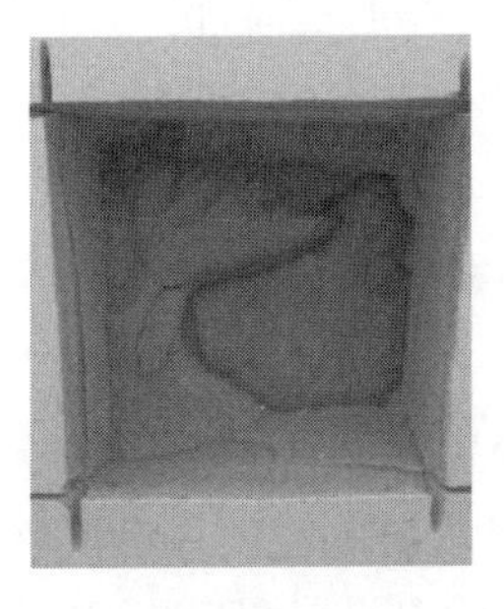
Photo of pull strength of external thermal insulation system based on the polyurethane composite insulation board and the adhesive polystyrene granule plaster/Average pull strength: 0.10MPa	Photo of pull strength of external thermal insulation system based on the polyurethane composite insulation board and the thin plaster/Average pull strength: 0.12MPa

4.3.2.4 Summary

(1) Due to deformation of the polyurethane sprayed on the site and shrinkage in this long process, cracking may occur in the curing stage. It is recommended, before surface construction, to keep the sprayed polyurethane paint finish system static until the polyurethane deformation is fully stabilized.

(2) In the weathering resistance test, it is risky to simply conduct point-and-frame bonding and thin plastering in the polyurethane composite insulation board system. Although cracking may be alleviated with the double-layer glass fiber mesh and reinforcing bars, there are still a large number of cracks in this system. When the inorganic insulation mortar (a kind of rigid material with high strength) is applied as a transition layer, the thermal stress can be reduced to some extent, but the strain generated in this system cannot be absorbed promptly and the stress cannot be released, which will easily result in cracking.

(3) Regardless of the on-site spraying of polyurethane or use of the polyurethane board, the thermal stress can be properly dispersed if an appropriate layer of adhesive polystyrene granule is used on the surface of the insulation layer. At the same time, the adhesive polystyrene granule layer has the effects of deformation absorption and flexible layer-by-layer

change, which can reduce the deformation rate difference of adjacent materials, thus greatly decreasing the cracking of the anti-cracking surface mortar. With the thickness of adhesive polystyrene granule increasing, the above effect will become better.

(4) The double-component surface mortar with higher flexibility and cracking resistance is superior to the dry-mixed surface mortar in the external thermal insulation system.

4.3.3 Analysis Report of Weathering Resistance Test of Extruded Polystyrene (XPS) Board Insulation System

4.3.3.1 Test Purpose

The XPS board was used as the main insulation material in the 11th weathering resistance test. The weathering resistances of the external thermal insulation systems involving different structures of one insulation material (XPS board) were compared to analyze the impact of the full bonding, point-and-frame bonding, thin plastering, mineral binder as well as EPS granule transition and its thickness on the weathering resistance of the external thermal insulation system based on the XPS board.

4.3.3.2 System Structure and Material Selection

Refer to Table 4-3-9 for the system structure and material selection.

Structure and Material Selection of External Thermal Insulation System based on XPS Board Table 4-3-9

System	Structure						
	Base course wall	Interface layer	Bonding layer	Insulation layer	Leveling layer	Anti-cracking layer	Finish layer
Paint finish system based on point-and-frame bonded and thin-plastered XPS board (hereinafter referred to as the point-and-frame bonding system)	Concrete wall	None	5mm XPS bonding mortar	60mm XPS board	None	Dry-mixed surface mortar + alkali-resistant glass fiber mesh	Waterproof flexible putty+paint
Paint finish system based on "LB type" adhesive polystyrene granule and pasted XPS board (hereinafter referred to as the "LB type" system)	Concrete wall	Interface mortar	15mm adhesive polystyrene granule	50mm XPS board	None	Dry-mixed anti-cracking mortar+alkali-resistant glass fiber mesh +elastic primer	Waterproof flexible putty+paint
Paint finish system based on "LBL type" adhesive polystyrene granule and pasted XPS board (hereinafter referred to as the "LBL type" system 1)	Concrete wall	Interface mortar	15mm adhesive polystyrene granule	50mm XPS board	10mm adhesive polystyrene granule	anti-cracking mortar + alkali-resistant glass fiber mesh + elastic primer	Waterproof flexible putty+paint
Paint finish system based on "LBL type" adhesive polystyrene granule and pasted XPS board (hereinafter referred to as the "LBL type" system 2)	Concrete wall	Interface mortar	15mm adhesive polystyrene granule	60mm XPS board	30mm adhesive polystyrene granule	anti-cracking mortar + alkali-resistant glass fiber mesh + elastic primer	Waterproof flexible putty+paint

Note: The XPS boards (600mm×450mm) with two holes were applied in the "LBL type" and "LB type" system. The fireproof interface agent was applied on two sides of XPS boards, and 10mm joints were left between boards, and filled and compacted with the adhesive polystyrene granule. The XPS boards of 600mm×900mm were applied by means of point-and-frame bonding in the point-and-frame bonding XPS board system, without joints, but the fireproof interface agent was applied on two sides of XPS boards.

4.3.3.3 Weathering Resistance Test Record and Analysis

1. Cracking and hollowing records

Cracking did not occur to the wall in the curing stage but occurred in the weathering resistance test, but cracks are significantly different from each other. Refer to Table 4-3-10 for details.

Cracking and Hollowing Record of Weathering Resistance Test of External Thermal Insulation System based on XPS Board　　Table 4-3-10

Category	Before the Test	During the Test	After the Test
Point-and-frame bonding system	No cracking or hollowing	Thick cracks (through type) appeared in the 32nd heat-rain cycle. Total number of cracks: 6; no hollowing	No crack expansion
"LB type" system	No cracking or hollowing	Thin cracks appeared in the 20th heat-rain cycle. Total number of cracks: 6; no hollowing	No crack expansion
"LBL type" system 1	No cracking or hollowing	Cracks appeared from the 2nd heat-rain cycle, and expanded and increased later. Total number of cracks: 4; no hollowing	No crack expansion
"LBL type" system 2	No cracking or hollowing	Total number of cracks: 0; no hollowing	No expansion

2. Temperature curve record and analysis

Figure 4-3-12 shows the temperature curves of the external surfaces of walls with four XPS board systems, recorded by the software of the weathering resistance tester in one steady period of the heat-rain cycle test.

Figure 4-3-13 shows the curves of the air temperature inside the box and the external surface of the exterior finish, obtained by means of numerical simulation in the same period of the heat-rain cycle of four XPS board systems. It can be seen in this figure that the change trend of the image of numerical simulation is basically consistent with that of the test results.

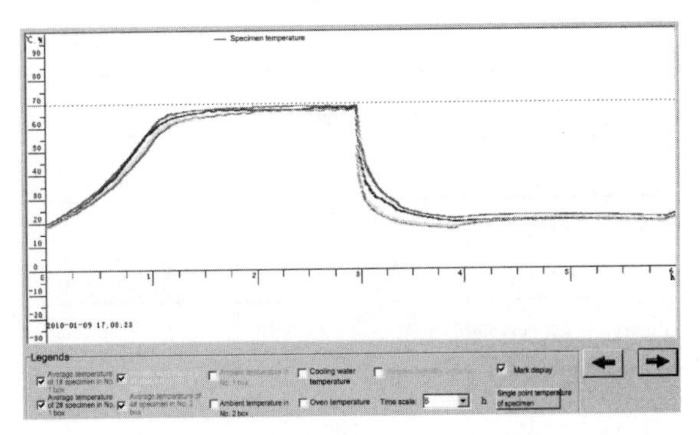

Figure 4-3-12　Temperature Recorded by Tester in One Steady Period of Heat-rain Cycle

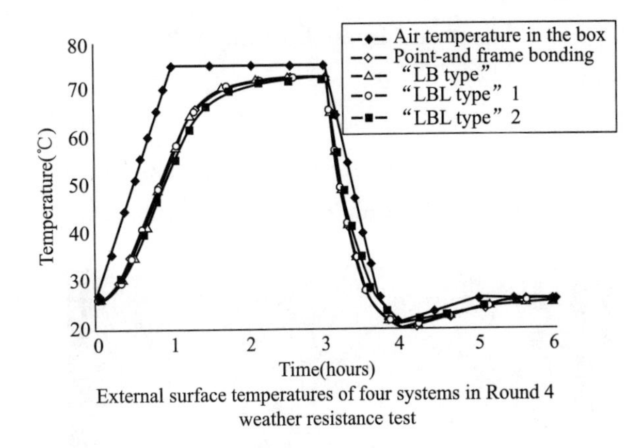

Figure 4-3-13　Temperature of External Surface of Each Wall with XPS Board System in One Period of Heat-rain Cycle

As shown in Figure 4-3-13, the temperature of the exterior finish of the "LBL type" system rises (drops) less slowly than that of the other three systems. This is because the thermal conductivity of the adhesive polystyrene granule leveling layer is higher than that of the XPS board but lower than that of the surface mortar (see Table 4-3-11 for the thermal conductivity of insulation materials), thus achieving good transitional effects. It is conducive to the slowdown of change in the temperature of the anti-cracking layer and exterior finish and reduction of the probability of cracking at low temperature (temperature cracks of the external thermal insulation mainly appear in the anti-cracking layer and finish layer).

Thermal Conductivity of Insulation Materials

Table 4-3-11

Insulation Material	Thermal Conductivity [W/(m·K)]	Multiple of Difference from Anti-cracking Mortar
Anti-cracking mortar	0.93	1
Adhesive polystyrene granule mortar	0.075	12.4
XPS board	0.030	31.0

3. Test results

(1) Point-and-frame bonding XPS board system

Refer to Figure 4-3-14 for the wall after the weathering resistance test. Thick cracks through the wall occurred in board gaps after this test.

Figure 4-3-14 After the Weathering Resistance Test of Point-and-frame Bonding XPS Board System

(2) "LB type" system

Refer to Figure 4-3-15 for the wall after the weathering resistance test. There were minor cracks in windows after this test and then local cracks in board gaps.

(3) "LBL type" system 1

Refer to Figure 4-3-16 for the wall after the weathering resistance test. There were minor cracks in windows after this test, and then short, thin and irregular local cracks.

(4) "LBL type" system 2

Refer to Figure 4-3-17 for the wall after the weathering resistance test. There was no cracking or hollowing after this test.

Figure 4-3-15 After the Weathering Resistance Test of "LB Type" System

Figure 4-3-16 After the Weathering Resistance Test of "LBL Type" System 1

Figure 4-3-17 After the Weathering Resistance Test of "LBL Type" System 2

The weathering resistance test results of different structures of the XPS board system are significantly different from each other, provided that the performance indicators of the material conform to the standards. There are long through cracks in board gaps of the point-and-frame bonding system; the weathering resistance of the "LB type" system is far superior to that of the point-and-frame bonding system, but there are short and thin cracks instead of through cracks; the cracks of the "LBL type" system 1 are thinner and shorter than those of the "LB type" system; and the "LBL type" system 2 has excellent resistance to weathering, so there was no cracking in this system.

4. Analysis of test results

The weathering resistance test results show that, with the thickness of adhesive polystyrene granule plaster increasing outside the insulation board, the weathering resistance of the external thermal insulation system is greatly improved. The XPS board is quite different from the surface mortar in the linear expansion coefficient and elastic modulus. Because of the deformation rate difference between adjacent materials, the thermal and wet stress will be concentrated. When the stress exceeds the strength of surface mortar, cracking will occur to the insulation system. Therefore, the 30mm insulation mortar should be applied outside the XPS boards to form a composite insulation layer, followed by thin plastering, to avoid direct contact between the XPS boards and surface mortar. With the transition layer, the deformation rate different between adjacent materials will be decreased, thus synchronizing the deformation of structural layers and reducing the shear stress caused by the deformation rate difference.

The joints of XPS boards should be treated properly. The results of comparison between the point-and-frame bonding and thin plastering system and the "LB type" system show that if there are joints between insulation boards, surface and through cracks can be reduced. The sub-elastic adhesive polystyrene granule layer, with the deformation amount between those of the mortar and XPS board, is able to absorb some deformation and release part of stress caused by board deformation. Meantime, its stability is helpful to restrict and decrease the board deformation and reduce the probability of cracking.

(1) Thermal stress mechanism

Wang Zhaojun, et al. from the School of Materials Science and Engineering of Beijing University of Technology analyzed the thermal deformation of EPS and XPS boards. Their deformation is shown in Figure 4-3-18. The EPS board is subject to shrinkage deformation within the temperature range from 30℃ to 100℃ and the XPS board to expansion deformation within the temperature range from 45℃ to 106℃. Refer to Figure 4-3-19 for the expansion deformation of the XPS board 10min after the ambient temperature rises from 30℃ to 105℃.

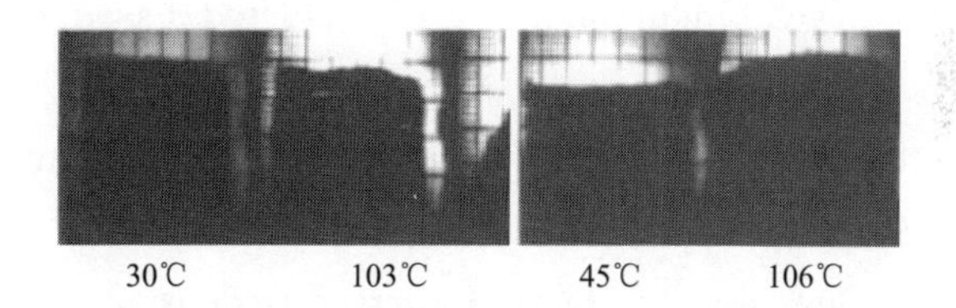

Figure 4-3-18 Thermal Deformation Process of EPS and XPS Boards (left: EPS board; right: XPS board)

Figure 4-3-19 Dimensional Changes of XPS Board in 10min under Ambient Temperature of 30℃ to 105℃

Figure 4-3-20 shows the extrusion deformation caused by thermal expansion of XPS boards of a wall on one construction site in summer.

When heated, the XPS board expands while the EPS board shrinks. This indicates that their thermal deformation mechanisms are different from each other. Therefore, the thin-plastered

Figure 4-3-20 Stress Caused by Excessive Temperature Difference between Internal and External Surfaces of XPS Board

XPS board system must not be specified by directly using the requirements of the thin-plastered EPS board system. Instead, the targeted study and development should be conducted based on the actual conditions of XPS boards.

(2) Analysis of stress generated by the surface temperature difference of XPS board

Due to its high (surface) density and strength, the XPS board has large stress in the case of deformation or under the temperature difference. For the point-and-frame bonded and thin-plastered XPS board system, the linear expansion coefficient of the surface mortar greatly varies from that of the XPS board as the difference in their thermal conductivities is 31 times. The temperature of the external surface of the outer wall varies significantly (up to 50℃) in summer. The thermal stress is generated in the event of a sudden shower after long-term exposure to the sunlight, which will lead to the difference in the deformation rates of adjacent materials as well as the thermal expansion and contraction of such materials. After the instable thermal expansion and contraction in a long time, there will be shearing forces between the polymer mortar and XPS board, which affects the bonding strength. Both cracking and hollowing occur when the bonding strength of the surface is less than the thermal stress (Figure 4-3-21).

Figure 4-3-21 Cracking and Hollowing of Surface Mortar Layer

The shear stress is generated at the interface between different materials subject to thermal deformation under constraints. The premise of significant relative deformation is that there are great differences in the temperature and thermal expansion coefficients of two materials, and the essential condition for high shear stress is great constraints. In the thin-plastered XPS board system, the temperature of the inner side of the XPS board varies little, so the thermal stress of each interface is low. The temperature of the outer side of the XPS board varies much, resulting in great differences in the thermal expansion coefficient of the surface mortar interface of the XPS board. Because of its large elastic modulus, the XPS board has strong constraints to the surface mortar, which leads to large shear stress of the interface. In addition, a large amount of stress is concentrated in board gaps, which causes the stress instability and cracking. The strength of the XPS board is twice that of the EPS board, and its elastic modulus is at least 20MPa. Under the same temperature difference, the XPS board can generate more stress than the EPS board.

The linear expansion coefficient of the XPS board is 0.07mm/(m·K) or more. That is, the expansion/contraction value per meter is 0.07mm/(m·K)×50K=3.5mm when the temperature rises or drops by 50℃ under normal conditions. The surface temperature of the outer wall is about 70℃ in summer. In the case of thin plastering, the anti-cracking layer and finish layer has no insulating effect, so the temperature of the outer side of the XPS board is about 70℃ and that of the inner side is basically 20～30℃. The significant temperature difference between the inner and external surfaces results in warping of the XPS board.

The deformation of the XPS board should be reduced, restricted and dispersed as much as possible.

If the adhesive polystyrene granule transition layer with insulating effects is applied between the XPS board and anti-cracking mortar, the temperature difference between the inner and external surfaces as well as the thermal deformation of the board can be reduced. In addition, the adhesive polystyrene granule layer with good volume stability and appropriate sub-elasticity is able to absorb part of deformation and release the stress, thus improving the cracking resistance of the surface.

If the base course wall is fully bonded, the deformation of the XPS board can be limited, reducing its deformation. If two holes are drilled at the center of the XPS board, the vapor permeability of the insulation system can be improved, and the deformation of the XPS board can be limited and reduced due to anchorage and restriction of the adhesive polystyrene granule in the two holes.

The deformation of the XPS board can be "divided" by reducing its size, in order to promptly release the stress, avoid the accumulation of deformation, and also reduce cracking risks arising from excessive deformation in board gaps.

4.3.3.4 Conclusions

(1) The strength and elastic modulus of the XPS board are greater than those of the EPS board. The principle of dimensional deformation of the former is opposite to that of the latter when heated. Therefore, the weathering resistance of the external thermal insulation system based on the XPS board system should be higher than that based on the EPS board.

(2) The deformation of the XPS board can be reduced, restricted and dispersed by plastering the insulation mortar outside the XPS board, applying the 10mm adhesive polystyrene granule in XPS board gaps, fully bonding the XPS board with the adhesive polystyrene granule, reducing the size of the XPS board, etc., to improve the weathering resistance of the external thermal insulation system based on the XPS board.

4.3.4 Comparative Analysis of Weathering Tests of External Thermal Insulation Systems based on EPS Board, XPS Board and Polyurethane

4.3.4.1 Test purpose

Typical insulation materials were analyzed in this report, namely, the polyurethane, EPS board and XPS board, involved in the 6th, 7th and 11th weathering resistance test, respectively. In all the tests, the paint finish was applied, the 4mm anti-cracking mortar layer used was compounded with the alkali-resistant glass fiber mesh, and the interface of the main insulation material was treated with the interface agent in the external thermal insulation system containing the adhesive polystyrene granule.

4.3.4.2 System Structure and Test Results

1. System with 30mm adhesive polystyrene granule leveling layer

Refer to Table 4-3-12 for the structure of the external thermal insulation system with the 30mm adhesive polystyrene granule leveling layer on the surface in the weathering resistance test. Figure 4-3-22 to 4-3-24 illustrate this system after the weathering resistance test, in which cracking did not occur.

Main Structural Layers Involved in Weathering Resistance Test **Table 4-3-12**

No.	Bonding Layer	Insulation Layer	Leveling Layer	Finish Layer	Remarks
6.1	—	40mm sprayed rigid polyurethane foam	30mm adhesive polystyrene granule	Paint	The polyurethane surface is not leveled.

Continued

No.	Bonding Layer	Insulation Layer	Leveling Layer	Finish Layer	Remarks
7. 1	15mm adhesive polystyrene granule	75mm EPS board	30mm adhesive polystyrene granule	Paint	—
11. 3	15mm adhesive polystyrene granule	60mm XPS board	30mm adhesive polystyrene granule	Paint	—

Figure 4-3-22 External Thermal Insulation System based on Sprayed Polyurethane (30mm adhesive polystyrene granule leveling layer) (No. 6. 1)

Figure 4-3-23 External Thermal Insulation System based on Pasted Polyurethane (30mm adhesive polystyrene granule leveling layer) (No. 7. 1)

2. System with 10mm adhesive polystyrene granule leveling layer

Refer to Table 4-3-13 for the structure of the external thermal insulation system with the 10mm adhesive polystyrene granule leveling layer on the surface in the weathering resistance test. Figure 4-3-25 to 4-3-27 illustrate this system after the weathering resistance test.

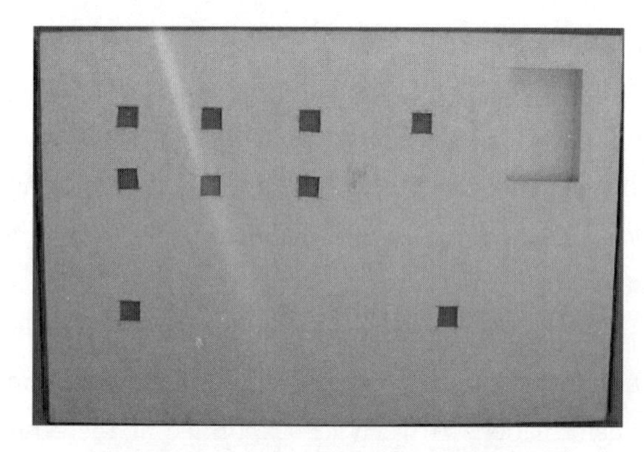

Figure 4-3-24 Pasted XPS Board System (30mm adhesive polystyrene granule leveling layer) (No. 11. 3)

3. System without adhesive polystyrene granule leveling layer

Refer to Table 4-3-14 for the structure of the external thermal insulation system without the adhesive polystyrene granule leveling layer on the surface in the weathering resistance test. Figure 4-3-28 to 4-3-33 illustrate this system after the weathering resistance test.

Main Structural Layers Involved in Weathering Resistance Test **Table 4-3-13**

No.	Bonding Layer	Insulation Layer	Leveling Layer	Finish Layer	Remarks
6. 2	—	40mm sprayed rigid polyurethane foam	10mm adhesive polystyrene granule	Paint	The polyurethane surface is not leveled
7. 2	15mm adhesive polystyrene granule	85mm EPS board	10mm adhesive polystyrene granule	Paint	—
11. 1	15mm adhesive polystyrene granule	60mm XPS board	10mm adhesive polystyrene granule	Paint	—

Main Structural Layers Involved in Weathering Resistance Test **Table 4-3-14**

No.	Bonding Layer	Insulation Layer	Leveling Layer	Finish Layer	Remarks
6.3	—	40mm sprayed rigid polyurethane foam	—	Paint	The polyurethane surface is leveled
6.4	—	40mm sprayed rigid polyurethane foam	—	Paint	The polyurethane surface is not leveled
7.3	15mm adhesive polystyrene granule	90mm EPS board	—	Paint	—
7.4	5mm bonding mortar	100mm EPS board	—	Paint	Rock wool type fire barrier
11.2	15mm adhesive polystyrene granule	60mm XPS board	—	Paint	—
11.4	5mm bonding mortar	60mm XPS board	—	Paint	—

Figure 4-3-25 External Thermal Insulation System based on Sprayed Polyurethane (10mm adhesive polystyrene granule leveling layer) (No. 6. 2)
(Cracks appeared from the 8th heat-rain cycle, and expanded and increased later. Total number of cracks: 8; no hollowing)

Figure 4-3-26 Pasted EPS Board System (10mm adhesive polystyrene granule leveling layer) (No. 7. 2)
(There was no cracking or hollowing in the heat-rain cycle. The 20mm long cracks appeared in the 45° lower right corner of the window in the 7th heat-cold cycle. Total number of cracks: 1; no hollowing)

4. 3. 4. 3 Analysis of test results

Table 4-3-15 illustrates the cracking of different adhesive polystyrene granule layers fully bonded on the surfaces made with three insulation materials. As shown in this table:

(1) For one insulation material, the number of cracks decreases with the thickness of the adhesive polystyrene granule layer increasing.

Total Cracks of Each System subject to Full Bonding in Weathering Resistance Test **Table 4-3-15**

Thickness of Adhesive polystyrene granule Layer	30mm	10mm	0mm
Sprayed polyurethane	0	8	20/31
EPS board	0	1	1
XPS board	0	4	6

(2) The external thermal insulation systems based on three insulation materials were compared

Figure 4-3-27 Pasted XPS Board System (10mm adhesive polystyrene granule leveling layer) (No. 11. 1)
(Cracks appeared from the 2nd heat-rain cycle, and expanded and increased later. Total number of cracks: 4; no hollowing)

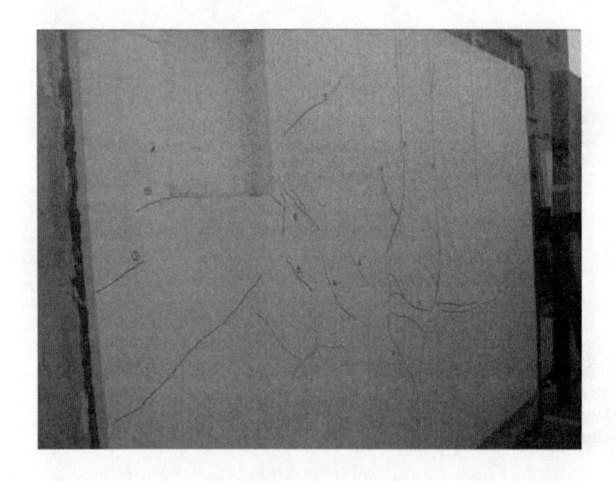

Figure 4-3-28 External Thermal Insulation System based on Sprayed Polyurethane and Thin Plaster (with the polyurethane surface leveled) (No. 6. 3)
(There were 15 cracks in the curing stage; and new cracks appeared from the 2nd heat-rain cycle, and expanded and increased later. Total number of cracks: 15; no hollowing)

Figure 4-3-29 External Thermal Insulation System based on Sprayed Polyurethane and Thin Plaster (with the polyurethane surface not leveled) (No. 6. 4)
(There were 20 cracks in the curing stage; and new cracks appeared from the 2nd heat-rain cycle, and expanded and increased later. Total number of cracks: 31; no hollowing)

Figure 4-3-30 Pasted EPS Board System (without the adhesive polystyrene granule leveling layer) (No. 7. 3)
(The upper right corner of the wall was damaged before this test; bubbles were produced mainly in the spot and zone form under the damaged part and in the window; and cracks appeared in the 62nd heat-rain cycle but did not expand. Total number of cracks: 1)

in parallel, with the 0mm and 10mm adhesive polystyrene granule leveling layer. The EPS board system has the fewest cracks, followed sequentially by the XPS board system and sprayed polyurethane system.

(3) When the 30mm adhesive polystyrene granule leveling layer is applied, there were no cracks on the surfaces of external thermal insulation systems based on these three insulation materials.

When a transition layer of insulation mortar is applied between the anti-cracking layer and insulation layer of the external thermal insulation structure, the temperature change rates of the anti-cracking layer and insulation layer can be reduced effectively, with the ambient temperature changing. This will decrease the thermal stress and strain on the surface of the insulation layer and improve the weathering resistance of this sys-

Figure 4-3-31 Point-and-frame Bonded and Thin-plastered EPS Board System (without the adhesive polystyrene granule leveling layer) (No. 7. 4)
(Cracks appeared from the 52nd cycle, and expanded and increased later. Total number of cracks: 6)

Figure 4-3-32 Point-and-frame Bonded and Thin-plastered EPS Board System (without the adhesive polystyrene granule leveling layer) (No. 11. 2)
(Thin minor cracks appeared in the 20th heat-rain cycle. Total number of cracks: 6; no hollowing)

tem.

The sub-elasticity of the adhesive polystyrene granule leveling layer (transition) falls between those of the organic insulation material and mortar. Due to flexible deformation of each layer, the external thermal insulation system is able to absorb the strain and release the stress, thereby avoiding cracking and improving the weathering resistance.

Refer to Table 4-3-16 for the results of the full bonding and point-and-frame bonding with

Figure 4-3-33 Point-and-frame Bonded and Thin-plastered EPS Board System (without the adhesive polystyrene granule leveling layer) (No. 11. 4)
(Thick through cracks appeared in the 32nd heat-rain cycle. Total number of cracks: 6; no hollowing)

the thin plaster. As shown in this table:

(1) In the thin plaster system without the adhesive polystyrene granule leveling layer on the surface, there are fewer cracks in the EPS board system subject to full bonding than that subject to point-and-frame bonding.

(2) There are six cracks in the XPS board systems subject to full bonding and point-and-frame bonding. In detailed, there are thin and short cracks in the case of full bonding, but through cracks in the case of point-and-frame bonding, which is more serious comparatively. Refer to Table 4-3-1 for details.

(3) In the case of full bonding, there are the fewest cracks in the EPS board system, followed sequentially by the XPS board system and sprayed polyurethane system.

As the insulation material can be wholly and uniformly bonded in the case of full bonding, the board deformation can be better restricted and reduced, thus decreasing the cracks.

4. 3. 4. 4 Conclusions

(1) Under the same structural conditions, the external thermal insulation system based on the EPS board is the most stable, followed sequentially by those based on the XPS board sys-

tem and sprayed polyurethane.

Total Cracks of Different Thin Plaster Systems subject to Full Bonding and Point-and-frame Bonding in Weathering Resistance Test

Table 4-3-16

System without Adhesive polystyrene granule Layer	Full Bonding	Point-and-frame Bonding
Sprayed polyurethane	20/31	No corresponding test
EPS board	1	6
XPS board	6(short and thin cracks)	6(through cracks)

(2) For one insulation material, the thicker the mineral binder and EPG granule leveling layer, the fewer cracks are produced in the weathering resistance test, and the more stable the external thermal insulation system is.

(3) Full bonding is superior to point-and-frame bonding, which can restrict the board deformation and improve the weathering resistance of the external thermal insulation system.

(4) The overall sequence of stability is EPS board, XPS board and sprayed polyurethane, so more reasonable targeted solutions should be provided depending on the insulation material, instead of just using one structure. The structure of the external thermal insulation systems based on the XPS board and polyurethane should be properly enhanced to meet their durability and safety requirements.

5. Study on Fire Resistance of External Thermal Insulation System

The fire safety of the external thermal insulation system is associated with the safety of people's life and property. It has long attracted the attention of industry insiders. The research, design, production and construction units of China's external insulation industry have been jointly conducting research on the fire safety of the external thermal insulation system since 2004.

In 2006, Beijing Zhenli High-tech Co., Ltd., Building Fire Prevention Research Institute of Beijing Building Research Institute of CSCEC, Science and Technology Development Promotion Center of the Ministry of Housing and Urban-Rural Development, Beijing Liujian Group, China Building Materials Academy Co., Ltd., Beijing Quality Supervision and Test Center of Fire Products, Beijing Building Design Standardization Office and Tsinghua University took the lead in applying for and undertaking the "Technology Research on Fire Protection Test Methods, Fire Rating Evaluation Standards and Architectural Applications of External Wall Insulation Systems" (06-K5-35), one scientific research project of the Ministry of Housing and Urban-Rural Development, and made the groundbreaking achievements suitable for China's national conditions. This project was formally accepted by experts in September 2007 and received high praise. Five conclusive suggestions were put forwards in this project.

(1) The fire safety of the external thermal insulation system should be an essential condition for applications of the external thermal insulation technology;

(2) The fire resistance of the overall structure is critical to the fire safety of the external thermal insulation system;

(3) Three key factors for fire protection of the external thermal insulation system are as follows: no cavity, fire partition and protective surface;

(4) The large-scale window fire test is an effective method to inspect the fire resistance of the external thermal insulation system;

(5) The classification of fire ratings of external thermal insulation systems and the provisions on their applicable building heights are effective means to improve the fire safety.

The experimental research in this project plays a guiding and promoting role in development of China's fire protection technologies for external thermal insulation. Later, with the active participation and strong support of associate organizations which are socially responsible and committed to fire protection of external thermal insulation, more than 40 large-sized model fire tests have been performed in three newly-built fire simulation test bases in Beijing, setting off a wave of large-scale researches on the fire safety of external thermal insulation systems in China. These researches have received the vigorous engagement of the governments and fire departments, extensive attention from all walks of life, and also high appraisals from foreign experts.

More than six years of hard work has witnessed the completion of a large number of fire tests and studies (including the cone calorimeter test, combustion shaft furnace test, window fire test and corner fire test), accumulation of a wealth of test data, and also phased achievements. In the fire tests, China's construction industry standard "Test Method for Fire-resistant Performance of External Wall Insulation Systems Applied to Building Facade" (GB/T 29416-2012) has been formulated and released based on digestion and absorption of foreign test techniques.

This chapter, by analyzing the fire safety of external thermal insulation systems, puts forward the concept and key elements of overall fire protection of external thermal insulation systems, and describes the test achievements made in large-, medium-and small-scale fire tests, and finally introduces the technology research on the fire rating classification and applicable building heights of external thermal insulation systems.

5.1 Fire Safety Analysis of External Thermal Insulation System

5.1.1 Status of Application of External Insulation Materials

From the perspective of combustibility, the insulation materials used in the outer walls of buildings can be divided into three categories: 1, inorganic insulation materials represented by mineral wool and rock wool, usually regarded as non-combustible materials; 2, organic-inorganic composite insulation materials represented by the adhesive polystyrene granule insulation mortar, usually regarded as flame-retardant materials; and 3, organic insulation materials represented by the polystyrene foam (including EPS and XPS boards), rigid polyurethane foam and modified phenolic resin, usually regarded as combustible materials. Refer to Table 5-1-1 for details.

5.1.1.1 Combustibility of Non-combustible Materials such as Rock Wool and Mineral Wool

The thermal conductivity of rock wool and mineral wool at room temperature (around 25℃) is usually 0.036—0.041W/(m・K). These materials are a kind of non-combustible inorganic silicate fiber. Although organic materials such as binders or adhesives used in the production process has some impact on the combustibility of related products, these materials are still regarded as non-combustible materials in general.

5.1.1.2 Combustibility of Adhesive polystyrene granule Mortar

The adhesive polystyrene granule mortar meeting the *Products for External Thermal Insulation Systems based on Mineral Binder and Expanded Polystyrene Granule Plaster* (JG/T 158-2013) is a kind of organic and inorganic composite insulation material, in which the volume of EPS granules accounts for about 80%. The adhesive polystyrene granule insulation mortar is a kind of flame-retardant material, with the thermal conductivity below 0.06W/(m・K) and combustibility of Class B_1; and the adhesive polystyrene granule pasting mortar is a kind of non-combustible material with the thermal conductivity below 0.08W/(m・K) and combustibility of Class A. When the adhesive polystyrene granule mortar is heated, internal EPS granules are usually softened and molten, but combustion will

Combustibility Rating and Thermal Conductivity of Insulation Materials **Table 5-1-1**

Material Name	Adhesive polystyrene granule	EPS Board	XPS Board	Polyurethane	Rock Wool	Mineral Wool	Foam Glass	Aerated Concrete
Thermal conductivity W/(m・K)	0.06	0.041	0.030	0.025	0.036—0.041	0.053	0.066	0.116—0.212
Combustibility rating	B1	B2	B2	B2	A	A	A	A

not occur. As EPS granules are encapsulated by inorganic materials, a closed cavity will be formed after melting. In this case, this insulation material will has lower thermal conductivity and involves slower heat transfer. However, there is hardly change in the volume of this insulation material during heating.

5.1.1.3 Combustibility of Organic Insulation Materials

Organic insulation materials are generally regarded as high-efficiency insulation materials, with low thermal conductivity. At present, three types of organic insulation material are mainly applied in China, namely, polystyrene foam (including EPS and XPS board), rigid polyurethane foam and modified phenolic resin board. Among them, polystyrene foam is a kind of thermoplastic material. It will first shrink and melt in the presence of fire or heat, followed by burning, but there is almost no residue after burning. The rigid polyurethane foam and modified phenolic resin board are thermosetting materials. In the presence of fire or heat, they hardly melt but burn into charcoal, with little change in the volume. Normally, the organic insulation materials for building insulation should have the combustibility of Class B_2 at least.

5.1.1.4 Status Quo at Home and Abroad

External thermal insulation systems have been used in Europe and the United States for decades. They are technologically mature, and a lot of research on their fire safety has been done. Currently, the external thermal insulation system based on the EPS plaster still plays an important role. Figure 5-1-1 shows the market shares of various external thermal insulation systems in the German market in 2006, of which EPS systems accounted for 87.4% and rock wool systems for 11.6%. The results of exchange with German External Insulation Association in 2010 are as follows: the EPS systems accounted for about 82% of market shares, rock wool systems for about 15% and other systems for about 3% to 4%.

From 2008 to 2009, Beijing Uni-construction

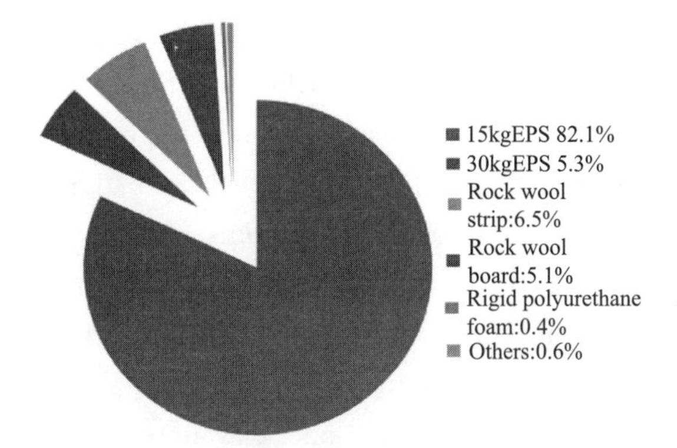

Figure 5-1-1 Market Shares of External Thermal Insulation Systems in the German market in 2006

Group conducted a survey of 43 projects ($1256000m^2$ in total) under construction in Beijing. Results show that organic insulation materials accounted for 97% in external wall insulation applications. Figure 5-1-2 illustrates the shares of materials applied in the external thermal insulation systems in Beijing.

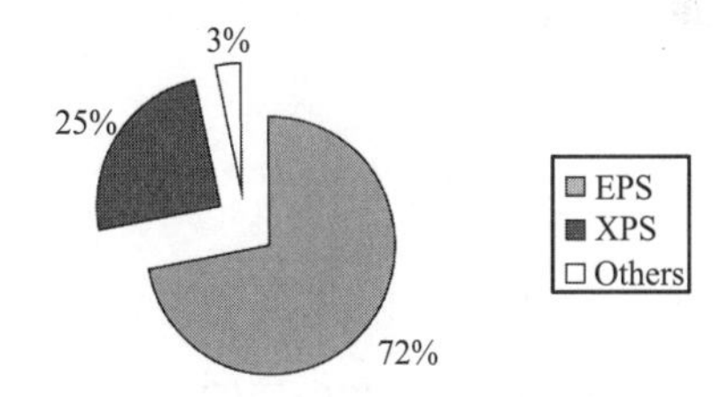

Figure 5-1-2 Shares of External Insulation Materials Used in Beijing

It can be seen that the application of insulation materials in China is almost the same as that in foreign countries. Although organic insulation materials are flammable, they are still widely applied at home and abroad. For technical and economic reasons, there are currently no high-efficiency substitutes. Organic insulation materials will still be the mainstream products in China's building insulation market at present and in a certain period of future.

5.1.2 Fire Accident Analysis of External Thermal Insulation System

Despite of their good insulation performance, light weight and mature applications, organic insulation materials are combustible, which may bring about fire risks. Fire accidents related to combustible materials for external thermal insulation have occurred from time to time in recent years. According to a survey conducted in Beijing, more than 90% of fires of external thermal insulation systems occurred in the construction phase, as a result of sparks generated in welding or careless use of flame. Certainly, the nonconformity of combustibility of some insulation materials to related standards is also one reason of fire. It is known that there are no such construction fires abroad. This means that there are some problems in the construction site management in China.

Although the results of current surveys indicate that most of fires occurs during construction of external thermal insulation systems, the fire protection work should focus on the improvement of fire resistance of existing external thermal insulation systems and eliminate fire hazards in the use of buildings, in terms of the importance and long-time application.

This is mainly based on the following considerations:

(1) There should be few people involved in the fire during construction, few flammable and explosive materials in buildings, but a number of escape ways and little difficulty in rescue. In the event of a fire during building use, the safety of personnel and property and the rescue capabilities of firefighters will face major challenges. Most of deaths caused by fires occur in the stage of building use.

(2) The construction period of the external thermal insulation system is short, ranging from one or two months to half a year. The service life of a building is usually more than 50 years, so that if the external thermal insulation system should be over 25 years. In terms of the importance for safety of personnel and property and the long-time use of buildings, it is particularly essential to reduce or avoid fires during building use.

(3) It is indicated in fire accidents on the construction site that organic insulation materials applied at present may lead to fire, and also bring fire hazards to external thermal insulation systems. Therefore, this sounds the alarm for us to attach importance to the fire safety of external thermal insulation systems.

5.1.3 Fire Safety Solution of External Thermal Insulation System

Fire protection problems of external thermal insulation systems should be solved by means of attaching great importance, conducting scientific research, and taking appropriate actions, instead of sweepingly reduce and even prohibit organic insulation materials or forcing out external thermal insulation systems due to fire accidents.

In order to prevent fire during building use, first, the external thermal insulation system must not be ignited; secondly, once the external thermal insulation system is ignited, the flame should be prevented from spreading. This is the basic idea to solve fire protection problems of the external thermal insulation system.

The external thermal insulation system is a non-bearing insulation structure attached to the outer wall. In accordance with China's fire protection codes, the fire resistance limits of outer walls of buildings are 1-3h. If the insulation layer is made of combustible foam, the burning time will not exceed the specified fire resistance limit after the external thermal insulation system is ignited, so the structural performance of the outer wall will not be damaged. However, the entire outer wall is enveloped by the external thermal insulation system that covers fire zones between different layers of a building. If not stopped by the external thermal insulation system, the flame may spread into the building through windows in the outer wall, causing fire in the stage of building use. In this sense, the main fire protection function of the external thermal insulation system is to prevent the flame from spreading. Especially in China's large and medium-sized cities, where there are a large number of high-rise buildings. This is different from foreign countries where low-rise and multistoried buildings play a dominant role. It is necessary to solve the problem of flame spreading after ignition of the external thermal insulation system. Two possibilities should be considered in case that a fire occurs to the external thermal insulation system.

First, if a fire occurs in a building, the flame may spread out of the window or hole, making the external thermal insulation system.

Secondly, the adjacent objects and external thermal insulation system may burn.

In both cases, once the external thermal insulation system is burning, the flame must be stopped from spreading into other floors, or into a building through window openings of other floors. This is the basic concept of China's current studies on the fire safety of external thermal insulation systems.

In summary, the fire safety of an external thermal insulation system should be studied from the following two aspects.

(1) Ignitability: inspect whether the insulation material or external thermal insulation system can be ignited or lead to combustion in the presence of fire or kindling.

(2) Flame spreading performance: inspect whether the insulation material or the external thermal insulation system is able to spread the flame in the event of combustion or fire.

The ignitability and flame spreading performance of the external thermal insulation system are based on the flame retardant properties of its material. It is stipulated in the national standards GB/T 10801. 1-2002 and GB/T 10801. 2-2002 that the expanded polystyrene (EPS) and extruded polystyrene (XPS) boards should meet the requirements of Class B_2 combustibility, and the oxygen index of the EPS board should not be less than 30%. The same provisions for the EPS board are also available in the *External Thermal Insulation Composite System based on Expanded Polystyrene* (JG 149-2003) and Technical Specification for External Thermal Insulation on Walls (JGJ 144-2004). The *Technical Code for Rigid Polyurethane Foam Insulation and Waterproof Engineering* (GB 50404) specifies that the combustibility of the rigid polyurethane foam should reach Class B_2 at least, and its oxygen index should not be less than 26%. In September 2009, the Ministry of Public Security and the Ministry of Housing and Urban-Rural Development jointly released the *Provisional Rules for Fire Protection of External Thermal Insulation System and Exterior Wall Finish of Civil Buildings*, stipulating that "the combustibility of the external insulation material of civil buildings should be of Class A, and must not be below Class B_2". Apparently, Class B_2 is the minimum requirement for organic insulation materials. At the same time, the fire protection management of the construction site should be enhanced while materials are stored in the baked state or pasted on walls, in order to ensure the fire safety during construction.

So, will the fire protection problems of external thermal insulation be solved by improving the combustibility of organic insulation materials? Relevant data shows that, under current technological conditions, if the flame retardant performance of organic insulation materials is improved, production costs will be greatly increased. Some flame retardants often lead to more smoke and higher toxicity of flue gas in the process of burning suppression, which may cause greater hazards. In addition, the combustibility rating of one insulation material does not fully represent its burning state in real fires. For example, some flame-retardant materials can burn fiercely if appropriate conditions are met.

It is found in the test that if coated with interface mortar, the fire resistance of the EPS board or rigid polyurethane foam will be improved during storage and construction, and the ignitability and flame spreading performance will also be better. If fire partitions are constructed after the EPS board or rigid polyurethane foam coated with interface mortar is applied on the wall, better fireproof effects will be achieved, as shown in Figure 5-1-3. The organic insulation material can be protected as it is isolated from a small fire via interface mortar. The above measures have some effects of fire prevention during storage and construction of combustible insulation materials, but cannot guarantee the fire safety in the event of large-scale and continuous fires.

Figure 5-1-3 Effect of Blast Burner on EPS Board Coated with Interface Mortar on Wall

A layer of cement-based mortar can be applied on the surface of the combustible insulation board to prevent ignition by a small fire on the construction site. This is an example of application of modern material compounding theories. All materials have their own advantages and shortcomings. It is impossible and will also be very difficult to find or develop an ideal material with good insulation properties, fire resistance and compatibility to ensure the excellent performance and durability of the external thermal insulation system. The simplest solution is compounding materials with respective advantages in accordance with the application requirements. Steel, like combustible materials, does not have fire resistance. Once it is compounded with high-efficiency fire-retardant materials, however, it can be applied in construction of skyscrapers that are much higher than concrete buildings. In this sense, there is no need to deliberately seek the insulation material of high combustibility. Instead, the solution to fire protection of external thermal insulation systems should focus on the improvement of the overall fire resistance and implementation of fireproof measures based on the building height. In fact, EPS boards of Class B_2 are widely applied in external thermal insulation systems in Europe and America, involving appropriate structural measures. During renovation of four 18-storey old residential buildings in Asian Sports Village in Beijing, the 100mm thin-plastered EPS board system were pasted based on the joint design of German and Chinese experts, and 200mm high rock-wool fire protection beams were added in each window to meet the fire safety requirements.

Obviously, the ultimate goal is to improve the overall fire resistance of the external thermal insulation system to solve fire protection problems in the stage of building use. Therefore, what is important is to take effective fireproof measures to improve the overall fire resistance of the external thermal insulation system, and properly test and evaluate the fire resistances of external thermal insulation systems with different structures.

5.1.4 Key Factors Affecting Fire Safety of External Thermal Insulation System

The international practice for fire protection of external thermal insulation systems is as follows: if the insulation material has high fire resistance, there are relatively low requirements for the protective layer and structural measures; otherwise, appropriate structural measures should be taken, and the protective layer should meet relatively high requirements. As a whole, the fire resistance of the insulation material should be balanced with the structural measures and protective layer. Accordingly, fireproof structural measures should be implemented as a main means at present to guarantee the fire safety of external thermal insulation systems in China. It is also an effective technical means suitable for China's national conditions and current applications of external thermal insulation systems. The insulation material (subject to cladding with inorganic materials), protective layer and fireproof structure should be regarded as a whole when the fire safety of an external thermal insulation system is assessed.

Fire hazards usually result from the released heat, so the fundamental of fire protection of the external thermal insulation system is three ways of heat transfer, namely, heat conduction, heat convection and heat radiation. In the presence of heat, combustible substances in the external thermal insulation system will burn and the flame will spread to other parts. As long as the three ways of heat transfer are cut off, combustible materials will not be ignited or the flame will not spread after combustible materials are ignited. Therefore, fireproof structural measures of the external thermal insulation system have two functions: preventing or slowing down direct fire attack to the external thermal insulation system in the involved area; and more importantly, preventing the flame from spreading through the external thermal insulation system itself. According to the existing research results, it can be concluded that the non-cavity structure connecting the insulation layer and base course wall, the protective layer covering the insulation layer and the fireproof structure separating the external thermal insulation system to prevent flame spreading, which are often referred to as "three structural fireproof elements", can effectively avoid the ignition of the external thermal insulation system and spreading of flame in this system.

The non-cavity structure is able to limit the heat convection in the external thermal insulation system; the increase of protective layer thickness can significantly reduce heat radiation of the external flame onto the internal insulation material; and the fire partition structure can effectively suppress heat conduction and prevent flame spreading, including fire compartments, fire barriers and fire protection beams.

The principles of these three structures are shown in Figure 5-1-4.

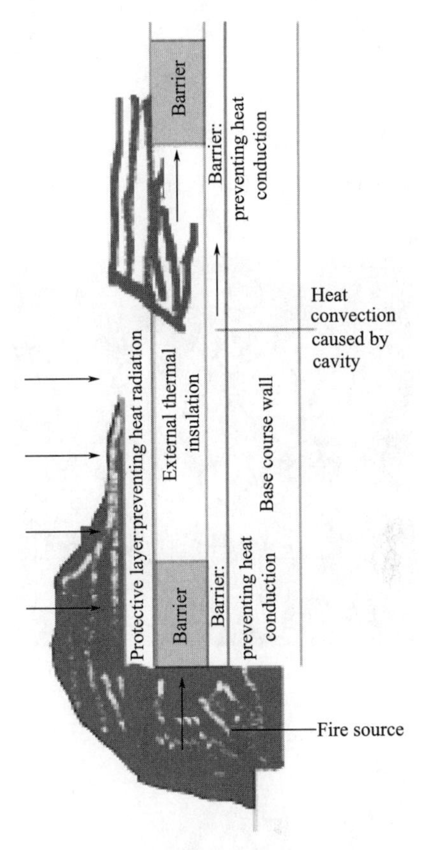

Figure 5-1-4 Thermal Insulation Effects of Three Structural Measures

Figure 5-1-5 to 5-1-7 show the non-cavity structure, fire partition structure and fireproof protective surface structure, respectively.

Figure 5-1-5 No-cavity Structure

Figure 5-1-6 Fire Partition Structure

Figure 5-1-7 Fireproof Protective Surface Structure

5.1.5 Focus of Fire Protection Research on External Thermal Insulation System

As mentioned above, the focus of fire protection of the external thermal insulation system is to improve its overall fire safety. Only the excellent fire resistance and reasonable structure of the external thermal insulation system can guarantee the conformity of this system to fire safety requirements, which is of extensive and practical significance for engineering applications. Therefore, importance should be attached to the effective fireproof structural measures for higher fire resistance of the whole external thermal insulation system and the testing and evaluation of external thermal insulation systems with different structures in the current research.

The following three aspects should be studied at the same time to solve the fire protection problems of the external thermal insulation system.

(1) By reference to advanced foreign technologies, develop and study external thermal insulation systems with high fire resistance, to provide more choices for their engineering applications. This is also the future direction of the external thermal insulation industry.

(2) Based on the foreign experience, establish a fire test method suitable for China's national conditions, scientifically evaluate the fire safety of China's external thermal insulation systems by means of testing, and further regulate the external insulation market, to prevent fires in the external thermal insulation system.

(3) Carefully investigate the causes of fire in the external thermal insulation system, develop the targeted technical and management measures for the fire safety of external thermal insulation construction, and formulate, release and rigorously enforce relevant standards to prevent the external thermal insulation system from fires. Top priority should be given to this aspect.

5.2 Fire Test of External Insulation Material and System

The earliest experimental conclusion is that the increase in the thickness of the fireproof protective surface can improve the fire resistance of the external thermal insulation system. It was found in the cone calorimeter test and combustion shaft furnace test, which have an extensive experimental basis and are highly accepted in China.

5.2.1 Cone Calorimeter Test

5.2.1.1 Principle of Cone Calorimeter Test

The combustibility refers to the ability of a material to react to a fire. Essentially, it is a comprehensive performance involving the amount and rate of heat release, ignitability of a material, and toxicity of flue gas released in combustion. The modern fire science research shows that the heat released by burning combustibles is the most important factor of fire hazards. It is not only decisive to the fire change, but also often controls the occurrence and evolution of other fire hazards. For a long time, fire science researchers have been seeking one accurate and easy-to-use test method to assess the heat released in a fire, especially the heat release rate.

The small-scale cone calorimeter test is a continuous process based on the principle of calorimetric oxygen consumption, in which the actual fire conditions of a material are simulated, and the parameters such as the ignitability, heat release, smoke and toxic gas are determined. The cone calorimeter, named after its cone heater, is an easy-to-use small combustibility tester, which was first designed on a rigorous scientific basis in the history of fire test technology. It is an important technological progress in the field of fire science and engineering and research. In the cone calorimeter test, all products of combustion are collected and treated with an exhaust pipe to fully mix gases, followed by determination of the mass flow and composition. At least the volume fraction of O_2 should be measured. For more accurate results, the volume fractions of CO and CO_2 should also be measured. The oxygen consumption (mass) in the combustion process can be calculated, with which the heat release rate can be obtained.

The ignition time, heat release, smoke and toxic gas generation of a material should be determined in the cone calorimeter test. Under the stable heat radiation of the cone furnace, the properties of the test material or product will change physically or chemically, relying on their ability to react to a fire. For reference, the amount of radiation heat received by a black non-combustible material at 750℃ is 62kW/m^2, but the amount of heat radiation in a real fire is normally 20-150kW/m^2.

The radiation energy provided by the cone calorimeter is 0-100kW/m^2, and usually 50kW/m^2 in the international practice. This is basically the same as that in a real fire.

Theoretically, the cone calorimeter test, designed on the basis of the room corner test and the principle of oxygen consumption calorimeter, is a small-scale scientific and reasonable fire simulation test. In terms of the practicality and popularity, it can be used as a routine test method for the fire resistance of the external thermal insulation system of a building. The model of the cone calorimeter test is shown in Figure 5-2-1.

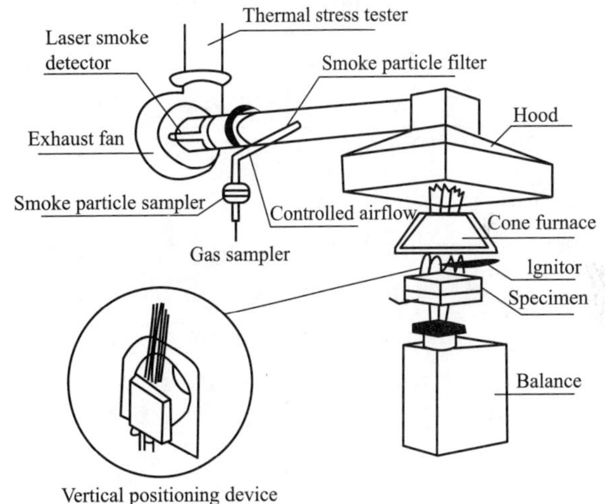

Figure 5-2-1 Principle Model and Physical Diagram of Cone Calorimeter Test

The principle of oxygen consumption is the ratio of the mass of oxygen consumed to the amount of heat generated during combustion.

Generally, the net heat of combustion of a material is proportional to the oxygen consumed in combustion. This relationship can be expressed as the heat of approximately 13.1×10^3kJ corresponding to the consumption of oxygen of 1kg. For most combustibles, this value changes within about $\pm5\%$. According to this principle, specimens are burnt in the air with the preset external radiation, the oxygen condensation and exhaust gas flow of combustion products are measured, and the related heat amount and heat release rate are determined with the above data.

At present, the small-scale cone calorimeter test and large-scale room corner test are internationally recognized, with the calculation as follows:

1. Calibration constant (C) for oxygen consumption analysis

$$C=\frac{10.0}{(12.54\times10^3)\times1.10\sqrt{\frac{T_d}{\Delta P}\frac{1.105-1.5X_{O_2}}{X^0_{O_2}-X_{O_2}}}} \quad (5\text{-}2\text{-}1)$$

Where, C —calibration constant for oxygen consumption analysis, $m^{1/2}kg^{1/2}K^{1/2}$;

T_d — absolute temperature of gas in the orifice-plate flow meter, K;

ΔP — pressure difference of the orifice-plate flow meter, Pa;

X_{O_2} —oxygen concentration, %;

$X^0_{O_2}$ —initial oxygen concentration, %.

The value "10.0" is equivalent to methane of 10kW, "12.54×10^3" is the $\Delta h_c/r_0$ value of methane (Δh_c is the net heat of methane combustion, in kJ/kg; r_0 is the stoichiometric ratio of oxygen to fuel mass); and "1.10" is the molecular weight ratio of oxygen to air.

2. Heat release rate $q(t)$

First, the time lag correction should be conducted to the oxygen concentration.

$$X_{O_2}(t)=X_{O_2}(t+t_d) \quad (5\text{-}2\text{-}2)$$

Where, $X_{O_2}(t)$ —oxygen concentration after the delay time correction, %;

X_{O_2} —oxygen concentration before the delay time correction, %;

t —time, s;

t_d —delay time of the oxygen analyzer, s.

The heat release rate $q(t)$ is calculated from the equation (5-2-3):

$$q(t)=\frac{\Delta h_c}{r_0}\times1.10\times C\sqrt{\frac{\Delta p}{T_d}}\frac{X^0_{O_2}-X_{O_2}}{1.105-1.5X_{O_2}} \quad (5\text{-}2\text{-}3)$$

Where, $q(t)$ —heat release rate, kW;

Δh_c —net heat of combustion of a material, kJ/kg;

r_0 —stoichiometric ratio of the oxygen to material by mass.

The value of $\Delta h_c/r_0$ is 13.10×10^3 for ordinary samples, and the specific value should be used in calculations if the value of Δh_c is known.

The heat release rate $q''(t)$ per unit area can be calculated from the equation (5-2-4):

$$q''(t)=q(t)/A_s \quad (5\text{-}2\text{-}4)$$

Where, $q''(t)$ —heat release rate per unit area, kW/m^2;

A_s —exposed surface area of the specimen, m^2.

3. Average effective heat of combustion ($\Delta h_{c,af}$)

$$\Delta h_{c,af}=\frac{\sum q(t)\Delta t}{m_i-m_f} \quad (5\text{-}2\text{-}5)$$

Where, $\Delta h_{c,af}$ —average effective heat of combustion, kJ/kg;

m_i —initial mass of the specimen, kg;

m_f —remaining mass of the specimen, kg.

4. Light absorption parameter (k) of smoke

The light absorption parameter (k) of smoke is determined as follows via the laser detection system in the cone calorimeter test:

$$k=\frac{I}{L}\ln\frac{I_0}{I} \quad (5\text{-}2\text{-}6)$$

Where, I —laser beam intensity;

I_0 —laser beam intensity in the absence of smoke;

k — light absorption parameter of smoke, m^{-1};

L —laser beam path, m.

5. Specific light absorption area $\sigma_{f(avg)}$

$$\sigma_{f(avg)} = \frac{\sum V_i k_i \Delta t_i}{m_i - m_f} \qquad (5\text{-}2\text{-}7)$$

Where, $\sigma_{f(avg)}$ —specific light absorption area;

V —volume flow rate of the exhaust pipe, m^3/s.

The following standards were applied in the cone calorimeter test: *Standard Test Method for Heat and Visible Smoke Release Rates for Materials and Products Using an Oxygen Consumption Calorimeter* (ASTM E 1354), *Reaction-to-fire Tests-Heat Release, Smoke Production and Mass Loss Rate-Part 1: Heat Release Rate (Cone Calorimeter Method)* (ISO 5660-1); and *Test Method of Heat Release Rate of Building Materials* (GB/T 16172), equivalent to the standard ISO 5660-1.

5.2.1.2 Contract Test I

In order to explore their fire resistance, the reaction-to-fire test was conducted to the external thermal insulation systems based on the thin-plastered EPS board, rock wool, and mineral binder and EPG granule, respectively. In the test process, the actual state of the external insulation material in the presence of fire was simulated with specimens consists of a 5mm polymer surface mortar layer receiving radiation, a middle 50mm insulation layer and a 10mm cement mortar base course wall, and the side faces of specimens were sealed with the 5mm polymer surface mortar, referred to as closed specimens. To compare the difference of performance in the test and actual application, specimens with side faces exposed were produced, that is, the side face are not sealed with surface mortar, referred to as open specimens. The size of each specimen is 100mm×100mm×65mm. Test conditions were set as follows: the amount of radiation energy was $50kW/m^2$; the flow rate of the exhaust pipe was $0.024m^3/s$; specimens were kept horizontal; and specimen shields and metal meshes were not used. Open specimens were used as observation specimens, while closed specimens as test specimens.

1. Specimens of external thermal insulation system based on the EPS board

The specimens were composed of the 10mm cement mortar base course wall + 50mm EPS board+5mm polymer plaster mortar (compounded with alkali-resistant glass fiber mesh).

The EPS board of the open specimen melted and shrank after 2s, and the polymer surface mortar (compounded with alkali-resistant glass fiber mesh) layer was attached to the cement mortar base course wall after 105s. The middle EPS board layer disappeared, and there were only a few black sinters.

The edges and corners of the closed specimen were cracking. The flue gas out of cracks were ignited at 52s and burned for about 70s. When the specimen shell was knocked off, nothing except sinter residues was found in this specimen.

2. Specimens of external thermal insulation system based on the adhesive polystyrene granule

The specimens were composed of the 10mm cement mortar base course wall+50mm adhesive polystyrene granule insulation mortar+5mm polymer surface mortar (compounded with alkali-resistant glass fiber mesh).

The open specimen was not ignited during the test. After the test, it was found that the surface of the insulation layer close to the heat radiation surface became slightly darker, the thickness of color change was about 3-5mm, the thickness of the insulation layer did not change greatly, and there were no other significant changes.

The closed specimen was not ignited during the test, without cracking or significant changes. When the specimen shell was knocked off after the test, it was found that the insulation layer became slightly darker due to heat radiation, the thickness of color change was about 3-5mm, and there were no other significant changes.

3. Specimens of external thermal insulation system based on the rock wool

The specimens were composed of the 10mm cement mortar base course wall + 50mm rock wool board+5mm polymer surface mortar (compounded with alkali-resistant glass fiber mesh).

The open specimen was not ignited during the test. After the test, it was found that the rock wool board close to the heat radiation surface became slightly darker, the thickness of color change was about 3-5mm, and the thickness of the rock wool board increased a little (as a result of thermal expansion). There were no other significant changes during and after the test.

The closed specimen was not ignited during the test, without cracking or significant changes. When the specimen shell was knocked off after the test, no significant changes of rock wool were found.

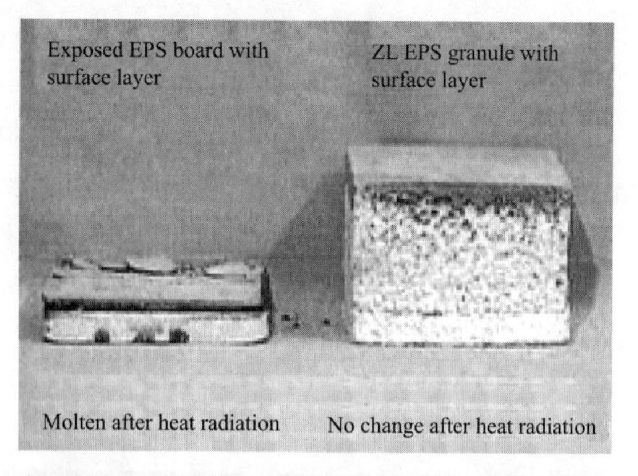

Figure 5-2-2 Specimens of Different Insulation Materials after Reaction to Fire

Refer to Figure 5-2-2 for specimens of different insulation materials after reaction to the fire.

The test results of reaction-to-fire performances of different insulation materials are shown in Table 5-2-1, in which No. 1, 2 and 3 specimens are for the external thermal insulation systems based on the EPS board, adhesive polystyrene granule, and rock wool, respectively.

Results of the reaction-to-fire performance tests show that:

(1) The specimens of the external thermal insulation system based on the adhesive polystyrene granule did not burn, the thickness of the insulation layer did not change greatly, only the insulation layer close to the heat radiation surface became slightly darker, and the thickness of color change was about 3-5mm. This is because the combustible EPS granule was coated with the non-combustible inorganic cementing material, and a cavity supported by the inorganic cement ma-

Reaction-to-fire Performance Test Results **Table 5-2-1**

Specimen	Ignition Duration (s)	Heat Release Rate(kW/m^2)		Effective Combustion Heat(MJ/kg)		Total Heat (kJ)	CO			CO_2		
		Peak	Average	Peak	Average		Peak(g/g)	Average(g/g)	Total(g)	Peak(g/g)	Average(g/g)	Total(g)
1	64	108.6	6.0	16.4	3.2	49.9	0.0525	0.0067	0.083	0.0848	0.111	1.38
2	Not ignited	0.9	0.0	0.2	0.0	0.2	0.0021	0.0013	0.027	0.032	0.022	0.46
3	Not ignited	0.5	0.0	0.2	0.0	0.2	0.0080	0.0029	0.027	0.099	0.049	0.46

Note: 1. External thermal insulation system based on the EPS board; 2. External thermal insulation system based on the adhesive polystyrene granule; and 3. External thermal insulation system based on the rock wool.

terial was formed as a result of hot melting and shrinkage of the EPS granule close to the heat source of intense heat radiation. The structure was kept stable as this layer will not be deformed in a certain period. At the same time, this layer is able to prevent the material below from heat, thus achieving good fire stability.

(2) The test results of the external thermal insulation system based on the rock wool show that specimens were not ignited, the rock wool board close to the heat radiation surface became slightly darker, the thickness of color change was about 3mm, and the thickness of the rock wool board increased a little. This is because rock wool is a kind of Class A non-combustible material with good fireproofing effects. The minor expansion of the rock wool board was caused by volatilization of organic additives (about 4%, such as the binder and water repellent) used in extrusion of rock wool into boards.

(3) The test results of the external thermal insulation system based on the EPS board show that this system shrank and melted fast in the presence of high-temperature radiation, and burned in the case of a fire. That is, this system tends to be damaged in the event of a fire (open flame or high-temperature radiation).

In conclusion, the external thermal insulation system based on the thin-plastered EPS board has poor fire resistance. If point bonding (pasting area: normally less than 40%) does not meet the standards, there will be a communicating air layer. In the event of a fire, shrinkage and melting of the EPS board will lead to the rapid formation of a "fire duct", making the fire spread fast. With a large amount of smoke generated in combustion, the visibility will decline greatly, resulting in psychological panic and difficulties in escape, and affecting the rescue works of firefighters. This system also has poor volume stability in the presence of a high-temperature heat source. Especially when the tile finish is applied, this system may be easier to damage in the event of a fire, and bring greater safety hazards to escape and rescue. This problem becomes more prominent with the building height increasing.

5.2.1.3 Contrast Test II

The adhesive polystyrene granule composite insulation system was used in this test to simulate the actual state of a wall in the presence of fire. Insulation materials included the rigid polyurethane foam, EPS board and XPS board, respectively involving flat and tank-type specimens, as shown in Figure 5-2-3*a* and 5-2-3*b*. Specimens (100mm×100mm×60mm) were coated with the 10mm refractory mortar (waterproof mortar based on the adhesive polystyrene granule) or cement mortar; and the size of the core made of insulation material was 80mm×80mm×40mm. Ordinary cement mortar specimens (100mm × 100mm × 35mm) were used as contrast specimens. Refer to Table 5-2-2 for specimen codes.

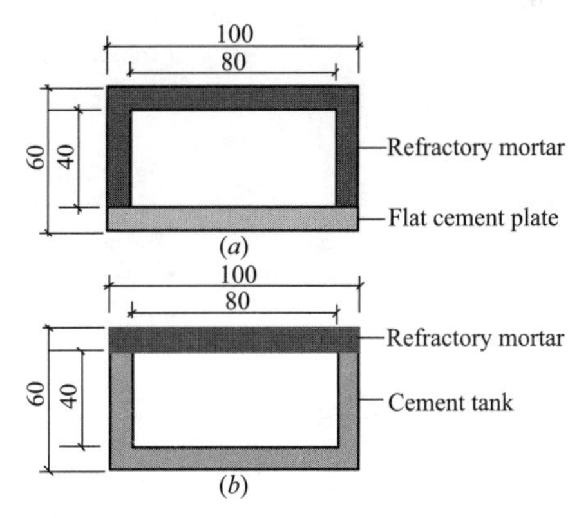

Figure 5-2-3 Schematic Diagram of Specimens of Adhesive polystyrene granule Composite Insulation System

(*a*) flat cement specimen; (*b*) tank-type cement specimen

Specimen Codes of Adhesive polystyrene granule Composite Insulation System in Cone Calorimeter Test

Table 5-2-2

Specimen Code	Insulation Layer	Structure Category	Number of Specimens
AP	Polyurethane	Flat specimen	6
AU	Polyurethane	Tank-type specimen	6

Continued

Specimen Code	Insulation Layer	Structure Category	Number of Specimens
BP-1(1st group)	EPS	Flat specimen	5
BU-1(1st group)	EPS	Tank-type specimen	6
BP-2(2nd group)	EPS	Flat specimen	6
BU-2(2nd group)	EPS	Tank-type specimen	6
SP	XPS	Flat specimen	5
SU	XPS	Tank-type specimen	6
C	Ordinary cement mortar	Uniform specimen	6

The state of the specimens of the adhesive polystyrene granule composite insulation system is the same as that of ordinary cement mortar specimens in this test.

Figure 5-2-4 shows the specimens of the adhesive polystyrene granule composite insulation system after reaction to the fire.

Ignitability: the specimens of the adhesive polystyrene granule composite insulation system and the ordinary cement mortar specimens were not ignited. Refer to Table 5-2-3 for the test results.

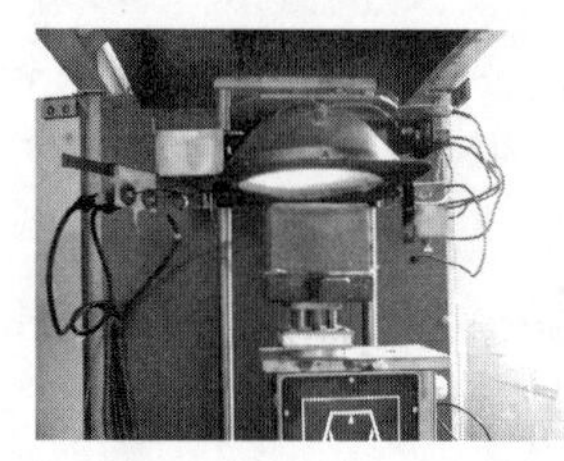

Figure 5-2-4 Specimens of Adhesive polystyrene granule Composite Insulation System after Reaction to Fire

Ignitability Results of Cone Calorimeter Test **Table 5-2-3**

Specimen Code	No. 1 Specimen	No. 2 Specimen	No. 3 Specimen	No. 4 Specimen	No. 5 Specimen	No. 6 Specimen	Average
AP	Not ignited	Not ignited	Not ignited	Not ignited	Not ignited	Not ignited	Not ignited
AU	Not ignited	Not ignited	Not ignited	Not ignited	Not ignited	Not ignited	Not ignited
BP-1(1st group)	Not ignited	Not ignited	Not ignited	—	Not ignited	Not ignited	Not ignited
BU-1(1st group)	Not ignited	Not ignited	Not ignited	Not ignited	Not ignited	Not ignited	Not ignited
BP-2(2nd group)	Not ignited	Not ignited	Not ignited	Not ignited	Not ignited	Not ignited	Not ignited
BU-2(2nd group)	Not ignited	Not ignited	Not ignited	Not ignited	Not ignited	Not ignited	Not ignited
SP	Not ignited	Not ignited	Not ignited	Not ignited	Not ignited	—	Not ignited
SU	Not ignited	Not ignited	Not ignited	Not ignited	Not ignited	Not ignited	Not ignited
C	Not ignited	Not ignited	Not ignited	Not ignited	Not ignited	Not ignited	Not ignited

(2) Heat release performance: the test results show that the peak heat release rates of specimens of the adhesive polystyrene granule composite insulation system are basically the same as those of ordinary cement mortar specimens, but the process averages of heat release rate and total amounts of heat of the former are slightly smaller than those of the latter. The heat release performance of the adhesive polystyrene granule composite insulation system can be regarded as identical to that of the ordinary cement mortar. The test results are show in Table 5-2-4.

Heat Release Performance Results of Cone Calorimeter Test

Table 5-2-4

Specimen Code	Heat Release Rate/(kW/m^2)			Total Amount of Heat /(MJ/m^2)
	Peak		Process Average	
	Range	Average		
AP	2.0-5.0	3.4	1.3	1.8
AU	3.1-6.0	4.2	1.4	1.8
BP-1(1st group)	3.1-3.9	3.5	1.4	1.8
BU-1(1st group)	1.3-7.2	3.8	0.8	1.0
BP-2(2nd group)	3.3-4.9	4.1	1.1	1.6
BU-2(2nd group)	4.2-5.6	5.0	1.2	1.6
SP	2.5-5.0	3.7	1.2	1.5
SU	2.6-5.4	3.4	1.1	1.4
C	3.0-5.6	3.9	2.0	2.4

(3) Smoke: the test results show that the light absorption parameters of smoke generated by specimens of the adhesive polystyrene granule composite insulation system are the same as those of ordinary cement mortar specimens, both of which are close to the baseline value. The average specific light absorption area of the specimens of the adhesive polystyrene granule composite insulation system is greater than that of ordinary cement mortar specimens, because the mass loss of the former was smaller than that of the latter in the late period of testing and part of the specific light absorption area is borne in the non-combustion mass loss of the latter. However, the total amount of smoke generated by the specimens of the adhesive polystyrene granule composite insulation system is basically identical to that of ordinary cement mortar specimens. The test results are shown in Table 5-2-5.

(4) CO: the test results show that the measured CO value of the specimens of the adhesive polystyrene granule composite insulation system is slightly greater than that of ordinary cement mortar specimens, both of which are close to the baseline value. The average CO production ratio and total amount of CO of the former are basically identical to those of the latter. The test results are shown in Table 5-2-6.

Smoke Results of Cone Calorimeter Test **Table 5-2-5**

Specimen Code	Mass Loss /g	Light Absorption Parameter of Smoke		Specific Light Absorption Area/(m^2/kg)		Total Amount of Smoke/m^2
		Peak	Average	Peak	Average	
AP	31.6	0.2	0.0	121	32	1.0
AU	33.5	0.2	0.0	171	28	0.9
BP-1(1st group)	15.4	0.2	0.0	78	17	0.5
BU-1(1st group)	14.1	0.2	0.0	265	38	1.1
BP-2(2nd group)	33.0	0.0	0.0	11	3	0.0
BU-2(2nd group)	30.4	0.0	0.0	18	5	0.1
SP	34.4	0.1	0.0	89	9	0.3
SU	36.9	0.3	0.0	121	25	0.9
C	32.1	0.2	0.0	72	17	0.5

CO Results of Cone Calorimeter Test **Table 5-2-6**

Specimen Code	Mass Loss /g	$CO/\times10^{-6}$		CO/(kg/kg)		Total Amount of CO /mg
		Peak	Average	Peak	Average	
AP	31.6	2	2	0.007	0.002	50
AU	33.5	3	2	0.012	0.002	64
BP-1(1st group)	15.4	5	4	0.476	0.004	118
BU-1(1st group)	14.1	4	2	0.763	0.002	67
BP-2(2nd group)	33.0	2	2	0.008	0.003	51
BU-2(2nd group)	30.4	2	2	0.011	0.004	50
SP	34.4	2	2	0.008	0.001	49
SU	36.9	2	2	0.003	0.001	49
C	32.1	2	2	0.005	0.001	44

(5) CO_2: the test results show that the measured CO_2 value of the specimens of the adhesive polystyrene granule composite insulation system is slightly greater than that of ordinary cement mortar specimens, both of which are close to the baseline value. The average CO_2 production ratio and total amount of CO_2 of the former are basically identical to those of the latter. The test results are shown in Table 5-2-7.

Analyzed from the above test results, the reaction-to-fire performance of the adhesive polystyrene granule composite insulation system is basically identical to that of the ordinary cement mortar, so the adhesive polystyrene granule can be used as a Class A non-combustible materials.

Table 5-2-8 compares the impact of the thickness of the protective layer on the combustibility of specimens. The test results show that: when the protective layer was 10mm, specimens were not ignited in the cone calorimeter test, the peak heat release rate was less than $10kW/m^2$, and the total amount of heat was $5MJ/m^2$, basically identical to those of ordinary cement mortar. When the protective layer is less than 5mm, specimens were ignited in the cone calorimeter test.

CO_2 Results of Cone Calorimeter Test **Table 5-2-7**

Specimen Code	Mass Loss/g	CO_2/%		CO_2/(kg/kg)		Total Amount of CO_2 /mg
		Peak	Average	Peak	Average	
AP	31.6	0.002	0.002	0.118	0.027	847
AU	33.5	0.005	0.004	0.358	0.054	1761
BP-1(1st group)	15.4	0.018	0.005	1.302	0.042	1368
BU-1(1st group)	14.1	0.016	0.007	2.063	0.080	2250
BP-2(2nd group)	33.0	0.002	0.002	0.141	0.056	848
BU-2(2nd group)	30.4	0.002	0.002	0.174	0.061	836
SP	34.4	0.004	0.002	0.133	0.025	830
SU	36.9	0.002	0.002	0.049	0.023	834
C	32.1	0.002	0.002	0.077	0.023	723

Impact of Protective Layer Thickness on Combustibility of Specimen **Table 5-2-8**

Insulation Material	Protective Layer Thickness(mm)	Ignition Time (s)	Heat Release Rate(kW/m^2)		Total Amount of Heat(kJ)
			Peak	Average	
PU	Color plate	2	280.9	71.6	349.0
PU	3	50	112.5	34.1	153.7
PU	5	65	101.0	11.4	129.1
PU	10	Not ignited	4.2	1.4	1.8
EPS	5	995	24.6	6.9	8.6
EPS	10	Not ignited	5.0	1.2	1.6
XPS	10	Not ignited	3.7	1.2	1.5
XPS	10	Not ignited	3.4	1.1	1.4
C	—	Not ignited	3.9	2.0	2.4

5.2.1.4 Summary

When the protective layer is 5mm and the non-combustible insulation material or the non-flammable insulation material without flame spreading capability is used, the specimens are not ignited in the cone calorimeter test, the peak heat release rate is less than 10kW/m^2, and the total amount of heat is less than 5MJ/m^2. These results are basically the same as those of ordinary cement mortar. If the combustible insulation material is applied, however, specimens are ignited in the cone calorimeter test, and the peak heat release rate is greater than 100kW/m^2.

When the protective layer is 10mm and the combustible insulation material is used, the specimens are not ignited in the cone calorimeter test, the peak heat release rate is less than 10kW/m^2, and the total amount of heat is less than 5MJ/m^2. These results are basically the same as those of ordinary cement mortar.

If a combustible insulation material is used, its combustibility can be improved by increasing the thickness of the protective layer.

5.2.2 Combustion Shaft Furnace Test

5.2.2.1 Test principle

The combustion shaft furnace test is a test method used in German standards to assess the combustibility of building materials. It is a kind of medium-scale model fire test. The application of this test method is equivalent to that in German standards. Test equipment includes the combustion shaft furnace and control instruments. The purpose of the shaft furnace test is to examine the impact of the thickness of the protective layer of the external thermal insulation system on flame spreading, as well as changes of the combustible insulation material applied in the presence of a fire. Relevant standards include the *Fire Behaviour of Building Materials and Elements-Part 1: Classification of Building Materials-Requirements and Testing* (DIN 4102-1: 1998), *Fire Behaviour of Building Materials and Elements-Part* 15: *"Brandschacht"* (DIN 4102-15: 1990), *Fire Behaviour of Building Materials and Elements-Part* 16: *"Brandschacht" Tests* (DIN 4102-16: 1998), and *Test Method of Difficult-flammability for Building Materials* (GB/T 8625-2005).

In the shaft furnace test, specimens (190mm × 1000mm, four in each group) were vertically fixed on the specimen support, forming a vertical square equivalent flue with the internal dimensions of 250mm × 250mm. That is, the clearance between each two parallel specimens was 250mm among four specimens.

In the shaft furnace test, the rectangular burner was horizontally mounted at the center of the equivalent flue at the lower ends of specimens, and the specimen parts of about 4cm from their lower ends were involved in a fire. The methane gas with the purity greater than 95% was used, the combustion power was stable at about 21kW, and the flame temperature was about 900℃. The standard test time is 10min.

In accordance with Article 7.1 of the *Test Method of Difficult-flammability for Building Materials* (GB/T 8625-2005), the shaft furnace test results should meet the following two requirements:

(1) The average remaining length of specimens should not be less than 150mm after combustion, and the remaining length of each specimen must not be zero.

(2) The average flue gas temperature measured with five thermocouples in each test must not exceed 200℃.

In line with Article 7.2 of the *Test Method of Difficult-flammability for Building Materials* (GB/T 8625-2005), the materials that are accepted in the combustion shaft furnace test and conform to the requirements of the *Classification on Burning Behaviour for Building Materials* (GB 8624-1997), *Test Method of Flammability for Building Materials* (GB/T 8626-1988) and *Test Method for Density of Smoke from Burning or Decomposition of Building Materials* (GB/T 8627-1999) can be identified as flame-retardant building materials.

The remaining length of specimens and flue gas temperature inside the exhaust pipe are measured in the shaft furnace test to inspect the flame retardancy of a material, which can be regarded as the flame spreading and heat release performance of a building material or component. For the external thermal insulation system, the shaft furnace test can be conducted to examine the reaction-to-fire of various layers. As specimens are only 100cm high, however, the fire resistance of the overall structure of the external thermal insulation system cannot be assessed in the shaft furnace test.

In the combustion shaft furnace test, temperature measuring points of the insulation layer, exposed to the protective layer, were set at intervals of 20cm along the height centerline of specimens, as shown in Figure 5-2-5 and 5-2-6. The flame power applied was a constant, and the thermocouples 5 and 6 corresponded to specimen parts involved in the fire.

Figure 5-2-5 Combustion Shaft Furnace Test Equipment

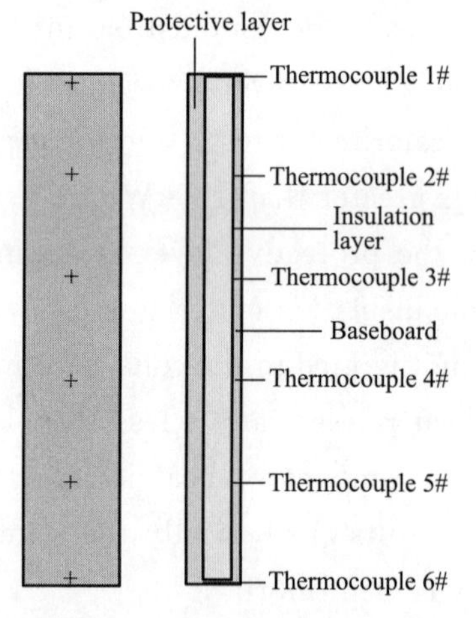

Figure 5-2-6 Specimen and Thermocouple Measuring Point of Combustion Shaft Furnace Test

5.2.2.2 Test Results

In the combustion shaft furnace test, the EPS board, XPS board and rigid polyurethanes foam were used as insulation materials, and the protective layers (thickness: 5-45mm) of specimens were made of the adhesive polystyrene granule or cement mortar. The specimen numbers and layer structures are listed in Table 5-2-9.

Specimen Number and Layer Structure of Combustion Shaft Furnace Test **Table 5-2-9**

Specimen No.	Material of Insulation Layer	Material of Protective Layer	Thickness of Protective Layer/ mm	Thickness of Anti-cracking Layer+ Finish Layer/ mm	Thickness of Insulation Layer/ mm	Base course wall Thickness/ mm
EPS-5	EPS board	Adhesive polystyrene granule	0	5	30	20
EPS-15			10	5	30	20
EPS-25			20	5	30	20
EPS-35			30	5	30	20
EPS-45			40	5	30	20
XPS-5	XPS board	Adhesive polystyrene granule	0	5	30	20
XPS-15			10	5	30	20
XPS-25			20	5	30	20
XPS-35			30	5	30	20
XPS-45			40	5	30	20
PU-5	PU	Adhesive polystyrene granule	0	5	30	20
PU-15			10	5	30	20
PU-25			20	5	30	20
PU-35			30	5	30	20
PU-35			30	5	30	20
PU-45			40	5	30	20
EPS-20/30-1	EPS board	Adhesive polystyrene granule	20/30	5	30/40	20
EPS-20/30-2			20/30	5	30/40	20
EPS-10/20-3		Cement mortar	10/20	5	30/40	20
EPS-10/20-4			10/20	5	30/40	20

As per the *Test Method of Difficult-flammability for Building Materials* (GB/T 8625-2005), the combustion power of methane gas is about 21 kW and the flame temperature is about 900℃. The flame was applied for 20min.

The maximum temperature of each measuring point of specimens is shown in Table 5-2-10.

Maximum Temperature of Each Measuring Point of Specimen **Table 5-2-10**

Category	No.	Measuring Point/ mm					
		0	200	400	600	800	1000
Flat EPS board	EPS-5	314.7	438.4	323.9	246.5	177.8	127.7
	EPS-15	143.0	280.0	202.1	95.9	96.8	95.6
	EPS-25	145.0	194.2	99.0	100.4	97.2	40.6
	EPS-35	97.2	99.0	98.6	99.5	84.3	41.6
	EPS-45	48.8	56.5	31.5	28.4	26.8	26.3
Slotted EPS board	EPS-20/30-1	150.7	168.9	114.1	91.5	95.2	91.5
	EPS-20/30-2	119.5	222.1	159.5	100.6	98.6	78.3
	EPS-10/20-3	164.0	257.8	221.4	137.6	98.8	68.8
	EPS-10/20-4	187.7	165.5	143.5	130.1	107.2	91.3

Continued

Category	No.	Measuring Point/ mm					
		0	200	400	600	800	1000
Flat XPS board	XPS-5	258.3	439.3	264.1	185.3	199.7	170.1
	XPS-15	155.0	225.7	206.6	96.1	84.7	48.3
	XPS-25	53.0	86.1	51.3	50.2	42.3	22.8
	XPS-35	48.1	49.5	54.1	39.7	34.1	32.9
	XPS-45	44.4	32.4	40.9	31.5	27.5	28.2
Flat PU board	PU-5	453.0	566.9	428.8	216.4	121.8	81.3
	PU-15	92.2	386.5	330.1	95.4	91.3	74.4
	PU-25	102.5	192.2	91.1	94.7	94.0	71.8
	PU-35	96.3	95.0	45.1	95.9	34.5	31.2
	PU-35	60.0	64.6	56.0	52.7	75.1	46.0
	PU-45	59.3	67.9	73.2	41.3	30.8	28.4

Note: Refer to Figure 5-2-12 for specimen classification.

Refer to Figure 5-2-7 to 5-2-10 for comparison of the maximum temperature of each measuring point of specimens.

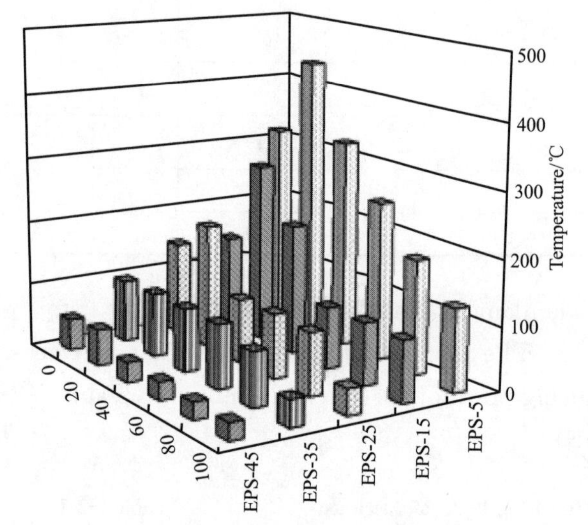

Figure 5-2-7 Comparison of Maximum Temperature of Flat EPS Board Specimens

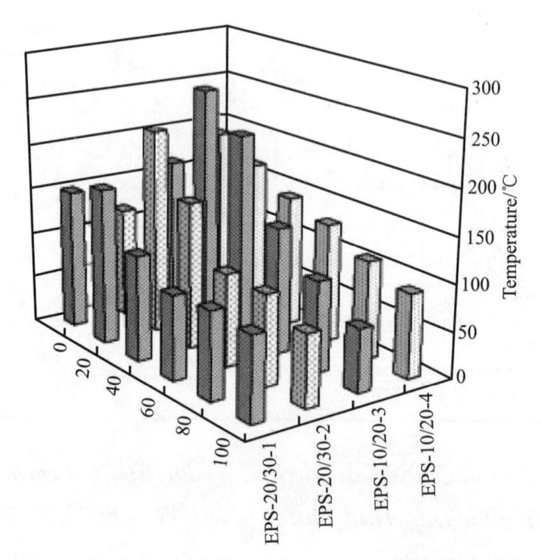

Figure 5-2-8 Comparison of Maximum Temperature of Slotted EPS Board Specimens

Refer to Figure 5-2-11 for the curve of each measuring point of specimens, and Figure 5-2-12 for the analysis of each specimen.

5.2.2.3 Summary

(1) The temperature of each measuring point of specimens drops with the thickness of the protective layer increasing. The thickness of the protective layer determines the ability of the external thermal insulation system to withstand a fire.

(2) The burning height of the insulation layer increases with the thickness of the protective layer decreasing. The insulation layer of the thin-plastered EPS board specimen with no fireproof protective layer was completely burnt, and the burn area of the protective layer of the thin-plastered rigid polyurethane foam specimen was about 65cm. When the protective layer of the adhesive polystyrene granule specimen is more than 30mm (thickness of the anti-cracking layer and finish layer: 5mm), the organic insulation material was not damaged under test conditions (the flame of 900℃ was applied on the bottom surfaces of specimen for 20min).

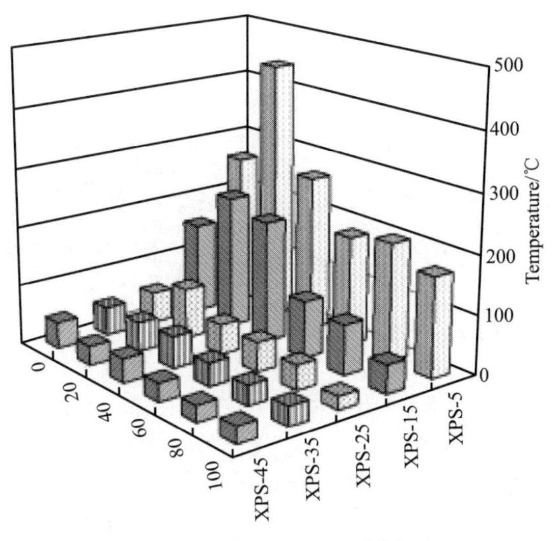

Figure 5-2-9 Comparison of Maximum Temperature of Flat XPS Board

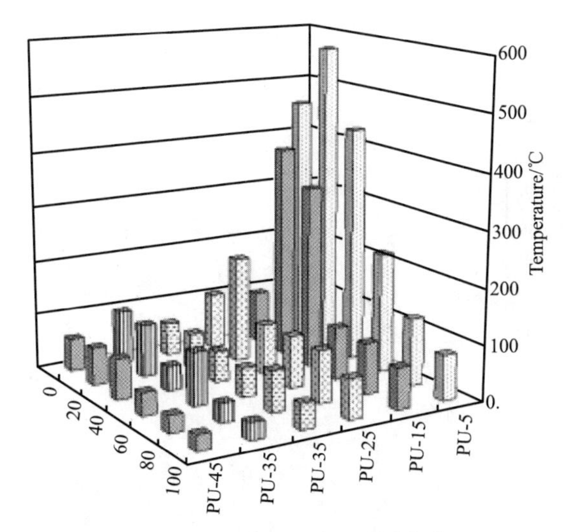

Figure 5-2-10 Comparison of Maximum Temperature of Flat PU Board

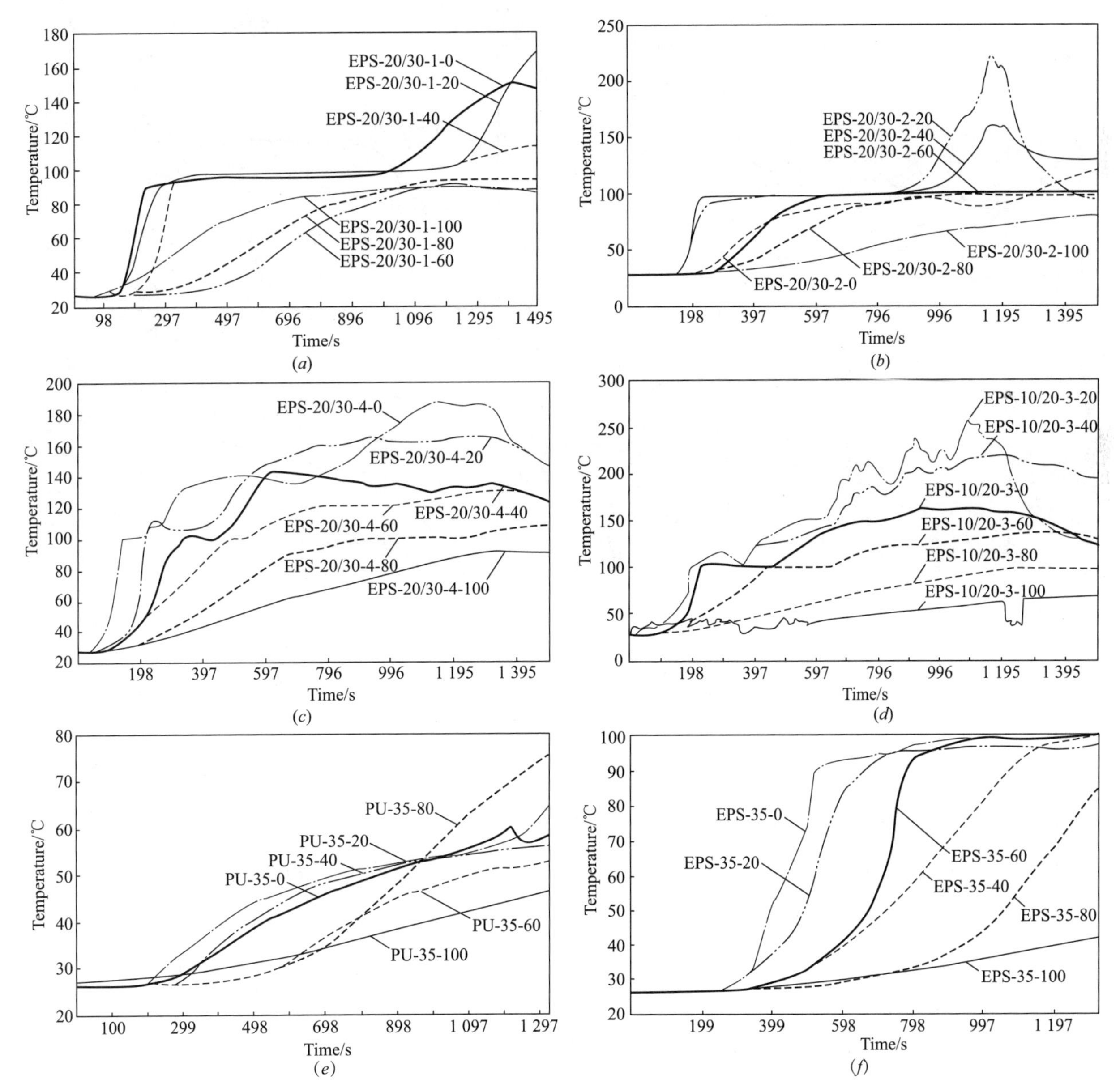

Figure 5-2-11 Curve of Each Measuring Point of Specimens (I)

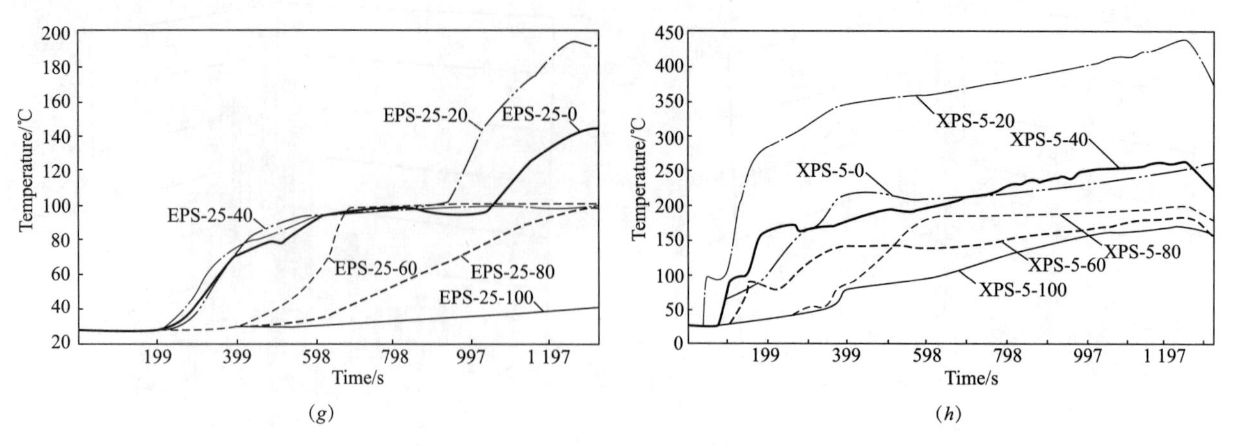

Figure 5-2-11 Curve of Each Measuring Point of Specimens (I) (Continued)

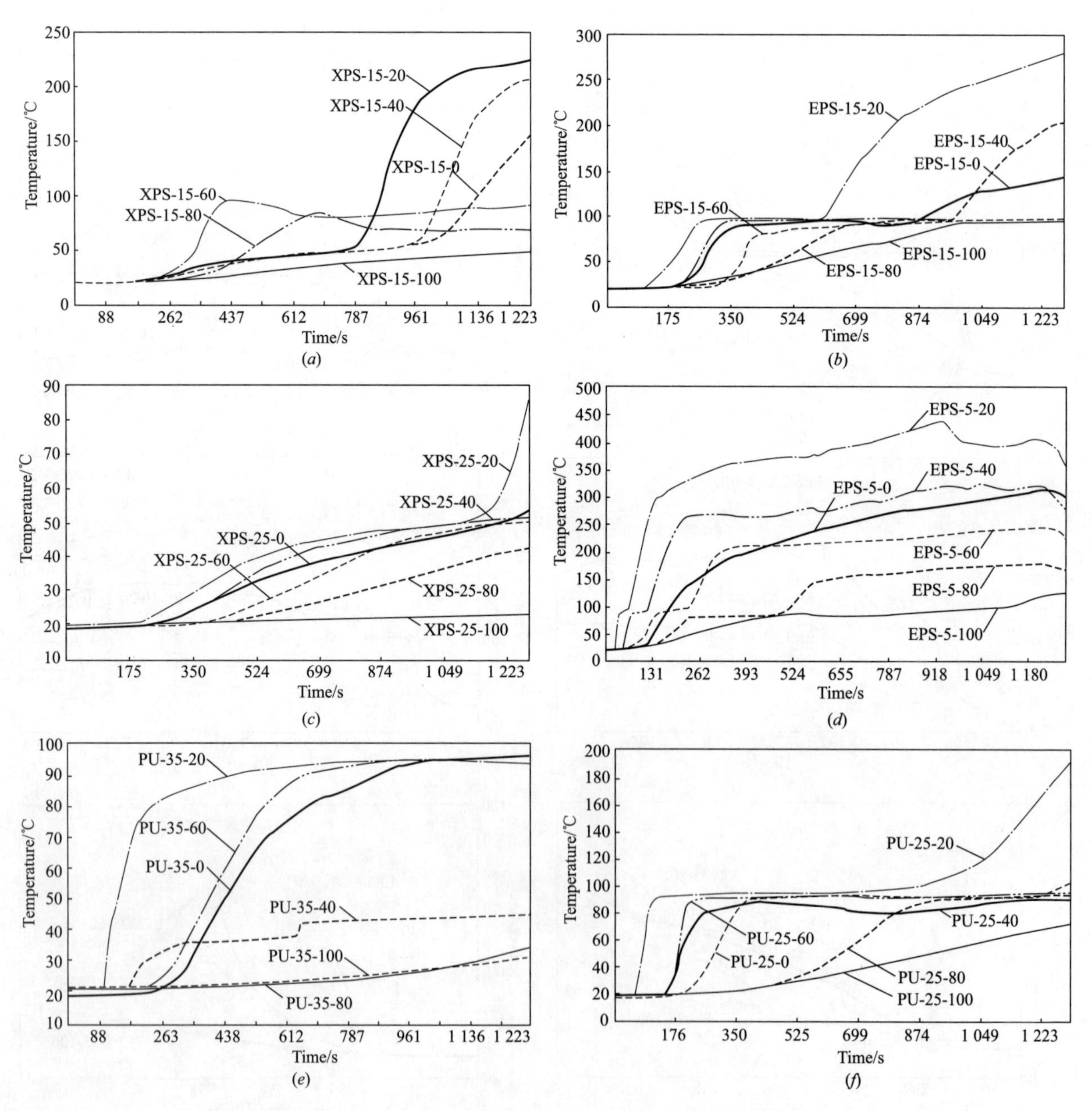

Figure 5-2-12 Curve of Each Measuring Point of Specimens (II)

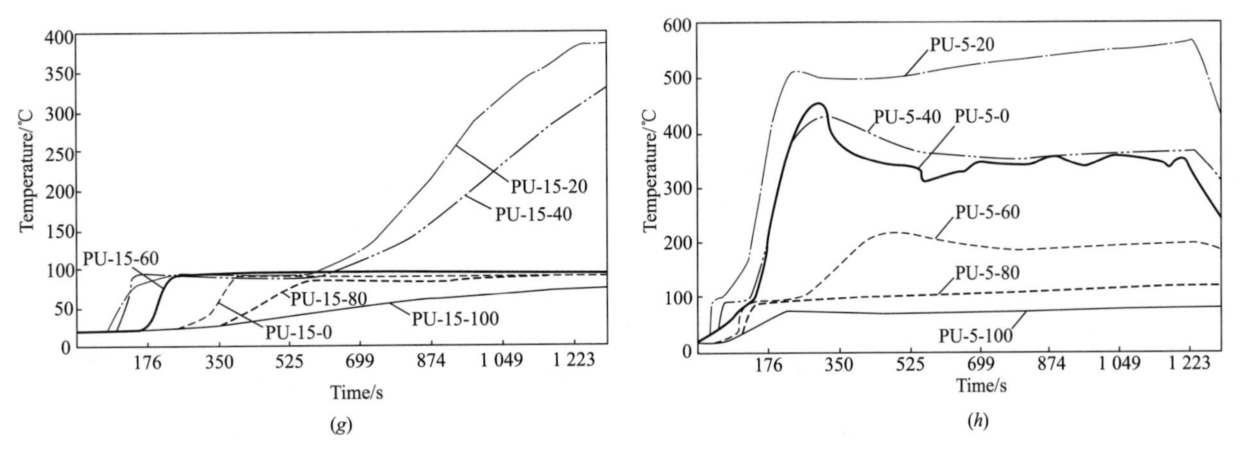

Figure 5-2-12 Curve of Each Measuring Point of Specimens (II) (Continued)

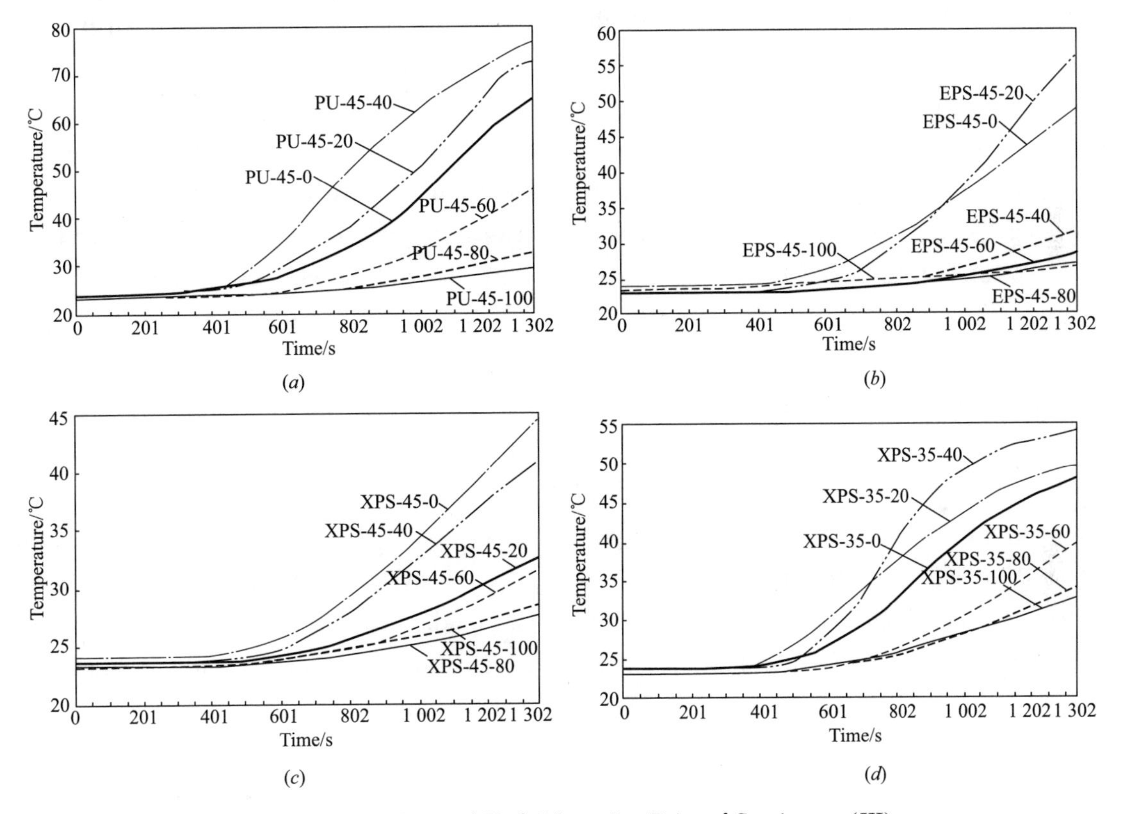

Figure 5-2-13 Curve of Each Measuring Point of Specimens (III)

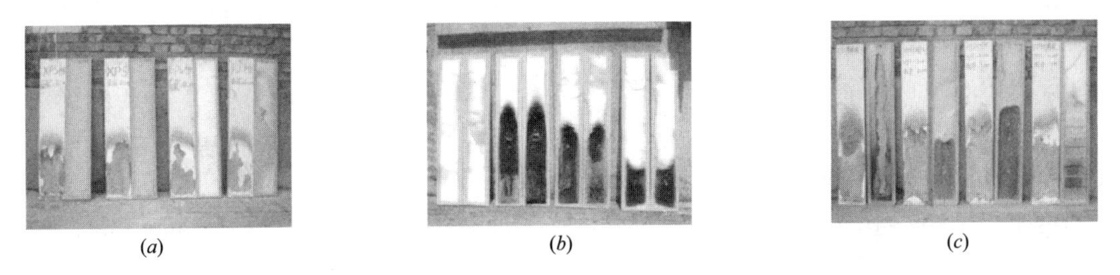

(a) (b) (c)

Figure 5-2-14 Analysis of Specimens

(*a*) XPS-35, XPS-45, EPS-45 and PU-45 from left to right; (*b*) PU-35, PU-5, PU-15 and PU-25 from left to right; (*c*) EPS-5, XPS-25, XPS-15 and EPS-15 from left to right;

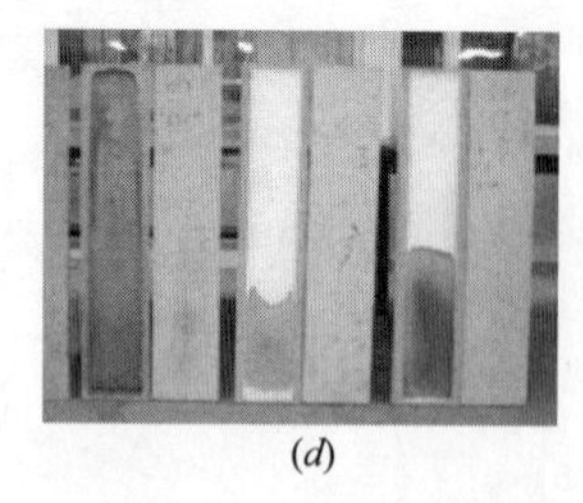
(*d*)

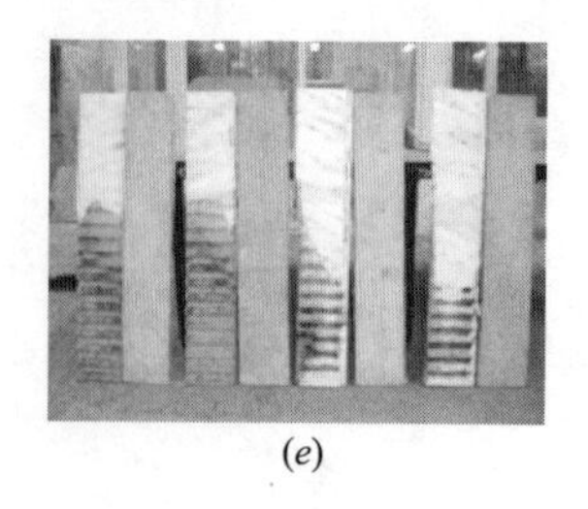
(*e*)

Figure 5-2-14 Analysis of Specimens (Continued)
(*d*) XPS-5, EPS-35 and EPS-25 from left to right;
(*e*) EPS-20/30-1, EPS-20/30-2, EPS-10/20-3 and EPS-10/20-4 from left to right

(3) The specimen itself can be regarded as a separate compartment of the external thermal insulation system. When its joint is wide enough and its material has high fire resistance, namely, with the thick enough protective layer, the compartment is able to prevent the flame from spreading. That is, the insulation layer of the specimen is not completely destroyed. The insulation layer of the thin-plastered EPS board specimen was fully burnt in this test, so the ability of its compartment structure to prevent the flame from spreading needs to be verified in the large-scale model test.

(4) Given the same thickness, the adhesive polystyrene granule system has better protective effects on the organic insulation material than cement mortar. On one hand, the adhesive polystyrene granule are a kind of insulation material with low thermal conductivity, while cement mortar has high thermal conductivity. The transfer of external heat into the former is slower than that into the latter. In addition, it takes a long time for the organic insulation material on the inner side to reach the melting and shrinkage temperature. As a closed cavity is formed after melting of the EPS granule, the thermal conductivity of the adhesive polystyrene granule becomes lower, making heat transfer slower. On the other hand, mortar will crack when heated, thus making the heat transferred faster into the mortar and the organic insulation material reach the melting and shrinkage temperature more quickly.

5.3 Large-scale Model Fire Test of External Thermal Insulation System

At present, the fire safety of external thermal insulation systems of buildings is assessed based on the fire test in China. Therefore, it is critical to select a correct and reasonable test method to objectively and scientifically evaluate the fire safety of external thermal insulation systems.

The small-and medium-scale test methods apply only to simulation of a specific aspect of the combustion process, instead of fully reflecting the combustion status of the external thermal insulation system. Comparatively, the large-scale test method is closer to combustion conditions of a real fire, and has a certain correlation with actual fire conditions. As it is impossible to wholly simulate and reproduce the actual combustion process under laboratory conditions, however, all tests fail to provide comprehensive and accurate fire test results, but can only be used as a reference for behavioral characteristics of a fire.

With the development of fire science, a consensus on the development direction of fire test methods and technologies has been gradually reached. That is, the best and most useful test method should be well associated with real fire scenarios, and its results can be applied in the simulation calculations of real fires as well as the engineering design of fire safety. The test method should also be a performance-based reaction-to-fire test method. Most of prevailing standards in China specifies the combustibility of insulation materials, involving small-scale tests and at most medium-scale tests. This indicates that the stipulated test conditions are far different from real fires. Therefore, the related test results can be used for comparison of individual indicators, but hardly reflect the performance of materials in real fires.

Currently, the fire safety of external thermal insulation systems should be assessed based on large-scale fire test results in China. By means of demonstrating and analyzing foreign standards, the large-scale UL 1040 wall corner fire test and BS 8414-1 window fire test were finally selected to examine external thermal insulation systems. So far, 10 wall corner fire tests and 32 window fire tests have been completed. The available test results show that the window fire test model is closer to the state of the external thermal insulation system in most real fires and can be used as a main means for future experiments.

5.3.1 Introduction to Fire Test Methods

5.3.1.1 UL 1040 Wall Corner Fire Test

The *Fire Test of Insulated Wall Construction* (UL 1040: 2001) is a standard of the Underwriters Laboratories of the United States. The attack of external fire to a building is simulated in the test process to examine the fire resistance of the external thermal insulation system. This test has the following advantages: the model covers the external thermal insulation system including the fire compartment; the spreading of flame in the horizontal or vertical direction of the external thermal insulation system can be observed; and the test state can fully reflect the overall resistance of the external thermal insulation system to a real fire.

The model applied in the UL 1040 wall corner fire test consists of two walls mounted in a right angle to form a large corner (6.10m × 6.10m × 9.14m). The roof is covered with a non-combustible inorganic sheet. A real building can be represented in this test, with the test model shown in Figure 5-3-1.

Figure 5-3-1 UL 1040 Test Model

The insulation system is installed on both walls that are connected to the roof as per actual conditions. The fire source is a timber stack (1.22m × 1.22m × 1.07m) consisting of 12 layers of battens (347 ± 4.54kg). Flame spreading in the vertical and horizontal direction of the external thermal insulation system can be observed in this test.

Temperature measuring points are located above the timber stack, on the surface of the wall with the external thermal insulation system and also in the atmospheric environment. Moreover, video recording is conducted from different angles during the test.

Criteria for the conformity of UL 1040 test results:

(1) During the test, surface burning shall not extend beyond 18 feet (5.49m) from the intersection of the two walls.

(2) Post-test observations shall show that the combustive damage of the test materials within the assembly diminishes at increasing distance from the immediate fire exposure area.

5.3.1.2 BS 8414-1 Window Fire Test

The British standard BS 8414-1: 2002 "Fire Performance of External Cladding Systems-Part 1: Test Method for Non-loadbearing External

Cladding Systems Applied to the Face of the Building" is mainly implemented to inspect the flame spreading range in the longitudinal direction of the external thermal insulation system.

The BS 8414-1 Window fire test provides an assessment method for the fire resistance of the non-bearing cladding system applied on the face of a building and exposed to external flame under controlled conditions, the rain shield above the cladding system, and the external thermal insulation system. In this test, the impact of flame out of the window or hole on the external thermal insulation system in the presence of a typical external fire or flashover in a building. This test method also applies to examination of the ignition and combustion of the external thermal insulation system in the event of fire or kindling, as well as the ability of this system to prevent the flame from spreading after combustion or fire. That is, the resistance of the external thermal insulation system to external fire attack or its overall fire resistance can be assessed simultaneously.

The BS 8414-1 window fire test, which has the same advantages as the wall corner fire test, can be conducted to examine the fire resistance of a whole system including the external thermal insulation system. In terms of the probability of real fire attach to buildings, however, this test method is of more universal significance. Figure 5-1-2 illustrates the principle of outward spreading of an indoor fire along the external thermal insulation system from the window of a building. If the external thermal insulation system is able to prevent the flame from spreading, the fire will not spread.

Figure 5-3-2 shows the window fire test model.

Figure 5-3-2 Window Fire Test Model

The flame spreading and its probability should be taken into consideration in the performance standard and classification based on the test method specified in the standard BS 8414-1: 2002. The system performance will be evaluated in accordance with the following three criteria: external fire spreading, internal fire spreading and mechanical properties.

(1) External fire spreading

If the temperature of the external thermocouple mounted outside the horizontal line 2 exceeds 600℃ for more than 30s within 15min, external fire spreading will occur.

(2) Internal fire spreading

If the temperature of the external thermocouple mounted inside the horizontal line 2 exceeds 600℃ for more than 30s within 15min, internal fire spreading will occur.

(3) Mechanical properties

There are no failure standards for mechanical properties. Instead, some details such as system damage, debris, delamination or flame fragments will be included in the test report. The failure of mechanical properties will be used as part of comprehensive risk assessment.

In order to transform the standard BS 8414-1 into a fire resistance test method suitable for external thermal insulation systems in China, 12 organizations, including Beijing Building Research Institute of CSCEC, jointly applied for the formulation of the product standard "Fire Test Method for External Thermal Insulation System of Building" of the construction industry. The draft for approval has been completed, involving the following details related to the standard BS 8414-1.

(1) The external thermal insulation system

is mounted on the building base surface of the test model in the manner specified by the client. The attack of flame out of the window or hole to the external thermal insulation system after flashover in a room is simulated to examine the damage to this system and determine its flame spreading performance.

(2) The wall body of the test model consists of main and auxiliary walls. A combustion chamber is provided below the main wall. The test model is composed of aerated concrete blocks (not less than 600kg/m^3), in which the wall body is 8400mm high, the main wall is 3500mm wide and the auxiliary wall is 2500mm wide. The auxiliary wall is perpendicular to the main wall and at a distance of (250±10) mm away from the opening edge of the combustion chamber. The external dimensions of the opening of the combustion chamber are: (2, 000±100) mm high and (2, 000±100) mm wide; and its internal dimensions are: (2, 300±50) mm high, (2, 000 ±100) mm wide and (1, 050 ± 50) mm deep. The horizontal lines at 2, 500mm and 5, 000mm above the top of the opening of the combustion chamber are respectively defined as the horizontal baseline s 1 and 2. External and internal thermocouples are located above the horizontal baseline s 1 and 2 based on the specific structure of the external thermal insulation system. The measuring points of the external thermocouple shall extend beyond the external surface of the external thermal insulation system by (50±5) mm. The measuring points of the internal thermocouple shall be located at 1/2 of the thickness of each combustible layer. If there is cavity in the external thermal insulation system, a thermocouple shall be installed at 1/2 of the thickness of the cavity layer. If a layer is less than 10mm, the thermocouple may not be used. The test model and thermocouple location are shown in Figure 5-3-3.

(3) Specimens shall involve all parts of the external thermal insulation system, and be

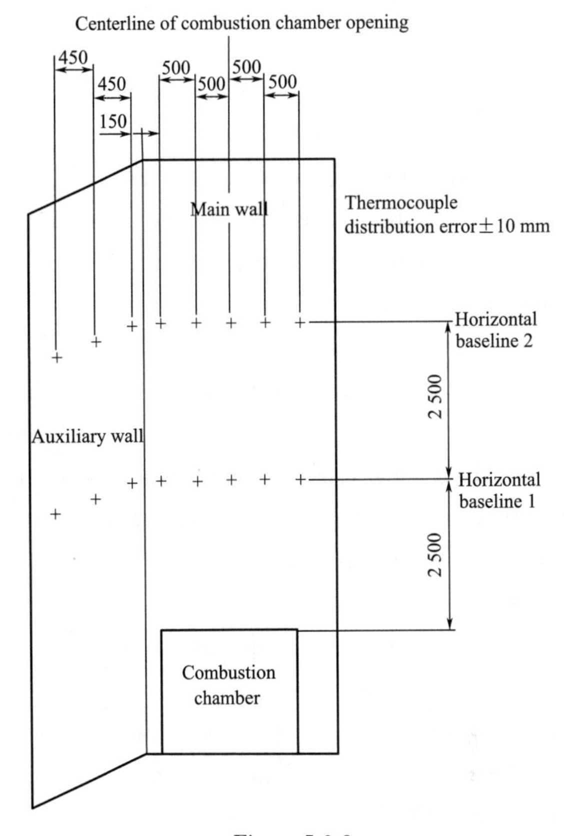

Figure 5-3-3

mounted in accordance with the installation requirements of this system. The thickness of specimens shall not exceed 200mm, and their width and height shall completely cover main and auxiliary walls of the model. The specimen in the corner formed by the main and auxiliary walls shall be connected tightly or installed as per the requirements of the client. The edges of specimens and peripheries of the opening of the combustion chamber shall be protected based on the actual application or the requirements of the client. If there is horizontal deformation joints in a real external thermal insulation system, horizontal joints of the specimen shall be set at intervals specified by the client, at least one at (2400±100) mm above the opening of the combustion chamber. If there are vertical deformation joints in a real external thermal insulation system, vertical joints of the specimen shall be set at intervals specified by the client, one of which shall be located at the upper extension of the opening centerline of the combustion chamber, with the allowable deviation of ±100mm from this centerline. If there are horizontal fire barriers in a real external thermal insulation system, horizontal fire barriers of the specimen shall be set in line with the requirements of the client, and the highest fire barrier shall be located below the horizontal baseline 2, with the upper edge at least 100mm away from the horizontal baseline 2.

(4) The time of change in the combustion state and mechanical properties of specimens shall be recorded during the test. The specimens shall be cooled and checked for damage such as cracking, melting, deformation and delamination after this test, but excluding blackening or fading. The following records shall be made (covering objects may be removed as per the inspection requirements): (i) flame spreading range in the vertical and horizontal direction of the specimen surface; (ii) flame spreading and damage in the vertical and horizontal direction of each middle layer; (iii) flame spreading and damage in the vertical and horizontal direction inside a cavity (if any); (iv) burning or stripping of the external surfaces of specimens; and (iv) details of any collapse or partial collapse of specimens.

(5) The flame spreading performance of the external thermal insulation system based on the test data are determined according to the following criteria. The external thermal insulation system will be regarded without the flame spreading performance if two requirements below are met simultaneously: (i) the temperature rise of any thermocouple inside the horizontal baseline 2 does not exceed 500℃, or the duration of the temperature rise of such a thermocouple above 500℃ is less than 20s; and (ii) inspection results show that the vertical combustion height of each combustible layer does not extend above the horizontal baseline 2 by 300mm. Otherwise, the external thermal insulation system shall be regarded with the flame spreading performance.

The formulation team members and review experts all agreed that the external thermal insulation system should be assessed properly based on China's national conditions.

5.3.2 Window Fire Test

5.3.2.1 Test Summary

Up to now, 32 window fire tests have been completed, summarized as follows.

Based on structural measures, the 32 tests are classified as follows: 3 tests of the fire barrier based on the rock wool, 4 tests of the fire barrier of the rigid polyurethane foam, 1 test of the fire barrier based on the phenolic resin, 2 tests of the fire protection beam based on the rock wool, 1 test of the rock-wool fire barrier of the foam cement cornice, 2 tests of the tile finish, 10 tests of the thin plaster, and 9 tests of the thin plaster. As per the types of insulation materials, these tests are classified into 19 tests of the external thermal insulation system based on the EPS board (15 tests of the system based on the thin plaster and 4 tests of the system with the thick protective layer), 8 tests of the external thermal insulation system based on the rigid polyurethane foam (3 tests of the system based on the thin plaster and 5 tests of the system with the thick protective layer), 3 tests of the external thermal insulation system based on the XPS board (1 test of the system based on the thin-plastered Class B_2 XPS board, 1 test of the system based on the thin-plastered Class B_1 XPS board and 1 test of the system based on the thick-plastered Class B_2 XPS board), and 2 tests of the system based on the modified phenolic board.

Table 5-3-1 shows the test details, and Table 5-3-2 to 5-3-8 summarize the test states based on the insulation material and system protection.

List of Window Fire Tests **Table 5-3-1**

No.	System Name	Test Date	Test Location	System Structure Characteristics				Fire Barrier (or Fire Protection Beam)	Flame Spreading Performance
				Insulation material	Protection layer type	Bonding method	Fire partition		
1	External thermal insulation system based on the adhesive polystyrene granule pasted EPS board	February 2,2007	Beijing-Zhenli	EPS	Thick plaster	No cavity	Compartment	—	None
2	External thermal insulation system based on the thin-plastered EPS board	April 14, 2007	Beijing-Zhenli	EPS	Thin plaster	With cavity; bonding area: ≥40%	None	—	Not evaluated
3	External thermal insulation system based on the thin-plastered EPS board	May 29, 2007	Beijing-Zhenli	EPS	Thin plaster	With cavity; bonding area: ≥40%	None	—	Yes
4	External thermal insulation system based on the thin-plastered rigid polyurethane foam composite board	May 30, 2007	Beijing Tongzhou	PU	Thin plaster	With cavity; bonding area: ≥40%	None	—	None
5	External thermal insulation system based on the sprayed rigid polyurethane foam plaster	July 16, 2007	Beijing Tongzhou	PU	10	No cavity	None	—	None
6	External thermal insulation system based on the poured rigid polyurethane foam	September 6,2007	Beijing Tongzhou	PU	Thin plaster	No cavity	None	—	None
7	External thermal insulation system based on the expanded glass bead, fireproof insulation mortar and composite EPS board	November 13, 2007	Beijing Tongzhou	EPS	Thick plaster	With cavity; bonding area: ≥40%	None	—	None
8	External thermal insulation system based on the thin-plastered EPS board-fire barrier of rigid polyurethane foam	April 23, 2008	Beijing Tongzhou	EPS	Thin plaster	With cavity; bonding area: ≥40%	None	Fire barrier of rigid polyurethane foam	None
9	External thermal insulation system based on the thin-plastered EPS board-rock-wool fire barrier	October 7,2008	Jingyeda	EPS	Thin plaster	With cavity; bonding area: ≥40%	None	Rock-wool fire barrier	None

Continued

No.	System Name	Test Date	Test Location	System Structure Characteristics				Fire Barrier (or Fire Protection Beam)	Flame Spreading Performance
				Insulation material	Protection layer type	Bonding method	Fire partition		
10	External thermal insulation system based on the thin-plastered EPS board-fire barrier of rigid polyurethane foam	October 21, 2008	Beijing Tongzhou	EPS	Thin plaster	With cavity; bonding area: ≥40%	None	Fire barrier of rigid polyurethane foam	None
11	External thermal insulation system based on the thin-plastered EPS board-phenolic fire barrier	November 11, 2008	Jingyeda	EPS	Thin plaster	With cavity; bonding area: ≥40%	None	Phenolic fire barrier	None
12	External thermal insulation system based on the thin-plastered EPS board-rock-wool fire protection beam	March 18, 2009	Jingyeda	EPS	Thin plaster	With cavity; bonding area: ≥40%	None	Rock-wool fire protection beam	Yes
13	External thermal insulation system based on the thin-plastered EPS board-rock-wool fire protection beam	March 18, 2009	Jingyeda	EPS	Thin plaster	With cavity; bonding area: ≥40%	None	Rock-wool fire protection beam	None
14	External thermal insulation system based on the thin-plastered EPS board-foam cement cornice/ rock-wool fire barrier	March 18, 2009	Jingyeda	EPS	Thin plaster	With cavity; bonding area: ≥40%	None	Foam cement cornice and rock-wool fire barrier	Yes
15	External thermal insulation system based on the thin-plastered EPS board	April 13, 2009	Jingyeda	EPS	Thin plaster	With cavity; bonding area: ≥40%	None	—	Yes
16	External thermal insulation system based on the thin-plastered EPS board-fire barrier of rigid polyurethane foam	June 3, 2009	Beijing Tongzhou	EPS	Thin plaster	With cavity; bonding area: ≥40%	None	Fire barrier of rigid polyurethane foam	None
17	External thermal insulation system based on the high-strength refractory plant fiber composite insulation board and cast-in-situ polyurethane foam	August 12, 2009	Jingyeda	PU	Thick protective layer	No cavity	None	—	None

Continued

No.	System Name	Test Date	Test Location	System Structure Characteristics				Fire Barrier (or Fire Protection Beam)	Flame Spreading Performance
				Insulation material	Protection layer type	Bonding method	Fire partition		
18	External thermal insulation system based on the thin-plastered XPS board-rock-wool fire barrier	August 12, 2009	Jingyeda	XPS	Thin plaster	With cavity; bonding area: ≥40%	None	Rock-wool fire barrier	None
19	External thermal insulation system based on the thin-plastered rigid polyurethane foam composite board	August 20, 2009	Beijing Tongzhou	PU	Thin plaster	With cavity; bonding area: ≥40%	None	—	None
20	External thermal insulation system based on the EPS board and tile finish	September 3, 2009	Jingyeda	EPS	Thick protective layer; tile finish	With cavity; bonding area: ≥40%	None	—	None
21	Curtain wall insulation system based on the sprayed rigid polyurethane foam	November 22, 2009	Beijing Tongzhou	PU	Thick plaster	The insulation layer is fully bonded with the base course wall, but there are cavities in the curtain wall. *	There are no fire partitions inside the insulation layer, but the cavities of the curtain wall are isolated by rock wool barriers.	Rock-wool fire barrier	None
22	External thermal insulation system based on the adhesive polystyrene granule pasted and thin-plastered EPS board	November 26, 2009	Beijing-Zhenli	EPS	Thin plaster	No cavity	Compartment	—	None
23	External thermal insulation system based on the thin-plastered EPS board-fire barrier of rigid polyurethane foam	February 3, 2010	Beijing Tongzhou	EPS	Thin plaster	With cavity; bonding area: ≥40%	None	Fire barrier of rigid polyurethane foam	Not evaluated
24	External thermal insulation system based on the adhesive polystyrene granule pasted XPS board	March 23, 2010	Beijing-Zhenli	XPS	Thick plaster	No cavity	Compartment	20cm adhesive polystyrene granule in window	None
25	External thermal insulation system based on the thin-plastered EPS board	May 13, 2010	Jingyeda	EPS	Thin plaster	With cavity; bonding area: ≥40%	None	—	None

Continued

No.	System Name	Test Date	Test Location	System Structure Characteristics				Fire Barrier (or Fire Protection Beam)	Flame Spreading Performance
				Insulation material	Protection layer type	Bonding method	Fire partition		
26	External thermal insulation system based on the EPS board and tile finish	May 13, 2010	Jingyeda	EPS	Thick protective layer; tile finish	With cavity; bonding area: ≥40%	None	—	None
27	Curtain wall insulation system based on the phenolic thin plaster-aluminum sheet	June 23, 2010	Beijing-Zhenli	PF	Thin plaster	With cavity; bonding area: ≥40%	None	—	Yes
28	External thermal insulation system based on the sprayed rigid polyurethane foam and thick plaster	September 2, 2010	Beijing Tongzhou	PU	Thick plaster	No cavity	None	—	None
29	Curtain wall insulation system based on the phenolic thick plaster (compartment structure)-aluminum sheet	September 10, 2010	Beijing-Zhenli	PF	Thick plaster	No cavity	Yes	Adhesive polystyrene granule partition	None
30	External thermal insulation system based on the thin-plastered XPS board	October 28, 2010	Jingyeda	XPS(Class B_1)	Thin plaster	With cavity; bonding area: ≥40%	None	—	Yes
31	External thermal insulation system based on the thin-plastered EPS board-rock-wool fire barrier	October 28, 2010	Jingyeda	EPS	Thin plaster	With cavity; bonding area: ≥40%	None	Rock-wool fire barrier	None
32	External thermal insulation system based on the rigid polyurethane foam insulation board and thick plaster	November 5, 2010	Beijing Tongzhou	PU	Thick plaster	With cavity; bonding area: ≥40%	None	—	None

Notes: 1. EPS-expanded polystyrene board; XPS-extruded polystyrene board, PU-rigid polyurethane foam, and PF-modified phenolic board.

2. No. 2 and 23 tests in this table were conducted for demonstration only, mainly introducing the window fire test method to leaders and experts. As the wind speeds applied in these tests did not meet the test standards, the test results were not evaluated.

Test Summary of External Thermal Insulation System based on Thin-plastered EPS Board and Not Fire Barrier

Table 5-3-2

Test System	3. External thermal insulation system based on the thin-plastered EPS board	15. External thermal insulation system based on the thin-plastered EPS board
Burnt state of insulation layer after testing		

Test Summary of External Thermal Insulation System based on Thin-plastered EPS Board and Fire Partition

Table 5-3-3

Test System	8. External thermal insulation system based on the thin-plastered EPS board-fire barrier of rigid polyurethane foam	9. External thermal insulation system based on the thin-plastered EPS board-rock-wool fire barrier	10. External thermal insulation system based on the thin-plastered EPS board-fire barrier if rigid polyurethane foam	11. External thermal insulation system based on the thin-plastered EPS board-phenolic fire barrier
Burnt state of insulation layer after testing				
Test System	12. External thermal insulation system based on the thin-plastered EPS board-rock-wool fire protection beam	13. External thermal insulation system based on the thin-plastered EPS board-rock-wool fire protection beam	14. External thermal insulation system based on the thin-plastered EPS board-foam cement cornice/ rock-wool fire barrier	16. External thermal insulation system based on the thin-plastered EPS board-fire barrier of rigid polyurethane foam
Burnt state of insulation layer after testing				

Continued

Test System	22. External thermal insulation system based on the adhesive polystyrene granule pasted and thin-plastered EPS board	25. External thermal insulation system based on the thin-plastered EPS board	31. External thermal insulation system based on the thin-plastered EPS board-rock-wool fire barrier
Burnt state of insulation layer after testing			

Test summary of System based on EPS Board and Thick Protective Layer Table 5-3-4

Test System	1. External thermal insulation system based on the adhesive polystyrene granule pasted EPS board	7. External thermal insulation system based on the expanded glass bead, fireproof insulation mortar and composite EPS board	20. External thermal insulation system based on the EPS board and tile finish	26. External thermal insulation system based on the EPS board and tile finish
Burnt state of insulation layer after testing				

Test summary of System based on Rigid Polyurethane Foam and Thin Plaster Table 5-3-5

Test System	4. External thermal insulation system based on the thin-plastered rigid polyurethane foam composite board	6. External thermal insulation system based on the poured rigid polyurethane foam	19. External thermal insulation system based on the thin-plastered rigid polyurethane foam composite board
Burnt state of insulation layer after testing			

Test Summary of System based on Rigid Polyurethane Foam and Thick Protective Layer **Table 5-3-6**

Test System	5. External thermal insulation system based on the sprayed rigid polyurethane foam plaster	17. External thermal insulation system based on the high-strength refractory plant fiber composite insulation board and cast-in-situ polyurethane foam	21. Curtain wall insulation system based on the sprayed rigid polyurethane foam	28. External thermal insulation system based on the sprayed rigid polyurethane foam and thick plaster	32. External thermal insulation system based on the rigid polyurethane foam insulation board-thick plaster
Burnt state of insulation layer after testing					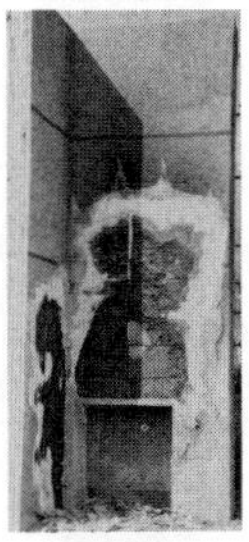

Test Summary of Curtain Wall Insulation System based on Phenolic-Aluminum Sheet **Table 5-3-7**

Test System	27. Curtain wall insulation system based on the phenolic thin plaster-aluminum sheet	29. Curtain wall insulation system based on the phenolic thick plaster(compartment structure)-aluminum sheet
Burnt state of insulation layer after testing		

Test Summary of External Thermal Insulation System based on XPS Board **Table 5-3-8**

Test System	18. External thermal insulation system based on the thin-plastered XPS board-rock-wool fire barrier	24. External thermal insulation system based on the adhesive polystyrene granule pasted XPS board	30. External thermal insulation system based on the thin-plastered XPS board
Burnt state of insulation layer after testing			

Table 5-3-9 lists the temperature of the measuring points and burning range of the insulation layer in the window fire test of the external thermal insulation system.

Window Fire Test Results of External Thermal Insulation System **Table 5-3-9**

No.	Maximum Temperature (℃) of Measuring Point of Combustible Insulation Layer on Horizontal Baseline 2	Burning Height of Combustible Insulation Layer	Judgment of Flame Spreading Performance
1	<500	No obvious burning	None
2	—	—	Not evaluated
3	>500	Complete burning	Yes
4	<500	10cm above the horizontal baseline 2	None
5	<500	5cm above the horizontal baseline 2	None
6	<500	10cm below the horizontal baseline 2	None
7	<500	No obvious burning	None
8	<500	Below the horizontal baseline 2	None
9	<500	Below the horizontal baseline 2	None
10	>500	Complete burning	Yes
11	<500	Below the horizontal baseline 2	None
12	<500	To the model top	Yes
13	<500	Below the horizontal baseline 2	None
14	>500	Lower edge of the highest fire barrier	Yes
15	>500	To the model top	Yes
16	<500	Below the horizontal baseline 2	None
17	<500	Below the horizontal baseline 2	None
18	<500	Below the horizontal baseline 2	None
19	<500	15cm above the horizontal baseline 2	None
20	<500	Below the horizontal baseline 2	None
21	<500	Horizontal baseline 1	None
22	<500	Below the horizontal baseline 2	None
23	—	—	Not evaluated
24	<500	Horizontal baseline 1	None
25	<500	Below the horizontal baseline 2	None
26	<500	Below the horizontal baseline 2	None
27	>500	To the model top	Yes
28	<500	Below the horizontal baseline 2	None
29	<500	Below the horizontal baseline 2	None
30	<500	To the model top	Yes
31	<500	Lower edge of the highest fire barrier	None
32	<500	Below the horizontal baseline 2	None

5.3.2.2 Analysis of Test Results

1. External thermal insulation system based on the EPS board

(1) The results of No. 3 and 5 tests show that the external thermal insulation system based on the thin-plastered EPS board without structural measures has the flame spreading performance. A single external thermal insulation system based on the thin-plastered EPS board involves fire risks, but there is difference between specific performances of the above two systems, which may be associated with the construction quality.

(2) The results of No. 9 and 31 tests show that the external thermal insulation system based on the thin-plastered EPS board and the rock-wool fire barrier does not have the flame spreading performance. With the 20cm rock-wool fire barrier, the flame is effectively stopped from spreading vertically upwards. Especially in No. 31 test, the EPS board was 20cm thick, with sufficient combustible materials. It can be concluded that the 20cm rock-wool fire barrier is able to suppress flame spreading.

(3) The results of No. 12 and 13 tests show that the external thermal insulation system based on No. 13 thin-plastered EPS board and rock-wool fire protection beam does not have the flame spreading performance, while No. 12 system has. Their fire resistance in the test process was superior to that of the external thermal insulation system based on the thin-plastered EPS board. Accordingly, the rock-wool fire protection beam has some impact on the prevention of the flame from spreading vertically upwards, but is less effective than the fire barrier. The failure in No. 12 test may be caused by pasting of the insulation boards first and then fire protection beams in their joints. The fire protection beams may not be fully bonded in the above process, resulting in cavities behind them. With the lesson learnt from this failure, construction was conduct bottom-up in No. 13 test, in which fire protection beams are fully bonded, thereby achieving significant effects of flame suppression. This once again proves that construction factors have important impact on the fire resistance of the external thermal insulation system.

(4) The results of No. 8 and 16 tests show that the external thermal insulation system based on the thin-plastered EPS board and fire barrier of rigid polyurethane foam does not have the flame spreading performance. With the 30cm fire barrier of rigid polyurethane foam, the flame is effectively prevented from spreading vertically upwards and limited below the first fire barrier. The results of No. 10 test show that the external thermal insulation system has the fire spreading performance, which is obviously related to the construction quality. During No. 10 test, the glass fiber mesh in the surface mortar was obliquely broken 17 minutes and 20 seconds, and the suspended surface was ignited from the upper part of its edge 17 minutes and 55 seconds after ignition and collapsed 11s later. Therefore, the fire barrier can be regarded with relative effects of flame suppression. The precondition for effective prevention of flame spreading is that the construction quality of the external thermal insulation system meets the relevant technical standards.

(5) The results of No. 11 test show that the external thermal insulation system based on the thin-plastered EPS board and phenolic fire barrier does not have the flame spreading performance. With the 30cm phenolic fire barrier, the flame is effectively prevented from spreading vertically upwards. The results of No. 14 test show that the external thermal insulation system based on the thin-plastered EPS board and foam cement cornice/ composite rock-wool fire barrier has the fire spreading performance, in which the rock-wool fire barrier has significant effects of flame suppression. If the rock-wool fire barrier is mounted unreasonably, flame will spread in the external thermal insulation system. The results of No. 22 test show that the external thermal insulation system based on the adhesive polystyrene granule pasted and thin-plastered EPS board does not have the flame spreading performance.

(6) For the EPS board system subject to protection with a thick protective layer, four tests were conducted, namely, No. 1 test (system with the adhesive polystyrene granule insulation mortar), No. 7 test (system with the expanded glass bead and fireproof insulation mortar), No. 20 test (system with the wire mesh and tile finish) and No. 26 test (system with the glass fiber mesh and tile finish). The test results show that the external thermal insulation system

with the thick protective layer does not have the flame spreading performance under test conditions.

2. External thermal insulation system based on the rigid polyurethane foam

The results of No. 4, 6 and 19 tests show that the external thermal insulation system based on the rigid polyurethane foam and thin plaster but without structural measures does not have the flame spreading performance. It can be concluded that the external thermal insulation system based on the rigid polyurethane foam and thin plaster is not able to spread fire, so the fire barrier is not needed in this system.

For the rigid polyurethane foam insulation system subject to protection with a thick protective layer, five tests were conducted, namely, No. 5 test (system based on the sprayed rigid polyurethane foam plaster), No. 17 test (system based on the high-strength refractory plant fiber composite insulation board and cast-in-situ polyurethane foam), No. 21 test (curtain wall insulation system based on the sprayed rigid polyurethane foam), No. 28 test (system based on the sprayed rigid polyurethane foam and thick plaster), and No. 32 test (system based on the rigid polyurethane foam insulation board and thick plaster). The test results show that the above system has no flame spreading performance, but excellent reaction-to-fire performance. It can be concluded that there are no fire risks if a thick protective layer is applied in the rigid polyurethane foam system.

3. External thermal insulation system based on the XPS board

Three tests were conducted to the external thermal insulation system based on the XPS board, namely, No. 18 test (system based on the thin-plastered XPS board and rock-wool fire barrier), No. 24 test (system based on the adhesive polystyrene granule pasted XPS board), and No. 30 test (system based on the thin-plastered Class B_1 XPS board).

Among them, the system involved in No, 18 test does not have the flame spreading performance, as the rock-wool fire barrier has effects of flame suppression; the system in No. 24 test also has no flame spreading performance, in which the damaged area of the XPS board under test conditions was well controlled due to the thick protective layer of adhesive polystyrene granule and the fully bonded no-cavity structure; and the flame tended to spread in No. 30 test under test conditions as a result of fireproofing measures such as the fire barrier, although the XPS board of Class B_1 was used.

As shown by the test results, the external thermal insulation system with a fire barrier will not have the adequate fire safety, even if the XPS board of Class B_1 is used. The maximum temperature inside the insulation layer above the horizontal line 2 in No. 30 test was 446℃, greatly exceeding that (235℃) in No. 18 test (system based on the XPS board of Class B_2 and the rock-wool fire barrier). It is recommended to install the fire barrier or take other fireproof structural measures in the external thermal insulation system based on the thin-plastered Class B_1 XPS board. It can be seen that improving the combustibility of the XPS board does not actually bring about the desired fireproofing effects, but may greatly increase the costs and has unexpected effects on the thermal conductivity, dimensional stability and other technical indicators of the material. At present, a more effective and economical method is to use the fire barrier that has significant effects.

4. Curtain wall system based on phenolic-aluminum sheet

Two tests have been conducted so far to the curtain wall system based on the phenolic-aluminum sheet, namely, No. 27 and 29 tests.

During No. 27 test, the 7cm phenolic insulation board was applied in the curtain wall system based on the phenolic thin plaster and composite aluminum sheet, involving the point-and-frame bonding (bonding area: ⩾40%), other fireproof structures were not used in this system, and the system was seriously destroyed under test conditions, so it can be concluded that this system has the flame spreading performance. During No. 29 test, the phenolic insulation board was pasted in a no-cavity manner, subject to full bonding (bonding area: 100%), compartments were built in the insulation board, the surface of the insulation board was plastered with the 1cm adhesive polystyrene granule, the keel of the curtain wall was sealed with the adhesive polystyrene granule, and this system had high fire safety under test conditions, so this system does not have the flame spreading performance. The comparison of combustion in the above two tests is show in Figure 5-3-4.

(*a*) (*b*)

Figure 5-3-4 Comparison of Combustion State of Curtain Wall System based on Phenolic-Aluminum Sheet

(*a*) burnt state in No. 27 test; (*b*) burnt state in No. 29 test

5.3.2.3 Summary

The fire safety of an external thermal insulation system is affected by its material and structure. Currently, the application of combustible insulation materials is undoubtedly the status quo that cannot be changed. The external thermal insulation system based on the thin-plastered EPS board, which is widely applied in China and the world as a whole, has certain fire safety. This is also beyond doubt due to such a wide range of applications. Our research should focus on further improvement of the fire safety of the existing external thermal insulation systems.

In addition to the improvement of the combustibility of organic insulation materials, fireproof structural measures should also be an essential part of the research on improvement of the fire resistance of external thermal insulation systems. Seen from the current applications of external thermal insulation systems in China, the system structure is a key factor affects its fire safety. The overall fireproof structure is an innovative way for energy efficiency and fire safety of buildings in China. Its principle is the same as that of applying the fireproof coating and protective plates to protect steel structures. If overall fireproof measures are taken to steel structures, their applicable height will greatly exceed that of concrete buildings.

1. Isolation measures

Isolation measures can help to reduce heat transfer in the insulation material to some extent, thus decreasing the risk of ignition of adjacent materials. They are taken mainly to isolate heat conduction in the insulation system. The following isolation measures have been verified in tests: fire barrier in the insulation layer, fireproof structure (fire protection beam) in the door/window opening, compartment structure of the system itself, etc.

The fire barrier is a belt-shaped fireproof structure that is mounted horizontally or vertically to prevent the flame from spreading in the external thermal insulation system. The fire protection beam is a kind of fire isolation measure taken in the door or window opening. Similar to the fire barrier, the fire protection beam is a belt-shaped fireproof structure mounted horizontally on the upper edge of the door or window opening. It extends the vertical edge of the door or window opening for a certain length under normal conditions. The compartment structure is a fireproof structure in which the insulation material is separated from other insulation boards via inorganic insulation mortar. Compartment joints should be wide enough.

The fire barrier is applied to prevent the flame from spreading inside the external thermal insulation system. The fire protection beam mainly has the function of preventing or reducing the external flame attack to the combustible insulation material inside the external thermal insulation system. In this sense, the fire barrier and fire protection beam should be able to maintain the stability of their fireproof structure as well as the essential stability of the protective surface of the insulation system in the presence of a fire. The fundamental stability and enough fire resistance should be maintained under fire conditions, in order to guarantee the essential stability of the overall fireproof structure of the fire barrier. Meanwhile, if the protective surface of the external thermal insulation system is essentially kept stable, the flame will be effectively prevented from spreading therein. The rock wool and inorganic insulation material can be used as a fire barrier. The mineral wool, which melts and shrinks when heated, is a kind of non-combustible insulation material, but must not be used as the fire barrier.

The results of window fire tests show that the acting height of the flame from the window onto the wall surface generally reaches the horizontal line 1 above the window (that is, the flame height is about 2.5m), so the fire barrier (fire protection beam) at the window is not enough to prevent the flame attack to the external thermal insulation system. Significant effects of flame suppression cannot be achieved until fire compartments are built in the insulation material. The distance between the fire barrier and window should be properly increased. If the fire barrier is directly mounted at the window, its isolation effect will be weak. Additionally, the longer the intervals between fire compartments, the poorer the ability to prevent flame spreading is. This needs to be fully considered during formation of relevant regulations.

The results of No. 8, 9, 11, 13, 14, 16, 18, 22, 24, 29 and 31 fully demonstrate the effectiveness of isolation measures to prevent flame spreading.

2. Closed cavity

There may be enough air in the cavity for combustion of the insulation material and spreading of the flame in the external thermal insulation system. The fire occurrence and spreading depend on air, so the cavity is conducive to flame spreading. The through cavity and closed cavity have different effects on the fire safety of the external thermal insulation system. The large through cavity is not conducive to the fire safety. The through cavity has the destructive effect similar to the "stack effect" in high-rise buildings. That is, heat convection is caused in the insulation system, thereby resulting in flame spreading. Through cavities should be avoided as much as possible to reduce fire risks of external thermal insulation systems.

In particular, it should be pointed out that the cavity of the pasted insulation board system should be closed. In the presence of a fire, the thermoplastic insulation material may shrink, melt and even burn, leading to the formation of cavities or communication of closed cavities, with adverse effects on the fire resistance of the external thermal insulation system. This has a lot to do with the construction quality of the external thermal insulation system.

In addition to the pasted insulation board system, cavities formed by surface bulging and involving greater hazards were also found in the other systems under fire conditions in the tests. In the event of holes caused by surface bulging, hot air will flow into the external thermal insulation system. This will aggravate the combustion of combustible components inside the system, and even result in flashover under test conditions.

Therefore, the reasonable construction process and conforming construction quality are important factors to guarantee the fire safety of the external thermal insulation system.

The cavity closure plays a more important role in the curtain wall system. The results of No. 21, 27 and 29 tests fully demonstrate the effectiveness of cavity closure.

3. Fireproof protective layer

The fireproof protective layer here includes the plaster layer and finish layer. The plaster layer is mainly composed of surface mortar, and its thickness and quality stability directly determine the fire resistance of the layered structure of the external thermal insulation system. The finish layer mainly consists of the finish paint and face brick. If the finish paint is applied and its thickness is less than 0.6mm or its mass per unit area is less than $300g/m^2$, its impact on the fire resistance of the external thermal insulation system may be ignored. The material, structure and construction quality of the protective layer have different effects on the fire resistance. The stability of the protective layer in a fire affects the overall reaction-to-fire performance of the external thermal insulation system. The damaged state and degree of the insulation material are influenced by the thickness of the protective surface of the external thermal insulation system. If the thickness of the protective layer is increased, heat radiation to the internal combustible insulation material through the surface of the external thermal insulation system will be reduced.

The fireproof protective layer can be applied to effectively reduce the peak heat release rate, improve the flame spreading performance and enhance the fire resistance of the external thermal insulation system. The results of the cone calorimeter test conducted in the laboratory show that, with an appropriate protective layer, the reaction-to-fire data of various external thermal insulation systems are basically identical to those of ordinary cement mortar specimens, so the test specimens of external thermal insulation system will not be ignited. In the combustion shaft furnace test, the burning height of the insulation layer increased with the thickness of the protective layer decreasing. When the protective layer is more than 30mm, the organic insulation material was not damage under the shaft furnace test conditions (the flame of 900℃ was applied on the bottom surfaces of specimens for 20min).

That is, the thicker the protective layer on its surface, the less likely the insulation material is to be ignited and destroyed.

The results of No. 1, 5, 7, 17, 20, 21, 26, 28 and 32 sufficiently demonstrate the effectiveness of a thick protective layer.

5.3.3 Wall Corner Fire Test

Table 5-3-10 to 5-3-12 and Figure 5-3-5 to 5-3-7 summarize and analyze the tests based on the system structure, test process, system state after testing, and test results.

System Structure Features of Wall Corner Fire Test **Table 5-3-10**

No.	System Name	System Structure Features				
		Combustibility rating of insulation material	Thickness (mm) of insulation material	Thickness (mm) of protective layer	Bonding method	Fireproof structural measure
1	External thermal insulation system based on the thin-plastered EPS board	B_2	80	Thin plaster	Point-and-frame bonding; bonding area: ≥40%	None

Continued

No.	System Name	System Structure Features				
		Combustibility rating of insulation material	Thickness(mm) of insulation material	Thickness(mm) of protective layer	Bonding method	Fireproof structural measure
2	External thermal insulation system based on the adhesive polystyrene granule pasted EPS board	B_2	60	Thick plaster and 10mm adhesive polystyrene granule leveling layer	Full bonding	No cavity + fire compartment + fireproof protective surface
3	Curtain wall system based on the adhesive polystyrene granule pasted EPS board aluminum sheet	B_2	70	Thick plaster and 20mm adhesive polystyrene granule leveling layer	Full bonding	No cavity fire compartment + fireproof protective surface
4	Curtain wall system based on the rock wool board-aluminum sheet subject to point bonding and anchorage	A	80	—	Point bonding and anchorage	—

Comparison of Burning Widths in Wall Corner Fire Test **Table 5-3-11**

	1	2	3	4
System	Curtain wall system based on the adhesive polystyrene granule pasted EPS board aluminum sheet	Curtain wall system based on the rock wool board aluminum sheet subject to point bonding and anchorage	External thermal insulation system based on the adhesive polystyrene granule pasted EPS board	External thermal insulation system based on the thin-plastered EPS board
Fireproof structural measure	No cavity + fire compartment + fireproof protective surface	—	No cavity + fire compartment + fireproof protective surface	None
Burning width/ m	0	0	2.4	6.1

Comparison of Burning Areas in Wall Corner Fire Test **Table 5-3-12**

	1	2	3	4
System	Curtain wall system based on the adhesive polystyrene granule pasted EPS board-aluminum sheet	Curtain wall system based on the rock wool board-aluminum sheet subject to point bonding and anchorage	External thermal insulation system based on the adhesive polystyrene granule pasted EPS board	External thermal insulation system based on the thin-plastered EPS board
Fireproof structural measure	No cavity + fire compartment + fireproof protective surface	—	No cavity + fire compartment + fireproof protective surface	None
Burning area/ m^2	0	0	About16	54

Figure 5-3-5 System States during and after the Test

Notes: 1. The system based on the thin-plastered EPS board (with no fireproof structure) was applied on the left sides of test walls in the figures (*a*), (*b*) and (*c*), and the system based on the adhesive polystyrene granule pasted EPS board on the right sides.

2. The curtain wall system based on the pasted EPS board and aluminum board was applied on the left sides of test walls in the figures (*d*), (*e*) and (*f*), and the curtain wall system based on the anchored rock wool board and aluminum sheet on the right sides.

(*a*) During the test; (*b*) After the test; (*c*) Damage to the insulation layer;
(*d*) During the test; (*e*) After the test; (*f*) Damage to the insulation layer

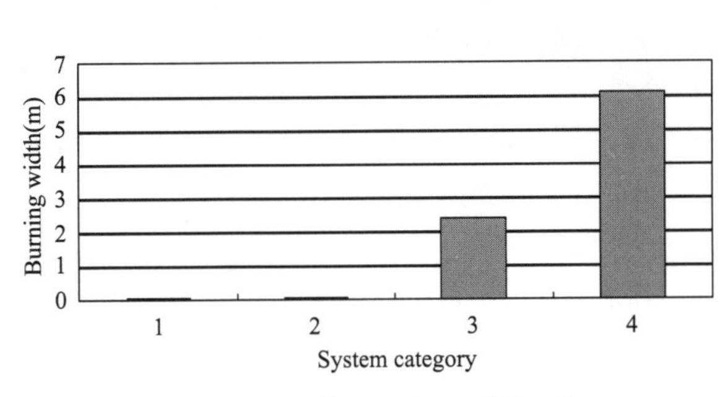

Figure 5-3-6 Comparison of Burning Widths in Wall Corner Fire Test

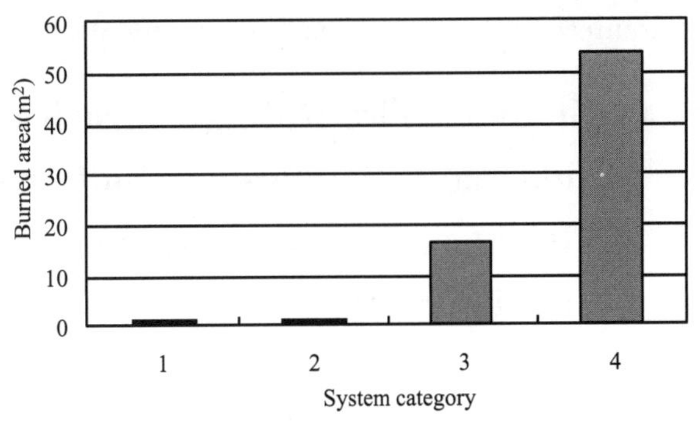

Figure 5-3-7 Comparison of Burning Areas in Wall Corner Fire Test

In the wall corner fire test, the burning area of the insulation system based on the adhesive polystyrene granule pasted EPS board was relatively large, as this system was tested at the same time with the system based on the thin-plastered EPS board and flashover occurred to the latter. Therefore, the fire attack to the pasted board system far exceeded the acting intensity of the test fire, but the flame did not spread in the pasted board system even under such harsh conditions. The adhesive polystyrene granule pasted EPS board in the curtain wall system did not burn during and after the test, but had the performance equivalent to the rock wool insulation system. This further demonstrates the advantages of fireproof structural measures.

5.3.4 Summary

The following conclusions are drawn based on a large number of data from the wall corner fire tests and window fire tests.

(1) The enough fire safety is an important part of the technological research on the external thermal insulation system, and a necessary condition and prerequisite for the application of external thermal insulation systems.

(2) The fire resistance of the overall structure is essential for the fire safety of the external thermal insulation system, so the solution to the fire safety of the overall structure is of practical significance and application value.

(3) Three key elements for fire protection of the external thermal insulation system are the no-cavity structure, fire compartment and fireproof protective surface. The results of a large number of tests as well as the study and application of structural measures prove that fire safety of the external thermal insulation system can be fully guaranteed with organic insulation materials.

5.4 Study on Fire Rating Classification and Applicable Building Height of External Thermal Insulation System

5.4.1 Key Factors of Fire Rating Classification

5.4.1.1 Combustibility Rating of Insulation Material

Because of its great impact on the fire resistance of the external thermal insulation system, the combustibility of an insulation material should be stipulated, which is a basic consensus of experts at home and abroad. For organic insulation materials, it is originally required in Germany that the EPS board should meet Class B_1 requirements. At present, the European Association of External Thermal Insulation intends to lower this requirement to Class E (equivalent to Class B_2 in German standards) of the standard EN13501. The prevailing requirements in China are as follows: minimum Class B_2 for the EPS board (oxygen index: ≥30%), XPS board and rigid polyurethane foam (oxygen index: ≥26%). They are minimum requirements for insulation materials and must be rigorously enforced to reduce the probability of fire.

5.4.1.2 Heat Release Rate of Insulation System

It can be seen from the test results that the peak heat release rate and the total amount of heat are key technical indicators for examination of the fire resistance of the external thermal insulation system, and has an internal relationship with the flame spreading performance. Essentially, the heat release rate is directly associated with the type of the insulation material and the thickness of the protective layer, the latter of which is also a key factor affecting the fire resistance of the external thermal insulation system. Hence, the peak heat release rate, which can be directly measured in the cone calorimeter test, is a main technical parameter for evaluation of the overall fire safety of the external thermal insulation system.

5.4.1.3 Flame Spreading Performance of Insulation System

The insulation system not only contains an insulation material, but also includes the materials of the anti-cracking surface layer and finish layer, in which the minimum unit is a continuous product monomer. In addition to the combustibility of the insulation material, the overall fire resistance should be assessed comprehensively in actual applications. This is of more practical significance in the study on the fire resistance of the external thermal insulation system. Fire hazards of an external thermal insulation system result from flame spreading. The monomer combustion test method adopted in China's new classification standard GB 8624-2012 was derived from the room wall corner fire test in the standard ISO 9705. It applies to interior decoration materials of buildings, based on the heat release rates of materials in the presence of a fire and involving relatively small specimens. However, the flame spreading performance of organic insulation materials is not fully taken into consideration, and the state of the external thermal insulation system in the test fire differs from that in an actual fire. Obviously, the combustibility rating determined by this test method cannot be used as a basis to evaluate the fire safety of the external thermal insulation system.

Through adequate investigations, we selected the UL 1040 wall corner fire test method in the United States and the BS 8414-1 window fire test method in the United Kingdom to examine the fire resistance of the external thermal insulation system, and obtained the initial classification indicators based on the test data.

5.4.2 Study on Fire Rating Classification and Applicable Building Height

5.4.2.1 Basis of Fire Rating Classification

The fire rating classification was studied based on the achievements of the "Technology Research on Fire Protection Test Methods, Fire Rating Evaluation Standards and Architectural Applications of External Wall Insulation Systems", a research and development project (06-k5-35) established in early 2006 and accepted in September 2007 under the 2006 Schedule of the Ministry of Housing and Urban-Rural Development for Science and Technology Project Planning. By reference to relevant foreign standards and test methods and specific conditions in China, it was proposed that the overall fire resistance of the external thermal insulation system is critical to the fire safety of external thermal insulation engineering. The project team obtained a large number of data by conducting cone calorimeter tests, combustion shaft furnace tests, large-scale window fire tests and wall corner fire tests, and through the analysis and study, made achievements in the fire rating classification and applicable building height of the external thermal insulation system.

The attacks of indoor and outdoor fires to a building can be simulated simultaneously in the wall corner fire test and window fire test, especially in the latter. The model dimensions cover various fireproof structures, the test state fully reflects the resistance of an external thermal insulation system to a real fire, and the test results can be applied to evaluate the overall ignitability and flame spreading performance of the external thermal insulation system. From the perspective of actual fire attacks to buildings, the window fire test has more universal significance.

The Test Method for Fire-resistant Performance of External Wall Insulation Systems Applied to Building Façade (GB/T 29416-2012) was implemented on October 1, 2013. It converts the British Standard BS 8414-1: 2002 into a more operational fire test method suitable for external thermal insulation systems in China. As China's first standard on the large-scale fire test method, this standard provides a scientific and unified test method for evaluation of the fire safety of the external thermal insulation system, which plays an active role in studying and improving the fire resistance of external thermal insulation systems in China, promoting the technological progress of external thermal insulation systems and guiding the healthy development of the external thermal insulation industry.

5.4.2.2 Fire Rating Test Method and Indicators

1. Test method

By summarizing the experience abroad, the fire resistance of the external thermal insulation system of a building should be evaluated in two aspects. First, its ignitability should be examined, namely, ignition and peak heat release rate in the presence of fire or kindling. The smoke and toxic gas release with great impact on escape in the event of a fire should also be taken into account at the same time. Such performance indicators can be inspected in the cone calorimeter test. Secondly, the flame spreading performance of the external thermal insulation system should be assessed, namely, the ability to stop the flame from spreading and resistance to external fire attacks or fire resistance in the case of combustion or fire. In principle, the test method in this project was selected depending on the real external thermal insulation systems (including fireproof structures) and correlation with a real fire. All such tests must be conducted within a large scale.

The fire ratings of buildings are different because of their distinctive features. The state of an external thermal insulation system in a large-scale model fire test can fully reflect its overall resistance to a real fire. Corresponding criteria should be formulated based on building categories, which is of universal significance.

As analyzed above, two important indicators are adopted in the fire rating classification standard: peak heat release rate obtained in the cone calorimeter test and flame spreading performance gained in the large-scale model fire test.

2. Fire rating test indicators

The indicators of fire rating classification are listed in Table 5-4-1.

Fire Rating Test Indicators of External Thermal Insulation System **Table 5-4-1**

Fire Rating	Combustibility of Insulation Material	Reaction-to-fire of External Thermal Insulation System	
		Peak heat release rate(kW/m^2)	Flame spreading performance(℃)
Ⅰ	Non-combustible	≤5 (Conventional non-combustible materials such as cement mortar cannot be ignited in the test, and the material of the protective layer is mainly specified.)	T2≤300(As a non-combustible material is used in the insulation layer, the material or thickness requirements for the protective layer are reduced.)
Ⅰ	Flame-retardant or combustible	≤5(If an organic material is applied in the insulation layer, the material or thickness of the protective layer will affect the peak heat release rate of the external thermal insulation system. The peak heat release rate of this system should be the same as that of cement mortar. The material or thickness of the protective layer is stipulated.)	T2≤200 and T1≤300 (An organic insulation material is applied, so the fire resistance of the external thermal insulation system is specified to prevent the insulation layer of L2 and L1 will not burn.)
Ⅱ		≤10(This is a critical value of combustibility. The external thermal insulation system should also meet this requirement for fire safety.)	T2≤300 and T1≤500(The insulation layer of L2 will not burn, and that of L1 will not burn drastically.)
Ⅲ		≤25(Although the overall external thermal insulation system may not meet the incombustibility requirements in the event of a fire, its combustion is limited. That is, minor combustion is allowed. In this case, the whole system is not completely non-combustible.)	T2≤300(The insulation layer of L2 should not burn, but there are no requirements for L1.)
Ⅳ		≤100(This is a critical value for evaluation of the fire resistance of the entire system.)	T2≤500(The insulation layer of L2 should not burn, but there are no requirements for L1.)

Notes: 1. The external thermal insulation system will not be regarded with the corresponding fire rating until the test and structural requirements are met at the same time.

2. The peak heat release rate is obtained in the cone calorimeter test, with heat radiation of $50kW/m^2$.

3. The flame spreading performance is determined in the window fire test equivalent to the test method in the standard BS 8414-1: 2002. T1 and T2 are the temperature of any measuring point of the insulation layer on the horizontal lines 1 and 2, respectively (see Section 5.3).

5.4.3 Study on Reaction-to-fire and Applicable Building Height of External Thermal Insulation System

The applicable building heights of external thermal insulation system of different fire ratings can be further divided based on the study results and China's national conditions.

There are multistoried buildings, small high-rise buildings and high-rise buildings in cities in China. High-rise buildings play a dominant role in large-and medium-sized cities with a high degree of modernization, where the density of population and buildings are higher than those of similar cities in other countries. In China's highly modernized cities, fire rescue ladders range from 50m to 60m. In this context, the fire rating should be subdivided according to building heights, and its classification based on the building height of 22m in German standards is slightly rough. Therefore, the fire rating classification should be based on building categories, and specific applicable building heights should be determined.

Refer to Table 5-4-2 to 5-4-4 for the reaction-to-fire and applicable building height of the external thermal insulation system.

Reaction-to-fire Requirements for External Thermal Insulation System of Non-curtain-wall Type Residential Building

Table 5-4-2

Building Height H(m)	Reaction-to-fire		
	Peak heat release rate (kW/m^2)	Window fire test	
		Horizontal baseline temperature(℃)	Burning area(m^2)
H≥100	≤5	T2≤200 and T1≤300, or T2≤300(using the insulation material of Class A combustibility)	≤5
60≤H<100	≤10	T2≤300 and T1≤500	≤10
24≤H<60	≤25	T2≤300	≤20
H<24	≤100	T2≤500	≤40

Reaction-to-fire Requirements for External Thermal Insulation System of Non-curtain-wall Type Public Building

Table 5-4-3

Building Height H(m)	Reaction-to-fire				
Non-curtain-wall type public building	Peak heat release rate(kW/m^2)	Window fire test		Wall Corner Fire Test	
		Horizontal baseline temperature(℃)	Burning area (m^2)	Burning width (m)	Burning area (m^2)
H≥50	≤5	T2≤200 and T1≤300, or T2≤300(using the insulation material of Class A combustibility)	≤5	≤1.52	≤10
24≤H<50	≤10	T2≤300 and T1≤500	≤10	≤3.04	≤20
H<24	≤25	T2≤300	≤20	≤5.49	≤40

Reaction-to-fire Requirements for External Thermal Insulation System of Curtain-wall Type Building

Table 5-4-4

Building Height H(m)	Reaction-to-fire				
Curtain-wall type building	Peak heat release rate (kW/m^2)	Window fire test		Wall Corner Fire Test	
		Horizontal baseline temperature(℃)	Burning area (m^2)	Burning width (m)	Burning area (m^2)

Continued

Building Height H(m)	Reaction-to-fire				
H≥24	≤5	T2≤200 and T1≤300, or T2≤300 (using the insulation material of Class A combustibility)	≤5	≤1.52	≤10
H<24	≤10	T2≤300 and T1≤500	≤10	≤3.04	≤20

5.4.4 Fireproof Structure and Applicable Height of External Thermal Insulation System

5.4.4.1 Thin-plaster External Thermal Insulation System with Organic Insulation Material

Based on the large-scale fire test of the external thermal insulation system with the fire barrier, the following provisions on the fireproof structure and applicable building height are put forward to guarantee the fire safety of the external thermal insulation system with the organic insulation material.

(1) When the external thermal insulation system with the Class B_2 insulation material is applied in buildings, the horizontal fire barrier should be mounted as per the requirements of Table 5-4-5.

Fireproof Design Requirements for Thin-plaster External Thermal Insulation with Class B_2 Insulation Board

Table 5-4-5

Building Height	Combustibility of Insulation Material	Use of Horizontal Fire Barrier (Class B_2 insulation material)
60m≤H<100m	Class B_2 or above	On each floor
24m≤H<60m	Class B_2 or above	Once every two floors
H<24m	Class B_2 or above	On the first floor

(2) The protective layer of the external thermal insulation system should be made of noncombustible or flame-retardant material. The thickness of the first protective layer should not be less than 6mm, and that of other layers should not be less than 3mm.

(3) The fire barrier (height: not less than 200mm) should be fully bonded with the base course wall via the Class A insulation material.

5.4.4.2 Insulation Slurry Type External Thermal Insulation System and Others

The results of study on three fireproof structures, namely the fire compartment, fireproof protective surface and no-cavity fully bonded structure, show that the following fireproof structural measures apply to the external thermal insulation engineering, and should conform to the requirements of Table 5-4-6 in order to ensure the fire safety.

(1) Fire compartments should be built in the insulation layer, and the area surrounded by fire compartments should not exceed 0.3m^2. The material width of fire compartments should not be less than 10mm. Insulation mortar without flame spreading performance should be applied.

(2) An appropriate protective layer consisting of the fireproof leveling layer, plaster layer and finish layer should be applied on the external surface of the insulation layer. The fireproof leveling layer should be constructed with insulation mortar that has no flame spreading performance.

(3) A no-cavity structure should be constructed. The gap between the insulation layer and finish layer and others in the curtain-wall type buildings should be sealed with the flame-retardant insulation material on each floor.

5.4.4.3 *Code for Fire Protection Design of Buildings* (GB 50016-2014)

The *Code for Fire Protection Design of Buildings* (GB 50016-2014) was implemented on May 1, 2015. The combustibility ratings and applicable heights of external insulation materials in

Fireproof Structural Measures of External Thermal Insulation Engineering **Table 5-4-6**

Type of External Thermal Insulation System	Fireproof Structural Measure			Applicable Building Height(m)		
				Non-curtain-wall type building		Curtain-wall type building
	Fire compartment	Thickness(mm)of fireproof leveling layer	Cavity bonded with base course wall	Residential building	Public building	
Insulation mortar system	Not used	—	No cavity	Not restricted	Not restricted	Not restricted
Non-meshed cast-in-situ system	Not used	≥10	No cavity	<24	N/A	N/A
		≥15		<60	N/A	
		≥20		<100	<24	
		≥25		Not restricted	<50	
		≥30		—	Not restricted	
Meshed cast-in-situ system	Not used	≥20	No cavity	<100	<24	N/A
		≥25		Not restricted	<50	
		≥30		—	Not restricted	
Pasted EPS board system	Used	—	No cavity	<24	N/A	N/A
		≥10		<60	N/A	
		≥15		<100	<24	
		≥20		Not restricted	<50	<24
		≥25		—	Not restricted	<100
Sprayed PU system	Not used	≥10	No cavity	<24	N/A	N/A
		≥15		<60	N/A	
		≥20		<100	<24	
		≥25		Not restricted	<50	<24
		≥30		—	Not restricted	<100
Anchored rock wool board system	Not used	—	—	Not restricted	Not restricted	Not restricted

Note: When the face brick finish is applied, the thickness of the fireproof protective layer may be reduced by 10mm at most based on the minimum requirements in this table.

this standard are listed in Table 5-4-7. Some fireproof structural measures are adopted in this standard, but the fireproof protective layer and fully-bonded no-cavity structure are not. The combustibility ratings of insulation materials are restricted rigorously to ensure the fire safety of external thermal insulation systems.

Combustibility and Applicable Height Classification of External Thermal Insulation Materials in *Code for Fire Protection Design of Buildings* (GB 50016-2014) **Table 5-4-7**

Building Type	Building Height H(m)	Minimum Combustibility Class of Insulation Material	Thickness (mm) of Protective Layer	Width (mm) of Fire Barrier	Minimum Combustibility Class of Material of Exterior Finish Layer
Residential building	H>100	A	—	—	A
	100≥H>27	B_1	1st floor: 15 Other floors: 5	Each floor: 300 Fire durance of door/window: 0.5h	H>50, A H≤50, B_1
	H≤27	B_2			

Continued

Building Type		Building Height H(m)	Minimum Combustibility Class of Insulation Material	Thickness (mm) of Protective Layer	Width (mm) of Fire Barrier	Minimum Combustibility Class of Material of Exterior Finish Layer
Public building	Densely populated place	Any height	A	—	—	H>50, A H≤50, B_1
	Others	H>50	A	—	—	A
		50≥H>24	B_1	1st floor: 15 Others: 5	Each floor: 300 Fire durance of door/window: 0.5h	B_1
		H≤24	B_2			

Note: There are no fire resistance requirements for doors and windows on the outer wall of the residential building with the Class B_1 insulation material and the height of less than 27m as well as the public building with the Class B_1 insulation material and the height of less than 24m.

6. Impact of Wind Load on External Thermal Insulation System

The external thermal insulation system is a non-bearing structure attached on the base course wall, so importance must be attached to the safety of connection between this system and base course wall. They are connected mainly in two ways: 1, the external thermal insulation system is directly pasted on the base course wall via the binder; and 2, the external thermal insulation system is secured on the base course wall via the binder and bolts. There are a number of engineering cases in which the external thermal insulation system falls off as a result of low bonding strength, small bonding area, false bonding, etc. Analysis results show that there are mainly three causes.

(1) The dead load of the external thermal insulation system is greater than the shear strength of its material or interface.

(2) The negative wind pressure perpendicular to the wall with the external thermal insulation system involving a cavity structure is greater than the tensile strength of the material or interface of this system.

(3) The materials of layers of the external thermal insulation system vary in the thermal expansion and contraction with the temperature changing and in the wet expansion and dry contraction with the moisture content changing, resulting in mutual constraints (or with deformation constrained by the base course wall). The thermal stress and wet expansion stress generated are greater than the shear strength of the material or interface of the external thermal insulation system.

Engineering accidents caused by falling of the external thermal insulation system threaten the life and property, so this system must be reliably connected with the base course wall.

This chapter includes the calculation of the wind load of the external thermal insulation system with a cavity structure, and analysis of its safety based on the prevailing relevant standards.

6.1 Causes of Positive and Negative Wind Pressure

The wind load of a building refers to the negative or positive pressure of wind formed by the airflow on the surface of this building. It is related to the wind properties (speed and direction), local landform and surrounding environment, building height and shape, etc. The pressure generated by the wind load on the building is not distributed uniformly. The internal air pressure is higher than the external air pressure on the crosswind and lee side, so the external thermal insulation system here is subject to thrust from inside to outside, i. e. negative wind pressure; and the windward side is subject to thrust from the external thermal insulation system to base course wall, i. e. positive wind pressure. For the external thermal insulation system with a cavity, if the air pressure inside the cavity is higher than the external air pressure, thrust will be generated from the cavity to this system, i. e. negative wind pressure (Figure 6-1-1); and if the air pressure inside the cavity is lower than the external air pressure, thrust will be generated from this system to cavity, i. e. positive wind pressure (Figure 6-1-2).

If the positive wind pressure is applied on the external thermal insulation system, the insulation board will be bent, the bonding point between the insulation board and bonding mortar

Figure 6-1-1 Schematic Diagram of Negative Wind Pressure

Figure 6-1-2 Schematic Diagram of Positive Wind Pressure

will be squeezed. If the negative wind pressure is applied on the external thermal insulation system, thrust will be generated from the cavity to this system. When the negative wind pressure is higher than the bonding strength between the bonding mortar and base course wall or insulation board, the external thermal insulation system will fall off. In the presence of negative wind pressure, the external thermal insulation system may be destroyed instantaneously or in a gale (i. e. within a short period of time). The common cases with external insulation systems blown off are caused by the negative wind pressure. As shown in Figure 6-1-3, the negative wind pressure often occurs on two sides parallel to the wind direction as well as the lee side. The maximum negative wind pressure on two sides of a building is the most likely to lead to damage. This chapter mainly analyzes the damage caused by the negative wind pressure to external thermal insulation systems.

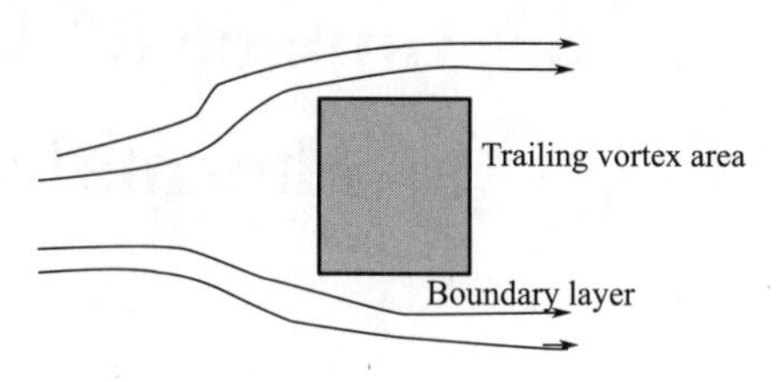

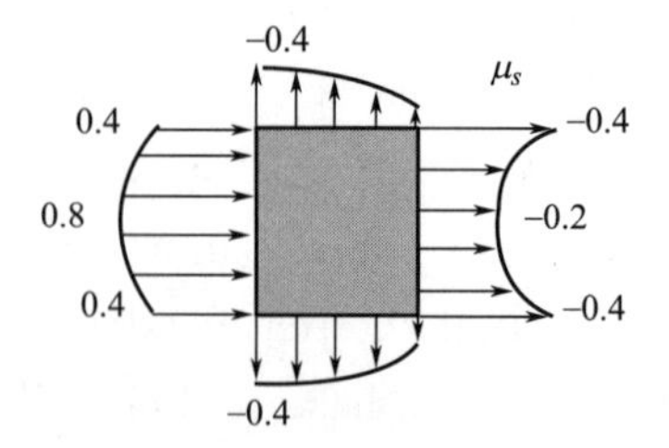

Figure 6-1-3 Distribution of Wind Pressure Generated by Airflow through Building Plane

The wind load increases at the same time with the building height, so special attention must be paid to the impact of the wind load on the external thermal insulation system of a high-rise building. The annual prevailing wind direction of one region can be determined through the wind rose diagram shown in Figure 6-1-4, and thus the area where the positive or negative wind pressure is prone to occur can be inferred.

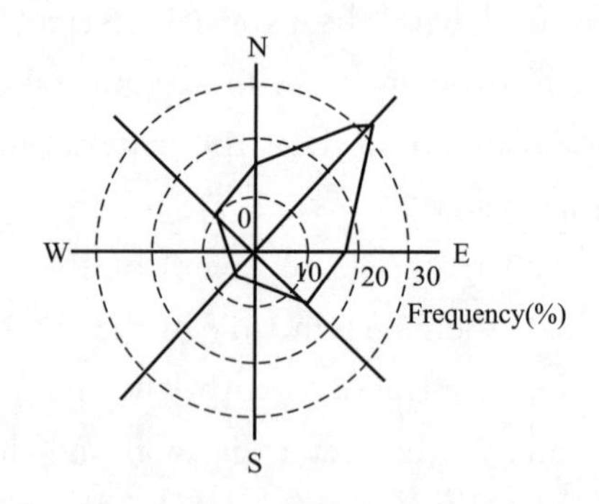

Figure 6-1-4 Example of Wind Rose Diagram

The wind pressure on the external thermal insulation system is calculated with that of the envelope in the *Load Code for the Design of Building Structures* (GB 50009—2012):

$$w_k = \beta_{gz}\mu_{sl}\mu_z w_0 \quad (6\text{-}1\text{-}1)$$

Where, w_k —standard wind load, kN/m^2;

β_{gz} —gust factor at the height Z;

μ_{sl} —local shape factor of the wind

load;

μ_z —height variation factor of the wind pressure;

w_0 —basic wind pressure, kN/m^2.

The sign of the shape factor of local wind pressure corresponds to the positive or negative wind pressure. A specific value of the shape factor of local wind pressure, involving complex considerations, needs to be determined in the wind tunnel test. Generally, the values in the *Load Code for the Design of Building Structures* (GB 50009—2001) can be applied, as shown in Figure 6-1-5.

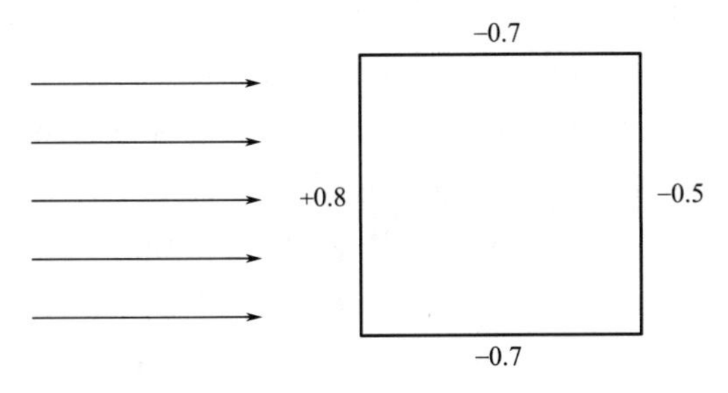

Figure 6-1-5 Shape Factor of Local Wind Pressure of Square Closed House

6.2 Factors Related to Wind Pressure

Which factors are associated with wind pressure? It can be seen in the equation (6-1-1) that the wind pressure is associated with the geographical environment, construction environment, building shape, etc. The basic wind pressure in inland areas is usually lower than that in coastal areas. For example, the basic wind pressure with the 100-year return period is 0.6kN/m^2 in Beijing, 0.35kN/m^2 in Chengdu, 2.2kN/m^2 in the Paracel Islands of Hainan, and 2.3kN/m^2 (more than 6 times that of Chengdu) in Ilan of Taiwan. The height variation factor and gust factor of the wind pressure are related to the building environment. For example, the height variation factor (gust factor) of wind pressure is 1.17 (1.69) at a height of 5m and 2.40 (1.46) at a height of 100m on the coast and lake bank (ground roughness: Class A); and 0.74 (2.30) at a height of 5m and 1.70 (1.60) at a height of 100m in cities (ground roughness: Class C) with dense buildings. It should also be multiplied by a correction factor if a building is located on a hill or at the bottom of a valley. The shape factor of local negative wind pressure is as follows (the design wind load differs from the actual value): −1.0 for the wall surface, −1.8 for the corner edge, −2.2 for part of a roof (ridge with the slope between its periphery and the roof more than 10°), and −2.0 for the cornice, appentice, sun louver and other protruding components. If high-rise building clusters are close to each other, cluster effects arising from mutual interference of wind pressure should be taken into account. In this case, the shape factor of local wind pressure needs to be multiplied by a mutual interference increase factor.

Therefore, the external thermal insulation system with cavities do not apply to the regions and high-rise buildings (especially high-rise building clusters close to each other) with high basic wind pressure along coasts or lake banks in China.

6.3 Engineering Case with External Thermal Insulation System Blown-off

The external thermal insulation system based on the thin-plastered EPS board is widely applied at present. It is a kind of insulating and decorating system composed of the expanded polystyrene board (EPS board), adhesive (sometimes supplemented with anchor bolts), surface mortar, alkali-resistant mesh and paint outside an outer wall.

The external thermal insulation structure involved in the *External Thermal Insulation Composite Systems based on Expanded Polystyrene* (GB 29906—2013) is shown in Table 6-3-1.

Basic Structure of External Thermal Insulation System based on Thin-plastered EPS Board Table 6-3-1

Base course wall(i)	Basic Structure				Schematic Diagram
	Bonding layer(ii)	Insulation layer(iii)	Protective layer		
			Plaster layer(iv)	Finish layer(v)	
Concrete or masonry wall	Adhesive (anchor bolt)	EPS board	Surface mortar compounded with glass fiber mesh	Coating material	(i) (ii) (iii) (iv)(v)

Note: Anchor bolts may be used as auxiliary fixings for EPS boards when required in the engineering design.

It is common that the external thermal insulation system is blown off as a result of material problems or violations in the construction process of actual projects. Damage often occurs in the following two positions: 1, interface between the bonding layer and base course wall (Figure 6-3-1); 2, interface between the bonding layer and EPS board (Figure 6-3-2).

Figure 6-3-1 Engineering Case (I) of Damage Caused by Negative Wind Pressure
(damage to interface between the bonding layer and base course wall)

Figure 6-3-2 Engineering Case (II) of Damage Caused by Negative Wind Pressure
(damage to interface between the point bonding layer and EPS board)

As shown in these figures, the damage in these engineering cases are caused by the nonconformity of the bonding strength between the bonding mortar and base course wall or EPS board to relevant standards and regulations. It is stipulated that the tensile bonding strength between the base course wall and adhesive of the thin-plaster system shall not be less than 0. 3MPa in Article 6. 1. 6 of the *Technical Specification for External Thermal Insulation on Walls* (JGJ 144-2004), and that the bonding area of the EPS board must not be smaller than 40% in Article 6. 1. 7. In accordance with *the External Thermal Insulation Composite Systems based on Expanded Polystyrene* (GB/T 29906—2013), the tensile strength of the EPS board and the tensile bonding strength between the binder and EPS board shall not be less than 0. 10MPa, and damage shall mainly occur in the EPS board. The external thermal insulation system may be blown off in the event of the inadequate bonding strength and area between the bonding mortar and base course wall or EPS board.

There are two main reasons in terms of construction: (1), poor adhesion of the base course wall with bonding mortar, poor bonding between the bonding mortar and EPS board, or nonconformity of construction; and (2), unreasonable point bonding. Most of external thermal insulation systems destroyed by wind are subject to point bonding only (see Section 6. 4. 3 for details).

6.4 Calculation of Negative Wind Pressure and Wind Pressure Safety of External Thermal Insulation System

6.4.1 Calculation of Negative Wind Pressure and Wind Pressure Safety Factor of External Thermal Insulation System

How much is the negative wind pressure in fact? Given a 50m high building with an external thermal insulation system based on the thin-plastered EPS board (bonding area: 40%) in the Paracel Islands under the maximum negative wind pressure, with the ground roughness of Class A, then $\beta_{gz}=1.51$, $\mu_s=-1$ (internal and external corner: $\mu_s=-1.8$), $\mu_z=2.03$, $w_0=2.2\text{kN/m}^2$ (basic wind pressure). The negative wind pressure can be obtained by substituting these values in the equation (6-1-1):

$w_{k1}=1.51\times(-1)\times2.03\times2.2\text{kN/m}^2=6.74\text{kN/m}^2$ (internal and external corner: 12.1kN/m^2)

Provided the full conformity of the external thermal insulation system to the standards GB/T 29906—2013 and JGJ 144—2004, the destructive force per unit area (destructive force onto the EPS board) is:

$$F=0.1\times40\%\ \text{MPa}=40\text{kN/m}^2$$

The negative wind pressure per unit area of the external thermal insulation system is associated with the contact area between the air in the cavity and the EPS board, calculated as follows:

$w_{k2}=w_{k1}\times(1-40\%)=4.05\text{kN/m}^2$ (internal and external corner: 7.28kN/m^2)

The safety factor of the external thermal insulation system against the negative wind pressure is:

$s_1=F/w_{k1}=40/4.05=9.9$ (internal and external corner: $s_2=40/7.28=5.5$).

From this point of view, the wind pressure safety of an external thermal insulation system can be guaranteed if it is constructed in accordance with the standards and regulations. This system may be blown off in the case of inadequate bonding area, unreasonable bonding method, problems in anchor bolt installation, etc.

The calculation results show that when the bonding area is 10.8%, the negative wind pressure (10.8kN/m^2) per unit area in the internal and external corner is just identical to the bonding strength (10.8kN/m^2) of the external thermal insulation system, and if the bonding area is larger than 10.8%, the external thermal insulation system will not be blown off. The critical value of bonding strength of the external thermal insulation system against the negative wind pressure varies in different regions. The bonding area corresponding to this critical value can be obtained by the above calculation method.

It can be seen from the engineering case shown in Figure 6-3-2 that a number of EPS boards and adhesive subject to point bonding fell off under the negative wind pressure, and the EPS boards on the upper part fell from the point bonding positions. Although the bonding area in this project exceeded 10%, the external thermal insulation system finally fell off as a result of connecting cavities formed by unreasonable point bonding.

6.4.2 Connecting Cavity

As mentioned above, when the bonding area of the external thermal insulation system is 10.8%, the negative wind pressure (10.8kN/m^2) is just the same as the bonding strength (10.8kN/m^2) of this system. If the bonding area is larger than 10.8%, the external thermal insulation system will not be blown off.

In actual projects, however, the bonding area of 10.8% is not a solution to prevent the insulation board from being blown off. In the case of unreasonable point bonding, connecting cavities will be formed between the base course wall and insulation layer, and the external thermal insula-

Wind Pressure Safety Factors of External Thermal Insulation Systems with Different Bonding Areas

Table 6-4-1

Bonding Area	Negative Wind Pressure(kN) per Unit Area	Negative Wind Pressure(kN) per Unit Area (Internal and External Corner)	Allowable Destructive Force(KN) per Unit Area	Negative Wind Pressure Safety Factor	Negative Wind Pressure Safety Factor(Internal and External Corner)
30%	4.72	8.47	30	6.4	3.5
20%	5.39	9.68	20	3.7	2.1
10%	6.07	10.89	10	1.6	0.9

tion system is more likely to fall off under the negative wind pressure.

If there are connecting cavities in a project, as shown in Figure 6-4-1, the negative wind pressure F per unit area will be uniformly applied on each bonding point in the external thermal insulation system with insulation boards subject to point bonding. The most unfavorable bonding point with the low bonding strength or small bonding area is destroyed first under the negative wind pressure (applied as a whole) generated by connecting cavities. This will further reduce the bonding area and exacerbate the damage caused by the negative wind pressure, thus damaging all bonding points one by one, gradually eliminating the bonding strength and eventually result in large-scale falling.

Therefore, it is specified in the regulations on the pasted insulation board that the bonding area must not be less than 40%, and the method of point-and-frame bonding shall be adopted to avoid large connecting cavities.

6.4.3 Bonding Area and Safety Factor

Assuming that the above project is built in the Paracel Islands, with the bonding area less than 40% but other indicators conforming to the standards, the wind pressure safety factor in Table 6-4-1 can be used to its external thermal insulation system.

According to the design requirement that the wind pressure safety factor is not less than 5%, the bonding area between the bonding mortar and EPS board must not be less than 40%. If there is the basic wind pressure and high-rise buildings are

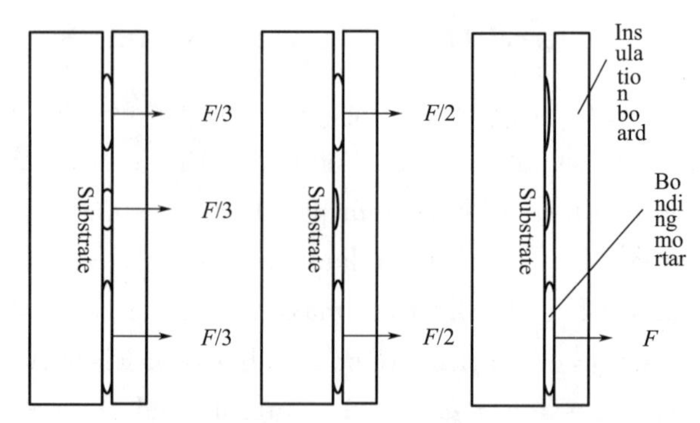

Figure 6-4-1 Schematic Diagram of Damage Caused by Negative Wind Pressure of Connecting Cavity

close to each other, the bonding area should be increased to improve the wind pressure safety factor of the external thermal insulation system. At the same time, the bonding strength and construction quality of the material should be guaranteed.

It is thought that mechanical fixing can be conducted as a protective measure (reinforcement with connectors and anchor bolts as shown by the schematic diagram of Figure 6-3-2) in the case of insufficient bonding area, but this is not rational for the following three reasons. First, the tensile strength of a single anchor bolt is only 0.3kN, which is negligible compared with the bonding strength (40kN/m^2) per unit area. Secondly, the anchor bolt is directly secured on the insulation board, so the part fixed by the anchor bolt may be sunken compared with its surrounding area, and the negative wind pressure is borne by the local EPS board near the anchor bolt. If the negative wind pressure exceeds the strength of the EPS board, the EPS board will be destroyed, finally making the external thermal insulation system fall off within a large scale. (3) If the anchor bolt to be directly secured on the surface of the insulation board is knocked with excessive force or just into a cavity, the EPS board may be damaged. In this case, the anchor bolt may lead to damage to the external thermal insulation system instead of reinforcement.

6.4.4 Rational Structure

According to the analysis of the impact of wind pressure on the external thermal insulation system, it is found that there are two essential conditions for damage under the wind pressure. Firstly, the negative wind pressure is prone to occur. Secondly, there are connecting cavities. In order to improve the resistance of the external thermal insulation system against the wind load, it is recommended to use the following three structures.

(1) Fully bonded structure

The bonding area is 100%, involving no cavity. The bonding strength is greater than 50kN/m^2, which is sufficient to overcome the wind pressure and prevent the insulation layer from falling off. As shown in Figure 6-4-2, there are no cavities between the boards and base course wall in the external thermal insulation system based on the adhesive polystyrene granule pasted EPS board and that based on the non-meshed cast-in-situ EPS board, both of which have a fully bonded structure.

(2) Closed small cavity structure

The unit area corresponding to the negative wind pressure is reduced in this structure. The bonding area of each closed bonding unit is larger than 40%, and the bonding strength is greater than 40kN/m^2, which is sufficient to overcome the wind pressure. As the bonding strength per unit area is greater than the destructive force generated by the negative wind pressure, the insulation layer will not fall off under the negative wind pressure. As shown in Figure 6-4-3, the closed small cavity structure is made with small boards (600mm×450mm), exhaust holes are located in the middle of insulation boards, and the joints around insulation boards are fully filled with mortar to form closed cavities.

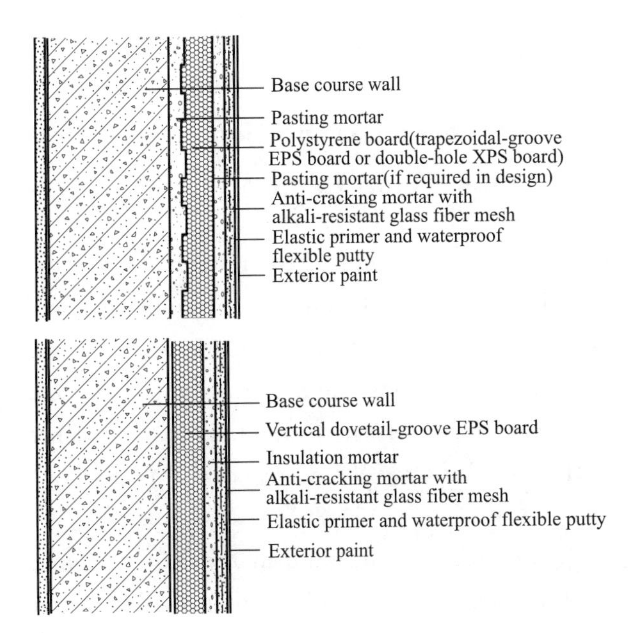

Figure 6-4-2 Fully Bonded No-cavity Structure

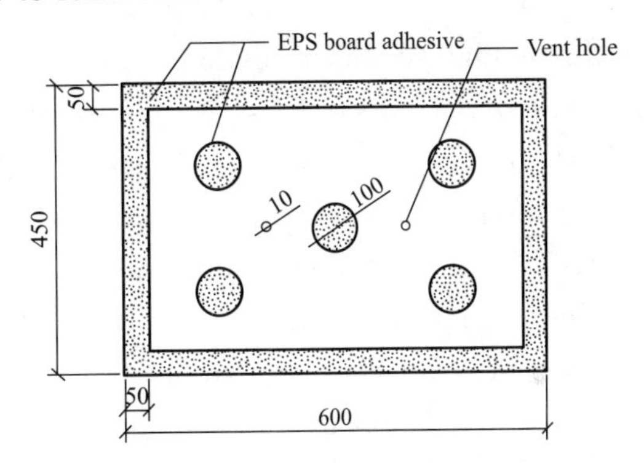

Figure 6-4-3 Closed Small Cavity Structure

(3) Strip type bonded structure

As shown in Figure 6-4-4, the adhesive is applied with a toothed trowel in one direction, the bonding area of insulation boards is large, and there are small cavities of the bonded insulation boards, similar to the fully bonded structure.

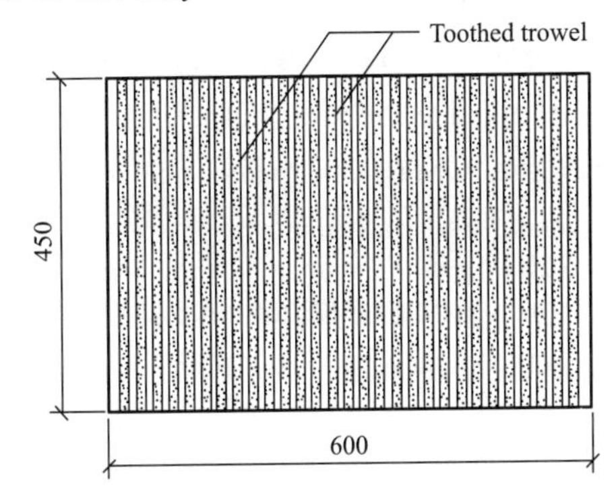

Figure 6-4-4 Strip Type Bonded Structure

6.5 Safety of Rock-wool External Thermal Insulation System under Wind Load

With the fire rating requirements of the external thermal insulation system improved and relevant policies issued in recent years, some Class A non-combustible materials with excellent properties have attracted attention. In accordance with the *Code for Fire Protection Design of Buildings* (GB 50016—2014):

The external insulation material of the outer wall in a densely populated building shall meet the requirements of Class A combustibility.

If there is no cavity between the external thermal insulation system and the base course wall or finish layer in a residential building (more than 100m high), the insulation material shall meet the requirements of Class A combustibility.

If the height of a building (except the residential building and densely populated building) is more than 50m, the insulation material shall meet the requirements of Class A combustibility.

Therefore, the consumption of Class A non-combustible insulation materials is bound to increase significantly. Rock wool, a kind of Class A material with good insulation effects, will be widely promoted and applied. In rock-wool external thermal insulation systems available at present in China, the rock wool boards are connected to the base course wall by means of bonding and anchorage. Because of the cavities in such systems and low strength of rock wool boards themselves, there are doubts on the wind load resistance of these systems. To address this issue, this section discusses the wind load resistance of the rock-wool external thermal insulation system.

6.5.1 Calculation of Standard Load of Negative Wind Pressure

In accordance with Article 8.1.1 of the *Load Code for the Design of Building Structures* (GB 50009—2012), the standard load of negative wind pressure can be calculated from:

$$w_k = \beta_{gz} \mu_{s1} \mu_z w_0 \qquad (6\text{-}5\text{-}1)$$

Where, w_k —standard wind load, kN/m^2;

β_{gz} —gust factor at the height of Z (related to the ground roughness);

μ_{s1} —shape factor of local wind pressure, −1.0 for the wall surface and −1.4 for the wall corner;

μ_z —height variation factor of the wind pressure (related to the ground roughness);

w_0 — basic wind pressure, kN/m^2. The basic wind pressure with the 100-year return period in Beijing is 0.50kN/m^2.

A high-rise residential building in Beijing was taken as an example, with the wall height of 100m and ground roughness of Class C. A wall surface of a closed rectangular flat room was studied. The calculation results of the standard load calculation results of its negative wind pressure are listed in Table 6-5-1 and 6-5-2. It can be seen that there is more severe negative wind pressure in corners, so the standard loads of negative wind pressure in Table 6-5-2 are used in the following calculations.

With reference to European technical data requirements, the safety factor is taken as 3 (ratio of the design value to standard value of wind load: 1.5) if the external thermal insulation system and base course wall are connected mainly by means of anchorage, and 10 (ratio of the design value to standard value of wind load: 1.5) if they are connected mainly by means of bonding.

6.5.2 Fixing of Rock Wool Board

(1) Fixing by main anchorage and auxiliary bonding

In accordance with the current national standard GB/T 25975 "*Rock Wool Products for Exterior Insulation and Finish Systems* (*EIFS*)", the tensile st-

Load Calculation of Negative Wind Pressure on Wall Surface of Closed Rectangular Flat Room in Beijing

Table 6-5-1

Height(m)above Ground	Height Variation Factor of Wind Pressure	Gust Factor at Height of Z	Shape Factor of Local Wind Pressure(wall surface)	Basic Wind Pressure (kN/m^2)in Beijing	Standard Load(kN/m^2) of Negative Wind Pressure
20	0.74	1.99	1.0	0.5	0.74
50	1.1	1.81	1.0	0.5	1.00
80	1.36	1.73	1.0	0.5	1.18
100	1.5	1.69	1.0	0.5	1.27

Load Calculation of Negative Wind Pressure on Corner of Closed Rectangular Flat Room in Beijing

Table 6-5-2

Height(m)above Ground	Height Variation Factor of Wind Pressure	Gust Factor at Height of Z	Shape Factor of Local Wind Pressure(corner)	Basic Wind Pressure (kN/m^2) in Beijing	Standard Load(kN/m^2) of Negative Wind Pressure
20	0.74	1.99	1.4	0.5	1.03
50	1.1	1.81	1.4	0.5	1.39
80	1.36	1.73	1.4	0.5	1.64
100	1.5	1.69	1.4	0.5	1.77

rength of rock wool boards is classified into TR7.5, TR10 and TR15. If the rock wool boards and base course wall are connected by means of bonding and anchorage, because of low tensile strength of rock wool boards, the bearing capacity of anchor bolts shall be included in the calculation of the wind load bearing capacity, but the bearing effect of bonding shall not.

Based on the standard maximum wind loads (w_k) at various heights in Beijing (Table 6-5-2), the theoretical bonding strength of rock wool boards should reach the values in Table 6-5-3 if the external thermal insulation system based on the rock wool board is mainly bonded, or subject to 100% full bonding or 40% point-and-frame bonding. The bonding strength of rock wool board subject to 40% point-and-frame bonding should be 2.5 times that subject to 100% full bonding.

It can be seen from Table 6-5-3 that the bonding strength of rock wool boards subject to40% point-and-frame bonding must be at least 25.75kPa to meet the load requirements under the negative wind pressure. The tensile strength of rock wool boards is only 7.5-15kPa, which does not meet the wind load requirements. The method of 100% full bonding is not operable. That is, bonding is completely not suitable for horizontal-wire rock wool boards. Hence, the external thermal insulation system based on the rock wool board should be fixed mainly by means of anchorage.

Calculation Results of Theoretical Bonding Strength of Rock Wool Board **Table 6-5-3**

Height(m)above Ground	Standard Load(kN/m^2) of Negative Wind Pressure	Safety Factor	Bonding Strength(kPa) of Rock Wool Board with Bonding Area of 100%	Bonding Strength(kPa) of Rock Wool Board with Bonding Area of 40%
20	1.03	10	10.3	25.75
50	1.39	10	13.9	34.75
80	1.64	10	16.4	41.00
100	1.77	10	17.7	44.25

(2) Fixing with anchored wire mesh/glass fiber mesh

The design value of wind load is usually 1.5 times of its standard value. Taking into account the inappropriate anchor bolt, excessive drilling diameter, improper insertion of anchor bolts, insufficient drilling depth, incorrect relative locations of anchor plates and wire meshes/glass fiber meshes, ageing and other potential impact, the safety factor (γ) based on the design value of wind load is taken as 2. That is, the total safety factor is taken as 3.

The anchorage force of anchor bolts secured on the base course wall is evenly transmitted onto rock wool boards through the wire mesh/glass fiber mesh. Calculated in line with Beijing's local standard DB11/T 1081—2014 "*Technical Specification for External Thermal Insulation Composite Systems based on Rock Wool*", the number of anchor bolts per unit area is calculated as follows:

$$N_A \geqslant w_d \gamma / f_0 \qquad (6\text{-}5\text{-}2)$$

Where, N_A —number of anchor bolts per unit area, pcs./m^2;

w_d —maximum design value of wind load at the corresponding height, equal to 1.5 times of the standard value (w_k), kN/m^2;

γ —safety factor, $\gamma = 2$;

f_0 —standard tensile strength of a single anchor bolt.

In accordance with Article 6.2 of the *Anchors for Fixing of Thermal Insulation Composite Systems* (JG/T 366—2012), the standard tensile strength of single anchor bolts of an autoclaved aerated concrete base course wall (Class E) shall not be less than 0.3kN, and that of an ordinary concrete base course wall (Class A) shall not be less than 0.6kN.

As shown in Table 6-5-4, when the rock wool board is fixed via the anchored wire mesh/glass fiber mesh, few anchor bolts are needed in ordinary concrete of the base course wall with higher strength, and the safety and workability of the anchorage system can be achieved at the same time. For the base course wall of lower strength, a large number of anchor bolts are required, so the external thermal insulation system based on the rock wool board is not applicable.

(3) Fixing by direct anchorage with anchor bolt through diffuser plate

As the rock wool board has low pull strength, the failure in anchorage caused by its deformation or damage under the wind pressure must be taken into consideration during direct anchorage with anchor bolts. As shown in Figure 6-5-1, assuming that the number of anchor bolts per square meter is N_A, the rock wool board is fixed with anchor bolts through diffuser plates, and the area B of the diffuser plates corresponding to the area $1/N_A$, then the rock wool board in the area B must be able to withstand the negative wind pressure in the area $1/N_A$, without damage. In this case, two essential conditions must be met to prevent damage under the negative wind pressure.

Calculation Results of Number of Anchor Bolts of Rock Wool Board Fixed by Anchored Wire Mesh/Glass Fiber Mesh

Table 6-5-4

Height(m) above Ground	Standard Load(kN/m^2) of Negative Wind Pressure	Safety Factor	Required Anchorage Strength(kPa)	Number of Anchor Bolts for Class ABase course wall	Number of Anchor Bolts for Class E Base course wall
20	1.03	3	3.09	6	11
50	1.39	3	4.17	7	14
80	1.64	3	4.92	9	17
100	1.77	3	5.31	9	18

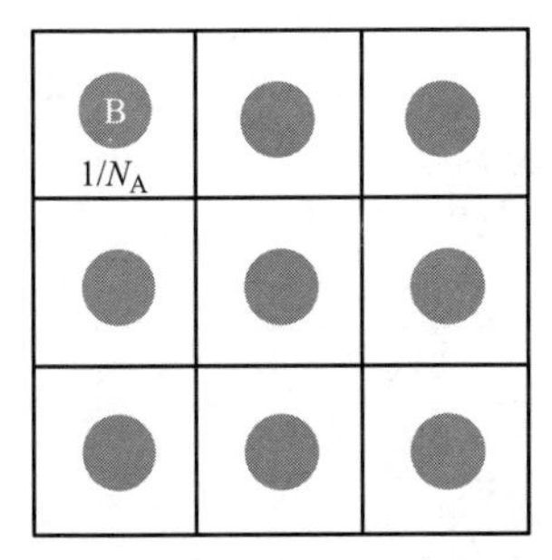

Figure 6-5-1 Schematic Diagram of Anchor Bolt Diffuser Plate Area and Effective Anchorage Area

Condition 1: the anchorage force per unit area is greater than or equal to the maximum design value of wind load per unit area at the corresponding height. The number of anchor bolts per unit area is:

$$N_{A} \geqslant w_{d}\gamma/f_{0} \quad (6\text{-}5\text{-}3)$$

Condition 2: the rock wool at the anchor bolt must not be damaged under negative pressure. The strength of rock wool is:

$$F_{0} \geqslant w_{d}\gamma \times 1\mathrm{m}^{2}/(N_{A} \times B) \quad (6\text{-}5\text{-}4)$$

The derived equation is:

$$N_{A} \geqslant w_{d}\gamma/(B \times F_{0}) \quad (6\text{-}5\text{-}5)$$

According to the calculation results in Table 6-5-5 6-5-6, the tensile strength of the rock wool board is 15kPa. Provided that the base course wall is constructed with ordinary concrete, the diameter of the anchor bolt diffuser plate is 100mm, and the building height is 20m, 14 anchor bolts with diffuser plates are needed per square meter, and the diffuser plate area should account for more than 21% of the board area in order to prevent the rock wool at anchor bolts from damage under negative pressure.

The calculation results demonstrate that even if TR15 rock wool boards are directly anchored, a large number of anchor bolts are needed. Direct anchorage is completely not suitable for TR7.5 and TR10 anchor bolts of lower strength.

6.5.3 Fixing of Rock Wool Strip

The rock wool strip is a kind of rock wool product which is made by cutting from a rock wool board and bonded onto the base course wall with fibers perpendicular to the wall surface. Due to the tensile strength greater than 80kPa, the rock wool strip can be secured on the outer wall by means of bonding (main) and anchorage (auxiliary). For safety reasons, the bonding area between the rock wool strip and base course wall should be larger than 70%. Full bonding is preferred to improve the resistance of the external thermal insulation system to the wind load.

6.5.4 Fixing of Reinforced Vertical-fiber Rock Wool Composite Board

The reinforced vertical-fiber rock wool composite board is a deep processing product of rock wool, in which the insulation core is composed of strips (pieces) made by cutting rock wool ($100\mathrm{kg/m^3}$), and the protective layer is formed by applying the inorganic insulation mortar and alkali-resistant glass fiber mesh on four surfaces in the length direction of the body.

Coated with the inorganic insulation mortar and alkali-resistant glass fiber mesh, the reinforced vertical-fiber rock wool composite board becomes an integrated bearing unit. The tensile strength is minimum 0.10MPa in the direction vertical to the board surface, and up to 0.2MPa at the center, more than 10 times that of the ordinary rock wool board.

(1) Fixing by bonding (main means)

The wind load at 100m in Beijing fully meets the requirements for the bonding strength corresponding to the bonding areas of 100% and 40% in Table 6-5-3. Both full bonding and point-and-frame bonding are applicable, thus fundamentally solving the problems of bonding as a main method of application of rock wool in high-rise buildings.

(2) Fixing with anchored wire mesh/glass fiber mesh

If fixed with the anchored wire mesh/glass fiber mesh, the reinforced vertical-fiber rock wool board is connected to the anchor bolts and base course wall via the whole wire mesh/glass fiber mesh, which has little relation with the strength

Calculation Results of Number of Anchor Bolts for Anchorage of Rock Wool Board **Table 6-5-5**

Height(m)above Ground	Standard Load (kN/m^2)of Negative Wind Pressure	Safety Factor	Required Anchorage Strength(kPa)	Number of Anchor Bolts per Square Meter under Condition 1	Number of Anchor Bolts per Square Meter under Condition 2
20	1.03	3	3.09	6	14
50	1.39	3	4.17	7	19
80	1.64	3	4.92	9	22
100	1.77	3	5.31	9	24

Note: The above results are obtained when the base course wall is constructed with ordinary concrete, the radius of the anchor bolt diffuser plate is 0.07m and the tensile strength of the rock wool board is 15kPa.

Calculation Results of Diffuser Plate Area per Square Meter under Conditions 2 **Table 6-5-6**

Height(m)above Ground	Number of Anchor Bolts under Condition 2	Diffuser Plate Area(m^2)per Square Meter
20	14	0.22
50	19	0.29
80	22	0.34
100	24	0.37

of rock wool itself. The values in Table 6-5-4 can be applied.

(3) Fixing by direct anchorage with anchor bolt through diffuser plate

When anchor bolts are directly anchored on the reinforced vertical-fiber rock wool board, the calculation should be conducted with the equations 6-5-3 and 6-5-4. The calculation results are shown in Table 6-5-7 below.

According to the results in Table 6-5-7, the reinforced vertical-fiber rock wool composite board can be anchored, and the number of anchor bolts under the condition 1 will apply. Anchorage is operable as a small number of anchor bolts are needed.

6.6 Summary

There are two essential conditions for damage caused by the wind pressure. Firstly, the negative wind pressure is prone to occur. Secondly, there are connecting cavities. The external thermal insulation system with cavities, which meets the standards and construction requirements, has the conforming wind pressure safety. In the case of nonconformity of construction to the standards or failure in material and quality control, the external thermal insulation system with cavities will be blown off.

The safety is affected in the following cases:

(1) There are large connecting cavities in the bonding layer.

(2) The bonding area is overly small as a result of no re-examination and supervision.

(3) The base course wall and insulation board are not bonded securely because of failure to inspect the base course wall or inadequate strength of bonding mortar.

The strength of the rock wool board is lower than that of the organic insulation board, so the external thermal insulation system based on the rock wool board is more susceptible to negative wind pressure. In actual applications of rock wool, therefore, the problems arising from its low strength can be solved in the structural design, and a stable bonding unit can be formed by means of deep processing, to improve the stability of rock wool boards on a wall.

(1) When the anchor bolt is directly applied on the rock wool board for reinforcement, the latter may be damaged (the rock wool board is likely to be damaged especially when the anchor

Calculation Results of Number of Anchor Bolts for Anchorage of Reinforced Vertical-fiber Rock Wool Composite Board

Table 6-5-7

Height(m) above Ground	Standard Load(kN/m^2) of Negative Wind Pressure	Safety Factor	Required Anchorage Strength(kPa)	Number of Anchor Bolts per Square Meter under Condition 1	Number of Anchor Bolts per Square Meter under Condition 2
20	1.03	3	3.09	6	4
50	1.39	3	4.17	7	6
80	1.64	3	4.92	9	7
100	1.77	3	5.31	9	7

Note: The above results are obtained when the base course wall is constructed with ordinary concrete, the radius of the anchor bolt diffuser plate is 0.05m and the tensile strength of the rock wool board is 15kPa.

bolt is knocked into a cavity).

(2) If used, the anchor bolt must be secured on the wire mesh or glass fiber mesh, with the force evenly distributed on the rock wool surface, in order to ensure the wind load resistance of rock wool.

(3) The rock wool strip subject to full bonding can meet the wind load requirements.

(4) When the reinforced vertical-fiber rock wool composite board made by deep processing of rock wool is coated with the glass fiber mesh and inorganic insulation mortar, the tensile strength of rock wool will be improved greatly, and both point-and-frame bonding and full bonding are applicable during rock wool construction. Meanwhile, the reinforced vertical-fiber rock wool composite board anchored meets the wind load resistance requirements. It is a solution to the safety and workability of rock wool in high-rise buildings.

7. Study on Seismic Performance of External Thermal Insulation System

7.1 Seismic Requirements for External Thermal Insulation System

7.1.1 Seismic Analysis of External Thermal Insulation System

The external thermal insulation system attached to the surface of an outer wall mainly bears its dead load as well as the wind load, seismic action and temperature effect applied directly on it, but free from the load and seismic action on the main structure. It should have the appropriate deformation capacity to adapt to the displacement of the main structure. That is, when the main structure is displaced under a significant earthquake load, there will be no excessive internal stress or unbearable deformation in the external thermal insulation system. Normally, an external thermal insulation system is composed of an insulation layer, plaster layer and finish layer, most of which are made of flexible material to adapt to the displacement of the structure. In the case of minor lateral displacement of the main structure, the external thermal insulation system can absorb the impact of such a displacement through flexible deformation. However, the external thermal insulation system is a composite system secured on a structural wall by means of bonding or mechanical anchorage, so attention should be paid to the connection of functional layers of this system as well as that between the main structure and this system in the event of an earthquake. Both the connection of functional layers and that between this system and main structure should be conducted reliably, to withstand the dead load of this system and prevent this system from falling under the wind load and seismic action. All connections must be properly able to adapt to the displacement so that the external thermal insulation system will not be damaged in the case of horizontal displacement of the main structure.

According to the results of seismic analysis of the external thermal insulation system, the seismic fortification zone and non-seismic fortification zone should be separated. The wind load, gravity load and temperature load should be taken into account in the non-seismic fortification zone, and the seismic action in the seismic fortification zone.

7.1.2 Basic Requirements for Seismic Performance of External Thermal Insulation System

It is stipulated in the *Code for Seismic Design of Buildings* (GB 50011-2010) that the basic seismic fortification goals for buildings are as follows: the main structure must not be damaged or may be used without repair in the event of frequent earthquakes below the local seismic fortification intensity; the structure can be put into operation after general repair in the event of earthquakes equivalent to the local seismic fortification intensity; collapse or serious damage endangering the life must not occur in the case of rare earthquakes above the local seismic fortification intensity.

The seismic performance of an external thermal insulation system attached as a whole to a structure is closely related to that of the structure. There are two types of damage to external thermal insulation systems in the Wenchuan Earthquake.

(1) The wall of a building with poor seismic performance is greatly displaced or deformed,

and the diagonal cracks or cross diagonal cracks on the external thermal insulation system of the wall are consistent with the structural deformation.

(2) The entire external thermal insulation system falls off under the combined effect of horizontal and vertical forces, as a result of large weight of the system itself, poor bonding between the system and base course wall and inadequate effective anchorage depth of anchor bolts.

Cracking of the external thermal insulation system causes economic losses, while falling may result in casualties. The external thermal insulation system is closely associated with the outer wall of a building. If a building is subject to minor seismic damage but its external thermal insulation system cracks seriously, huge economic losses will be caused. If a building is subject to severe seismic damage and the outer wall collapses, good seismic performance of its external thermal insulation system will be of no significance. Accordingly, the following seismic fortification goals are put forward for external thermal insulation systems: the external thermal insulation system must not be damaged or may be used without repair in the event of frequent earthquakes below the local seismic fortification intensity; small-scale cracking is permissible in the case of earthquakes equivalent to the local seismic fortification intensity, but the system should be applicable to use after general repair; and the system must not fall off in the case of rare earthquakes above the local seismic fortification intensity.

7.2 Seismic Calculation of External Thermal Insulation System

The external thermal insulation system is a non-structural component of a building. In accordance with the *Code for Seismic Design of Buildings* (GB 50011—2010), the calculation of seismic action of non-structural components should meet the following requirements:

(1) Seismic forces should be applied at the center of gravity of each component and part, and the horizontal seismic force may be in any horizontal direction.

(2) Under normal circumstances, the seismic action arising from the gravity of a non-structural component can be calculated by the equivalent lateral force method; and for non-structural components supported on different floors or two sides of a seismic joint, the effect caused by relative displacement between support points in an earthquake should also be taken into account, in addition to the seismic action arising from the gravity of non-structural components.

(3) When the natural vibration period of the system of a building attachment (including brackets) is more than 0.1s and its gravity exceeds 1% of that of the floor concerned, or the gravity of a building attachment exceeds 10% of that of the floor concerned, the seismic design of the overall structural model should be conducted, or the floor spectrum method specified in Appendix M.3 of this Code should be applied for calculation. If a device is inflexibly connected with a floor, they should be directly regarded as one mass point during analysis of the entire structure to obtain the seismic action of this device.

Non-structural components requiring seismic verification roughly include:

(1) Grade 7-9 earthquakes: curtain walls fundamentally made of brittle material and connections of curtain walls;

(2) Grade 8 and 9 earthquakes: supports and connections of suspended heavy objects, as well as anchors of overhanging billboards and similar components;

(3) Bearing parts of heavy-duty trademarks, signs, signals and overhanging decorative frameworks on high-rise buildings;

(4) Grade 8 and 9 earthquakes: supports and connections of cultural relics display cabinets of Class B buildings;

(5) Grade 7-9 earthquakes: anchors of elevators and lifting equipment, as well as elevator components and anchors of high-rise buildings;

(6) Grade 7-9 earthquakes: supports, bases and anchors of building attachment with the gravity exceeding 1.8kN or the natural vibration period of its system greater than 0.1s.

Normally, a simplified calculation method is acceptable, i.e. equivalent lateral force method. The additional internal force generated by the relative displacement between supports should be taken into account at the same time. If a floor and a device connected to it in a rigid manner are regarded as one mass point in the calculation and analysis of the entire structure, additional calculation in the floor spectrum method may be omitted.

7.2.1 Calculation Method for Horizontal Seismic Action of External Thermal Insulation System

In accordance with the *Code for Seismic Design of Buildings* (GB 50011—2010), the horizontal seismic action of an external thermal insulation system can be calculated in the equivalent lateral force method.

If the equivalent lateral force method is used, the standard value of horizontal seismic action should be calculated from:

$$F = \gamma \eta \xi_1 \xi_2 \alpha_{max} G \quad (7\text{-}2\text{-}1)$$

Where, F —standard value of horizontal seismic action at the center of gravity of the non-structural component in the most unfavorable direction;

γ —function coefficient of the non-structural component, determined based on the fortification category and use requirements of a building in relevant standards, taken as 1.4;

η —category coefficient of the non-structural component, determined based on the material performance and other factors in relevant standards, taken as 0.9;

ξ_1 —coefficient of the connection status, 2.0 for prefabricated components, cantilever components, flexible systems as well as any device with its support point below the center of mass, and 1.0 for others;

ξ_2 —location coefficient, 2.0 for the top of a building and 1.0 for its bottom, subject to linear distribution in the height direction;

α_{max} —maximum seismic action coefficient, determined according to the provisions on frequent earthquakes in Article 5.1.4 of the *Code for the Seismic Design of Buildings* (GB 50011—2010);

G —gravity of the non-structural component, including the gravity of personnel concerned, media in containers and pipes and articles in cabinets during operation.

7.2.2 Example of Seismic Calculation of External Thermal Insulation System

The seismic resistance was calculated with the external thermal insulation system based on the adhesive polystyrene granule pasted EPS board and face brick finish as an example. The relationship between its tensile bonding strength (the adhesive polystyrene granule layer was regarded the weakest in this system, so its tensile bonding strength was taken as that of the system) and seismic force was determined, and the seismic force (F) in a Grade 9 rare earthquake was calculated.

The structure of the above-mentioned system is as follows: concrete wall + adhesive polystyrene granule pasted EPS boards + adhesive polystyrene granule layer + anti-cracking mortar (compounded with hot-dip galvanized welded wire mesh and anchor bolts) + face bricks bonded by face brick bonding mortar + caulking material of face bricks (Table 7-2-1).

Material Properties **Table 7-2-1**

Structural Form	Material Name/Serial Number	Thickness d (mm)	Density ρ (kg/m^3)
External thermal insulation system based on the adhesive polystyrene granule pasted EPS board and face brick finish	Concrete wall 1	200	2300
	Adhesive polystyrene granule pasting mortar 2	15	320
	EPS board 3	60	20
	Adhesive polystyrene granule pasting mortar 4	10	320
	Anti-cracking mortar 5	8	1300
	Bonding mortar 6	5	1500
	Face brick finish 7	7	2000

Calculated based on the above structure, the gravity (G) of the external thermal insulation system is:

$$G = \sum_{i=2}^{7} \rho_i h_i g \qquad (7\text{-}2\text{-}2)$$

Where, i —serial number of the material;

ρ_i —material density, kg/m^3;

h_i —material thickness, m;

g —gravity acceleration, $10m/s^2$;

G —gravity (vertically downward) of the external thermal insulation system, N/m^2.

The gravity per square meter of the external thermal insulation system is: $G = 411N/m^2$.

The following parameters are used as per the *Code for Seismic Design of Buildings* (GB 50011—2010):

$\gamma = 1.4$, $\eta = 0.9$, $\xi_1 = 2.0$, $\xi_2 = 2.0$, $\alpha_{max} = 1.40$

The above parameters are maximum values under the most unfavorable circumstances, and the corresponding horizontal seismic force is $F \approx 2,900N/m^2$. When the external thermal insulation system fully meets the tensile bonding strength ($\geqslant 0.1MPa$) stipulated in the *Products for External Thermal Insulation Systems based on Mineral Binder and Expanded Polystyrene Granule Plaster* (JG/T 158—2013) (as the face brick finish was cut only to the anti-cracking mortar surface during measurement, so the tensile strength of the paint finish is taken here as that of this system), the destructive force per unit area of this system is $E = 100kN/m^2$. The safety factor of this system under such seismic action is:

$$s = E/F = \frac{100 \times 10^3}{2900} \approx 34$$

From the calculation results, it can be seen that the tensile strength of this system is much greater than the horizontal seismic action in the case of a Grade 9 rare earthquake. Thus, seismic verification is not needed as long as material properties meet the standards, the external thermal system based on the adhesive polystyrene granule conforms to the quality requirements and its tensile bonding strength is in line with the provisions.

The adhesive polystyrene granule mortar layer here is equivalent to a flexible transition between the external thermal insulation system and base course wall. Similar to a measure for vibration isolation and energy absorption, the adhesive polystyrene granule mortar layer is able to absorb the seismic energy transmitted through the wall, thus reducing the seismic action to the external thermal insulation and enhancing its seismic performance. Moreover, this layer is able to absorb some structural deformation. When full bonding is adopted, the safety factor will increase, and the possibility for the external thermal insulation system to fall in an earthquake will be reduced.

7.3 Seismic Test of External Thermal Insulation System

The seismic action of an external thermal insulation system can be tested by the vibrating table or quasi-static method. The test principles, apparatuses and instruments of these two methods differ from each other.

7.3.1 Vibrating Table Test

7.3.1.1 Test Principle

Install an external thermal insulation system on a vibrating table, enter a seismic wave of specific waveform into the seismic simulation vibrating table, and observe the seismic response of each part of this system under the simulated seismic action.

7.3.1.2 Test Equipment

1. Seismic simulation vibrating table

This table should be provided with three directions and six degrees of freedom, and output various simulated seismic waves as needed.

(1) Mounting wall: for installation of an external thermal insulation system. Generally, the wall should be able to form the desired angle of total displacement according to the design requirements.

(2) Specimen components: subject to self-inspection by the manufacturer. The insulation system should be installed in line with the design requirements.

(3) Specimen: full-sized.

2. Test instrument

(1) The frequency response, range and resolution of the test instrument shall meet the *Specification for Seismic Test Method of Buildings* (JGJ/T 101—2015).

(2) The test instrument shall be calibrated before testing.

(3) The test data should be collected and recorded by the data collection computer system.

(4) The measuring sensor shall have good resistance to mechanical impact, and be lightweight and small-sized to facilitate installation and dismantling. The connecting wire of the sensor shall be a shielded cable. The temperature rise, output impedance and output frequency of the measuring instrument shall match with those in the data collection system.

7.3.1.3 Measuring Point Layout

Acceleration sensors should be mounted in major parts of the specimen base course wall and external thermal insulation system, and strain gauges in the corresponding parts of the latter.

7.3.1.4 Test Procedures

(1) Install the specimen.

(2) Install sensors such as accelerator sensors and stain gauges.

(3) Input white noise (0.07~0.1) g, and test dynamic properties of the specimen, such as the natural vibration frequency, vibration mode and damping ratio.

(4) Input the seismic wave, and gradually increase the acceleration from 0.07g based on the fortification intensity of 0.5. Record the details of seismic response of the specimen under various conditions.

(5) Stop the test when the acceleration increases to the expected value or the specimen begins to fail. Carefully inspect each part of the specimen, and record damage in details.

(6) Dismantle the specimen.

7.3.1.5 Test Data

Test data should include:

(1) Maximum acceleration response at each measuring point of the specimen under various conditions;

(2) Maximum displacement and strain at each measuring point of the specimen under various conditions.

7.3.1.6 Test Report

The test report should include:

(1) Specimen name, type and dimensions;

(2) Manufacturer and client;

(3) Elevation, plane, section and node details of the specimen;

(4) Type and material properties of the external thermal insulation system;

(5) Test standards, equipment and instruments;

(6) Characteristics of the seismic wave;

(7) Dynamic properties, acceleration response, displacement response, strain and damaged part of the specimen under various conditions;

(8) Test purpose, signatures of test personnel, etc.

7.3.2 Quasi-static Test

7.3.2.1 Test Principle

The quasi-static test is a test method in which a load is repeatedly applied in low cycles on the specimen under appropriate load or deformation control, until the specimen is damaged from the elastic stage. In this test, the structure or its component is loaded and unloaded repeatedly in the positive and negative direction, to simulate the force and deformation during reciprocating vibration in an earthquake. As the vibrating results of the structure are obtained in a static method, this method is known as the quasi-static or pseudo-static test.

7.3.2.2 Test Equipment

The test equipment is a general term for various devices making the structure or its component under the expected force. The quasi-static test equipment is mainly composed of:

(1) Loading device: distributing the load applied onto the test structure;

(2) Support device: accurately simulating the actual force or boundary conditions of the test structure or component;

(3) Observation device: including the instrument stand and observation platform for installation of various sensors;

(4) Safety device: preventing safety accidents or damage arising from the destroyed specimen.

The quasi-static test equipment for the external thermal insulation system is shown in Figure 7-3-1.

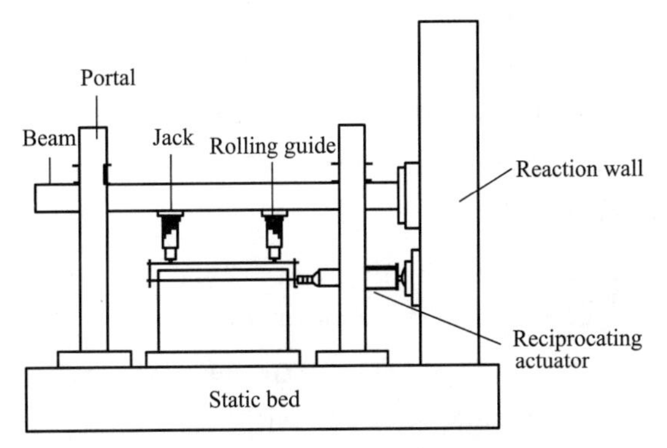

Figure 7-3-1 Schematic Diagram of Quasi-static Test Equipment

7.3.2.3 Measuring Point Layout

The layout of measuring points in the quasi-static test of the external thermal insulation system is shown in Figure 7-3-2. That is, five measuring points are distributed evenly on the centerline of one side of a wall in the height direction, to measure the maximum displacement of the top of the wall and also the lateral displacement curve.

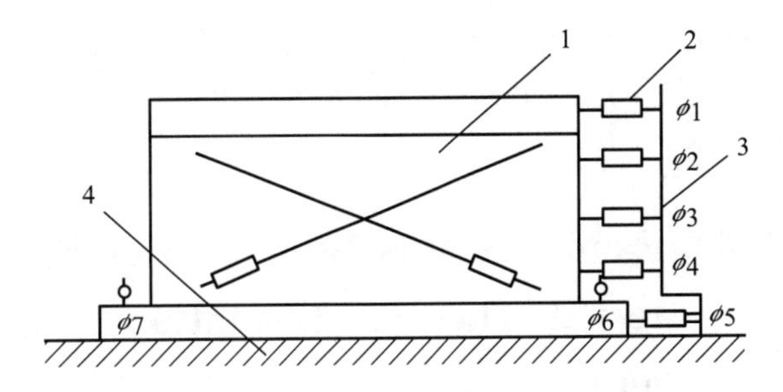

Figure 7-3-2 Measuring Point Layout for Measurement of Lateral Displacement and Shear Deformation of Wall

1: specimen; 2: displacement meter; 3: instrument stand on the test bed; and 4: test bed

7.3.2.4 Test Procedures

(1) Install the specimen.

(2) Install sensors such as strain gauges and displacement meters.

(3) Control the loading system (interlayer displacement angle: 1/1, 000 to 1/40).

(4) Stop the test when the interlayer displacement angle reaches the expected value or the specimen begins to fail. Carefully inspect each part of the specimen, and record damage in details.

(5) Dismantle the specimen.

7.3.2.5 Test Data

Test data should include:

(1) The load-deformation curve of the wall,

i. e. its resilience curve;

(2) Lateral displacement of the wall, mainly involving the lateral deformation of the specimen in the horizontal direction under the low-cycle repeated load.

7.3.2.6 Test Report

The test report should include:

(1) Specimen name, type and dimensions;

(2) Manufacturer and client;

(3) Elevation, plane, section and node details of the specimen;

(4) Type and material properties of the external thermal insulation system;

(5) Test standards, equipment and instruments;

(6) Characteristics of the loading system;

(7) Displacement response, strain and damaged part of the specimen under various conditions;

(8) Test purpose, signatures of test personnel, etc.

7.4 Examples of Seismic Test of External Thermal Insulation System

7.4.1 Vibrating Table Test of External Thermal Insulation System based on Adhesive polystyrene granule Pasted EPS Board and Tile Finish

7.4.1.1 Test Purpose

The seismic test under the fortification intensity in Beijing was conducted to two external thermal insulation systems based on the adhesive polystyrene granule pasted EPS board and tile finish on a concrete base course wall, to verify their damage in an earthquake, study the feasibility of their applications in high-rise buildings and analyze their seismic performance.

Compared with the paint finish, the tile finish has higher stain resistance and color durability. Tiles widely applied in exterior finishes of buildings, and there is also a great demand for applications in external thermal insulation systems. It is necessary to study the technology of tile finishing in external thermal insulation systems as well as the feasibility of applications in high-rise buildings in China. With cooperation of the Engineering Seismic Research Institute of the Beijing Building Research Institute of CSCEC, the Railway Engineering Research Institute of China Academy of Railway Sciences and other units, the seismic test plan and procedures for the external thermal insulation system with the tile finish have been formulated. On September 10, 2005, the seismic test was conducted to the external thermal insulation system based on the adhesive polystyrene granule pasted EPS board and tile finish in the Engineering Structure Test Center of Shijiazhuang Tiedao University. Given that the tile finish is flexibly connected to the main structure and the twisting force on the main structure is hardly transmitted to the finish layer, the seismic test was conducted with the seismic wave perpendicular to the tile finish layer. The most representative sinusoidal beat wave with the largest destructive force to the exterior finish was applied, making this seismic test more realistic and representative.

7.4.1.2 Specimen

1. Structural design

The insulation is composed of adhesive polystyrene granule pasted EPS boards. In the anti-cracking protective layer, the galvanized lead wire meshes at four corners are combined with anti-cracking mortar and tied by galvanized lead wires in tail fastener holes of the structural wall, to improve the bonding strength of the surface layer, insulation layer and structural layer as well as the safety and reliability of the entire structure.

Four reinforcing fasteners (42mm long) are used per square meter. According to the test results, the breaking tension of fasteners is 7kN separately and 28kN in total and that of lead wires for tying is 2kN. Thus the total tensile strength is 8kN per square meter with four mechanical fixing points, which was 4-5 times greater than Grade 8 seismic force at 100m of a high-rise building.

2. Model design and fabrication

Prior to the test, the specimen (width: 1.3m; height: 1.2m; thickness: 0.16m) with angle irons secured in holes was fabricated with C30 concrete, based on the test plan designed by the Engineering Seismic Research Institute of Beijing Building Research Institute of CSCEC and the Railway Engineering Research Institute of China Academy of Railway Sciences in line with the *Code for Seismic Design of Buildings* (GB 50011-2001), and also the cast-in-situ concrete structure used widely at present in high-rise buildings. The specimen model is shown in Figure 7-4-1.

The specimen structure is shown in Figure 7-4-2. Surace A (from inside to outside): C30 reinforced concrete wall, interface mortar, 15mm pasting mortar, 60mm EPS board, 10mm pasting mortar, 8mm anti-cracking mortar+hot-dip galvanized welded wire mesh and plastic anchor bolt, 5mm face brick bonding mortar, tile (45mm × 95mm), and face brick pointing material, to verify the seismic action of the concrete wall after EPS board bonding and leveling with the insulation mortar and bonding of the tile. Surace B (from inside to outside): C30 reinforced concrete wall, interface mortar, 15mm pasting mortar, 60mm EPS board, 8mm anti-cracking mortar+hot-dip galvanized welded wire mesh and plastic anchor bolt, 5mm face brick bonding mortar, tile (45mm × 95mm), and face brick pointing material, to verify the seismic action of the concrete wall after EPS board bonding with insulation mortar, thin plastering and tile bonding.

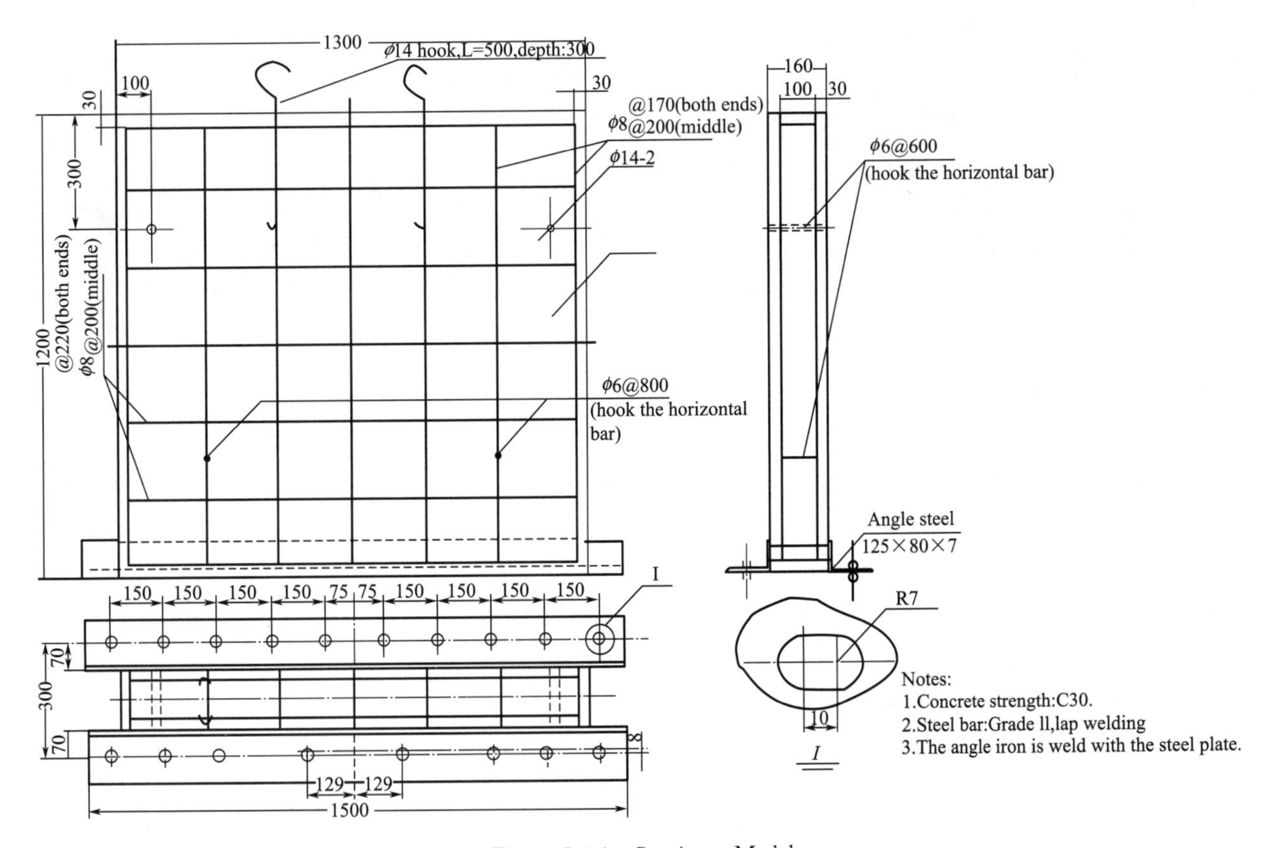

Figure 7-4-1 Specimen Model

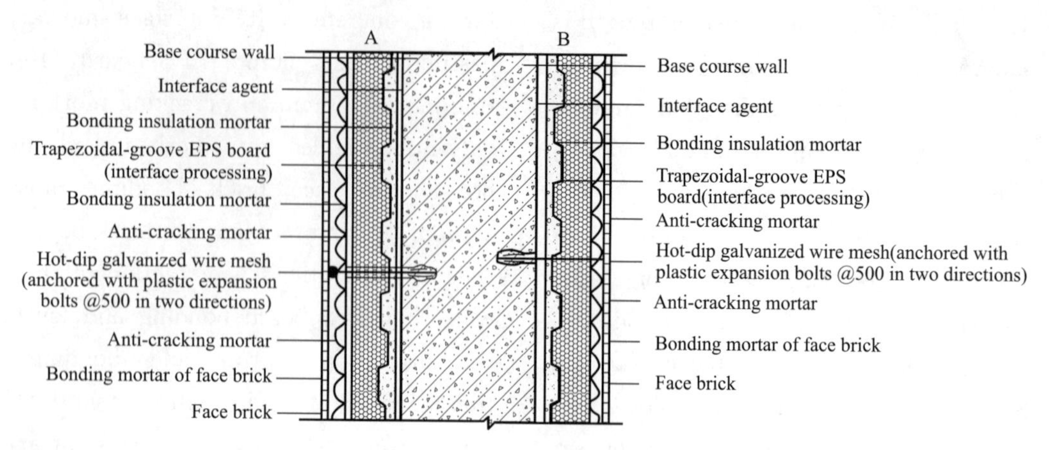

Figure 7-4-2 Specimen Structure

The test was conducted after the specimen was fixed onto and then reliably connected to the vibrating table. The connection between the specimen and vibrating table is shown in Figure 7-4-3.

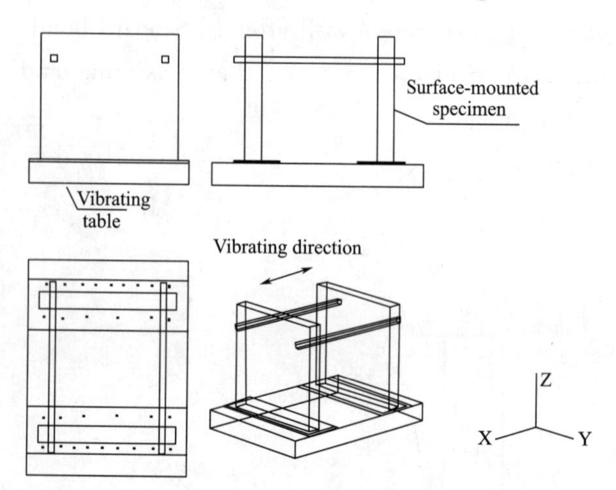

Figure 7-4-3 Connection of Specimen and Vibrating Table

3. Loading and testing procedures

Starting from the fortification intensity of Grade 8 and earthquake acceleration of 0.2g in Beijing, the seismic load was applied at five levels (increasing 0.1g respectively), namely, 0.2g (once), 0.3g (1.5 times), 0.4g (twice), 0.5g (2.5 times) and 0.6g (3 times). Considering that the horizontal seismic wave perpendicular to a building surface is the most destructive to a non-structural load-bearing material, the horizontal sinusoidal best wave was used (Fig. 7-4-4). Each vibration lasted for more than 20s and involved more than 5 beat waves. This test was completed within various frequency bands based on seismic response spectra of buildings in different regions and positions and with reference to the seismic influence coefficient in Section 5.1.4 of the *Code for Seismic Design of Buildings* (GB 50011—2010). The test frequency was divided by 1/3 octaves into 0.99Hz, 1.25Hz, 1.58Hz, 2.00Hz, 2.50Hz, 3.13Hz, 4.00Hz, 5.00Hz, 6.30Hz, 8.00Hz, 10.0Hz, 12.5Hz, 16.0 Hz, 20.0Hz, and 32.0Hz.

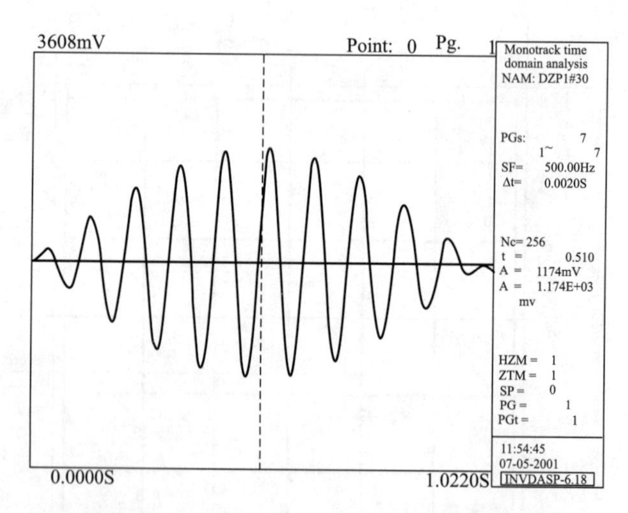

Figure 7-4-4 Sinusoidal Beat Wave

7.4.1.3 Test Results and Analysis

1. Test results

The specimen was subjected to a 10-hour two-cycle vibration test. When the acceleration rose to 0.5g in the first cycle, part of the reinforced concrete base course wall fell off and cracked. There was no cracking, damage or falling on the insulation material and finish layer of both surfaces (A and B), and tiles did not fall off or become loose.

Following the seismic test, the pulling test was conducted to tiles. The measured bonding strength of tile adhesive was 0.73MPa, which fully meets the tile bonding requirements.

2. Result analysis

Tile bonding on an external thermal insulation surface is different from that on a solid concretebase course wall. The load capacity of the insulation layer, bonding strength of the tile adhesive and the flexible deformation capacity against the violent movement under seismic effects must be taken into consideration during tile bonding on the external thermal insulation surface. Due to their flexible connection via an insulation material, the base course wall and finishing tiles cannot be regarded as a whole under the stress. Instead, they vary from each other in the presence of the stress, so the flexible tile adhesive compatible with the insulation material should be used in order to form a flexible gradient system. In the above seismic test, the tensile bonding strength after tile bonding with the selected adhesive was 0.40-0.80MPa, the compressive-to-flexural strength ratio was less than 3.0, and the elastic modulus was less than 6, 600MPa, thus achieving the appropriate flexibility and the technical requirements for the flexible gradient and layer-by-layer release of deformation. As the deformation of the tile adhesive was less than of anti-cracking mortar but more than that of tiles, the deformation difference between two materials that completely vary from each other in the mass, rigidity and thermal performance can be fully eliminated due to the deformation of the structure itself, thereby further ensuring that the stress generated by tiles like scales can be released separately and preventing tiles from falling off as a result of deformation under seismic actions.

The adhesive polystyrene granule mortar has excellent bonding and seismic performance, and its flexible structure is conducive to relief of seismic impact on the surface layer. If the appropriate elasticity of the tile adhesive is adopted, tiles will not crack or fall off in the event of rare earthquakes. The hot-dip galvanized wire mesh (12.7mm×12.7mm) was used instead of the alkali-resistant glass fiber mesh to enhance the safety and seismic resistance, so the tiles on the surface layer did not fall off under rare seismic actions. The seismic test results show that the maximum load of $60kg/m^2$ is acceptable for tiles bonded on the insulation layer. Therefore, the load of $60kg/m^2$ or less can be applied on the insulation layer.

After the devastating Wenchuan Earthquake, experts conducted investigations, demonstrating that external thermal insulation systems constructed as per the standards have normal seismic performance.

7.4.2 Quasi-static Test of Composite EPS Granule Self-Insulation of Wall

7.4.2.1 Test Purpose

The quasi-static test, in which the specimen was repeatedly loaded under appropriate load or deformation control until the specimen was destroyed from the elastic stage, was conducted to verify the seismic performance of the wall with composite EPS granule self-insulation.

7.4.2.2 Specimen

1. Specimen Structure and Model Design & Fabrication

In the composite wall concerned, the internal and external formworks was made ofreinforced vertical-fiber rock wool composite board and calcium silicate board, the middle lightweight self-insulation layer was made by pouring of EPS granule foamed concrete. It was connected to the main structure via the core column and tie beams. The basic structure of this wall is shown in Figure 7-4-5, and the layout of the core column and tie beams in Figure 7-4-6. The quasi-static test was conducted according to the method in the *Specification for Seismic Test Method of*

Buildings (JGJ 101), using the specimen of 2, 600mm×3, 400mm.

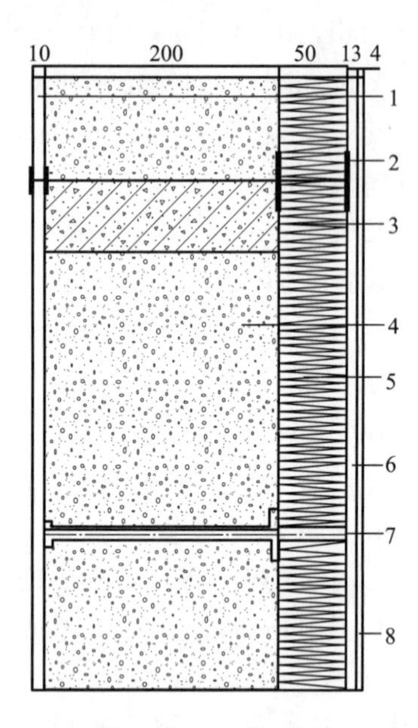

Figure 7-4-5 Schematic Diagram of Wall with Composite EPS Granule Self-insulation

1. Internal formwork (calcium silicate board); 2. Double "H" connector; 3. Concrete tie beam; 4. Self-insulated wall (EPS granule foamed concrete); 5. External formwork (reinforced vertical-fiber rock wool composite board); 6. Leveling layer of adhesive polystyrene granule mortar; 7. Through-wall pipe; 8. Anti-cracking protective layer

As shown in Figure 7-4-6, one core column (thickness: 200mm; width: 100mm; height: 3, 400mm) was set within 3, 400mm in the transverse direction of the wall, and three horizontal tie beams (thickness: 200mm; height: 100mm) bottom-up at intervals of 600mm within 2, 500mm in the longitudinal direction.

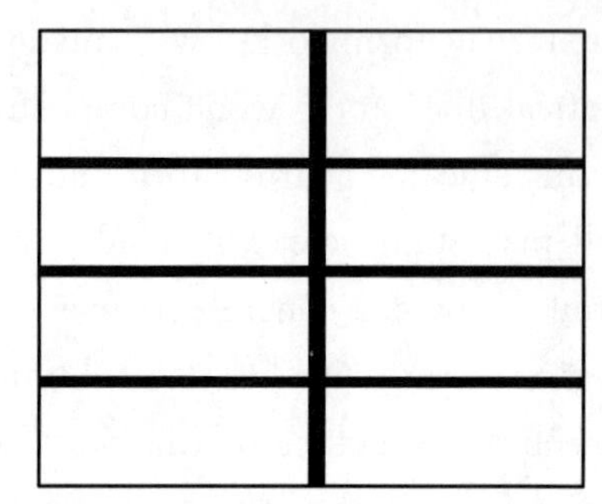

Figure 7-4-6 Layout of Structural Columns and Tie Beams of Wall

The core and tie beams were made with C20 concrete, and separately equipped with two steel bars (Φ16).

2. Loading and testing procedures

The controlled displacement loading method was used. The displacement loading levels are shown in Table 7-4-1.

Controlled Displacement Loading Levels and Main Phenomena during Test

Table 7-4-1

Controlled Loading Level	Phenomenon
±2.6mm (interlayer displacement angle: 1/1,000)	No cracks were found.
±3.3mm (interlayer displacement angle: 1/800)	No cracks were found.
±5.2mm (interlayer displacement angle: 1/500)	There were 0.02mm wide minor cracks on the inner wall, but no cracks on the outer wall.
±10.4mm (interlayer displacement angle: 1/250)	There were 0.02mm wide minor cracks on the outer wall.
±17.3mm (interlayer displacement angle: 1/150)	Cracks on two sides expanded, and the width of cracks on the outer wall was up to 0.5mm.
±26mm (interlayer displacement angle: 1/100)	Cracks expanded continuously. There were 1mm cracks in the vertical direction of the outer wall, and long horizontal cracks along connection joints of the inner wall. Squeezing and bulging occurred to some parts.
±32.5mm (interlayer displacement angle: 1/80)	Horizontal cracks on the inner wall expanded continuously. Through cracks appeared in the vertical direction of the outer wall, with the maximum width up to 10mm.
±52mm (interlayer displacement angle: 1/50)	There were a large number of cracks along the connection joints of the inner wall, with the maximum width up to 10mm, but no new cracks on the outer wall.
±65mm (interlayer displacement angle: 1/40)	Original cracks expanded continuously, but there were no new cracks.

7.4.2.3 Test Results and Analysis

1. Test results and conclusions

Main phenomena during the test are shown in Table 7-4-1. According to the test results:

(1) The wall cracks were 0.02mm wide before the interlayer displacement angle reached 1/250 (loaded displacement: 10.4mm).

(2) The wall material did not fall off and the wall did not collapse, except local squeezing and bulging, before the interlayer displacement angle reached 1/40 (loaded displacement: 65mm).

2. Result analysis

In the entire quasi-static test process, the self-insulated wall was reliably connected to the main structure through the core column and tie beams, and the wall deformation was always synchronized with the structure deformation, without collapse.

Figure 7-4-7 and 7-4-8 respectively show the cracks on the internal and external surface of the specimen and those on the inner surface, corresponding to the interlayer displacement angle of 1/50. It can be seen that cracks on the inner surface are mainly located in board gaps of the internal formwork (composed of calcium silicate boards). As calcium silicate boards are rigid, the stress will be finally transferred to and concentrated in board gaps in an earthquake, thereby leading to cracks. The core column on the outer wall cracked, and other parts were basically in good conditions.

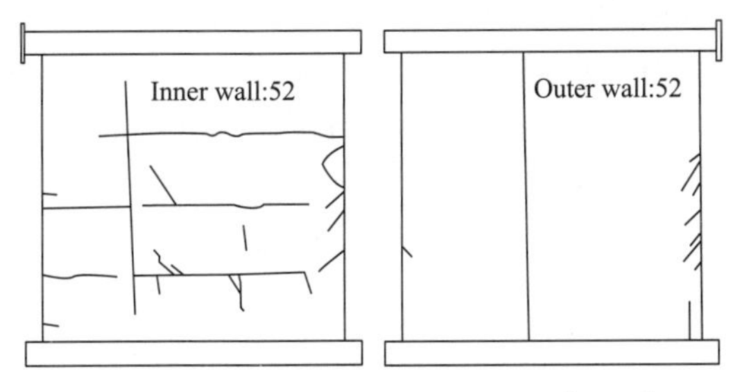

Figure 7-4-7 Schematic Diagram of Cracks on Internal and External Surfaces of Wall with Loaded Displacement of 52mm (1/50)

As shown in Table 7-4-1:

(1) When the interlayer displacement angle was 1/250, there were no cracks on the wall. This meets the requirement for the interlayer displacement angle (1/300) of steel structures.

(2) When the interlayer displacement angle was 1/40, except local squeezing and bulging, the wall material did not fall off and the wall did not collapse. This proves that the wall has excellent seismic performance.

Figure 7-4-8 Schematic Diagram of Cracks on Inner Surface of Wall with Loaded Displacement of 52mm (1/50)

The test results demonstrate that the reinforced vertical-fiber rock wool composite board is able to release a large amount of stress generated in the seismic test. As the EPS granule is used as the aggregate in the EPS granule foamed concrete, the wall constructed with such concrete has some quasi-elasticity and low elastic modulus, and the brittle failure does not occur extensively under the compressive stress. In addition, the wall material with the apparent density of less than 500kg/m^3 has light weight and good integrity. When an earthquake occurs, the wall suffers from small seismic forces, and the shock wave is transmitted slowly. The self-insulated wall has appreciable damping effects due to its long cycle of natural vibration and fast absorption of impact energy. The lightweight wall is conducive to the reduction of building weight and seismic impact.

The EPS granule composite self-insulated wall is poured integrally, in which the insulation layer has light weight, good flexibility and high seismic performance. When an earthquake occurs, its flexible structure is able to relieve the seismic impact on the finish layer, thus decreasing the peeling risk of the finish layer.

8. Safety of Bonded Face Brick of External Thermal Insulation System

Affected by climatic conditions, consumption levels and aesthetic habits in many regions of China, the outer wall with the face brick finish is popular with real estate developers and residents because of good decorative effect and high impact resistance as well as better stain resistance and color durability than the paint finish. The face brick finish is applied on more and more buildings in China. However, hollowing or falling occurs in a long time, including the so-called "tile rain", which inevitably lead to worries.

In order to meet the external thermal insulation requirements in the design standards for building energy efficiency, insulation layers of the majority of external thermal insulation systems are made of organic foamed plastics, with the lower density, strength and rigidity far lower than that of the base course wall material. With the development of building energy conservation and the formulation of higher design standards, the thickness of the insulation layer will also rise, and safety problems of face brick finishes in external thermal insulation systems will become increasingly prominent. This chapter analyzes the quality problems of face brick finishes in external thermal insulation systems, and introduces the ideas of corresponding solutions, as well as the testing, study, technical measures and guidance over the engineering practice.

8.1 Status Quo of Bonded Face Brick Finish of External Thermal Insulation System

8.1.1 Related Provisions for Bonded Face Brick Finish of External Thermal Insulation System

There are different provisions for finishes in the prevailing national and industry standards on external thermal insulation systems:

(1) The *External Thermal Insulation Composite Systems based on Expanded Polystyrene* (GB/T 29906—2013) stipulates that the paint finish shall be used in the thin-plastered EPS board system, but the face brick finish is listed in its informative appendix.

(2) The *Products for External Thermal Insulation Systems based on Mineral Binder and Expanded Polystyrene Granule Plaster* (JG/T 158—2013) specifies that the paint or face brick finish shall be used in the external thermal insulation system based on the adhesive polystyrene granule plaster. The face brick finish is significantly different from the paint finish, in which the system and anti-cracking plaster layer shall be reinforced with anchor bolts and reinforcement meshes. This standard not only provides the performance indicators of the tile adhesive and pointing material, but also defines the dimensions and mass per unit area of face bricks.

(3) The *Technical Specification for External Thermal Insulation on Walls* (JGJ 144—2004) stipulates that the paint finish shall be used in the thin-plastered EPS board system and the EPS board and cast-in-situ concrete system, but does not specify whether the facing brick finish is applicable to the adhesive polystyrene granule mortar system, EPS meshed board and cast-in-situ concrete system and the mechanically fixed EPS meshed board system.

(4) The *Technical Requirements of External Thermal Insulation with Expanded Polystyrene Panel for In-situ Concrete* (JG/T 228—2007) specifies that the paint finish or face brick finish shall be used in the cast-in-situ EPS board

system but the surfaces of EPS boards shall be leveled with the adhesive polystyrene granule waterproof mortar.

(5) The *Code for Acceptance of Energy Efficient Building Construction* (GB 50411—2007) stipulates that "the face brick finish should not be used in external thermal insulation projects."

The face brick finish of the external thermal insulation system is not involved in the national industry standards "Specification for Construction and Acceptance of Tapestry Brick Work for Exterior Wall" (JGJ 126) and "Test Standard for Bonding Strength of Building Facing Brick" (JGJ 110), some terms therein must be observed. For example, the exterior finish shall be designed specifically; and when the tensile strength of thebase course wall is lower than the bonding strength of the face brick finish of the outer wall, reinforcement measures must be taken, followed by the strength test of specimens; specimens of the face brick finish shall, prior to construction, be fabricated and accepted by the construction, design and supervision units in accordance with relevant standards; and the bonding strength of face bricks of the outer wall shall not be less than 0.40MPa.

8.1.2 Quality Problems of Bonded Face Brick of External Thermal Insulation System

On the contrary to "precautions" and "inappropriateness" in the above standards, face brick finishes are extensively applied in actual external thermal insulation projects, even in high-rise and super high-rise buildings, leading to engineering quality and safety problems. It should be pointed out that, in order to adapt to market needs, conduct a large number of tests and studies of face bricks used in external thermal insulation systems, take effective technical measures in strict accordance with relevant standards, make specimen walls on the site, and strengthen the quality control during construction to make the bonded face bricks stand the test of time. However, the face bricks of some external thermal insulation projects have quality problems of hollowing and falling off, which is mainly caused by the blindness and randomness in the design and construction. Face bricks are bonded on paint finishes of the majority of projects concerned, without the specimen fabrication and pull-out test before construction; resulting in hollowing and falling between the anti-cracking mortar surface and face brick finish, cracking along face brick joints, leakage of the thermal insulation system as well as large-scale falling of face bricks.

8.1.3 Study on Bonded Face Brick for External Thermal Insulation

Safety is the primary requirement for the face brick finish of the external thermal insulation system of an outer wall, especially on a high-rise building. Different from the rigid base course wall, the composite wall with the external thermal insulation system often exhibits the characteristics of a soft base course wall due to its low-density and low-strength built-in insulation layer. In order to bond face bricks thereon and ensure that the bonding strength is higher than 0.40MPa, the anti-cracking mortar layer on the insulation layer must be reinforced until its tensile strength is greater than the bonding strength. Meanwhile, as external forces such as the thermal stress, fire, water or vapor, wind pressure and seismic action are directly applied on the face brick finish, the weathering resistance and other related properties of the external thermal insulation system with the face brick finish must comply with the relevant standards.

Main differences between the face brick finish and paint finish of the external thermal insulation system:

(1) With the face brick weight and surface thickness increasing, the dead load of the face brick finish will rise;

(2) The deformation of face bricks with a large elastic modulus is inconsistent with that of

the anti-cracking mortar layer;

(3) Face bricks may easily fall off under the action of freeze-thaw cycles and natural forces;

(4) The thermal strain may lead to stress concentration at special nodes.

In order to ensure the engineering quality, the face brick finish of the external thermal insulation system should be studied from the following perspectives:

(1) Safety of the system;

(2) Reinforcement measures of the system;

(3) Properties of the system and related materials;

(4) Key points of construction of the system.

8.2 Study on Safety of Face Brick Finish System

The external thermal insulation system is a non-bearing structure that is composed of several functional layers and attached to a base course wall, so the connection between layers and that between the system and base course wall must be safe and reliable. Functional layers of the system are connected by directly bonding adjacent materials. The system and base course wall are connected mainly in two ways: 1, direct bonding with the adhesive; 2, bonding with the adhesive and anchor bolts. As the external thermal insulation system with the face brick finish increases, the safety of connection between the system and base course wall must be calculated and analyzed.

8.2.1 Calculation Model of Shearing and Tensile Forces Generated by Dead Load

In order to study and analyze the influence of the dead load of the external thermal insulation system with the face brick finish on the safety, a mechanical model was established, taking the thin-plastered EPS board system with a face brick finish as an example.

8.2.2 System Structure and Material Parameters

The basic structure of the EPS board system with the face brick finish is: concrete wall+EPS board bonded with mortar+anti-cracking mortar (with hot-dip galvanized welded wire mesh secured via anchor bolts) + face bricks bonded with adhesive+face brick pointing materials (Table 8-2-1).

8.2.3 Mechanical Model

The study object is an external thermal insulation system of an outer wall (1m×1m), with the bonding area of 40%. The following assumptions are made before model fabrication:

(1) As the actual locations of anchor bolts cannot be determined, their effects are not taken into consideration here;

(2) All the effects of EPS board bonding mortar are substituted with two bonding mortar areas ($A=0.20\text{m}^2$) centered at 1/4 and 3/4 (in the height direction) of EPS boards;

(3) Only the influence of the dead load is

Material Properties **Table 8-2-1**

Structural Form	Material Name	Thickness d (mm)	Density ρ (kg/m^3)
Point-bonded EPS board system with face brick finish	Concrete wall	200	2300
	EPS board bonding mortar	10	1500
Point-bonded EPS board system with face brick finish	EPS board	50	20
	Anti-cracking mortar	8	1300
	Bonding mortar	5	1500
	Face brick finish	7	2000

taken into account;

(4) The material of each layer is flat, and the wall is upright.

The mechanical model established is shown in Figure 8-2-1.

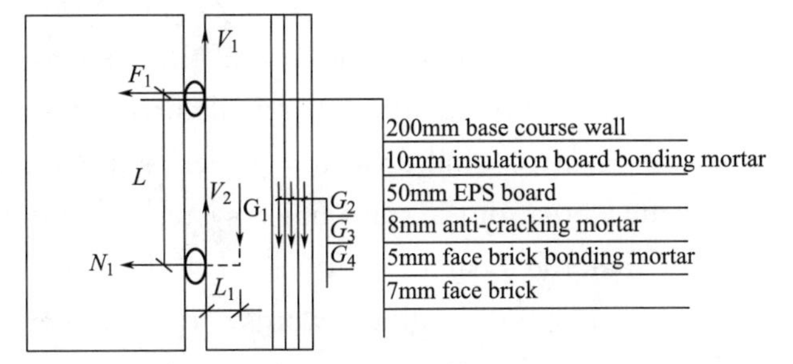

Figure 8-2-1 Mechanical Model of External Thermal Insulation System under Gravity

The face brick finish (excluding the EPS board bonding mortar) was analyzed as a whole. An equation set consisting of the equations (8-2-1) and (8-2-2) was used according to the balance of the force and moment of an object:

$$V_1 + V_2 = G_1 + G_2 + G_3 + G_4 \quad (8\text{-}2\text{-}1)$$

$$F_1 \times L = G_1 \times L_1 + G_2 \times L_2 + G_3 \times L_3 + G_4 \times L_4 \quad (8\text{-}2\text{-}2)$$

Where, V_1 and V_2 —upward shearing force applied by EPS board bonding mortar on the EPS board parallel to the wall surface, N;

G_1, G_2, G_3 and G_4—dead load of the EPS board, anti-cracking mortar, adhesive and face brick, respectively, N;

F_1 — tensile force applied by the EPS board bonding mortar close to the top on the EPS board, N;

L_1, L_2, L_3, L_4 and L —arm of the forces G_1, G_2, G_3, G_4 and F_1 with the connection point between the EPS board and its bonding mortar as the fulcrum, equal to $d_1/2$, $d_1+d_2/2$, $d_1+d_2+d_3/2$, $d_1+d_2+d_3+d_4/2$ and (3/4-1/4) ×1 (in m).

The equation (8-2-1) is a force balance equation, in which the right side means the gravity of the external thermal insulation system while the left side is the shearing force generated under the dead load of EPS board bonding mortar for system balance.

Where,

$$\begin{cases} G_1 = \rho_1 (S \times d_1) g \\ G_2 = \rho_2 (S \times d_2) g \\ G_3 = \rho_3 (S \times d_3) g \\ G_4 = \rho_4 (S \times d_4) g \end{cases}$$

S is the area of the studied object, taken as $1m^2$ here; ρ_1, ρ_2, ρ_3 and ρ_4 as well as d_1, d_2, d_3 and d_4 are the density and thickness of the EPS board, anti-cracking mortar, adhesive and face brick, respectively (in kg/m^3 and m); g is the acceleration of gravity, taken as $10m/s^2$ here.

The equation (8-2-2) is a moment balance equation, which is used for the following reason: the external thermal insulation system under the dead load tends to rotate around the EPS board and bonding mortar connection close to the bottom (right side of the equation-clockwise in Figure8-2-1), so a moment (right side of the equation-counterclockwise in Figure 8-2-1) to prevent such rotation is inevitable.

8.2.4 Calculation Results

The calculation results are:

$G_1 = 10N$, $G_2 = 104N$, $G_3 = 75N$, $G_4 = 140N$,

$L_1 = 0.025m$, $L_2 = 0.054m$,

$L_3 = 0.0605m$, $L_4 = 0.0665m$, $L = 0.5m$

Assuming $V_1 = V_2$, then:

$$V_1 = V_2 = 164.5N \quad (8\text{-}2\text{-}3)$$

$$F_1 = \frac{G_1 \times L_1 + G_2 \times L_2 + G_3 \times L_3 + G_4 \times L_4}{L} = 39.427N \quad (8\text{-}2\text{-}4)$$

The shearing force and tensile force applied by the dead load of the external thermal insulation system per unit area of the EPS board bonding mortar is:

$$w_1=\frac{V_1}{A}=0.822\ \text{kPa} \qquad (8\text{-}2\text{-}5)$$

$$w_2=\frac{F_1}{A}=0.197\ \text{kPa} \qquad (8\text{-}2\text{-}6)$$

If the material of larger mass in this system is farther away from the base course wall, the tensile force arising from the dead load will increase.

8.2.5 Dead Load Safety Factor

The above calculation model and method are also applicable to the no-cavity external thermal insulation system; and the calculation of shearing and tensile forces applied by the dead load on other interfaces. Comparing the value of w_2 in the equation (8-2-6) and the tensile bonding strength [$\sigma_2=0.10$ (MPa)] of the EPS board as well as the value of w_1 in the equation (8-2-5) with the compressive-shear bonding strength [$\sigma_1=0.10$ (MPa)] of the EPS board interface (assuming that the compressive-shear bonding strength of a material is equal to its tensile bonding strength), it can be seen that the compressive-shear bonding strength and tensile bonding strength of the EPS board are much greater than the shearing and tensile forces generated by the dead load. As shown by the equations (8-2-5) and (8-2-6), the shearing force arising from the dead load of the external thermal insulation system is far more than its tensile force.

The forces generated by the dead load of the external thermal insulation system are used in analysis of the safety factor. The shear safety factor α_1 and tensile safety factor α_2 of the EPS board (the shear strength and tensile strength of the system are represented by the strength of the weakest EPS board in the system) are $\alpha_1=\frac{\sigma_1}{w_1}=\frac{0.1\times 1000}{0.822}\approx 122$ and $\alpha_2=\frac{\sigma_2}{w_2}=\frac{0.1\times 1000}{0.197}\approx 508$, respectively. It can be seen that the compressive-shear bonding strength between the EPS board and bonding mortar is far more than the dead load of the face brick finish system, indicating that the system is safe enough in the vertical direction. A small amount of horizontal tensile force is generated by the dead load of the system, and the main force in the horizontal direction is still the negative wind pressure, which is almost the same as that of the paint finish system and indicates that the system is sufficiently safe in the vertical direction. In other words, the system safety is not affected by the additional weight of the face brick finish. Of course, the weight and dimensions of face bricks should be controlled, taking into account the combined effects of other factors. It is stipulated in relevant standards that the mass per unit area of face bricks bonded in the external thermal insulation system should not be greater than 20kg/m^2.

8.3 Study on Reinforced Structure of Face Brick Finish System

8.3.1 Necessity of Reinforced Structure

In order to ensure the bonding quality of face bricks, China's related standards stipulate that the bonding strength between the face bricks and base course wall shall not be less than 0.40MPa. When the tensile strength of a base course wall is lower than the bonding strength of face bricks, reinforcement measures must be taken. For the face brick finish system, the face brick bonding base course wall is the anti-cracking mortar layer on the insulation layer and its tensile strength should be higher than the face brick bonding strength (0.40MPa), so the system structure must be reinforced. The results of tests and studies show that the following reinforcements are applicable: 1, ensuring that the tensile bonding strength of anti-cracking mortar is greater than 0.5MPa; 2, laying a reinforcing mesh in the anti-cracking mortar layer. As a soft insulation layer is located under the anti-cracking mortar layer and horizontal tensile forces arising from face bricks to the anti-cracking mortar layer should be directly transferred to and borne by the base course wall, anchor bolts are used to connect the reinforcing mesh and base course wall to form a reinforced structure adaptive to the face brick finish and thus guarantee the system safety.

The studies on the reinforcing mesh and anchor bolt are described below.

8.3.1.1 Single-layer Glass Fiber Mesh

Currently, medium-alkali and alkali-resistant glass fiber meshes are mainly used in external thermal insulation. The alkali-free glass fiber has the lowest alkali metal oxide content, followed by the medium-alkali glass fiber and then alkali-resistant glass fiber (approximately ZrO_2 of 14.5% and TiO_2 of 6%). The ordinary glass fiber usually refers to the medium-alkali glass fiber, which is mainly composed of SiO_2 with good acid resistance but no alkali resistance. Through a large number of studies in China over years, it is determined that the ZrO_2 content is an important means for glass fibers to resist alkali erosion. The ZrO_2 content should be set reasonably; otherwise, its effects will not be obvious. In addition, ZrO_2 is a refractory material with the melting temperature above 1,600℃. The higher the ZrO_2 content, the more difficult it is for glass to be melted, and the higher technical requirements should be observed.

The durability of the glass fiber is manly affected by the following factors.

1. Fiber composition

The glass fiber composition is a prerequisite for the mesh durability. As mentioned earlier, the glass fibers containing ZrO_2 help to improve the alkali resistance of the mesh. It is thought in the literature that good alkali resistance can be achieved if the ratio of Na_2O/ZrO_2 is 1.0 to 1.2. When this ratio is decreased, the effect on the improvement of alkali resistance will not be obvious. When this ratio is increased, the alkali resistance will drop drastically.

2. Plastic coating quantity of glass fiber mesh

The plastic coating of the glass fiber mesh is used for protection of fibers from erosion by an alkali medium. The coating layer is formed by the following process: mortar absorption by the glass fiber mesh onto its surface, and then curing via baking, dehydration, chemical reaction for film formation, coiling, etc. The plastic coating quantity is not decisive for the alkali resistance of the mesh, but the plastic coating itself should be alkali-resistant. Most of plastic coatings of meshes in China are made of "acrylic acid+acrylic emulsion", "acetate emulsion+polyvinyl alcohol" or PVC emulsion.

3. Production technology of glass fiber mesh

The glass fiber has high tensile strength. The finer the fiber, the higher the strength will be. The fineness of warp filaments is generally 10.5-11.5μm, and that of weft filaments is 11.5-12.5μm. However, the glass fiber has poor shear properties, so the fiber surface may be easily worn or damaged due to the low equipment accuracy and high surface roughness during production. The plastic coating applied in the late period of production directly affects the strength of the glass fiber mesh.

4. External stress

The stress generated due to volumetric shrinkage of cement during hydration will lead to two kinds of component force on the glass fiber mesh in mortar: (1) tensile force parallel to fibers; and (2) force vertical to the fiber surface, resulting in bending deformation of fibers. Minor cracks on the fiber surface in the formation process will expand under the action of internal tensile and compressive stress, and finally the glass fiber mesh may be broken. In addition, the mesh has poorer mechanical properties in a humid environment than in a dry environment. This is because minor cracks are produced on the glass fiber surface due to temperature variations in the drawing process and expand in the presence of internal stress and external substances, especially with the evaporation and expansion of moisture, thus reducing the strength of the glass fibers and also the mesh.

5. Alkali resistance of glass fiber mesh

A large number of studies have conducted on the alkali corrosion of glass fibers in China. Theoretically, Ca $(OH)_2$ precipitated during the hydration of SiO_2 in the glass fiber with Portland cement is harmful to the silica-oxygen skeleton of fibers, making the glass fiber thinner, more brittle and less strong, and thus shortening the life of the glass fiber.

At present, there are many methods for alkali resistance testing of glass fiber meshes, such as the methods in the *Test Method for Determining Tensile Breaking Strength of Glass Fiber Reinforcing Mesh after Exposure to Sodium Hydroxide Solution* (GB/T 20102—2006, equivalent to the standard ASTME 98 of the United States), Appendix B of the *Glass Fiber Reinforcing Mesh-Part 2: Glass Fiber Reinforcing Mesh for Class PB EIFS* (JC 561.2—2006), Article A.12 of the *Technical Specification for External Thermal Insulation on Walls* (JGJ 144—2004), and Article 7.8.2 of the *Products for External Thermal Insulation Systems based on Mineral Binder and Expanded Polystyrene Granule Plaster* (JG/T 158—2013). Different alkaline media are specified in these methods. Table 8-3-1 lists the alkaline environments of various tests.

Refer to Table 8-3-2 for the contrast test of the alkali resistance of glass fiber meshes based on the above requirements.

The test results show that the alkali resistance of the glass fiber mesh varies with the alkali concentration and temperature. The alkali resistance drops the most at high temperature (5% NaOH solution), and its retention rate is the highest in the presence of cement paste. This means that 5% NaOH solution is the most corrosive to glass fibers at high temperature.

6. Reinforced structure of single-layer glass fiber mesh

It is stipulated in relevant standards of exter-

Domestic Relevant Standards on Alkali Resistance of Glass Fiber Mesh **Table 8-3-1**

Standard Code	Alkali Medium	Soaking Temperature/℃	Soaking time	Standard Code	Alkali Medium	Soaking Temperature/℃	Soaking time
GB/T 20102	5% NaOH solution	23	28d	Article A.12 of JGJ 144	Mixed solution	80	6h
Appendix B of JC 561.2	5% NaOH solution	80	6h	Article 7.8.2 of JG/T 158	Cement Paste	80	4h

Strength of Alkali-resistant Mesh in Different Alkali Environments **Table 8-3-2**

Solution Type		5% NaOH (80℃, 6h)	Mixed Solution (80℃, 6h)	Cement Paste (80℃, 4h)	5% NaOH (normal temperature, 28)	Mixed Solution (normal temperature, 28)	Cement Paste (normal temperature, 28)
Initial strength/ N	Warp	1480	1480	1480	1480	1480	1480
	Weft	1384	1384	1384	1384	1384	1384
Strength after alkali treatment/ N	Warp	1015	1070	1413	1133	1115	1432
	Weft	900	1043	1281	1065	1112	1211
Retention rate/%	Warp	68.6	72.3	95.5	76.6	75.4	96.2
	Weft	65.0	75.4	92.6	77.0	80.4	90.4

nal thermal insulation that the elongation at break of the alkali-resistant glass fiber mesh should be 3% to 5%. The maximum deformation (except thermal deformation) of the surface mortar (or anti-cracking mortar) with the elastic modulus of about 1, 000MPa and tensile strength of roughly 0.4MPa is 0.4MPa/1, 000MPa = 0.04%. The soft mesh is only able to disperse stress, instead of sharing tensile forces of the surface mortar (or anti-cracking mortar). It is impossible to make the alkali-resistant glass fiber mesh of which the elongation at break is about 0.04%. Moreover, rigid alkali-resistant glass fiber meshes may be easily broken during transportation.

Why is the mesh applicable to the paint finish? As paint is relatively soft (with low elastic modulus), there is little thermal stress in the paint finish, which can be completely borne by its constraint, namely, the surface mortar (or anti-cracking mortar). The mesh only needs to disperse the stress.

The elastic modulus of the face brick is much larger than that of the paint, so there is a lot of thermal stress on the face brick finish, which cannot be completely borne by its constraint, namely, the surface mortar (or anti-cracking mortar). In particular, quality problems such as cracking may occur in stress concentrated parts (window corners).

There is large internal thermal stress between the alkali-resistant glass fiber mesh and surface mortar (or anti-cracking mortar) at high or low temperature or in the case of abrupt changes in the temperature.

According to the test results, when a single-layer glass fiber mesh is used in the face brick finish, the tensile bonding strength of face bricks is related to the mass per unit area and the grid size of the glass fiber mesh. When the mass per unit area of the glass fiber mesh is increased, the tensile bonding strength of face bricks will rise a little. When the grid size of the glass fiber mesh is increased, the tensile bonding strength of face bricks tends to first rise and then drop.

8.3.1.2 Double-layer Glass Fiber Mesh

The construction methods of thermal insulation systems reinforced with the double-layer glass fiber meshes vary with manufacturers, and are generally divided into the method A (surface mortar+double-layer mesh+surface mortar) and method B (surface mortar+double-layer mesh+surface mortar+mesh+surface mortar). Test results show that the tensile bonding strength of various face bricks is greater than 0.4MPa, but it is lower than in the method B comparatively. Thus, the construction method has great impact on the tensile bonding strength of face bricks. In addition, the location of the glass fiber mesh in surface mortar also influences the tensile bonding strength of face bricks greatly. Therefore, the double-layer glass fiber mesh should be evenly distributed in the surface mortar layer.

The thermal expansion coefficients of glass fibers are (2.9×10^{-6} to 5×10^{-6})℃$^{-1}$, and those of wires are (11×10^{-6} to 17×10^{-6})℃$^{-1}$ (refer to the Mechanics of Composite Materials, compiled by Shen Guanlin and Hu Gengkai). The thermal expansion coefficient of cement mortar is not much different from that of the steel wire (refer to the section related to cement mortar among building materials). The alkali-resistant glass fiber mesh is located in the surface mortar (or anti-cracking mortar). Under normal conditions, their temperature values are the same, but their thermal deformation varies, resulting in thermal stress between them. The internal stress will be greater if the temperature is too high/low or changes too quickly.

The reinforced structure of the double-layer alkali-resistant glass fiber mesh has the following characteristics. Following the completion of the insulation layer, a layer of anti-cracking mortar (3~5mm) is applied on its surface, the first layer of alkali-resistant glass fiber mesh is pressed into the anti-cracking mortar at the same time, then the second layer of surface mortar and glass fiber mesh is constructed in the same way (total thickness: 6-10mm) to form the double-mesh anti-cracking protective layer, and finally face bricks are bonded.

Although the cracking resistance of the anti-cracking protective layer is effectively improved due to reinforcement with the alkali-resistant glass fiber mesh, the base course wall is not reinforced greatly in the presence of the face brick finish. The load of the face brick finish on the base course wall cannot be dispersed effectively, but is still applied on the insulation layer of lower strength.

In addition, the alkali-resistant glass fiber mesh just enhances the tensile strength in the parallel direction but has little impact on the strength in the vertical direction. The pull-out test results show that the damaged areas are concentrated on the surface of the glass fiber mesh, and the pull strength of the glass fiber mesh is relatively low, indicating that the glass fiber mesh is the weak part of this structure.

8.3.1.3 Galvanized Welded Wire Mesh

Galvanized welded wire meshes are divided into hot-dip galvanized welded wire meshes and cold-galvanized welded wire meshes based on the forming process. Hot galvanizing means immersion plating while cold galvanizing refers to electroplating. The galvanized welded wire mesh is made by electric welding of high-quality low-carbon steel wires with the precise automated mechanical technique, and has the flat surface, solid structure and high integrity. Even in the case of local cutting or pressure, the galvanized welded wire mesh will not become loose. Also, it has good corrosion resistance and other advantages that ordinary steel wires do not have.

At present, the following types of galvanized welded wire mesh are used in external thermal insulation systems in China: hot-dip galvanized welded wire mesh, cold-galvanized welded wire mesh, wire mesh welded first and then galvanized, and wire mesh galvanized first and then welded, which have different sizes and models. Table 8-3-3 shows specific requirements of hot-dip galvanized welded wire meshes in the *Products for External Thermal Insulation Systems based on Mineral Binder and Expanded Polystyrene Granule Plaster* (JG/T 158—2013).

1. Steel content of galvanized rectangular mesh

The rectangular mesh has significant impact on the anti-cracking protective layer. The performance of the entire anti-cracking protective layer varies with the weight of the rectangular mesh per unit volume. Under normal circumstances, it can be assessed based on the steel content.

The steel content refers to the weight of the rectangular mesh per unit volume in the anti-cracking protective layer, in kg/m^3. Theoretically, the higher the steel content, the more the strength of the anti-cracking protective layer will be enhanced, and the higher the bearing capacity will be. Specific operations are constrained by the costs, adaptability and other factors, so the steel content must not be over high.

Performance Indicators of Hot-dip galvanized Welded Wire Mesh **Table 8-3-3**

Item	Unit	Indicator	Item	Unit	Indicator
Wire diameter	mm	0.90±0.04	Tensile strength of spot weld	N	>65
Mesh size	mm	12.7×12.7	Weight of galvanized layer	g/m^2	>122

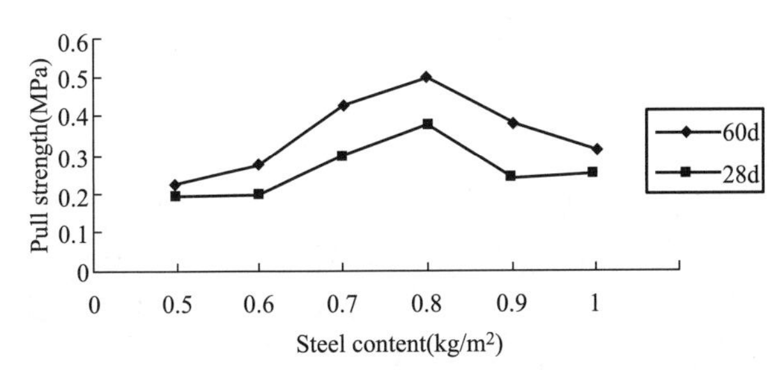

Figure 8-3-1 Impact of Steel Content on Pull Strength (pore size: 10mm×10mm to 20mm×20mm)

The test results of rectangular meshes (grid size: 10mm×10mm to 20mm×20mm) with different shapes and steel contents are shown in Figure 8-3-1. Given the same thickness of the anti-cracking protective layer, the pull strength of the system is low in the case of small steel content, indicating that the rectangular mesh does not reach the expected reinforcing effects. With the steel content rising, the rectangular mesh has better reinforcing effects and the pull strength grows gradually. When the steel content is 0. 8kg/m^2, the system has the peak pull strength. When the steel content continues to increase, the pull strength tends to drop.

Therefore, the steel content of the rectangular mesh should be 0. 8kg/m^2, which not only meets the needs for protection of the insulation layer and reinforcement of the anti-cracking protective layer, but also achieves excellent workability at appropriate costs.

2. Specifications of galvanized rectangular mesh

Reinforcement distribution is a main difference between the anti-cracking protective layer and reinforced concrete, and also an important condition to obtain the excellent performance of the anti-cracking protective layer. In the case of the same steel content, the distribution of reinforcing bars significantly influences the ultimate elongation, cracking strength, elastic modulus, creepage under long-term load, as well as bonding performance between materials of the anti-cracking protective layer. Therefore, it is of particular importance to determine the specifications of the rectangular mesh.

The rectangular mesh not only has favorable effects on the deformation and pressure suppression of the surrounding anti-cracking cement mortar under the stress, but also helps to reinforce the anti-cracking protective layer in the process of material combination. When the steel content is kept unchanged but the grid size is reduced, the wire diameter of the rectangular mesh will decrease, and the specific surface area of the rectangular mesh per unit area will increase, thereby enlarging the contact area between the rectangular mesh and anti-cracking cement mortar, enhancing the bond stress, and achieve better reinforcement effects on the anti-cracking protective layer. In the case of the same steel content, however, the smaller the grid size, the poorer the surface flatness is and the more difficult it is to lay the rectangular mesh. In this sense, the impact of other factors such as the construction adaptability should be taken into account when the specifications of the rectangular mesh are determined.

The test and analysis results (Figure 8-3-2) of the specific surface area coefficient KB of the rectangular mesh show that, when the steel content is 0. 8kg/m^2, the thickness of the anti-cracking protective layer is 5mm and the specific surface area of the rectangular mesh is 0. 46m^2/m^2, the rectangular mesh has good reinforcing effects on the anti-cracking protective layer, and the anti-cracking protective layer also has high pull strength.

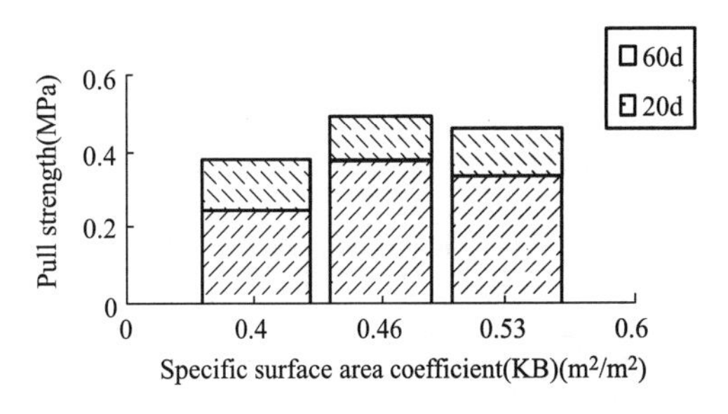

Figure 8-3-2 Relationship between Specific Surface Area and Pull Strength of Rectangular Mesh

3. Corrosion protection of rectangular mesh

As an important skeleton material of the anti-cracking protective layer, the durability of a rectangular mesh is related to both the durability of the anti-cracking protective layer, but also the durability and safety of the entire thermal insulation system. The rectangular mesh, a steel product, has general properties of steel. Because of the thermodynamic instability of steel, oxidation corrosion is an inevitable trend. Therefore, corrosion protection of the rectangular mesh is also an important issue to be studied and resolved in this system.

The typical research on the corrosion protection of the wire mesh abroad is the test conducted by Iranian Ramesht at Manchester Polytechnic in the United Kingdom, including the following procedures: pre-loading to generate minor cracks on the cement specimen (1 : 2 cement mortar with the water-cement ratio of 0.4) with the galvanized and non-galvanized welded wire mesh, wet curing for 28d, drying and soaking in NaCl solution (6%) alternately (once per hour) at 60℃, and corrosion testing after six months. The test results show that:

(1) If the wire mesh is tied tightly in the middle of the specimen to form a 9-12mm protective layer, the corrosion rate can be reduced significantly.

(2) Even minor pre-cracks will intensify the corrosion of the wire mesh, and the specimen surface subject to the tensile stress will be heavily corroded and destroyed.

(3) Although the galvanized and non-galvanized steel meshes are corroded to different degrees, the galvanized layer obviously has protective effects on the wire meshes. It is necessary to galvanize wire mesh to improve the durability of the cement structure.

The selection, layout and anti-corrosion treatment of the rectangular mesh are the same as those among the research results of foreign experts.

The corrosion of rectangular meshes made with different technologies under various alkali and salinity conditions are illustrated in Table 8-3-4 and 8-3-5. It can be seen from both tables that the hot-dip galvanized rectangular mesh is more corrosion-resistant than the cold-galvanized wire mesh. Main reasons thereof are differences in the galvanizing process and galvanized layer thickness. Under normal circumstances, the zinc layer thickness of 200μm can be easily obtained by means of hot galvanizing, but it is less than 10μm in the case of cold galvanizing, which does not meet the wire passivation needs with the pH value smaller than 13.3 and thus is not conducive to corrosion protection of the wire mesh. On the contrary, the corrosion resistance of the hot-dip galvanized welded wire mesh in cement mortar can be effectively enhanced with the thickening of the zinc layer.

Corrosion of Rectangular Meshes Made with Different Technologies under Different Alkali Conditions

Table 8-3-4

pH Value / Technology	7~9	9~11	11~13	≥13
Hot galvanizing	No rusting	No rusting	No rusting	No rusting
Cold Galvanizing	Serious rusting	Minor rusting	Minor rusting	Serious rusting

Corrosion of Rectangular Meshes Made with Different Technologies under Different Salinities Table 8-3-5

NaCl Value / Technology	3%	6%	9%	12%
Hot galvanizing	No rusting	No rusting	Minor rusting	Minor rusting
Cold Galvanizing	No rusting	Minor rusting	Minor rusting	Serious rusting

4. Tensile strength of rectangular mesh

The tensile strength of the rectangular mesh is composed of the tensile strength of welds and that of wires.

The tensile strength of welds represents the resistance of the rectangular mesh against the load in the vertical direction; and that of wires represents the resistance against the load parallel to the anti-cracking protective layer (main acting direction of the load).

The rectangular mesh with the wire diameter of 0.9mm and grid size of 12.7mm × 12.7mm was tested. The test results of the tensile strength of welds and wires are listed in Table 8-3-6.

Tensile strength of welds per unit area of the rectangular mesh: $F_H = 0.191 \times 81 \times 81 = 1253.2$ (kN);

Tensile strength of wires per unit area of the rectangular mesh: $F_W = 0.322 \times 81 = 26.1$ (kN).

As shown by the above data, the mechanical properties of the rectangular mesh satisfies the strength requirements of the system.

5. Reinforcement position of rectangular mesh

The reinforcement position of the rectangular mesh refers to the position of the anti-cracking protective layer therein. The location of the rectangular mesh in the anti-cracking mortar affects the anti-cracking protective layer, especially the isolation and protection of the insulation layer.

(1) When the rectangular mesh is directly exposed to the insulation layer or partially buried in anti-cracking mortar and partially exposed to the insulation layer, its reinforcement effect will decline. When the external force is applied on the anti-cracking protective layer, the insulation layer may be damaged easily.

(2) When the rectangular mesh is laid in the middle of anti-cracking mortar, the anti-cracking protective layer can be reinforced effectively and the insulation layer can be protected efficiently. When the external force is applied, damage may occur to and be absorbed by the anti-cracking protective layer.

(3) When the rectangular mesh is laid on the surface of the anti-cracking protective layer, the protection of the insulation layer will be improved, but the bond stress on the wire mesh is weak as a result of thin anti-cracking mortar on the surface of the rectangular mesh. When the external force is applied, the wire mesh surface may be damaged, and the tensile strength of the wire mesh is low.

Figure 8-3-3 shows the pull-out test data of different reinforcement forms, indicating that:

(1) When the rectangular glass fiber mesh is laid in the middle of anti-cracking mortar, the pull strength will high. When the rectangular glass fiber mesh is directly exposed to or close to the insulation layer, the pull strength will low, and the insulation layer may be damaged in the pull-out test. When the rectangular mesh is too close to the anti-cracking protective layer, the pull strength will fall between the above two values.

(2) In the absence of a rectangular mesh, the pull strength will increase from initial 0.10MPa to 0.22MPa with the thickening of the anti-cracking protective layer. If the thickness of the anti-cracking protective layer continues to increase, the pull strength will change little, and the damage in the pull-out test will be concentrated on the insulation layer, so the insulation layer will be damaged seriously.

Test Results of Tensile Strength of Welds and Wires of Hot-dip Galvanized Rectangular Mesh Table 8-3-6

Item \ S/N	1	2	3	4	Average
Tensile strength(N) of weld	195	187	192	190	191
Tensile strength(N) of wire	325	316	341	305	322

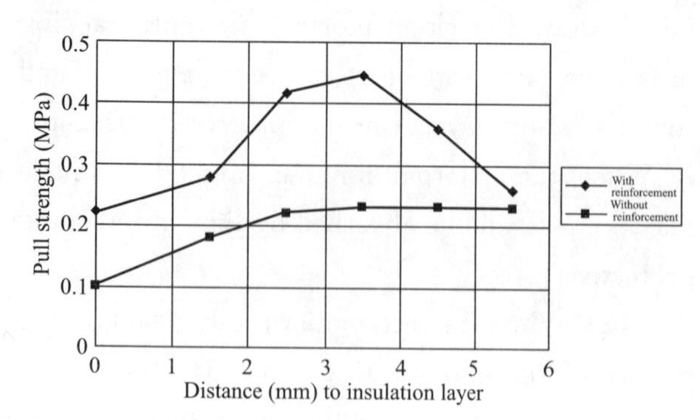

Figure 8-3-3 Pull-out Test Data of Different Reinforcement Forms

6. Reinforcement structure of galvanized rectangular mesh

Structural formation process: after the insulation layer is completed, applying a 2-3mm layer of anti-cracking mortar, laying a rectangular mesh and directly fixing it with plastic anchor bolts into anti-cracking mortar, then applying a 5-7mm layer of anti-cracking mortar to make the rectangular mesh inside anti-cracking mortar, and finally pasting face bricks. The pull-out test results of the reinforcement structure of the galvanized rectangular mesh are shown in Figure 8-3-4.

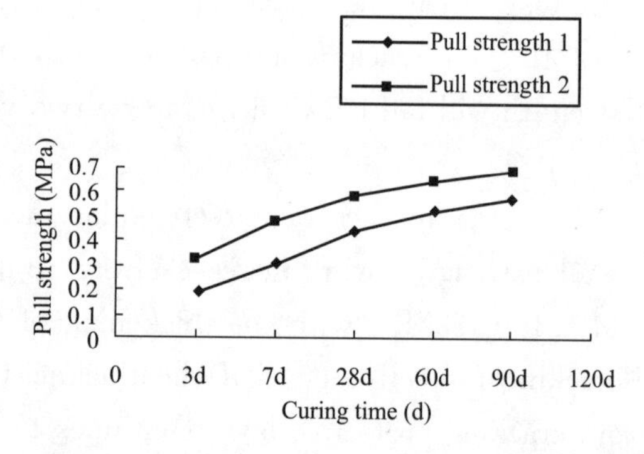

Figure 8-3-4 Pull-out Test Data of Reinforcement Structure of Galvanized Rectangular Mesh

Note: The curve of the pull strength 1 is based on the cement mortar substrate (1 : 3);
The curve of the pull strength 2 is based on the meshed anti-cracking mortar layer.

This structure is also able to protect the insulation layer through the rectangular mesh and transfer the surface load. Due to large bond stress between the rectangular mesh and anti-cracking mortar, the tensile strength in the horizontal and vertical direction will be improved, and the strength of the face brick bonding base course wall will be greatly enhanced.

The test results show that, in the presence of the face brick finish, the reinforcement structure of the galvanized rectangular mesh is superior to that of the alkali-resistant glass fiber mesh. The comparison of pull-out test results of both reinforcement structures are shown in Figure 8-3-5. If the galvanized rectangular mesh is used, its cracking resistance and the face brick requirements for the base course wall strength can be balanced effectively, and the tensile strength of the reinforcement system is 0.4MPa or more, thus meeting the stability, safety and durability needs of the thermal insulation system.

Figure 8-3-5 Comparison of Pull-out Test Results of Different Reinforcement Structures

8.3.1.4 Anchors

1. Anchorage mechanism of expansion bolt

The load is jointly supported via the friction

generated by pressing of the expansion parts of anchor bolts into borehole walls and the coordination of geometrical anchor bolt holes with the anchorage base course wall and borehole shapes.

2. Anchoring depth in base course wall

Anchorage is completed in two steps, namely, drilling and fastening. In order to avoid damage to the base course wall, a rotary drilling method should be adopted. Moreover, the drilling depth should be greater than the anchoring depth to guarantee the function of anchorage.

3. Base course wall protection in the anchorage process

Rotary drilling is preferred for the walls of low strength, such as hollow masonry walls, in order to avoid excessive drilling and prevent damage of external forces to the hollow masonry structure.

4. Anchor corrosion protection

Expansion bolts should be subject to corrosion protection and may be made of galvanized steel or stainless steel. Anchor bolts should be made of nylon plastic with high tensile strength and resistance to ageing, thermal deformation, cold, heat and high pressure.

5. Tensile strength of anchor

The tensile strength of expansion bolts used for anchorage is closely related to the bolt diameter. If the base course wall is composed of hollow bricks, the breaking load of each bolt should meet the requirements of Table 8-3-7.

Breaking Loads of Expansion Bolts with Different Diameters

Table 8-3-7

Bolt diameter(mm)	5	6	7	8
Breaking load(kN)	1.0	1.2	1.7	3.0

If bolts with the diameter of 7mm are used, the load (F_L) of each bolt will be 1.7kN or more. This system is designed based on at least four expansion bolts per square meter, so the reliable bearing capacity per unit area of expansion bolts should be: $F_L \geqslant 4\times1.7\geqslant6.8$ (kN).

8.4 Study on Related Materials of Face Brick Finish System

The results of observation on the construction site and studies in the laboratory show that there are two main forms of face brick falling: separate falling, indicating that the adhesive does not meet the face brick bonding requirements; falling with the adhesive, indicating that the adhesive does not meet the bonding requirements of the anti-cracking mortar layer; and cracking of the face brick pointing material as a result of extrusion, indicating that the deformation of the face brick finish cannot be absorbed by the pointing material. For this reason, related materials are tested.

8.4.1 Anti-cracking Mortar

8.4.1.1 Performance Indicators

In the face brick finish system for external thermal insulation, anti-cracking mortar should have both flexibility and appropriate strength. The test results show that, when the compressive-to-flexural strength ratio of anti-cracking mortar is less than or equal to 3.0 and its compressive strength is 10MPa or more, the anti-cracking protective layer has good cracking resistance and the base course wall strength meets the requirements for face brick bonding. Table 8-4-1 lists the performance indicators of anti-cracking mortar.

Performance Indicators of Anti-cracking Mortar

Table 8-4-1

Item		Unit	Indicator
Operable time		h	2
Tensile bonding strength	Initial value	MPa	≥0.7
	After soaking in water		≥0.5
	After freeze-thaw cycle		≥0.5
Compressive-to-flexural strength ratio		—	≤3.0

8.4.1.2 Thickness of Anti-cracking Mortar

The anti-cracking protective layer is an important part of the thermal insulation system, and plays a special role of "connection". That is, the insulation layer of low density and strength is organically combined with the face brick finish via the anti-cracking protective layer as a transition layer from the insulation layer unsuitable for face brick bonding to the protective layer with the appropriate strength and flexibility. At the same time, forces are transferred to the base course wall via anchor bolts. The test results show that the thickness of the anti-cracking mortar layer has significant impact on the protection of the insulation layer and also the pull strength of the thermal insulation system.

Figure 8-4-1 and Figure 8-4-2 illustrate the relationship between the thickness of the anti-cracking mortar layer and the pull strength of the thermal insulation system.

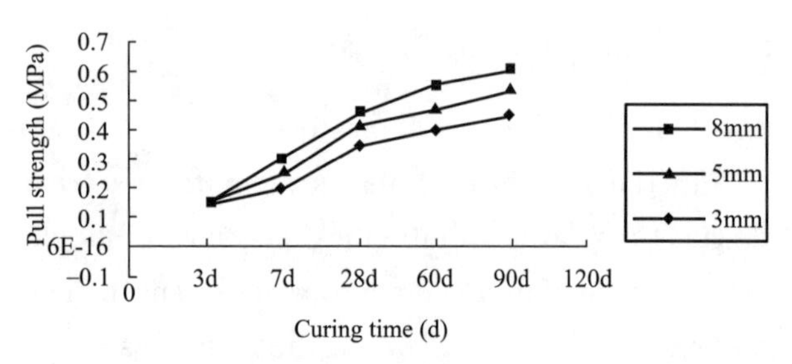

Figure 8-4-1 Relationship between Anti-cracking Mortar Thickness and Pull Strength

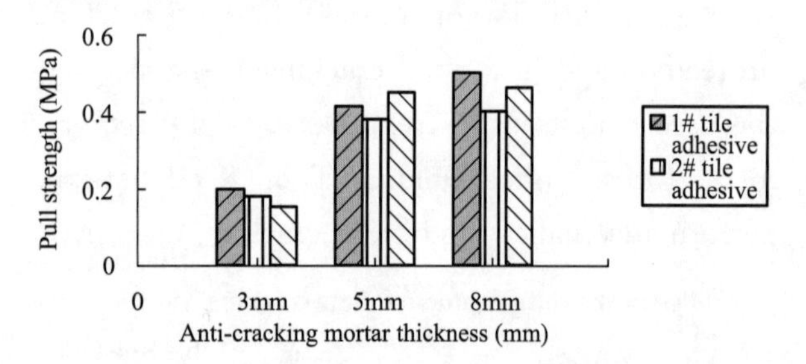

Figure 8-4-2 Relationship between Anti-cracking Mortar Thickness and Pull Strength (28d)

The test results show that, when the anti-cracking mortar thickness (*H*) is less than 5mm, the isolation and protection of the insulation layer will fail, and the damage in the pull-out test will be concentrated on the insulation layer. If the value of *H* is 5mm or more, especially 8mm or more, damage in the pull-out test will be concentrated in the anti-cracking protective layer, and the insulation layer will be protected effectively instead of damage in the presence of external forces. The 28d pull-out test results indicate that the pull strength of the thermal insulation system is 0.4MPa or more and damage occurs in the anti-cracking protective layer or bonding layer.

The thickness of the anti-cracking protective layer should be 10mm±2mm. If this layer is too thin, protection and reinforcement will fail; otherwise, engineering costs will increase. The reasonable cost-effective thickness is 10mm±2mm.

8.4.2 Face Brick Bonding Mortar

8.4.2.1 Performance Indicators

The conditions for face brick bonding on the surface of the external thermal insulation system are different from those on a solid concrete base course wall. The thermal expansion coefficient of face bricks varies greatly from that of the insulation layer. Accordingly, the thermal stress and deformation with the temperature change significantly. In addition to the weathering resistance, water resistance, ageing resistance and normal construction conditions, the internal stress caused by the deformation of two materials (different in the hardness and density) with the temperature must be taken into account when the face brick bonding mortar of the thermal insulation surface is selected. The binder used should be able to eliminate deformation variations of two materials (completely different in the mass, hardness and thermal properties) through its own deformation, in order to prevent the face bricks of high hardness, high density, large elastic modulus and low deformation performance from falling off the insulation layer of low hardness, low density, small elastic modulus and high deformation performance.

According to field surveys, the deformation rate of face brick bonding mortar should be more than 2‰ under the application conditions to prevent the thermal insulation system from cracking and eliminate the internal stress arising from temperature differences of materials. Considering that the adhesive cannot be directly pasted on the insulation layer but should be bonded with the anti-cracking protective layer, the deformation amount of face brick bonding mortar should be less than the thermal deformation of anti-cracking mortar but more than that of face bricks. Finally, the deformation amount of the adhesive corresponding to the 5mm anti-cracking protective layer is determined as 5‰ to 1%, which is less than that (5%) of anti-cracking mortar but more than that (1.5×10^{-6}/℃) of face bricks, thereby preventing face bricks from falling off in the case of thermal deformation.

The main performance of face brick bonding mortar is shown in Table 8-4-2.

Main Performance Indicators of Face Brick Bonding Mortar

Table 8-4-2

Item		Unit	Performance Indicator
Tensile bonding strength	Initial value	MPa	≥0.5
	After soaking in water		
	After thermal ageing		
	After freeze-thaw cycle		
	After 20min drying		
Transverse deformation		mm	≥1.5

8.4.2.2 Impact of Polymer-cement Ratio on Flexibility of Bonding Mortar

Flexibility is an important indicator of the face brick bonding material. There are many factors affecting the flexibility of the face brick bonding material, among which the polymer-cement ratio is the most influential. The ordinary cement bonding mortar containing no polymer has high strength but little deformation, and its compressive-to-flexural strength ratio is normally 5-8. When a base course wall is deformed in the presence of thermal stress, this kind of mortar for face brick bonding cannot be deformed to balance the base course wall deformation, but is often prone to hollowing or falling.

Given that the bonding strength is guaranteed, the flexibility of face brick bonding mortar of the external thermal insulation system should be improved to unify the face bricks and entire system, absorb external effects (especially those of thermal stress), and meet the face brick finish needs of the external thermal insulation system. Figure 8-4-3 shows the impact of the polymer-cement ratio on the compressive-to-flexural strength ratio, in which the compressive-to-flexural strength ratio 1 was measured after curing in water, the compressive-to-flexural strength ratio 2 after curing in a plastic bag, and the compressive-to-flexural strength ratio 3 after curing in water. It can be seen that the polymer-cement ratio has significant impact on the compressive-to-flexural strength ratio.

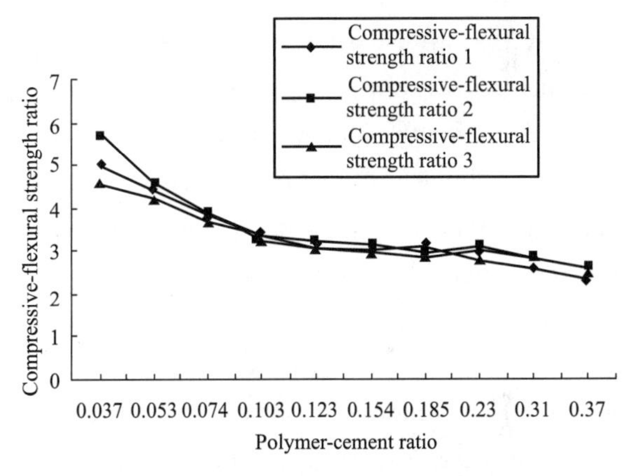

Figure 8-4-3 Relationship between Polymer-cement Ratio and Compressive-to-flexural Strength Ratio

(1) The cement mortar with the small polymer content has low flexibility and large compressive-to-flexural strength ratio.

(2) As the polymer-content ratio increases, the polymer-cement ratio will rise. When the polymer-content ratio is about 0.1, the compressive-to-flexural strength ratio will drop to be less than 3.5.

(3) If the polymer-cement ratio continues to increase to approximately 0.3, the compressive-to-flexural strength ratio will fluctuate between 3.0 and 3.5.

(4) When the polymer-cement ratio exceeds 0.3, the compressive-to-flexural strength ratio

will be less than 0. 3, which meets the requirements for flexible deformation.

Figure 8-4-3 also shows that the impact of one polymer-cement ratio on the compressive-to-flexural strength ratio varies with curing conditions.

(1) When the polymer-cement ratio is less than 0. 1, the compressive-to-flexural strength ratio is the highest after curing in a typical plastic bag, followed by that after curing in water and then that after curing in the air.

(2) When the polymer-cement ratio is 0. 1 to 0. 3, there are not significant differences in the effects of three curing methods on the compressive-to-flexural strength ratio.

(3) When the polymer-cement ratio is more than 0. 1, the compressive-to-flexural strength after curing in a plastic bag is the highest, followed by that after curing in the air and then that after curing in water.

Reasons for the above results are as follows: when the polymer-cement ratio is small, the cement performance plays a decisive role; when the polymer-cement ratio is 0. 1 to 0. 3, the effects of cement and polymer tends to be balanced; and when the polymer-cement ratio is more than 0. 3, the polymer has more obvious effects, though the bonding mortar still has the properties of the cement-based material. That is, the bonding mortar has clearly demonstrated the high flexibility and bonding performance of polymer, which meets the needs for face brick bonding of the external thermal insulation system.

8. 4. 2. 3 Impact of Curing Conditions on Bonding Performance

In general, appropriate curing measures are taken to a cement-based material after construction. Figure 8-4-4 illustrates the impact of curing conditions on face brick bonding mortar. Accordingly, the face brick finish was cured with water in seven consecutive days (twice per day) from 2h after bonding. The bonding strength of the tile adhesive is 20% more than that of the bonding mortar subject to no curing. The face brick bonding mortar used in the external thermal insulation system has been modified with polymers, and meets the bonding strength requirements even in the case of no curing. However, proper curing helps to achieve better bonding effects.

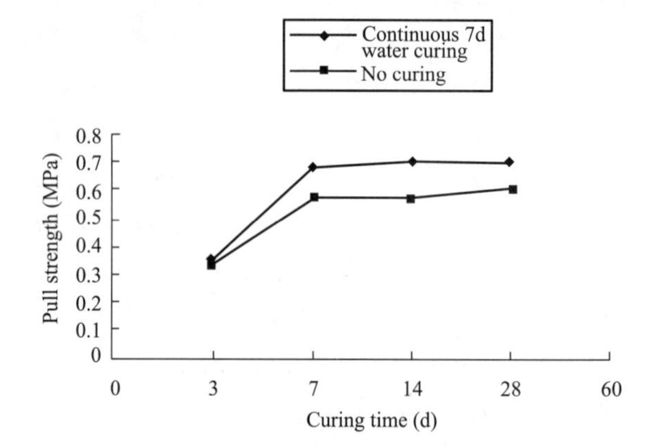

Figure 8-4-4 Impact of Curing Conditions on Performance Face Brick Bonding Mortar

8. 4. 2. 4 Impact of Application Life on Bonding Performance

Figure 8-4-5 shows the impact of the application life on the performance of bonding mortar.

According to Figure 8-4-5, the performance of face brick bonding mortar tends to decline greatly with the application life extended. If the face brick bonding mortar is used within the specified 4h, the tensile strength may reach 0. 4MPa. Otherwise, the tensile strength may abruptly drop to be less than 0. 2MPa, thus causing failure in face brick bonding.

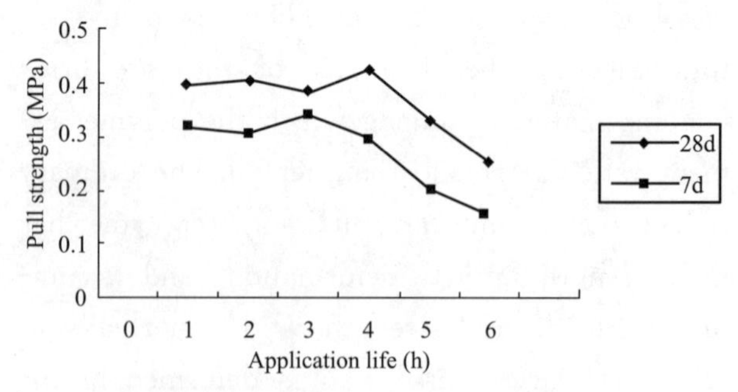

Figure 8-4-5 Impact of Application Life on Face Brick Bonding Mortar

8. 4. 2. 5 Impact of Water Absorption Rate of Face Brick on Performance of Bonding Mortar

The water absorption rate is an essential indicator of external face bricks. A low water absorption rate represents good sintering and bending performance, high strength and excellent resistance to wear, sudden heating/cooling and chemical corrosion, and vice versa.

Depending on the water absorption rate, face bricks for outer walls are classified as follow:

(1) $E \leqslant 0.5\%$;

(2) $0.5\% \leqslant E \leqslant 3\%$;

(3) $3\% \leqslant E \leqslant 6\%$;

(4) $6\% \leqslant E \leqslant 10\%$.

The water absorption rate of face bricks has great impact on the performance of face brick bonding mortar. That is, the effect of bonding mortar varies with the water absorption rate. This is mainly caused by the difference in the bonding mechanism. There are usually two mechanisms of bonding between the bonding mortar and face bricks.

(1) Physical and mechanical anchorage mechanism

In this mechanism, the adhesion of face bricks with bonding mortar is generated by permeation and filling of bonding mortar into pores and pits on the face brick surface, which seems the "grabbing" effect. Obviously, this mechanical is dominant for porous or rough materials. Dovetail-groove face bricks are based on this mechanism.

(2) Chemical bond mechanism

In this mechanism, the bonding mortar and face bricks are bonded through the Van der Waals forces among molecules or chemical bonds among reactive functional groups.

When the face brick has the low water absorption rate, good sintering performance and small void content, the physical and mechanical anchoring mechanism will be weakened. For the pure cement mortar relying mainly on physical and mechanical anchorage, the face brick bonding strength will be low. For the polymer-modified face brick bonding mortar, however, even smooth tiles can be bonded securely due to the Van der Waals forces between the functional groups of molecular chains of polymer and the molecules of the face brick surface as well as new valence bond combinations among some functional groups.

Figure 8-4-6 shows the performance of face brick bonding mortar on the face brick surface, in which the polymer content of the former and water absorption rate of the latter are different. Accordingly, when the water absorption rate (E) is 0.5% or less, the above three types of face brick bonding mortar have low bonding strength. The face brick bonding mortar A of large polymer content has the balanced effect on and high adaptability to face bricks with different water absorption rates.

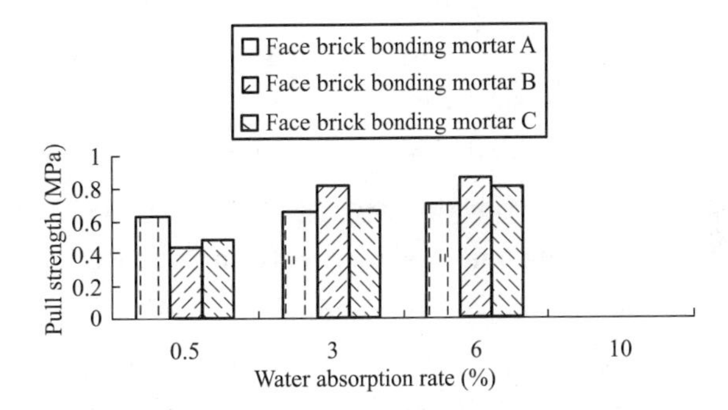

Figure 8-4-6 Impact of Water Absorption Rate of Face Brick on Bonding Performance

8.4.3 Pointing Material

8.4.3.1 Performance Indicators

The performance of the face brick pointing material should also meet the flexibility requirements, in order to effectively release the thermal stress and deformation of the face bricks and bonding material, and also prevent the face bricks of the finish layer from falling off. Meanwhile, the pointing material should have excellent water resistance. Table 8-4-3 lists the technical performance indicators of the face brick pointing material.

Main Performance Indicators of Face Brick Pointing Material

Table 8-4-3

Item		Unit	Performance Indicator
Shrinkage		mm/m	$\leqslant 3.0$
Flexural strength	Initial value	MPa	$\geqslant 2.50$
	After freeze-thaw cycle		$\geqslant 2.50$

Continued

Item	Unit	Performance Indicator
Water permeability(24h)	mL	≤3.0
Compressive-to-flexural strength ratio	—	≤3.0

8.4.3.2 Impact of Polymer-cement Ratio on Flexibility of Face Brick Pointing Material

The face brick pointing material is prepared by dry mixing of Portland cement (as a main cementing material), re-dispersed emulsion powder and other additives, with the compressive-to-flexural strength ratio of 3.0 or less. It has excellent workability, waterproofing resistance and efflorescence resistance.

The test and research results show that the compressive-to-flexural strength ratio of the face brick pointing material is greatly affected by the amount of re-dispersed emulsion powder. Figure 8-4-7 and Figure 8-4-8 show the relationship between the polymer-cement ratio and compressive-to-flexural strength ratio.

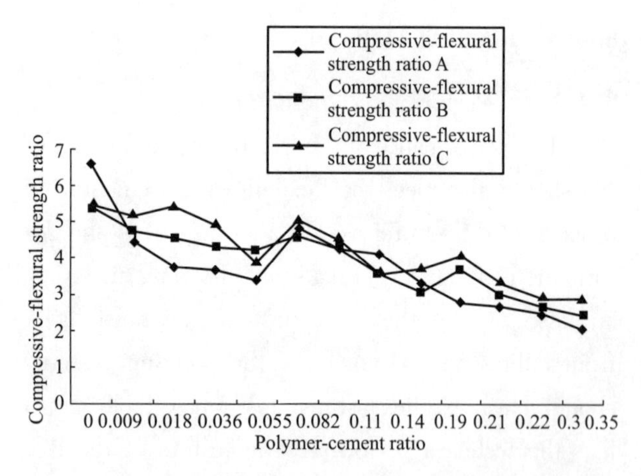

Figure 8-4-7 Relationship between Polymer-cement Ratio (low) and Compressive-to-flexural Strength Ratio

According to Figure 8-4-7 and Figure 8-4-8, the amount of re-dispersed emulsion powder has obvious impact on the compressive-to-flexural strength ratio of the face brick pointing material. With the polymer-cement ratio increasing, the compressive-to-flexural strength ratio declines rapid-

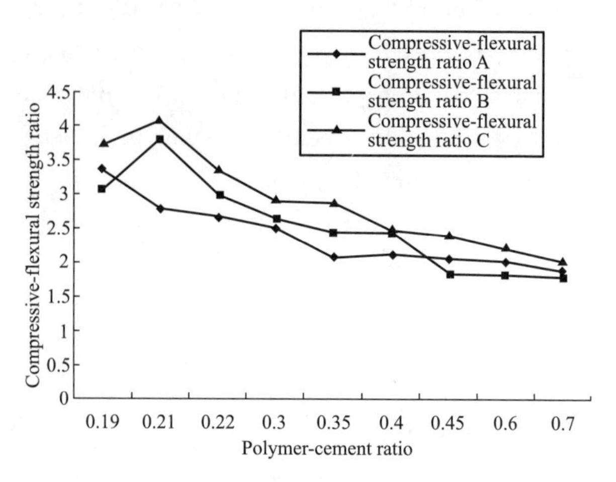

Figure 8-4-8 Relationship between Polymer-cement Ratio (high) and Compressive-to-flexural Strength Ratio

ly. When the polymer-cement ratio is about 0.3, the compressive-to-flexural strength ratio of the face brick pointing material will be less than 3.0; and when the polymer-cement ratio is more than 0.4, the compressive-to-flexural strength ratio tends to change smoothly.

8.4.4 Face Brick

Dovetail-groove face bricks with no release agent should be used in the face brick finish of the external thermal insulation system. In addition to conformity to relevant standards on face bricks of outer walls, such as the *Ceramic Tiles* (GB/T 4100) and *Ceramic Mosaic* (JC 456), the face brick should also meet the requirements of Table 8-4-4.

Performance Indicators of Face Bricks

Table 8-4-4

Item		Unit	Performance Indicator
Dimensions	Single area	cm^2	≤150
	Side length	mm	≤240
	Thickness	mm	≤7
Mass per unit area		kg/m^2	≤20
Water absorption rate	Class I, VI and VII climatic regions	%	0.5-3.0
	Class II, III, IV and V climatic regions		0.5-6.0

Continued

Item		Unit	Performance Indicator
Freezing resistance	Class I, VI and VII climatic regions	—	No damage in 50 freeze-thaw cycles
	Class II climatic regions		No damage in 40 freeze-thaw cycles
	ClassIII, IV and V climatic regions		No damage in 10 freeze-thaw cycles

Note: Climatic zones are divided based on Grade I requirements in the *Standard for Climatic Regionalization for Architecture* (GB 50178).

For the part with concentrated stress, especially a window corner (Figure 8-4-9), the results of face brick construction methods vary from each other.

Figure 8-4-9 shows the complete bonding of face bricks in the bottom left corner of a window, and special-shaped bricks in the bottom right corner. It is found after the weathering resistance test that special-shaped bricks cracked.

Figure 8-4-9 Face Brick System for External Themal Insulation after Weathering Resistance

8.4.5 Performance Requirements of External Thermal Insulation System with Face Brick Finish

The external thermal insulation system with the face brick finish should meet the basic requirements of general external thermal insulation projects, as well as the technical requirements listed in Table 8-4-5.

According to the *Specification for Construction and Acceptance of Tapestry Brick Work for Exterior Wall* (JGJ 126) issued by the Ministry of Housing and Urban-Rural Development, the pull strength of tiles measured in the pull-out test on the construction site shall not be lower than 0.4MPa, which is more than 100 times that of the maximum negative wind pressure at a height of 100m. Firstly, this value is determined based on the measurements of projects under different climatic conditions in Beijing, Harbin, Zhuhai, Henan and other places as well as the verifications in the laboratory, taking into account the climatic characteristics, critical values and probabilities of falling of face brick finishes on sites and in laboratories, water absorption rate and thermal deformation of face bricks, positive and negative wind pressure, typhoon effects, abrupt cooling/heating, as well as weathering. Secondly, two standards of the Housing Renovation Department of Japan's Ministry of Construction are used for reference. As long as the pull strength of face bricks on the construction site meets the standards, the connection safety of the face brick finish will be guaranteed.

Performance Indicators of External Thermal Insulation System with Face Brick Finish Table 8-4-5

Test Item		Performance Indicator
Weathering resistance	Appearance	No seepage cracks, efflorescence, hollowing or peeling
	Tensile bonding strength(MPa) between face brick and anti-cracking layer	≥0.4
Water absorption(g/m^2)		≤1000
Wet flow density [$g/(m^2 \cdot h)$] of vapor permeation		≥0.85

Continued

Test Item		Performance Indicator
Freeze-thaw resistance	Appearance	No seepage cracks, efflorescence, hollowing or peeling
	Tensile bonding strength(MPa) between face brick and anti-cracking layer	≥0.4
Water impermeability		No water permeation on the inner side of the anti-cracking layer

8.5 Construction and Engineering Case of Face Brick Finish of External Thermal Insulation System

8.5.1 Process

8.5.1.1 Process of Glass Fiber Mesh Reinforced Face Brick Finish

Insulation layer construction → applying of surface mortar → laying of covering and reinforcing mesh → applying of surface mortar → mesh laying→ applying of surface mortar → mesh laying→ anchorage with anchor bolts → face brick bonding → pointing.

8.5.1.2 Process of Wire Mesh Reinforced Face Brick Finish

Insulation layer construction → applying of anti-cracking mortar → laying of hot-dip galvanized rectangular mesh → anchorage with anchor bolts → applying of anti-cracking mortar → face brick bonding → pointing.

8.5.2 Key Construction Points

8.5.2.1 Key Construction Points of Glass Fiber Mesh Reinforced Face Brick Finish

Two layers of surface mortar should be applied during laying of the glass fiber mesh. That is, a layer (thickness: 1-2mm) of surface mortar, which is slightly larger than one glass fiber mesh, should be applied evenly on the EPS board surface with a stainless steel trowel. Then the glass fiber mesh should be immediately pressed into wet surface mortar. When mortar is dry enough to be touched, a second layer (thickness: 2-3mm) of surface mortar should be applied with the trowel, until the glass fiber mesh is fully covered. In this case, the glass fiber mesh will be located between two layers of surface mortar. The total thickness of surface mortar should be 4-6mm in the presence of a single layer of glass fiber mesh and 6-8mm in the presence of two layers of glass fiber mesh.

The glass fiber mesh should be laid bottom-up along an outer wall. If a door or window opening is encountered, a standard glass fiber mesh should be bonded in the 45° direction of each corner to prevent cracking. The lap joints of standard glass fiber meshes should be minimum 150mm, but reinforcement meshes must be butt-jointed tightly. If needed, the mesh turning width should not be less than 100mm. Cement paste should be applied on the mesh turning part of a window as well as the side face of the starting edge of the first layer. The surface mesh should be aligned and leveled with a guiding rule, and then compacted with mortar. The mesh part at the turning point should be extruded until there is mortar. The external and internal corners of the outer wall shall be directly overlapped for 200mm. The curved surface of the glass fiber mesh should face the wall and leveled with the trowel from the center to periphery, until the glass fiber mesh is completely embedded in surface mortar and there are no visible cracks. If the glass fiber mesh is still exposed, an appropriate amount of surface mortar should be applied for repair. There must be no dry lap joint between glass fiber meshes. Gaps between glass fiber meshes should be completely filled with surface mortar.

8.5.2.2 Key Construction Points of Meshed Face Brick Finish

The first layer (thickness: 2-3mm) of anti-cracking mortar should be applied on the insulation layer after acceptance. Then the hot-dip galvanized welded wire meshes (maximum length: no more than 3m) meeting the structural size requirements should be laid in sections. In order to ensure the construction quality of edges and corners, hot-dip galvanized welded wire meshes for edges and corners are folded into a right-angled shape via the wire mesh flattening machine as well as hydraulic shearing and formation machine before construction. The fixed folds must not be formed in the cutting or laying process. After unfolding, the hot-dip galvanized welded wire meshes should be laid flatly in the desired direction, clamped via U-type clips made of 14# wire onto the anti-cracking mortar surface, and then anchored on the base course wall via nylon expansion bolts. They should be distributed in a plum shape at intervals of 500mm in both directions, with the effective anchoring depth of 25mm or more. Partial unevenness can be resolved with U-type clips. The lap width of hot-dip galvanized welded wire meshes should not be less than 50mm, and the number of lap layers must not exceed 3, with lap joints secured via U-type clips, wires or expansion bolts. Hot-dip galvanized welded wire meshes at the starting and ending points of the inner side of windows and on the parapet and settlement joint should be secured onto the main structure via cement nails and gaskets or expansion bolts.

The hot-dip galvanized welded wire meshes laid should be inspected, following by the application of a second layer of anti-cracking mortar layer as the coating of these meshes. The anti-cracking mortar should be plastered evenly with the total thickness of 10mm±2mm, and its surface should meet the flatness and verticality requirements.

8.5.2.3 Face Brick Finish

The face brick finish should be constructed in accordance with the *Specification for Construction and Acceptance of Tapestry Brick Work for Exterior Wall* (JGJ 126). The thickness of the bonding mortar layer should be 3～5mm. Face brick joints should not be less than 5mm wide, and 20mm wide face brick joints should be set at intervals of six floors. When the side length of face bricks is more than 100mm, special angle tiles should be applied in external and internal corners. External corners should not be treated by connecting face bricks with edges processed into a 45° angle. The top drainage angle of a horizontal external corner should not be less than 3°. Top face bricks should be laid on elevation fabrics, and the face bricks of the bottom row of the elevation on those of the bottom plane. In addition, dripping structures should be built.

Face bricks should be 100% bonded with flexible bonding mortar. It is permitted to apply the tile adhesive on a wall surface with a serrated trowel, make face bricks compacted in the adhesive, and if necessary, remove the mortar area on the back. If used, paper tiles should be constructed as follows: applying a thin layer of bonding mortar first on the wall and then on paper tiles, and finally making paper tiles compacted in bonding mortar. The paper tiles should not be directly attached onto a wall surface via a thin layer of bonding area, as the bonding area cannot be guaranteed if the bonding mortar thickness does not meet the requirements.

8.5.2.4 Face Brick Pointing

The flexible and highly hydrophobic pointing powder should be used in the process of face brick pointing. Pointing should be conducted first in horizontal joints and then in vertical joints. The face brick joints should be recessed into the face brick surface for 2～3mm. After pointing is completed, the outer surface should be inspected and cleaned within a large area, to ensure the attractiveness.

8.5.3 Engineering Cases

8.5.3.1 Beijing Binduyuan Community

Beijing Binduyuan Community is located on the Maizidian North Road in Chaoyang District and at the interaction of Nongzhanguan West Road, with the Maizidian West Road on its western side, Nongzhanguan North Road on its southern side, and green belts, bungalows and irrigation channels on its northeastern side. It consists of 20-storey buildings on the ground, with a height of 61m and total construction area of 19, 043m^2. There are two tower buildings (in an L shape in the plan) in the east-west and north-south direction. The first floor is used for commercial purposes, the 2nd to 20th floors for ordinary residential purposes, and the basement for the garage and equipment rooms. The Grade 6 civil air defense is adopted on the 2nd floor on the ground. The adhesive polystyrene granule pasted EPS board system with the face brick finish is used, covering an area of approximately 10, 000m^2.

This project complies with relevant regulations. It passed the one-time acceptance after the completion.

8.5.3.2 Beijing Yongtai Garden Community

BeijingYongtai Garden Community was contracted by Tianhong Group, designed by Tianhong Yuanfang Design Institute and constructed by Beijing Urban Construction FirstGroup Co. ,Ltd.

This project covers a construction area of 50, 000m^2, and consists of six-storey buildings with a shear wall structure and eave height of 21.5m. The external thermal insulation system, with an area of 20, 000m^2 and energy conservation rate of 50%, was commenced in August 2004 and completed in November, 2004.

The outer walls of this project are insulated with the adhesive polystyrene granule pasted EPS board system containing the face brick finish. The EPS boards in the composite insulation layer consisting of the adhesive polystyrene granule as well as EPS boards are 50mm, the internal bonding layer based on the adhesive polystyrene granule is 20mm, and the external leveling layer is 10mm. The anti-cracking protective layer is composed of the anti-cracking mortar and hot-dip galvanized welded wire mesh, and anchored with plastic expansion bolts. The finish layer is formed by bonding face bricks with the dedicated bonding mortar (compressive-to-flexural strength ratio: less than 3). The entire system with no cavity has high resistance to the wind load, cracking and weathering, and meets the requirements for the thermal insulation and energy conservation as well as the safety of the face brick finish.

This project complies with relevant quality requirements, and passed the acceptance at a time after completion.

8.5.3.3 Qingdao Lucion Changchun Garden

QingdaoLucion Changchun Garden (Figure 9-3-4) was invested and built by Shandong Lushin Real Estate Co., Ltd., and located at No. 1, Yinchuan East Road, Qingdao, with a construction area of approximately 990, 000 m^2. It consists of 99 buildings in total, which are composed of the cast-in-situ concrete, meshed EPS board and framed shear walls filled with aerated concrete blocks.

8.6 Summary

Although priority should be given to the paint finish in the external thermal insulation system, the face brick finish that has many irreplaceable advantages is still extensively applied due to differences in aesthetic concepts, even in super high-rise buildings. Therefore, the external thermal insulation industry must pay attention to and guarantee the safety and quality of the face brick finish in external thermal insulation projects.

The simulation calculations show that the dead load increased by the face brick finish has little impact on the external thermal insulation system, and the connection between the system and base course wall is safe enough. Hollowing or falling occurs on the surface of the anti-cracking mortar layer and between the face bricks and adhesive, so the anti-cracking mortar layer must be reinforced. In addition, anchor bolts should be used to connect the anti-cracking mortar layer and base course wall, thereby transferring external forces to the base course wall. The hot-dip galvanized rectangular mesh is preferred for reinforcement in the anti-cracking mortar layer, to effectively protect the insulation layer. As there is high bond stress between the rectangular mesh and anti-cracking mortar, the tensile strength in the horizontal and vertical direction is enhanced, and the bonding strength between the face bricks and base course wall is greatly improved.

Materials of the face brick finish system for external thermal insulation should not only comply with relevant standards, but also have the appropriate thickness, flexibility and waterproof performance to meet respective functional needs. Certainly, the construction quality is essential for the functions of materials. Therefore, the construction method should be observed rigorously.

9. Technology Research on and Application of Adhesive polystyrene granule Composite External Thermal Insulation System

9.1 Development of and Research on Material of Adhesive polystyrene granule Composite Insulation System

9.1.1 Development of Adhesive polystyrene granule Mortar

9.1.1.1 Development of Adhesive polystyrene granule Mortar Abroad

Rhodius subordinated to Burgbrohl in Germany was awarded an invention patent of "insulation mortar" on October 16, 1968. The insulation mortar applies to industrial products and has been developed into a high-efficiency insulation material that can be directly applied on a wall. Germany has formulated the *Rendering Systems for Thermal Insulation Purposes Made of Mortar Consisting of Mineral Binders and Expanded Polystyrene (EPS) as Aggregate* (DIN 18550-3), which applies to the insulation mortar system made of inorganic binders and with the expanded polystyrene (EPS) granule (hereinafter referred to as the EPS granule) as the main aggregate. This standard stipulates the requirements and test methods for the bottom/surface insulation mortar and mortar system, the technical requirements for application and construction, as well as the procedures of inspection and marking. In its explanatory notes, it is stated that the insulation mortar system has been applied on outer walls of lightweight masonry buildings and existing buildings for 25 years; and through continuous research, the volume weight of insulation mortar has been reduced from 600kg/m^3 to 200-300kg/m^3, and the plaster thickness is up to 100mm (instead of 20mm), which helps to improve the thermal insulation performance of new buildings. The EPS granule insulation material is also widely applied in France, Austria, Italy, the former Yugoslavia and other countries, but has not been deeply developed.

9.1.1.2 In-depth Development of Adhesive polystyrene granule Mortar

1. Internal insulation material made of adhesive polystyrene granule mortar

In the early stage of building energy conservation in China, the energy-saving rate is 30%, and internal thermal insulation is mainly adopted for building walls. BeijingZhenli, absorbing European technological achievements, developed the adhesive polystyrene granule insulation mortar, which is promoted as an internal insulation material of outer walls.

The adhesive polystyrene granule insulation mortar is prepared with the adhesive polystyrene granule. The mineral binder is prepared and bagged in a factory, containing calcium hydroxide, amorphous silica, a small amount of Portland cement, additives such as polymer binder as well as water-retaining and thickening admixture, and a large number of fibers. The EPS granules are prepared by pulverizing recycled polystyrene boards to the appropriate size, followed by mixing and bagging. The insulation mortar is made by mixing one bag of mineral binder (25kg), one bag of EPS granules (200L) and water of 34-36kg. The quantities of such materials can be controlled easily as per the accurate ratio. The one-time plaster thickness is up to 40-60mm. This insulation material also has high bonding strength, with no slipping and little dry shrinkage.

The internal thermal insulation technology with the adhesive polystyrene granule has the advantages of high construction speed, easy control of quality, strong integrity, large rate of material utilization, no need for priming, etc. The waste EPS board is used as the insulation material of mortar, thereby fully utilizing the recycled resources and reducing white pollution. This solves the problems of the plaster gypsum and EPS granule insulation mortar and composite silicate insulation mortar as the internal thermal insulation material, such as instable insulation effects, thermal conductivity change, poor anti-slip performance, thin one-time plaster, small softening coefficient, falling at the dew point, "hollowing" "bulging" and "cracking", caused by the inaccurate mixing ratio during on-site plastering construction.

2. External thermal insulation material made of adhesive polystyrene granule

As the internal thermal insulation has inherent thermal defects, the external thermal insulation technology was developed and put into use in China in the late 1990s, to meet China's needs of increasing the energy-saving rate from 30% to 50%. Beijing Zhenli, through its efforts, developed the adhesive polystyrene granule insulation mortar suitable for external thermal insulation. This insulation material has been successfully applied in external thermal insulation projects.

The adhesive polystyrene granule insulation mortar has reliable insulation performance, high cracking resistance and Class B_1 combustibility, but low requirements for the flatness of a base course wall. It can be easily applied in various base course walls. The recycled EPS granule is used as its aggregate, thereby saving the energy and facilitating environmental protection. The results of testing, research and engineering practice show that the external thermal insulation system made of adhesive polystyrene granule insulation mortar has been improved constantly, applied widely for thermal insulation in severe cold, cold and hot-summer warm-winter regions, and listed as one of China's five major external thermal insulation systems in the *Technical Specification for External Thermal Insulation on Walls* (JGJ 144—2004). The industry standard *External Thermal Insulating Rendering Systems Made of Mortar with Mineral Binder and Using Expanded Polystyrene Granule as Aggregate* (JG 158—2004) has been formulated. Subsequently, a series of relevant standards including local standards, construction procedures, structural atlas and acceptance codes have been developed, which promotes the study on and application of adhesive polystyrene granule mortar.

With the application of the external thermal insulation system made of adhesive polystyrene granule, the roof insulation material, sloping roof insulation material and roof top insulation material made of adhesive polystyrene granule, the technology for application of the adhesive polystyrene granule mortar in high-rise buildings, and the face brick technology for external thermal insulation have been developed. The extensive application of adhesive polystyrene granule mortar provides a powerful technical support for China to realize the goal of cutting the building energy consumption by 50%. The external thermal insulation system based on the adhesive polystyrene granule is also rated as a state-level construction method.

By preliminarily summarizing the practice of application of adhesive polystyrene granule mortar, three major technical concepts of external thermal insulation have been established: the external thermal insulation of the outer wall is superior to its internal thermal insulation; the anti-cracking technology by changing the flexible deformation and releasing the stress layer by layer in the external thermal insulation system should be adopted; and the external thermal insulation system should have a no-cavity structure. They play an important role in enrichment and development of China's theoretical research on building wall insulation.

3. Adhesive polystyrene granule composite insulation technology

When the mineral binder and EPS insulation mortar is separately applied on an outer wall, its thickness should be 100mm to meet the requirements of building energy conservation (65%). In this sense, the mortar should be plastered repeatedly during construction, which is unreasonable economically and technologically. For this reason, the adhesive polystyrene granule composite insulation technology has been developed. The key point of this technology is to converter adhesive polystyrene granule insulation mortar into EPS granule pasting mortar, and combining the EPS granule pasting mortar with insulation boards into a "sandwich" type external thermal insulation system, thus greatly improving the thermal insulation performance and fire resistance. The EPS granule pasting mortar is the core of the adhesive polystyrene granule composite insulation technology. Compared with the adhesive polystyrene granule insulation mortar, the EPS granule pasting mortar has high bonding strength and Class A2 combustibility. It cannot only be used with insulation boards to build the adhesive polystyrene granule composite insulation layer, but also can be used as a plaster transition layer on insulation board surfaces. The adhesive polystyrene granule composite insulation system has higher weathering resistance and wind pressure safety than the thin-plastered insulation board system. It is able to effectively prevent flame from spreading during a fire. Due to good adaptability to construction, it meets the energy conservation (65% or greater) requirements as well as fire protection requirements of high-rise buildings.

In 2013, with theadhesive polystyrene granule composite insulation technology more and more mature, the industry standard *External Thermal Insulating Rendering Systems Made of Mortar with Mineral Binder and Using Expanded Polystyrene Granule as Aggregate* (JG 158—2004) was upgraded into the *Products for External Thermal Insulation Systems based on Mineral Binder and Expanded Polystyrene Granule Plaster* (JG 158—2013), providing a new technical plan for application of insulation materials such as EPS and XPS boards, as well as a reference for improvement of the weathering resistance of insulation materials such as polyurethane insulation boards, phenolic foam boards and rock wool boards. The results of large-scale weathering resistance test and engineering practice of the polyurethane composite board system prove that if light mortar is applied on its surface, its weathering resistance of the polyurethane composite board will be improved significantly, and surface hollowing and cracking will be reduced. The local standard *Technical Specification for External Thermal Insulation Construction based on Rigid Polyurethane Foam Composite Board with Plastering Lightweight Mortar* (DB11/T 1080—2014) was implemented on July 1, 2014.

The theory of external thermal insulation has been further developed currently, enriching technological application theories such as the temperature field theory related to the insulation layer location of the outer wall, waterproofing and vapor permeation theory of the insulation system, theory of no-cavity wind-resistant structure, theory of structural fire protection, technical route of layer-by-layer flexible release of stress, face brick structure of external thermal insulation system, etc. This has positive effect in guiding the research on and engineering application of external thermal insulation systems.

4. New progress of adhesive polystyrene granule mortar

Beijing took the lead in putting forward the building energy conservation rate by 75% and the development of passive residential buildings. The light wall based on the reinforced vertical-fiber rock wool board and EPS granule foamed concrete has been developed for steel structures as well as building with framed lightweight infilled walls. This new type of wall has low density, light weight and small elastic modulus. There is little seismic action applied on this type of wall. The shock wave is transmitted slowly through this type of wall. Due to long self-vibration period and fast absorption of impact energy, cracking and leakage of outer walls of steel structures in the current stage can be solved, thus improving the seismic performance and fire resistance of buildings.

With the promotion of building energy conservation in China, adhesive polystyrene granule mortar has been deeply studied and developed, extensive experience in application of external thermal insulation systems have been gained, and the technology and theory of external thermal insulation have been advanced. The technological summary will drive further development of adhesive polystyrene granule mortar.

9.1.2 Concepts of External Thermal Insulation Technology

The above-mentioned three major technical concepts are based on large quantities of scientific experiments and engineering practices during development of external thermal insulation and building energy conservation. Based on the rule that the safety and durability are primary contradictions between the external thermal insulation system and energy conservation, they reveal the relationship between the location of the insulation layer and the temperature field change of the outer wall. The technical route to control cracking of the insulation layer is put forward. Moreover, these three major technical concepts change conventional thoughts on building deformation, and involve some frontier issues related to energy conservation of building walls.

9.1.2.1 Superiority of external thermal insulation of outer wall to internal thermal insulation

1. With the external thermal insulation system, the service life of a building will be extended.

If the internal thermal insulation system is applied in an outer wall, the outer and inner walls will be kept in two different environments due to the insulation layer. That is, inner walls and slabs are located indoors, with the annual temperature difference of 10℃, but the outer wall of the envelope indoors, with the annual temperature difference of 60～80℃. As long as the ambient temperature changes by 10℃, wall concrete of 1/10,000 will expand or shrink. Deformations arising from two different temperature values vary from each other, making the structure always instable and further leading to cracks on the wall surface, damage to the waterproof structure of the roof along the outer wall, leakage of the waterproof structure of the basement, etc. These phenomena greatly shorten the life of structures, what is called "internal insulation syndrome". Its principle also applies to the sandwich insulation system and the thick rigid rendering on the surface of an insulation layer. When stones are pasted in a wet manner on an insulation layer, the outer side of the insulation layer is subject to similar deformation. In the early stage of the external thermal insulation practice, overhanging parts of buildings and structures, which are often neglected, include balconies, air conditioner pallets, drainage ditches and appentices. With no insulation layer, their deformation under thermal effects differs from that of the wall with external thermal insulation, so connections between such parts and the wall are prone to cracking and damage.

The temperature field of a building with no insulation layer greatly differs from that of a building with an insulation layer, so it is essential to study the impact of such a difference on the building. One key point of the external thermal insulation practice is to keep a building or structure in one temperature environment. In this case, the thermal deformation is mainly affected by the indoor temperature, to avoid variations in deformation of building parts under annual outdoor temperature differences, make the building stable and extend its life.

2. External thermal insulation is areasonable way to eliminate thermal bridges.

One of the causes of indoor heat loss is the impact of thermal bridges. The internal thermal insulation involves large thermal bridges, which reduces the energy efficiency. Because of heat loss, there is a significant difference in the temperature of the part with a thermal bridge and that with no thermal bridge. Infrared images show a temperature difference of 5℃ at connections between the inner and outer walls and between the outer wall and floor in winter. Condensation often occurs at thermal bridges.

The dew point is close to the internal surface of the outer wall in the internal thermal insulation practice, and the external surface of the external insulation layer of the outer wall in the external thermal insulation practice. Condensation usually occurs at thermal bridges of the internal thermal insulation system in the north in winter. The internal surface of the non-insulated outer wall is affected little by air-conditioning but the temperature of its external surface is high in the south in summer, so mildew often appears on the internal surface of the outer wall with no external thermal insulation.

3. Wall cracking is easier to control in the external thermal insulation system than the internal thermal insulation system.

Insulation boards in the internal thermal insulation system are prone to cracking, as internal insulation boards influenced by the indoor temperature are attached on the outer wall deformed because of the annual outdoor temperature difference. Under the deformation stress of the outer wall caused by annual temperature differences, internal insulation boards are bound to crack.

It is easier to control wall cracks in the external thermal insulation system than the internal thermal insulation system. The application of thorough external thermal insulation is equivalent to a "cotton-padded jacket" on the entire envelope of a building, so the base course wall is completely affected by the indoor temperature with little fluctuation, and the impact of resulting deformation can be ignored. Affected by the outdoor environment temperature is only the external surface. The premise of protection of the outer wall is to prevent hollowing, cracking and falling and ensure the safety and durability of the external thermal insulation system. They are exactly the problems to be solved.

9.1.2.2 Anti-cracking Technical Route of "Layer-by-layer Change and Flexible Release of Stress" for External Thermal Insulation

It is required to prevent five natural destructive forces (thermal stress, wind, water, fire and earthquake) and release the deformation stress via the technical route of allowing, limiting and inducing deformation, in order to ensure the safety of the external thermal insulation system. Among many contradictions, cracking is the primary one to be addressed. It is critical to study the causes of cracking and control cracks in the external thermal insulation system.

1. The stress is distributed more evenly in the seamless insulation material formed on the site than the prefabricated board.

The stress is concentrated in the joints of insulation boards and each board contracts and expands separately. When the external insulation boards are secured on a wall, the impact of their deformation on the filling material of joints must be fully considered. If the elastic deformation of the filling material is less than the expansion and contraction of boards, cracking will be inevitable in joints.

Insulation mortar of lower strength and little contraction has good integrity because of no joints after on-site plastering on a wall, without obvious contraction or stress concentration. If the flexible mortar and alkali-resistant glass fiber mesh are applied on a surface, the stress can be distributed evenly, which is conducive to cracking control.

When insulation mortar of high strength is applied without glass fiber mesh and flexible mortar, it will be difficult to control cracking. When ordinary cement mortar and insulation material of high strength such as perlite are applied, hollowing and cracking are likely to occur under the effect of temperature differences of one or two year (s), as thermal deformation of the material and wall are not synchronized and deformation stress is not released promptly and efficiently.

2. A flexible material system should be used to completely release stress and avoid two contradictory cycles of cement.

Cracking is common to plaster finishes of cement mortar because of two contradictory cycles of cement products, namely, fast increase in strength but slow volume shrinkage. There are successful engineering cases in the ancient times to avoid such contradictions. For example, the materials of volcanic ash series are adopted in large-scale artificial concrete pouring projects in the ancient Rome, with low strength meeting construction requirements in the early period and high strength meeting functional requirements in the late period, but there is not a large amount of heat of hydration. Due to little volume shrinkage, cracking did not occur in the early period, and related structures are not damaged in thousands of years. The long life of buildings left over from ancient times fully relies on the release of thermal stress. There are also no deformation joints in the Great Wall under harsh climatic conditions. The 25mm brick joints of the ancient Great Wall are filled with thick mortar and slaked lime, of which the strength increases slowly. The Great Wall may be regarded as a huge flexible body.

Falling of face bricks are caused by excessive strength but insufficient flexibility of bonding mortar. From the investigation and analysis of dozens of projects, falling of face bricks can be classified into three damaged parts (large-scale middle hollow section, edge/corner, and connection between the top parapet and roof slabs, damaged as a result of stress concentration and insufficient flexibility) and two fractured layers (face bricks fall separately in the case of a concrete base course wall and with mortar in the case of a brick masonry base course wall).

The results of the pull-out test of non-falling face bricks on the site involving falling of face bricks show that the bonding mortar has high strength and good flexibility, and its compressive-to-flexural strength ratio is greater than 8. For example, if flexible mortar with the compressive-to-flexural strength ratio of less than 3 is used, the deformation stress generated in expansion and contraction of face bricks can be released, and thus falling of face bricks can be avoided.

3. There should be a small difference in the thermal conductivity of adjacent materials.

The temperature rise rates of materials lead to differences in the speeds of thermal expansion and contraction. If there is a difference in the deformation rates and speeds of two adjacent materials, thermal stress will be generated at the interface.

It is found in severe cold areas that cracking usually occurs to the external surfaces of EPS boards as a result of insufficiently flexible or ageing resistance of anti-cracking mortar. There are often through cracks in joints of XPS boards on the anti-cracking mortar surface layer.

There are differences in the deformation rates and speeds of the EPS board and cement mortar outside the wire mesh. With the cement mortar of high strength expanding or contracting fast, the temperature of the adjacent EPS board rises or drops slowly. For a soft base course wall deformed slowly, the cement mortar layer of high strength is prone to cracking due to fast volume deformation with the temperature changing.

Materials of one wall differ in the thermal conductivity and heat transfer rate. For a framed light weight infilled wall, the difference in the deformation speeds of materials can lead to cracking. When the ambient temperature changes greatly, the temperature change rates of materials also vary from each others. The temperature change rate of reinforced concrete of the frame is about 8 times greater than that of the filling material such as aerated concrete. Affected by the temperature, the volume deformation of the latter is roughly 20% less than that of the former. Tensile stress is generated in the case of hot expansion arising from the deformation rate difference and volume deformation at the connection of different materials, and compressive stress in the case of cold contraction. Both the tensile and compressive stress at the connection of different materials of the framed lightweight infilled wall affected by the temperature will inevitably lead to cracks.

Hollowing usually occurs to the cement mortar plaster on the surface of aerated concrete blocks after one or two year (s) with temperature differences. It is also caused by the difference in the deformation speed and volume deformation of materials.

4. The deformation of the outer layer should be greater than that of the inner layer.

The surface temperature of the external thermal insulation system changes more than the external surface of the wall with no insulation layer. In summer, the impact of solar radiation per square meter of the surface of the external thermal insulation system is more significant than that on a non-insulated wall. As heat is isolated by the external insulation layer, the temperature rise rate of the external surface and its sudden temperature drop rate in a rain are higher than those of the surface of a non-insulated wall.

Deformation stress occurs frequently and rapidly to the surface of the external thermal insulation system.

The external thermal insulation system should be completely flexible, with the deformation of the outer layer greater than that of the inner layer. If the outermost layer is more flexible than the adjacent inner layer, their thermal deformation rates should be coordinated.

5. Deformation should be induced or its direction should be changed.

The thermal stress on the surface of the external thermal insulation system should be released promptly and thoroughly, to avoid damage to this system. The flexible reinforcement combined with flexible mortar containing fibers can be applied to promptly release deformation stresses. The technical route of allowing, limiting and inducing deformation and changing the direction of stress transfer should be implemented.

9.1.2.3 No-cavity Structure of External Thermal Insulation System

It is not acceptable to completely copy all foreign technologies in China's buildings with their own features. Full bonding of the insulation layer and base course wall with no cavity is one important measure to ensure the reliable connection between the external thermal insulation system and outer wall, and extend the engineering life of the external insulation layer. Under the combined effects of positive and negative wind pressure, the external thermal insulation system with connecting cavities is always in an instable state and may be blown off in a heavy wind, while that with no cavity is stable enough to effectively resist the wind load. Due to changes in the gas pressure inside connecting cavities under the wind pressure, the insulation layer will be instable. In addition, cavities may lead to fatigue of and damage to mortar in board gaps, thereby shortening the life of the insulation layer.

9.1.3 Study on Adhesive polystyrene granule Mortar

The adhesive polystyrene granule mortar is a kind of insulation mortar, which is prepared with the mineral binders consisting of redispersible adhesive powder, inorganic cement material, admixtures and fibers and the EPS granule as the main aggregate. It is divided into the adhesive polystyrene granule insulation mortar (hereinafter referred to as the insulation mortar) and the adhesive polystyrene granule pasting mortar (hereinafter referred to as the pasting mortar).

9.1.3.1 Impact of EPS Granule Morphology and Grade on Thermal Conductivity

Kou Xiurong from Suzhou University conducted an experimental study the impact of the EPS granule morphology and grade on thermal conductivity in the *Research on Adhesive polystyrene granule Insulation Materials*. The following EPS granules were used in the experiment: four types of granule prepared by foaming of expandable polystyrene and designated as E1, E2, E3 and E4 respectively; and four types of granule (smaller than 5mm) obtained by crushing waste polystyrene foam with a special crusher and designated as e1, e2, e3 and e4, respectively. With the portions of other components unchanged, the same amount of the above eight types of polystyrene granules were mixed, respectively. With the fluidity controlled to be (160±5) mm, the wet apparent density of the new insulation material, its dry apparent density after drying at 110℃ and its thermal conductivity at 25℃ were determined. The experimental results show that the properties of the insulation material prepared with crushed EPS granules is superior to that of the foamed EPS granule, and the insulation performance of the former is obviously better than that of the latter. Therefore, the grade of EPS granules has significant impact on the thermal conductivity.

9.1.3.2 Relationship of Apparent Density, Compressive Strength and Thermal Conductivity

1. Insulation mortar

Refer to Table 9-1-1 for the measured thermal conductivity of the specimen of adhesive polystyrene granule insulation mortar (average temperature: 318K).

The relationship between the dry apparent density and thermal conductivity of the specimen of the adhesive polystyrene granule insulation mortar is shown in Figure 9-1-1.

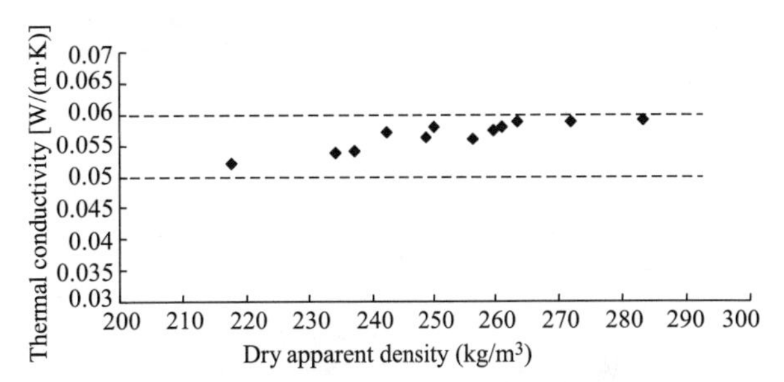

Figure 9-1-1 Relationship between Dry Apparent Density and Thermal Conductivity of Insulation Mortar

The following conclusion can be drawn from the above test data: the thermal conductivity of the adhesive polystyrene granule insulation mortar falls between those of the EPS board and mineral binder mortar. The dry apparent density of the insulation mortar is 200～300kg/m^3. When the average temperature of the test is 318K, its thermal conductivity is 0.05～0.06W/ (m·K).

Refer to Table 9-1-2 for the measured apparent density and compressive strength of insulation mortar, and Figure 9-1-2 for the relationship between the dry apparent density and compressive strength. The dry apparent density and compressive strength of the specimen are basically linearly proportional to each other. As the adhesive polystyrene granule insulation mortar is a kind of heterogeneous material, there may be deviations in test results.

Test Results of Apparent Density and Thermal Conductivity of Insulation Mortar **Table 9-1-1**

Dry apparent density, kg/m^3	283.2	271.9	263.5	261.0	259.5	256.4
Thermal conductivity, W/(m·K)	0.0590	0.0587	0.0587	0.0579	0.0573	0.0559
Dry apparent density, kg/m^3	250.0	248.7	242.3	237.0	234.1	217.5
Thermal conductivity, W/(m·K)	0.0580	0.0562	0.0571	0.0540	0.0538	0.0522

Test Results of Compressive Strength of Insulation Mortar **Table 9-1-2**

Dry apparent density, kg/m^3	206	212	218	229	233	236	238	244	248	252
Compressive strength, MPa	0.29	0.31	0.32	0.34	0.35	0.35	0.36	0.39	0.38	0.39

Test Results of Apparent Density, Thermal Conductivity and Compressive Strength of Pasting Mortar **Table 9-1-3**

Dry apparent density, kg/m^3	300	312	325	330	345	350	363	376	380	398
Thermal conductivity, W/(m·K)	0.065	0.067	0.066	0.068	0.068	0.066	0.069	0.074	0.075	0.076
Compressive strength, MPa	0.53	0.62	0.71	0.65	0.74	0.75	0.74	0.84	0.82	0.88

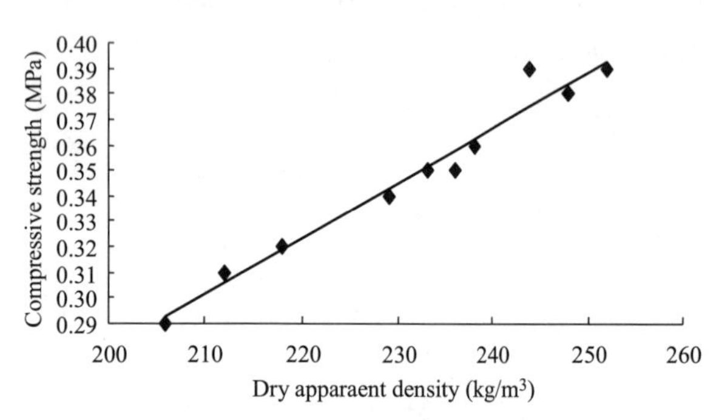

Figure 9-1-2 Relationship between Dry Apparent Density and Compressive Strength of Insulation Mortar

2. Pasting mortar

Refer to Table 9-1-3 for the test results of thermal conductivity and compressive strength of the specimen of adhesive polystyrene granule pasting mortar (average temperature: 318K).

Refer to Figure 9-1-3 for the relationship between the dry apparent density and thermal conductivity of the adhesive polystyrene granule pasting mortar.

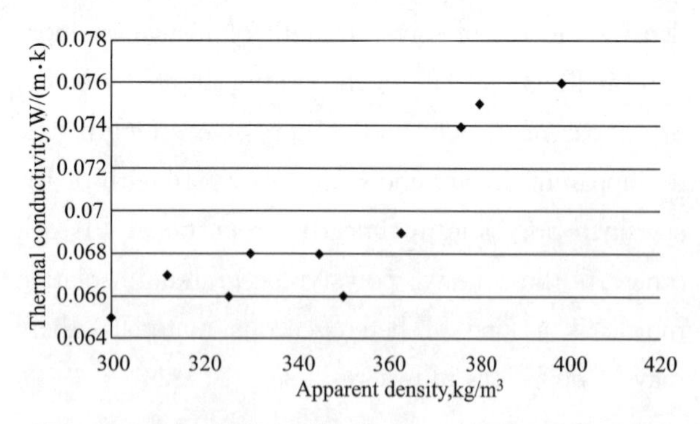

Figure 9-1-3 Relationship between Apparent Density and Thermal Conductivity of Pasting Mortar

Analyzed from the above experimental results, when the apparent density of the pasting mortar is less than $350kg/m^3$, its compressive strength is much greater than 0.3MPa and its thermal conductivity is lower than 0.08W/ (m·K).

It can be seen from the analysis of insulation and pasting mortar that the adhesive polystyrene granule mortar has good thermal insulation performance and fully meets the technical standards and engineering requirements.

9.1.3.3 Bonding Performance

The pasting mortar and insulation board should be reliably connected in a composite insulation structure. Therefore, the pasting binder was developed based on original insulating binders, and the bonding strength between the insulation board and pasting mortar prepared with different ratios of EPS granules was tested. When the ratio of the pasting binder to EPS granule is 25kg: 200L, the bonding strength between the pasting mortar and EPS board is 0.102MPa (28d), and the pasting mortar ($18kg/m^3$) layer is completely damaged, which does not meet the design requirements of the EPS board masonry layer. It is assumed here that the pasting binder proportion can be increased to enhance the bonding strength, with the EPS granule volume unchanged. The impact of mortar with different proportions of the pasting binder on its tensile strength (28d) was analyzed, as shown in Figure 9-1-4. The bonding strength is improved constantly with the pasting binder proportion increasing, but the magnitude of improvement drops gradually. When the amount of pasting binder is 35kg (bonding strength: 0.14MPa or higher), the tensile strength changes little. The increase in the pasting binder proportion also leads to the rise of thermal conductivity but drop of the insulation performance. The costs are increased greatly at the same time. Comprehensively, the best ratio of pasting binder and EPS granule is 40kg: 200L. When the pasting binder proportion is increased from 25kg to 40kg per bag, the tensile bonding strength and compressive strength of mortar will be greatly enhanced.

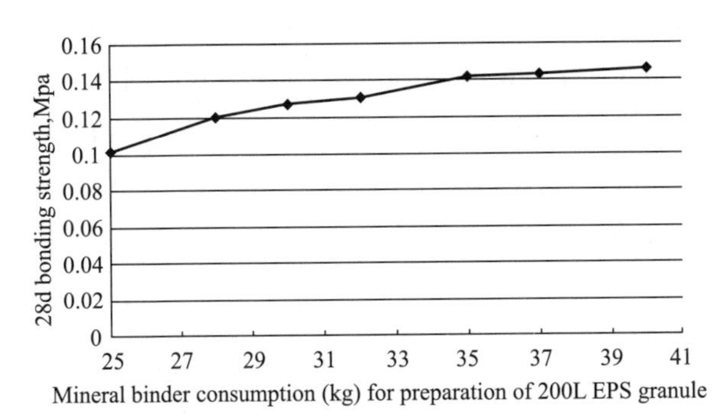

Figure 9-1-4 Impact of Mineral Binder Proportion on Bonding Strength

When an insulation layer is composed of the pasting mortar and insulation board, its shear strength, which is an important indicator of the EPS board masonry layer, can be used for measure the bearing capacity of the masonry layer and the bonding strength between two layers. The pasting mortar layer has high shear strength (up to 65kPa), bearing capacity and stability. This intuitively reflects the powerful load capacity of the insulation system. The test method is shown in Figure 9-1-5.

9.1.3.4 Durability

1. Impact of inorganic cementing material on durability

The fly ash-silica fume-lime-cement type cementation system is used as a cementing material in the adhesive polystyrene granule mortar, in which cement is substituted with calcium hydroxide, fly ash, amorphous silicon dioxide and other substances. A variety of polymer materials are used in inorganic cement materials. The substitution of the pure cement system with a composite

Figure 9-1-5 Load Test of Insulation Board Pasted with Mortar

cementing system is based on the complementary properties of various cementing materials, providing a solution to rapid growth of cement strength, long cycle of volume shrinkage and tendency to crack. The composite system guarantees good workability and reasonable use of cementing materials in the process of hydration, and involves a reasonable cycle of strength growth.

Cement is used to enhance the early strength and acts as a crystal nucleus in the hydration reaction of cementing materials. Lime is applied to provide a supersaturated Ca $(OH)_2$ environment to simulate the chemical activity of fly ash. An appropriate amount of compound sulfate is added for further activation. As a calcareous material, lime is engaged in hydration reaction with active SiO_2 and Al_2O_3 in fly ash and highly active SiO_2 in silica fume to generate gel. The C-S-H and C-A-H colloids are generated in chemical combination of SiO_2 and Al_2O_3 in fly ash with Ca $(OH)_2$. In the presence of gypsum, the C-A-H colloid will further change into stable ettringite.

The CHS gel is formed on the surface of fly ash particles and used for binding particles in the insulation system. The ettringite is filled in holes in the hardened cement paste to make these holes denser and gradually lead to micro-expansion, thus improving the performance of the hardened cement paste.

The hydration reaction mechanism of the cement-fly ash-sodium sulphate-slaked lime-silica ash cementation system can be summarized as follows. First, the hydration of cement occurs:

$2\ (3CaO \cdot SiO_2) + 6H_2O = 3CaO \cdot 2SiO_2 \cdot 3H_2O + 3Ca\ (OH)_2$

$2\ (2CaO \cdot SiO_2) + 4H_2O = 3CaO \cdot 2SiO_2 \cdot 3H_2O + Ca\ (OH)_2$

$3CaO \cdot Al_2O_3 + Ca\ (OH)_2 + 12H_2O = 4CaO \cdot Al_2O_3 \cdot 13H_2O$

Then calcium hydroxide reacts with sodium sulfate to produce calcium sulfate and sodium hydroxide, calcium sulfate reacts with hydrated calcium aluminate to produce ettringite, and the glass body interface in fly ash is corroded by sodium hydroxide and calcium hydroxide.

$Ca\ (OH)_2 + Na_2SO_4 = CaSO_4 + 2NaOH$

$3CaSO_4 + 4CaO \cdot Al_2O_3 \cdot 13H_2O + 20H_2O = 3CaO \cdot Al_2O_3 \cdot 3CaSO_4 \cdot 32H_2O + Ca\ (OH)_2$

$NaOH + Ca\ (OH)_2 + CaO \cdot nAl_2O_3 \cdot mSiO_2 + H_2O \rightarrow CaO \cdot xAl_2O_3 \cdot yH_2O + Na_2O \cdot xAl_2O_3 \cdot yH_2O + CaO \cdot xAl_2O_3 \cdot yH_2O + CaO \cdot xSiO_2 \cdot yH_2O + Na_2O \cdot xSiO_2 \cdot yH_2O$

Finally, when there is not adequate sodium sulfate, part of ettringite is transformed into monosulfur-hydrated calcium sulfate crystals:

$3CaO \cdot Al_2O_3 \cdot 3CaSO_4 \cdot 32H_2O + 2\ (3CaO \cdot Al_2O_3) + 4H_2O = 3\ (CaO \cdot Al_2O_3 \cdot CaSO_4 \cdot 12H_2O)$

In the cement-fly ash-sodium sulfate-slaked lime-silica ash cementation system, the formation of ettringite is a reason of enhancement of the compressive strength. To improve the compressive strength, the amount of sodium sulfate can be increased, but excessive sodium sulfate will lead to efflorescence. Silica fume can be added to effectively prevent efflorescence. In this sense, an appropriate amount of silica fume should be added at the same time with sodium sulfate. The flexural strength is improved as the boundary phase of hydrated calcium silicate and hydrated calcium aluminate produced in interfacial reaction between the spherical glass body in fly ash and calcium/sodium hydroxide, instead of the boundary phase of calcium hydroxide. With the sodium hydroxide, the rate of the interfacial reaction can be increased. The results of XRD and SEM analysis show that crystal phases such as mullite, quartz and magnetite in fly ash and spherical glass body are not engaged in reactions under cement hydration conditions. If activators are used, the rate of interfacial reaction of the spherical glass body will increase, changing the weak calcium hydroxide phase of the interface between the hydrated calcium silicate gel and spherical glass body into a hydrated calcium silicate phase, thus improving the system strength, especially the flexural strength.

For the weathering-resistant wall constructed with the adhesive polystyrene granule mortar in April 1998, the cementing material still has high strength without efflorescence in the natural environment, except the exposed EPS granules, as shown in Figure 9-1-6.

Figure 9-1-6 Weathering-resistant Wall Constructed with Adhesive polystyrene granule in 1984

2. Impact of organic polymer on durability

The redispersible polymer powder is a copolymer of vinyl acetate and other resins or a copolymer (polymer) containing ethylene, which has high water resistance, elasticity and bonding performance. It can be used for bonding surfaces that are difficult to bond with the organic material EPS granule. Meanwhile, ethylene has good flexibility and complete non-saponifying performance (polymer degradation under strong alkali conditions). The redispersible polymer powder will be redispersed into original emulsion when mixed with water on the site. As a polymer, it has the same film-forming and reinforcing effects as emulsion. The emulsion film formed is able to effectively cut off air circulation to form a number of closed voids, thus avoiding through holes and further reducing the thermal conductivity of the material. When filler particles are bonded with the polymer film as an organic binder, the tensile bonding strength and flexural strength of the material will be improved greatly, which can overcome the disadvantages caused by brittle inorganic materials. When the redispersible polymer powder is added in the cement material, therefore, the bonding, compressive and shear strength and deformability of the material will be enhanced significantly, with the flexural strength unchanged. Table 9-1-4 lists the data of performance comparison before and after the use of redispersible polymer powder in the insulation mortar.

As shown by the above data, when redispersible polymer powder is added, the flexural, compressive and tensile strength of insulation mortar will be greatly improved, thereby effectively enhancing the resistance to cracking strain. This means that the insulation mortar has higher overall performance and better cohesion.

9.1.3.5 Cracking Resistance

1. Impact of fiber on cracking resistance of insulation mortar

The cracking resistance of the adhesive polystyrene granule insulation mortar is affected by its own shrinkage and thermal stress, depending on the superposition of additives such as organic polymers, water-reducing agents and shrinkage reducers. It is also closely related to fiber compounding of the adhesive polystyrene granule insulation mortar. Fiber compounding is an important technical feature of such insulation mortar.

Based on the material, fibers are classified into metal fibers, inorganic fibers and organic fibers. Inorganic fibers are further divided into natural mineral fibers (such as chrysotile and acicular wollastonite), artificial mineral fibers (such as alkali-resistant glass fibers and alkali-resistant mineral wool), and carbon fibers. Organic fibers are further divided into synthetic fibers (polypropylene fibers, nylon fibers, polyethylene fibers, high modulus polyvinyl alcohol fibers, modified polyacrylonitrile fibers, aramid polyimide fibers and the like) and plant fibers.

Based on the elastic modulus, fibers are classified into high-elastic-modulus fibers (with the elastic modulus higher than that of the cement base course wall, such as steel fibers, asbestos, glass fibers, carbon fibers, high-elastic-modulus polyvinyl alcohol fibers, and aryl polyimide fibe-

Impact of Redispersible Polymer Powder on Properties of Insulation Mortar **Table 9-1-4**

Item	No Redispersible Polymer Powder	Redispersible Polymer Powder of 1%	Item	NoRedispersible Polymer Powder	Redispersible Polymer Powder of 1%
Compressive strength, MPa	0.33	0.41	Tensile strength, MPa	0.11	0.14
Flexural strength, MPa	0.10	0.20	Tensile bonding strength, MPa	0.08	0.13

rs) and low-elastic-modulus fibers (polypropylene fibers, nylon fibers, polyethylene fibers and most plant fibers).

Based on the length, fibers are classified into discontinuous short fibers and continuous long fibers.

When high-elastic-modulus fibers are added, the flexural (shear) strength of the adhesive polystyrene granule mortar will be enhanced. The higher the elastic modulus, the more stress is borne by fibers, thus forming a reliable space skeleton. Low-elastic-modulus fibers are conducive to the suppression of plastic cracking and reduction of dry shrinkage in the early period. Inorganic fibers feature high temperature resistance, physical and chemical stability, high wet dispersibility, low thermal conductivity and capacity, good thermal stability and high tensile strength, and can be well bonded with mortar. The thermal properties of organic fibers can compensate thermal expansion and contraction of cement-based materials. The fiber length has some impact on the fiber dispersibility and cracking resistance.

As inorganic and organic binders and fibers of different in the elastic modulus and length are compounded in the adhesive polystyrene granule mortar, the expansion of internal cracks can be prevented effectively, thereby improving the fracture toughness, tensile strength, cracking resistance as well as freeze-thaw resistance of the insulation mortar, and also extending its service life.

The adhesive polystyrene granule mortar contains anti-cracking fibers, which can effectively prevent the occurrence and development of cracks. In order to verify the impact of anti-cracking fibers on the cracking resistance of the mortar, the compressive strength, tensile bonding strength and cracking resistance were tested after 28d natural curing, with the mortar containing organic fibers. Refer to Table 9-1-5 for specific data.

From Table 9-1-5, it can be seen that if appropriate fibers are added, the compressive strength and tensile bonding strength of the insulation mortar will be improved, and its cracking resistance will become much better. However, the excessive use of fibers has a negative impact on the workability of fibers. Given the same conditions, fibers of high dispersibility have more obvious effects. Comprehensively, high-dispersibility of about 5‰ should be added.

2. Impact of EPS granule on cracking resistance

The EPS granule is a kind of organic aggregates with a very low elastic modulus but excellent toughness. It is able to absorb some energy by means of elastic deformation under external forces. When the mineral binders and EPS granules are fully combined and fibers are added, a spatial meshed structure will be formed in the insulation material, in which all components are evenly distributed in spatial grids. The external force applied on the insulation material can be dispersed in the three-dimensional space, and transferred to EPS granules via fibers and mineral binders. The EPS granules will be deformed to absorb energy. Thus, cracking of the insulation material will be reduced or eliminated, which helps to improve the cracking resistance.

Impact of Type and Quantity of Fibers on Properties of Pasting Mortar **Table 9-1-5**

Fiber Category	Quantity ‰	Compressive Strength	Tensile Bonding Strength	Cracking Resistance	Dispersibility
No fiber	0	0.54	0.11	With cracks	—
D	2.0	0.63	0.11	Without visible cracks	Normal
E	5.0	0.65	0.13	Without visible cracks	Normal
F	5.0	0.68	0.14	Without visible cracks	Good
F	8.0	0.61	0.12	Without visible cracks	Good

9.1.3.6 Fire Resistance

The fire resistance is an important feature of the adhesive polystyrene granule mortar, which contains the organic light aggregate coated with inorganic cementing materials. When the mortar is heated, EPS granules therein are softened and molten but do not burn (Figure 9-1-7). As the EPS granules are encapsulated with inorganic materials, closed cavities will be formed after melting. Then the thermal conductivity of this insulation material will be lower, and heat will be transferred more slowly, but the volume of this insulation material will hardly change.

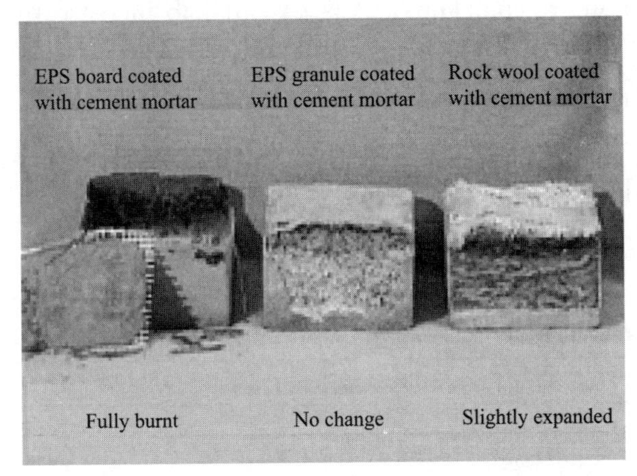

Figure 9-1-7 Comparison of EPS Board, Adhesive polystyrene granule Mortar and Rock Wool after Heating

For the leveling layer of adhesive polystyrene granule pasting mortar, as the fire resistance of pasting mortar itself does not meet Class A non-inflammability requirements and its thermal conductivity is much lower than cement mortar, a little heat is transferred slowly to the inside in the presence of fire or high temperature. That is, heat is concentrated on the surface of the pasting mortar leveling layer. Hence, the pasting mortar has more significant protective effects on internal insulation boards than cement mortar.

In the fire resistance test of the wall composed of the adhesive polystyrene granule mortar and rock wool boards, the volume of the adhesive polystyrene granule mortar changed little after 4h in the presence of fire, while EPS granules on the surface disappeared and those in deep parts were intact. This means that the temperature in deep parts of the mortar layer is below 70℃. It can be inferred that, when subjected to a fire, part of EPS granule foamed concrete on the mortar surface will become a heat-insulating honeycomb structure, thus preventing the heat from further transfer to the center of the wall and protecting insulation boards (Figure 9-1-8).

Figure 9-1-8 Fire Resistance Test of Wall Constructed with Adhesive polystyrene granule

9.1.3.7 Insulation Performance

The thermal insulation of an external envelope refers to the isolation of the roof and outer wall (especially the western wall) from heat via an insulation material, thus reducing the heat transferred to indoors and reducing the internal surface temperature of the envelope. As the outdoor comprehensive temperature changes periodically within 24h in summer, the insulation performance is measured by the attenuation multiple, total delay time and other indicators. The attenuation multiple refers to the ratio of the amplitude of the outdoor comprehensive temperature to that of the strength of the internal surface. The greater the attenuation multiple, the better the insulation performance is. The total delay time refer to the difference between the occurrence time of the maximum outdoor comprehensive temperature and that of the maximum internal surface temperature. The longer the delay time, the better the insulation performance is.

Due to the thermal capacity of a material during heating and cooling and the thermal resistance of the material layer during heat transfer, the temperature wave will be subject to attenuation and delay in the transmission process. This is why the insulation material of low thermal conductivity and large heat storage coefficient should be selected and the envelope should have the appropriate heat transfer coefficient based on insulation requirements. Compared with other insulation materials such as the EPS board and polyurethane, the adhesive polystyrene granule mortar has higher thermal capacity. Given the same thermal resistance, the delay time of maximum temperature is long because of the small temperature amplitude of the internal surface. The heat storage coefficient of the adhesive polystyrene granule mortar is greater than 0.95W/ (m^2 · K), and that of the EPS board is 0.36W/ (m^2 · K), of which the former is approximately three times that of the latter. Therefore, the adhesive polystyrene granule mortar has better insulation performance. In accordance with the *Design Standard for Energy Efficiency of Residential Buildings in Hot Summer and Cold Winter Zone* (JGJ 134—2010) and *Design Standard for Energy Efficiency of Residential Buildings in Hot Summer and Warm Winter Zone* (JGJ 75—2012), the external thermal insulation system with 20-40mm adhesive polystyrene granule insulation mortar meets the insulation requirements of a variety of outer walls.

If the EPS granule insulation mortar is applied as a transition layer, the surface temperature of insulation boards will drop greatly, and the occurrence time of maximum temperature will be delayed, thus significantly decreasing the thermal stress arising from surface deformation of insulation boards, relieving drastic fluctuations in the stress of insulation boards with the temperature changing, and achieving high weathering resistance of the insulation board system.

9.1.3.8 Impact of Water on Mortar Properties

1. Water resistance

The water resistance of the adhesive polystyrene granule mortar is expressed by the softening coefficient. As rigid hydrated calcium silicate and calcium aluminate are produced in chemical reaction of organic materials in water, the strength of this insulation material rises greatly in the middle and late period. The molecules of hydrated calcium silicate are densely arranged, which is conducive to waterproofing as well as increase of the softening coefficient. An important premise for application of the insulation mortar in the external thermal insulation system is that the solidified material must have appropriate water resistance, and will not be softened or fall off in the case of rain erosion. Thus, it is essential to increase the softening coefficient. The test results of the softening coefficient of the EPS granule insulation mortar in different ageing periods are listed in Table 9-1-6.

Comparison of Softening Coefficients of EPS Granule Insulation Mortar in Different Ageing Periods

Table 9-1-6

Ageing Period(d)	Softening Coefficient
28	0.55
90	0.71
180	0.74

On one hand, the activity of fly ash can be stimulated to increase the strength growth rate and water resistance. On the other hand, an appropriate portion of hydrophobic surfactant and redispersible powder can be added to increase the softening coefficient of the insulation mortar to more than 0.5 after 28d curing and 0.7 after 90d curing, basically reaching the expected design goals.

If the adhesive polystyrene granule mortar is exposed to water in a long time, water molecules will infiltrate into it, leading to the decrease of bonding strength of internal molecules and compressive strength of mortar, and also affecting the physical properties of mortar. Long-term exposure to water may result in loosening. Figure 9-1-9 shows the results of the softening coefficient after 96h complete immersion of the insulation mortar in water.

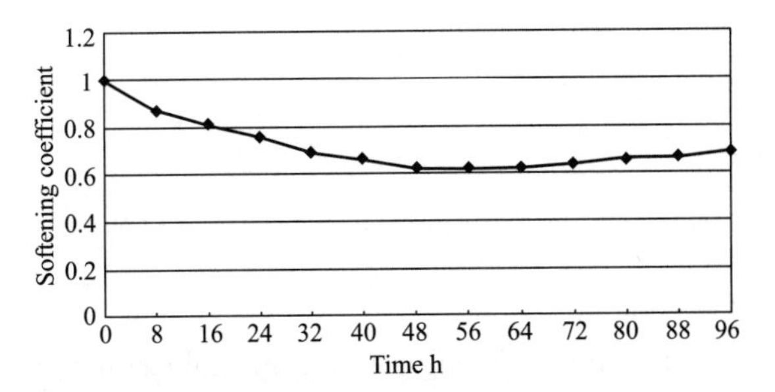

Figure 9-1-9 Trend of Change in Softening Coefficient of Adhesive polystyrene granule Mortar over Time

The test results show that the softening coefficient gradually decreases over time, namely, dropping with the amount of absorbed water rising. It tends to change gently after 40h and does not basically change after 48h. The impact on the physical properties of the mortar is still within an acceptable range when the softening coefficient is the lowest.

2. Impact of water on thermal conductivity

The property of a material to absorb moisture from the damp air is known as the hygro scopicity, expressed as the moisture content. Hygroscopicity is an important factor affecting the actual thermal conductivity of a material. The thermal conductivity of water is $\lambda = 0.5815$W/(m·K). If a material has high hygroscopicity, its thermal conductivity should be increased properly. The trace pre-dispersion technology with inorganic materials coated with organic ones is applied in the insulation mortar. When the hydrated silicate compound is produced in reaction of the mortar and water, the organic polymer material of high hydrophobicity will be evenly distributed on the surface of the silicate compound to form an organic protective film, preventing moisture infiltration. Accordingly, the insulation mortar has high water resistance and low water absorption capacity.

Refer to Table 9-1-7 and Figure 9-1-10 and 9-1-11 for the increase in thewater absorption rate and thermal conductivity of the specimen in a standard environment [temperature: (20±2)℃; humidity: (50±10)%].

Relationship between Water Absorption Rate and Thermal Conductivity of Adhesive polystyrene granule Mortar

Table 9-1-7

Equilibrium Time, d	0	1	3	8	24	30
Moisture content, %	0	1.31	2.40	3.63	3.83	3.84
Thermal conductivity, W/(m·K)	0.05174	0.05268	0.05371	0.05810	0.05931	0.05933
Thermal conductivity increase rate, %	0	1.82	3.81	12.29	14.63	14.67
Equilibrium Time, d	45	60	90	120	150	180
Moisture content, %	3.80	3.88	3.81	3.90	3.82	3.85
Thermal conductivity, W/(m·K)	0.05930	0.05945	0.05929	0.05948	0.05929	0.05937
Thermal conductivity increase rate, %	14.61	14.90	14.59	14.96	14.59	14.75

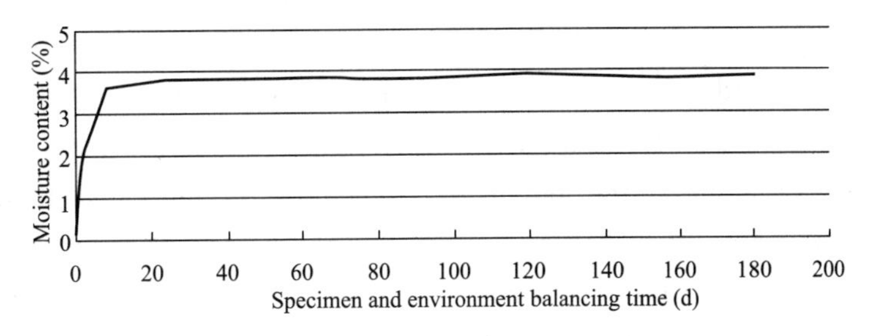

Figure 9-1-10 Relationship between Equilibrium Time and Moisture Content of Insulation Mortar

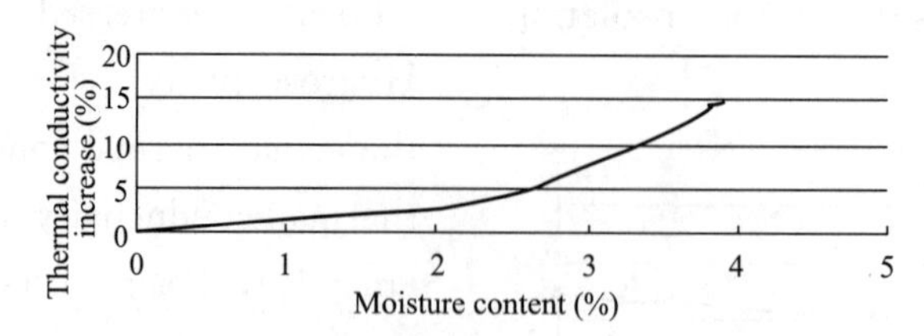

Figure 9-1-11 Relationship between Moisture Content and Thermal Conductivity Increase of Insulation Mortar

As shown in the moisture content figure, the moisture content of theadhesive polystyrene granule mortar generally rises over time in a standard environment. It increased greatly in the first 30 days and is gradually stabilized at roughly 4% after 30d. Therefore, the equilibrium moisture content of this material in the standard environment is 4%.

The thermal conductivity figure illustrates that the relationship between the moisture content and thermal conductivity increase seems a smooth curve. That is the thermal conductivity rises with the moisture content increasing. When the moisture content reaches the maximum value (4%), namely, the equilibrium moisture content, the thermal conductivity will increase to 15%.

The adhesive polystyrene granule mortar has the water absorption capacity and vapor permeability. That is, water vapor is discharged through micro-porous channels along with absorption of moisture from inside and also in the environment. The dynamic equilibrium can be maintained under appropriate temperature and humidity conditions, making the properties of the insulation mortar steady.

9.1.3.9 Summary of Study on Adhesive polystyrene granule Mortar

The adhesive polystyrene granule mortar has been profoundly studied and applied in China, promoting the formulation and improvement of relevant standards for materials of external thermal insulation systems. The external thermal insulation theory summarized and refined from the practice of adhesive polystyrene granule mortar further guides and advances the engineering practice of external thermal insulation and energy conservation.

The adhesive polystyrene granule composite insulation system has the advantages of good insulation performance, durability, adaptability to construction, fire resistance and cracking resistance. It is an external thermal insulation technology applying to China's national conditions, involving the mature application technology and excellent insulation effects at present, and making significant contributions to the building energy conservation and emission reduction in China.

The adhesive polystyrene granule mortar technology is not a single material technology, but a comprehensive technology including various material and construction technologies. The combination method greatly improves the stability of insulation boards and weathering resistance of this insulation system.

The adhesive polystyrene granule mortar with excellent properties and adaptability will be further developed during continuous research and application.

9.1.4 Study on Anti-cracking Protective Layer

The anti-cracking protective layer of the adhesive polystyrene granule composite insulation technology is composed of the anti-cracking polymer cement mortar and alkali-resistant glass fiber mesh layer, polymer emulsion elastic primer and anti-cracking waterproof flexible putty. It plays a key role for the cracking resistance of the whole insulation system. The flexible ultimate tensile deformation of the protective layer should be greater than the sum of its own deformation (dry shrinkage, chemical deformation, humidity deformation and thermal deformation) and the base course wall deformation under the most unfavorable conditions, in order to meet the cracking resistance requirements.

The performance indicators of materials in the anti-cracking protective layer should change layer by layer to meet the deformation requirements of each layer, fully release the stress of each layer, minimize the impact of stress on the system and thus avoid cracking. The anti-cracking structure of the anti-cracking protective layer is a guarantee for the cracking and weathering resistance of the adhesive polystyrene granule composite insulation technology, so the anti-cracking partition and centralized stress release structure are not needed in the composite insulation system.

9.1.4.1 Study on Anti-cracking Mortar

Anti-cracking mortar is a kind of polymer mortar, which is prepared with polymer, cement and sand as main materials and has excellent deformation resistance and bonding performance. It has important significance for the weathering resistance and service life of the external thermal insulation system. This section analyzes the main influencing factors and anti-cracking mechanism of anti-cracking mortar.

Based on the physical status, anti-cracking mortar can be divided into:

(1) Dry-mixed polymer mortar: prepared by accurately metering and mixing raw materials including cement, sand and redispersible polymer powder in factory, packed in a quantitative manner, and used after mixing with water on the construction site as per the proportion defined in the specifications.

(2) Anti-cracking polymer emulsion mortar: prepared by fully mixing the cement mortar cracking prevention agent, ordinary Portland cement as well as medium-sized and fine sand according to a specific proportion, and used for anti-cracking protection of the surface of an insulation layer. It has flexibility and cracking resistance after curing. When the mortar is used, the bonding performance and flexible cracking resistance of the anti-cracking protective layer can be improved. The cement mortar cracking prevention agent is a glue substance prepared by mixing admixtures and anti-cracking substances with elastic polymer emulsion.

1. Study on polymer in anti-cracking mortar

Polymers used in mortar mainly include emulsion and redispersible polymer powder.

(1) Polymer emulsion

Polymer emulsions used in mortar mainly include styrene-butadiene emulsion (SBR), polyacrylic acid ester (PAE), polyvinyl acetate (EVA), and styrene-acrylic emulsion (SAE). One or more emulsion (s) can be used in a complementary manner depending on the desired properties. In order to improve the properties of cement mortar, the polymer emulsion used must be well suited for cement hydration. The polymer to be applied must conform to the following requirements:

(i) There must be no adverse effects on cement hardening and bonding;

(ii) The polymer must not be hydrolyzed or destroyed the alkaline medium of cement;

(iii) When the temperature is 10℃ or below, the segments of polymer should be still flexible;

(iv) There must be no reverse reaction with various medium after reaction;

(v) The polymer must be available extensively with moderate prices.

(2) Redispersible polymer powder

The redispersible polymer powder is a kind of modified emulsion powder prepared by modifying, spraying and drying polymer emulsion with polyvinyl alcohol as a protective colloid. Normally, the redispersible polymer powder has good redispersibility. When exposed to water, it can be redispersed into emulsion with chemical properties close to those of the original emulsion.

(3) Effect of polymer in mortar

(i) Effects in wet mortar: improve the workability, fluidity, thixotropy, sag resistance, cohesion and water retention, and extend the open time.

(ii) Effects after curing:

a) Raise the tensile strength, and significantly improve the bonding strength between mortar and organic insulation boards;

b) Improve the flexural strength, cohesion, deformability and cracking resistance, and reduce the elastic modulus and depth of carbonation;

c) Increase the compactness and wear resistance, and reduce the water absorption, to achieve excellent hydrophobicity (with hydrophobic polymer powder).

(4) Anti-cracking mechanism of polymer emulsion

The results of a large number of theoretical studies and basic experiments prove that the use of polymer emulsion with high homogeneity, strength and flexibility is an effective way to guarantee the strength and flexibility of anti-cracking mortar. The polymer emulsion mainly has the following effects.

Activation effect: the polymer emulsion contains surfactants can be used to reduce water, disperse cement particles, improve the workability of mortar and decrease the water consumption, thereby reducing harmful pores such as capillary pores of cement and improving the compaction and impermeability of mortar.

Bridge bond effect: special bridge bonds are generated during exchange of active genes in polymer molecules and free ions such as Ca^{2+}, Al^{3+} and Fe^{2+} in cement hydration. Physical and chemical absorption occurs around cement particles, forming a continuous phase of high homogeneity. This helps to reduce the overall elastic modulus, improve the physical structure and internal stress of cement mortar, and enhance the ability to withstand deformation, thus greatly reducing the possibility of micro gaps. Even if there are micro cracks, their development can be restricted due to the bridge bond effect of polymer.

Filling effect: the polymer emulsion can quickly condense into a tough and dense film. When filled among cement particles, the polymer emulsion reacts with cement hydration products to form a continuous phase in gaps, thus blocking the passage to the outside.

In the case of water loss, the organic polymer emulsion becomes an elastic film with excellent adhesion and continuity, and cement absorbs water in the emulsion in the hardening process. The flexible polymer film and hardened cement body interpenetrate firmly into a solid and resilient waterproof layer. Thus, the anti-cracking mortar has excellent flexibility, considerable elasticity, low linear shrinkage rate and good bonding performance.

2. Study on fibers in anti-cracking mortar

Fibers can be applied to improve the resistance to cracking, impact load and flex. As stress is transferred through fibers, crack generation and expansion can be effectively controlled and reduced in the curing process, thereby improving the frost resistance and cycle thawing performance of the surface, decrease and eliminate minor cracks on the surface, and lower the air permeability. Three types of fiber were mixed at different volume fractions in mortar in this test, namely, fibers 1, 2 and 3, based on one formula, followed by 14d natural curing, as shown in Table 9-1-8.

Impact of Three Types of Fiber of the Same Volume Fraction on Mortar Properties **Table 9-1-8**

Fiber Type	Volume Fraction(‰)	Flexural Strength(MPa)	Compressive Strength(MPa)	Compressive-to-flexural Strength Ratio	Dispersibility
No fiber	0	1.67	3.10	1.86	—
Fiber 1	5	2.05	3.16	1.54	Poor
Fiber 2	5	2.03	4.30	2.12	Normal
Fiber 3	5	2.17	5.13	2.36	Good

When the fractions of three types of fiber are all 5‰ (by the anti-cracking agent), the fiber 1 has poor dispersibility in mortar, but the fiber 2 has normal dispersibility and the fiber 2 has good dispersibility. When the fiber 3 of the same mass was added into anti-cracking mortar, the compressive strength and flexural strength were greatly improved. This means that the cracking resistance rises with fibers added. The impact of the fiber fraction on mortar properties was examined with the 5mm long fiber 3, as shown in Table 9-1-9. It can be seen that, with the quantity of fibers increasing, the consistency of anti-cracking mortar drops a little, indicating that the fiber adversely affects the workability or mortar, while the compressive strength and flexural strength rise quickly, indicating that the flexibility and cracking resistance of mortar are enhanced significantly. Given that the use of excessive fibers leads to poor workability of anti-cracking mortar and limitations in improvement of its cracking resistance, the fiber fraction should be 5‰ (by the anti-cracking agent). The reason why mortar containing fibers has higher resistance to cracking is that the resistance of mortar to plastic shrinkage is improved greatly. In addition, fibers help to reduce stress concentration at the tips of cracks, which can prevent cracks from further expansion.

The anti-cracking mechanism of fibers is analyzed as follows:

Synthetic fibers can be added into cement mortar to enhance the tensile strength and significantly reduce the plastic micro-cracks arising from flow and shrinkage. The decrease or elimination of plastic cracks leads to the best long-term integrity of cement mortar. Fibers should be evenly distributed throughout cement mortar to achieve additional reinforcement and prevent shrinkage cracks. In this case, the width and length of potential cracks of cement mortar under loads can be minimized. This is a modern technology of micro reinforcement.

According to the available data, attention should be paid to the following fiber parameters in the engineering practice:

(1) Fineness (D): in the standard unit Tex or Dtex. The large fineness is conducive to stirring, while the small fineness to dispersion.

(2) Tensile strength (T): ratio of the specific gravity to fineness.

(3) Elastic modulus (E) and tensile strength: high tensile strength and large elastic modulus are helpful to inhibit the occurrence and expansion of cracks in cement mortar.

(4) Volume fraction (V_f): a large volume fraction is conducive to the ductility of cement mortar. In the event of an excessive volume fraction, however, the porosity of cement mortar will increase, making the strength decline.

(5) Length-to-diameter ratio: a large length-to-diameter ratio is conducive to the rise of remaining strength and elastic modulus.

(6) Specific surface area (SFS): meaning the total fiber area per unit volume of a composite material. It plays a decisive role in base course wall damage under the stress. The higher the degree of dispersion and the larger the area, the better effects will be achieved.

Impact of Fiber Fraction on Mortar Properties **Table 9-1-9**

Fiber Fraction(‰)	Flexural Strength(MPa)	Compressive Strength(MPa)	Compressive-to-flexural Strength Ratio
0	1.67	3.10	1.86
3	2.23	4.2	1.88
5	2.17	5.13	2.36
8	2.4	4.27	1.78
10	1.975	3.78	1.91

The above analysis results indicate that polymer emulsions and fibers are of great significance to improve the cracking resistance of anti-cracking mortar and reduce its elastic modulus as well as compressive-to-flexural strength ratio, thus guaranteeing that anti-cracking mortar is compatible with the insulation layer.

9.1.4.2 Study on Alkali-resistant Glass Fiber Mesh

Glass fiber meshes are mainly divided into the alkali-resistant glass fiber mesh (glass composition: ZrO_2 of 14.5%±0.8% and TiO_2 of 6%±0.5%), medium-alkali glass fiber mesh (content of alkali metal oxides: approximately 12%) and alkali-free glass fiber mesh (content of alkali metal oxides: no more than 0.5%).

With continuous deepening of building energy conservation, the glass fiber mesh as a key reinforcing material has been promoted fast in external thermal insulation systems, so its variety and quality are essential for development of external thermal insulation. The variety and quality of the glass fiber mesh suitable for an external thermal insulation system should be determined by analyzing the alkali-resistant breaking strength retention rates and alkali-resistant mechanisms of glass fibers, as well as the impact of the plastic coating quantity, medium and soaking time on the alkali resistance of the alkali-resistant mesh.

1. Comparison of alkali-resistant breaking strength retention rates of glass fiber meshes

The initial breaking strength (F_0) of the alkali-resistant, medium-alkali and alkali-free glass fiber meshes were tested in accordance with the *Reinforcements-Test Method for Woven Fabrics-Part 5: Determination of Glass Fiber Tensile Breaking Force and Elongation at Break* (GB/T 7689.5). Five identical specimens were completely soaked in ordinary Portland cement mortar (80℃) with the water-cement ratio of 10: 1 for a specific period of time (1d, 2d, 3d, 4d, 5d, 6d and 7d). Then these specimens were taken out, rinsed with clean water to remove cement mortar, and dried to the constant weight in an oven of (105±5)℃. Finally, the alkali-resistant breaking strength (F_1) was tested.

The alkali-resistant breaking strength retention rate is calculated from:

$$B=\frac{F_1}{F_0}\times 100\%$$

Where, B —alkali-resistant breaking strength retention rate, %;

F_1 —alkali-resistant breaking strength, N;

F_0 —initial breaking strength, N.

In order to keep the test conditions, plastic coating was not conducted on the surfaces of three glass fiber meshes. The test results are listed in Table 9-1-10.

According to the test results, the strength retention rate of the alkali-free glass fiber mesh with high initial strength dropped to 22% after 1d soaking in alkali liquor, and decreased over time to zero finally as a result of corrosion of alkali liquor. Thus, the alkali-free glass fiber mesh is not discussed here. This section focuses on the properties of the alkali-resistant and medium-alkali glass fiber meshes.

2. Comparison of properties of alkali-resistant and medium-alkali glass fiber meshes

(1) Systematic study on mesoscopic morphology and breaking strength.

Test Data of Alkali-resistant Breaking Strength Retention Rates of Glass Fiber Meshes (without plastic coating)

Table 9-1-10

—	Alkali-resistant Breaking Strength Retention Rate(%)						
	1d	2d	3d	4d	5d	6d	7d
Alkali-resistant fiber(ER-13)	75.0	67.3	57.8	51.6	46.5	40.8	40.2
Medium-alkali fiber(C)	45.1	41.2	35.0	20.2	20.4	10.2	—
Alkali-free fiber(C)	22.0	18.3	16.2	10.5	—	—	—

Figure 9-1-12 Changes in mesoscopic morphology of Alkali-resistant Glass Fiber Mesh after Soaking in Alkali Liquor

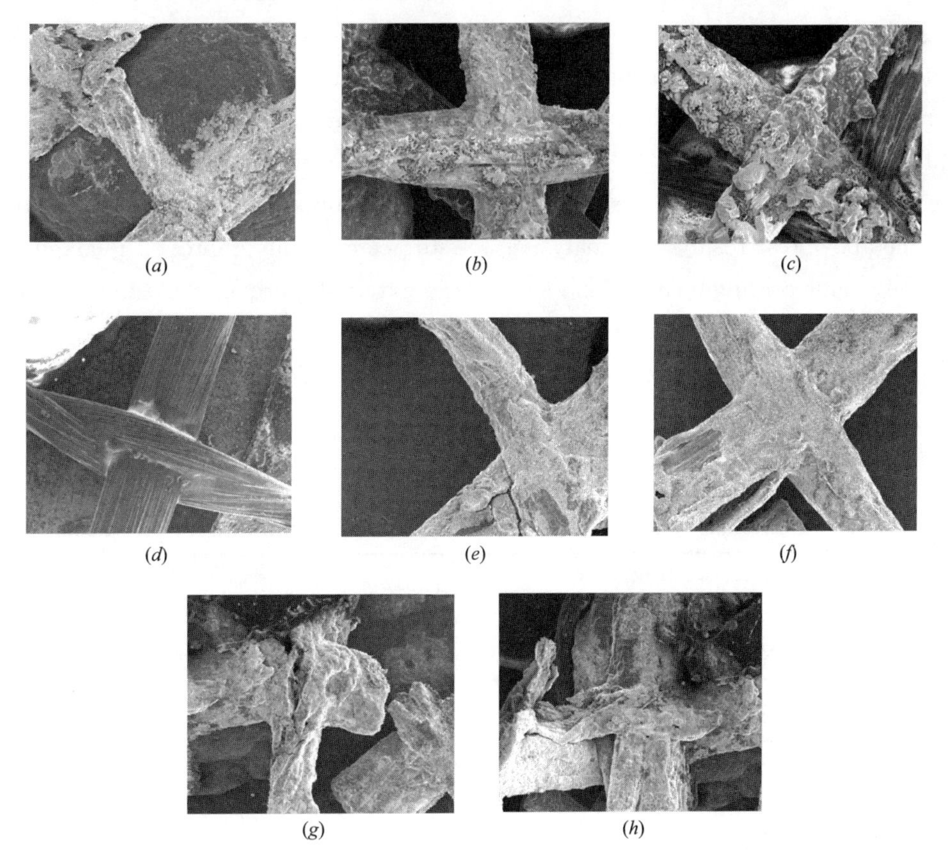

Figure 9-1-13 Changes in mesoscopic morphology of Medium-alkali Glass Fiber Mesh after Soaking in Alkali Liquor

(i) Test procedures

Mix Grade 42.5 ordinary Portland cement (0.2kg) and distilled water (2kg) batch by batch with a magnetic stirrer for 30 minutes, and keep the mixture for 30 minutes. Filter the mixture with the filter paper, and collect the filtrate. Transfer the filtrate into two clean beakers, respectively. Keep nine alkali-resistant glass fiber meshes (50mm×50mm) and nine medium-alkali glass fiber meshes (50mm×50mm) static in cement filtrate. Take out one alkali-resistant glass fiber mesh and medium-alkali glass fiber mesh respectively on the 1st, 2nd, 3rd, 5th, 7th, 10th, 15th, 21st and 28th day, and observe their mesomorphology. Then dry them and test their breaking strength.

(ii) Test results

a) Refer to Figure 9-1-12 for changes in the micromorphology of alkali-resistant glass fiber meshes after soaking in alkali liquor.

b) Refer to Figure 9-1-13 for changes in the micromorphology of medium-alkali glass fiber meshes after soaking in alkali liquor.

c) Refer to Table 9-1-11 for changes in the breaking strength of alkali-resistant and medium-alkali glass fiber meshes.

(2) The following conclusions are obtained by comparing the above test results and analyzing the impact of the plastic coating quantity and soaking time on the alkali resistance of these meshes as well as their actions in external thermal insulation systems.

(i) The glass fiber mesh composite anti-cracking mortar plays a key role in anti-cracking protection of the external thermal insulation system. Given that anti-cracking mortar cannot be isolated by the mesh with some strength, the correct way is to lay the glass fiber mesh (pore size: 3～6mm; alkali-resistant breaking strength: no less than 1000N/50mm) in high-flexibility mortar and close to the surface side.

(ii) The plastic coating material and quantity are of important significance for the alkali resistance of glass fiber meshes in the early stage, but the glass fiber variety is decisive for the long-term alkali resistance. In addition to the breaking strength, the alkali-resistant strength retention rate should be specified in order to ensure the long-term effects of glass fiber meshes. The alkali resistance of glass fiber meshes depends on the glass fiber variety as well as the plastic coating material and quantity. Moreover, the alkali resistance of alkali-resistant glass fiber meshes, especially the long-term alkali resistance, is superior to those of medium-alkali and alkali-free glass fiber meshes. Their fundamental difference lies in the presence of a zirconium-rich interface on the alkali-resistant glass fiber surface, which can reduce the concentration of hydroxide ions in alkali liquor, suppress their diffusion on the glass fiber surface and thus improve the alkali resistance. As an external thermal insulation system must have a life of 25 years at least, the glass fiber mesh used must be made by weaving alkali-resistant glass fibers, subject to plastic coating with alkali-resistant polymer materials.

Changes in Breaking Strength of Alkali-resistant and Medium-alkali Glass Fiber Meshes Table 9-1-11

Time (d)	Alkali-resistant Glass Fiber Mesh			Medium-alkali Glass Fiber Mesh		
	pH value of mother liquor	Breaking strength F(kN/m)	Alkali-resistant strength retention rate χ(%)	pH value of mother liquor	Breaking strength F(kN/m)	Alkali-resistant strength retention rate χ(%)
0	11.2	62.6	100	11.2	64.3	100
1	11.2	62.0	99.04	11.3	63.5	98.76
2	11.2	62.0	99.04	11.2	61.7	95.96
3	11.2	57.9	92.49	11.1	60.9	94.71

Continued

Time (d)	Alkali-resistant Glass Fiber Mesh			Medium-alkali Glass Fiber Mesh		
	pH value of mother liquor	Breaking strength F(kN/m)	Alkali-resistant strength retention rate χ(%)	pH value of mother liquor	Breaking strength F(kN/m)	Alkali-resistant strength retention rate χ(%)
5	11.1	58.0	92.65	11.1	60.2	93.19
7	11.2	57.5	91.85	11.1	59.8	93.00
10	11.1	57.3	91.53	11.0	54.7	85.07
15	10.9	57.1	91.21	11.0	54.6	84.52
21	10.9	56.8	90.73	10.9	53.1	82.20
28	10.9	56.3	89.94	10.8	52.7	81.96

(iii) To sum up, it is believed that the patented ZL alkali-resistant glass fiber meshes available currently on the market are relatively ideal for external thermal insulation systems. Their technical properties meet the requirements in Table 9-1-12.

9.1.4.3 Study on Polymer Elastic Primer

The polymer elastic primer is a kind of base coating, which is prepared with elastic waterproof emulsion, additives, pigments and fillers and has waterproofing and permeable effects. It is a key structural layer in the adhesive polystyrene granule composite insulation system, effectively preventing the inflow of liquid water and allowing the discharge of water vapor, avoiding damage caused by changes in three phases of water to the external thermal insulation system, and guaranteeing the durability of the external thermal insulation system.

1. Mechanism of film formation and action

The polymer emulsion elastic primer has good weathering resistance, as the emulsion with a fine film and small particle size is used as the film-forming substance, and the self-crosslinking pure acrylic emulsion is compounded with silicone acrylic emulsion. The polymer emulsion is a polymer or monomer dispersion system in water, involving complicated film formation. Polymers are dispersed as small (colloidal) particles, and move freely in the manner of Brownian move-

Properties of ZL Alkali-resistant Glass Fiber Mesh **Table 9-1-12**

Item		Unit	Indicator
Appearance		—	Conforming
Length & Width		m	50-100 and 0.9-1.2
Mesh center distance	Normal type	mm	4×4
	Reinforced type		6×6
Mass per unit area	Normal type	g/m^2	≥160
	Reinforced type		≥500
Breaking strength (warp and weft)	Normal type	N/50mm	≥1250
	Reinforced type	N/50mm	≥3000
Alkali-resistant strength retention rate(warp and weft)		%	≥90
Elongation at break(warp and weft)		%	≤5
Plastic coating quantity		g/m^2	≥20
Glass composition		%	Conforming to the standard JC 719, containing ZrO_2 of 14.5%±0.8%, TiO_2 of 6%±0.5%

ment. This dispersion system is not stable. When water evaporates or is absorbed by a porous material, the movement of polymers will be restricted and small particles will be gradually converged and eventually integrated. To form a continuous film without voids, small particles must be deformed. With water evaporating, the dispersed small (colloidal) particles are drawn together under the capillary pressure. Colloidal particles are deformed under the increasing pressure to form a continuous film with good water resistance and vapor permeability.

2. Impact of emulsion amount on polymer elastic primer

The polymer elastic primer is prepared by mixing one kind of acrylic emulsion and silicone acrylic emulsion as per the specific proportion, with assistance with additives and fillers. The proportion of polymer emulsions in the primer ranges from 20% to 60%. The performance indicators of the polymer elastic primer were tested, among which the breaking strength values are for reference only. The test results are listed in Table 9-1-13.

As shown by the performance indicators in Table 9-1-13, the polymer elastic primer is uniform with no agglomerate after mixing, and easy to brush during construction, but the drying time varies slightly. With the amount of polymer emulsion increasing, the elongation at break rises greatly. The conforming amount of emulsion is 40%. Taking into account the elongation at break, the amount of emulsion should be more than 50%.

9.1.4.4 Study on Waterproof Flexible Putty

With the large-scale development and application of the external thermal insulation technology, the early " rigid anti-cracking technical route" has been gradually transformed into the "technical route of layer-by-layer change and flexible prevention of cracking". Normally, the insulation layer has low density and strength and its outline dimensions are instable because of deformation arising from the temperature and humidity, the supporting plaster layer, putty layer and coating layer must be effectively adaptive. In addition to the general performance requirements, the putty used in the external thermal insulation system must adapt to dynamic stress changes of this system. The waterproof flexible putty is able to dissipate, release and balance the dynamic stress of the insulation system, while the rigid putty with insufficient flexibility may crack with the anti-cracking protective layer deformed. Figure 9-1-14 shows the contrast test results of rigid and flexible putty on one test wall.

Performance Indicators of Polymer Elastic Primer of Different Emulsion Amounts　Table 9-1-13

Item		Unit	Indicator	Emulsion of 20%	Emulsion of 30%	Emulsion of 40%	Emulsion of 50%	Emulsion of 60%
Status in container		—	Uniform with no agglomerate after mixing	Uniform with no agglomerate after mixing	Uniform with no agglomerate after mixing	Uniform with no agglomerate after mixing	Uniform with no agglomerate after mixing	Uniform with no agglomerate after mixing
Workability		—	No impact on brushing	No impact on brushing of two layers	No impact on brushing of two layers	No impact on brushing of two layers	No impact on brushing of two layers	No impact on brushing of two layers
Drying time	Surface drying time	h	≤4	75min	65min	65min	62min	55min
	Full drying time	h	≤8	205min	165min	130min	140min	135min
Elongation at break		%	≥100	150	283	420	562	694
Surface hydrophobicity		%	≥98	89	93.4	99.3	99.91	99.91
Breaking strength (for reference)		MP	≥0.5	0.14	0.24	0.46	0.58	0.64

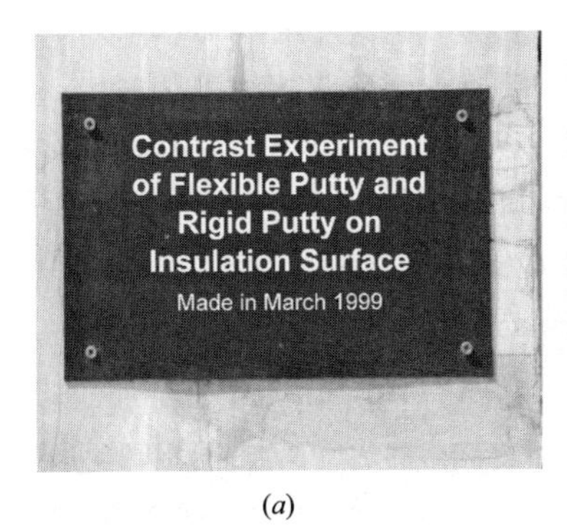

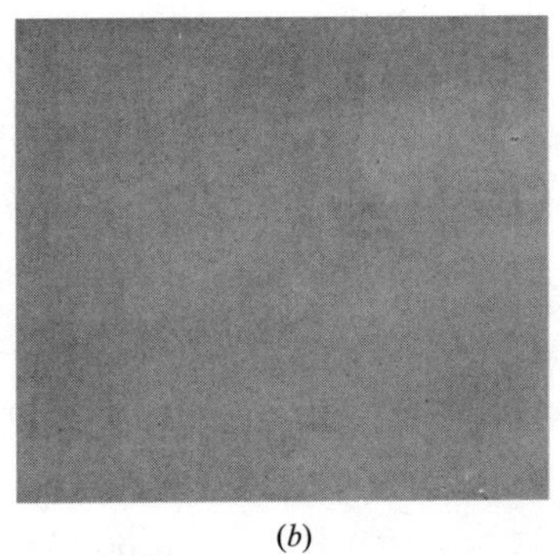

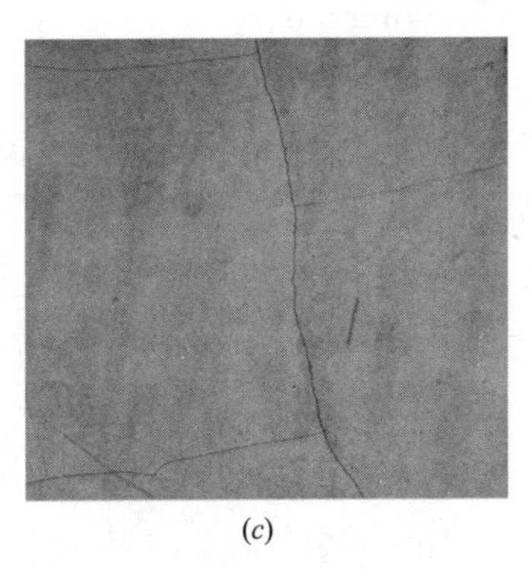

(*a*) (*b*) (*c*)

Figure 9-1-14 Comparison of Rigid and Flexible Putty

(*a*) putty comparison; (*b*) waterproof flexible putty; (*c*) rigid putty

1. Study on factors influencing waterproof flexible putty

(1) Test formula

Flexible emulsion (matching high Tg value and low Tg value): 25% to 35%;

Anti-cracking fiber: 0.2-0.5mm, 0.5% to 1.0%;

Methyl cellulose (solid powder): 0.1% to 0.5%;

Thixotropic agent (solid powder): 0.1% to 0.5%;

Defoamer: 0.1% to 0.3%;

Preservative: 0.1% to 0.2%;

Composite hydrophobic surfactant: 0.1% to 0.3%;

Additive mixture: 1% to 3%;

Quartz sand: 100 to 325 meshes, 25% to 40%;

Heavy calcium carbonate: 150 to 200 meshes, 30% to 50%;

Water: as needed.

(2) Preparation

Weigh raw materials according to the production formula, mix the emulsion and solids into the solution of appropriate concentration. Add the emulsion, defoamer, preservative, liquid cellulose and anti-cracking fibers into a mixing tank, and start the stirrer. Slowly add the additive mixture, thixotropic agent and surfactant, and stir them for 40-60min to form uniform mortar. Then pack the putty product.

During construction, the paste should be prepared with the putty and ordinary Portland cement (42.5) according to the proportion of 1:0.4 (by weight). The putty must be applied via a scraper or trowel within 2h after preparation. The thickness of putty applied at a time should be 1mm at least. It is allowed to apply putty in 2 to 4 batches.

(3) Performance test

The test should be conducted in accordance with the *Waterproofing Flexible Putty for Exterior Thermal Insulation Systems* (JG/T 229).

(4) Relationship between cement content and flexibility/bonding strength

The cement content is usually regarded as one main factor affecting the overall performance of putty. The flexibility and adhesion of the putty used in an external thermal insulation system should be unified. As the putty layer is used as a non-structural coating or covering material supporting the anti-cracking insulation layer, its compressive strength or elastic modulus generally needs to be lower than that of the base course wall, in order to better adapt to the base course wall deformation, coordinate its own deformation under external factors, and reduce the stress as well as potential cracking and peeling. That is what is called the unified flexibility and adhesion. The relationship between the cement content and flexibility/bonding strength here was obtained based on the test results of putty that is prepared with binders of four cement ratios (using heavy calcium carbonate as the rest) and putty glue of two emulsion contents.

According to the test results of the bonding strength and flexibility of putty specimens involving different cement contents, the following conclusions are drawn: when the emulsion content is 20%, the bonding strength of putty does not meet the standards (Table 9-1-14), regardless of the cement content; and when the emulsion content is 30% and the cement content is 25% to 30%, all indicators of the putty meets the standards (Table 9-1-15). The appropriate emulsion content and best cement content can be found to unify the flexibility and adhesion.

(5) Impact of emulsion content

The putty used in the external thermal insulation system must have good flexibility as well as high hardness, adhesion, alkali resistance and stain resistance, it is particularly important to select emulsion. Normally, emulsions of high and low T_g values are mixed, which not only meets the flexibility requirements, but effectively solves the problems of low rigidity, stain resistance and water resistance and poor adhesion to the base course wall. The putty properties corresponding to four emulsion contents were tested based on the cement content of 28% (Table 9-1-16).

According to the above test results, the putty prepared with the cement content of 25% to 30% and emulsion content of 30% to 35% has the best flexibility and bonding strength.

(6) Summary

The polymer emulsion, cement, filler and additives affect the properties of putty, such as the bonding strength, water absorption capacity, flexibility and alkali resistance. The waterproof flexible putty should have good workability and flexibility, without cracking in the case of bending of 10%. In addition to prevention of defects of the conventional putty, such as peeling and poor resistance to water and weathering, it should also adapt to the layer-by-layer change in and flexible release of dynamic stress and strain, in order to act as an ideal supporting material for the external thermal insulation paint finish system.

9.2 Analysis of and Study on Performance of Insulation Board

The insulation board for heat preservation and isolation is an important part of the external thermal insulation system. Thin-plastered EPS boards have been applied the most extensively in

Putty Properties Corresponding to Different Cement Contents (emulsion content: 20%) Table 9-1-14

Cement content, %	20	25	30	35
Bonding strength, MPa	0.33	0.43	0.49	0.47
Flexibility	Conforming	Conforming	Conforming	Nonconforming

Putty Properties Corresponding to Different Cement Contents (emulsion content: 30%) Table 9-1-15

Cement content, %	20	25	30	35
Bonding strength, MPa	0.48	0.61	0.71	0.69
Flexibility	Conforming	Conforming	Conforming	Nonconforming

Putty Properties Corresponding to Different Emulsion Contents Table 9-1-16

Emulsion content, %	20	25	30	35
Bonding strength, MPa	0.46	0.54	0.67	0.73
Flexibility	Nonconforming	Conforming	Conforming	Conforming

Europe and America, where the board forming and processing technology, insulation board pasting and reinforcement technology as well as supporting materials have been studied profoundly. After introduction into China, EPS boards are mostly applied in construction projects. Subsequently, organic boards such as XPS boards, polyurethane composite boards and phenolic insulation boards were developed, which have excellent insulation performance, but their other performance indicators are far different from those of EPS boards. Rock wool boards with Class A combustibility are mainly used in high-rise buildings, public buildings or certain wall structures. Insulation boards on an outer wall are subject to repeated effects of sudden cold/heat and freeze-thaw cycles, water vapor and other factors, with the insulation material expanding or contracting at the same time. The thin-plastered EPS board practice and its supporting materials are adopted during construction with other types of insulation board in thermal insulation of buildings. Are the rules of changes in deformation of various boards the same as each other? How the protective and surface layers are affected? Does the weathering resistance of the thermal insulation system vary? Starting from the analysis of basic properties of insulation boards and the testing of insulation boards affected by the heat and humidity, this section discusses the deformation and stress change of insulation boards on walls, and provides a solution to improvement of the deformation stress of the insulation board system, to reduce losses arising from the weathering resistance of the thermal insulation system.

9.2.1 Analysis of Basic Properties of Insulation Board

The EPS board is a white object, which is formed in a mold by heating and pre-foaming expandable polystyrene beads containing a volatile liquid foaming agent, and has a fine closed-cell structure. It features light weight, good insulation performance, low water absorption capacity, high resistance to low temperature, good workability, as well as some elasticity.

The XPS board is a rigid board formed by continuous extrusion and foaming polystyrene resin through a special process, and has an independent closed bubble structure. It is a kind of insulation material featuring high compressive strength, low water absorption rate, moisture resistance, air-tightness, light weight, corrosion resistance, long service life, low thermal conductivity, etc.

The polyurethane composite board is a kind of composite insulation board prefabricated in a factory, with rigid polyurethane foam [containing polyurethane rigid foam (PUR) and polyisocyanurate rigid foam (PIR)] as a core material. Its six sides are coated with cement-based polymer mortar of some thickness. The polyurethane composite board is characterized by low thermal conductivity, high tensile and compressive strength, good corrosion resistance, and high stability of physical and chemical properties.

The phenolic insulation board is a kind of closed-cell rigid foamed plastic, made of phenolic foam containing phenolic resin, foaming agent, curing agent and other additives. Phenolic foam has the characteristics of fire resistance, low smoke and low thermal conductivity.

Rock wool is a kind of inorganic fiber, produced by high-temperature melting and fibrillation, with natural rocks as main raw materials, such as basalt, gabbro, dolomite, iron ore and bauxite. It is characterized by non-combustibility and good thermal insulation performance.

In recent years, a number of national and industrial specifications on insulation boards have been issued, involving great differences in specific indicators. As shown in Table 9-2-1, differences in basic technical indicators will affect the applications of insulation boards in engineering. In particular, the external thermal insulation system is applied in a complex environment, so attention should be paid to thermodynamic indicators of boards. The raw materials used in board formation also play an essential role.

Comparison of Basic Properties of Insulation Materials **Table 9-2-1**

Item	Unit	EPS Board	XPS Board	Polyurethane Composite Board	Phenolic Board	Rock Wool Board	Adhesive polystyrene granule Insulation Mortar
Apparent density	kg/m^3	≥18	22-35	≥32	≥45	≥140	250～350
	Multiple	1	1.94	1.78	2.5	7.78	19.4
Thermal conductivity	W/(m·K)	0.039	0.03	0.024	0.033	0.04	0.06
	Thermal resistance multiple	1	1.3	1.625	1.18	0.975	0.65
Tensile strength perpendicular to board surface	MPa	≥0.10	≥0.20	≥0.10	≥0.08	≥0.0075	—
	Multiple	1	2	1	0.8	0.075	—
Compressive strength	kPa	≥100	≥200	≥150	≥100	≥40	≥300
	Multiple	1	2	1.5	1	0.4	3
Bending deformation	mm	≥20	≥20	≥6.5	≥4.0	—	—
	Multiple	1	1	0.325	0.2	—	—
Dimensional stability(70℃,2d)	%	≤0.3	≤1.0	≤1.0	≤1.0	≤1.0	≤0.3
	Multiple	1	3.33	3.33	3.33	3.33	1
Linear expansion coefficient	mm/(m·K)	0.06	0.07	0.09	0.08	—	—
	Multiple	1	1.17	1.5	1.33	—	—
Heat storage coefficient	W/(m^2·K)	0.36	0.54	0.27	0.36	0.75	0.95
	Multiple	1	1.5	0.75	1	2.1	2.6
Water absorption rate	%	≤3	≤1.5	≤3	≤7.5	—	—
	Multiple	1	0.5	1	2.5	—	—
Vapor permeability coefficient	ng/(Pa·m·s)	4.5	3.5	6.5	8.5	10	20
	Multiple	1	0.78	1.44	1.9	2.2	4.5
Elastic modulus	MPa	9.1	20	26	16.4	—	100
	Multiple	1	2.2	2.87	1.8	—	11
Poisson's ratio	—	0.1	0.28	0.42	0.24	—	—
Combustibility	—	Class C(B_1)or above		A		A	

Note: The multiple is a ratio to the corresponding indicator of the EPS board.

(1) Apparent density

There is a large difference in the apparent density of various insulation boards. The apparent density of the XPS board and polyurethane composite board is nearly twice that of the EPS board, and that of the phenolic board is 2.5 times, and that of rock wool is 7.78 times. The apparent density is an important factor affecting the thermal conductivity, compressive strength, bending deformation, etc.

The thermal conductivity of the EPS board drops with its apparent density rising. Both the porous structure of the EPS board and its application practice show that there is correlation between the apparent density and thermal conductivity within the optimal density range. When the apparent density of the EPS board is 18～25kg/m^3, its compressive strength is 70～150kPa.

When the apparent density of the polyurethane composite board is around 40kg/m^3, its thermal conductivity is the lowest. The insufficient apparent density will directly lead to a decline in physical properties, such as the strength and dimensional stability.

If the apparent density is too large, appropriate brackets should be applied during construction.

(2) Thermal conductivity

The polyurethane composite board has the lowest thermal conductivity. With the same energy-saving indicators, the board thickness decreases according to the following sequence: rock wool board, phenolic board, EPS board, XPS board, and polyurethane composite board. The greater the thermal resistance of an insulation material, the more slowly the heat is transferred to the inside. In this case, the heat will be excessively concentrated in the external surface, the temperature difference between both sides of the board and the deformation of the board itself will increase, and the cracking resistance of the protective surface will be more demanding.

(3) Compressive strength

The compressive strength refers to the compressive stress of a specimen during deformation of 10%. The compressive strength of the XPS board is twice that of the EPS board. When the same deformation occurs, the outer protective mortar needs to meet higher requirements with the board stress rising.

(4) Bending deformation

The principle of the bending deformation test is shown in Figure 9-2-1. The load was applied from the load head to specimens on two supports at a specific speed. It should be perpendicular to the specimens and at the center of two fulcrums. The load values corresponding to the specified deformation or the breaking load values of specimens were recorded. The bending deformation of the polyurethane composite board is 32.5% of that of the EPS board, and that of the phenolic board is only 20%. The bending deformation of the polyurethane composite board and phenolic board is far less than that of EPS and XPS boards. This indicates both the polyurethane composite board and the phenolic board are brittle and less flexible, and their ability of absorb internal stress and release deformation is far worse than that of the EPS board. When applied in the external thermal insulation system, they are more likely to crack and fall off.

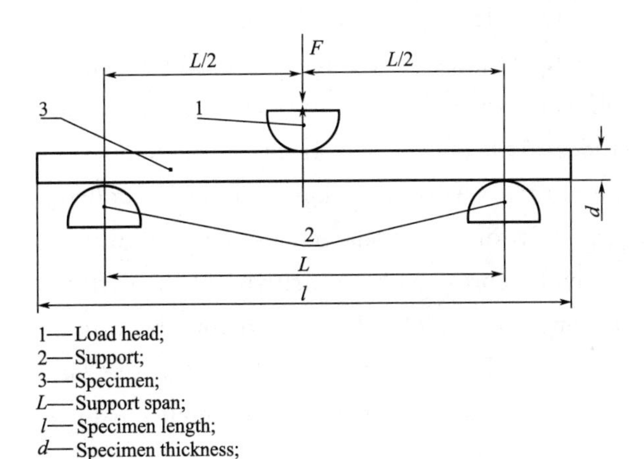

Figure 9-2-1 Schematic Diagram of Bending Deformation Test

(5) Dimensional stability

In the 48h test at 70℃, the dimensional change rate of the EPS board is 0.3%. The dimensional change rate of other insulation boards is 3.3 times that of the EPS board. That is, under the laboratory conditions, when the deformation of the EPS board is 3mm, that of other boards will be 10mm, so the volume of the EPS board is more stable in the presence of heat. Boards subject to significant thermal deformation are more prone to cracking under sudden hot and cold conditions.

(6) Linear expansion coefficient

The EPS board has the smallest linear expansion coefficient, while the polyurethane composite board has the largest. The linear expansion coefficient of the XPS board is 1.17 times that of the EPS board.

(7) Elastic modulus

The elastic modulus can be regarded as an indicator measuring the difficulty of a material in elastic deformation. The greater the elastic modulus, the larger stress is generated by elastic deformation. That is, the higher the stiffness, the less the elastic deformation is under the given stress conditions. The elastic modulus of the XPS board is 2.2 times that of the EPS board, and that of the polyurethane composite board is 2.87 times.

(8) Heat storage coefficient

In the thermodynamics, the heat storage coefficient refers to the ratio of the heat flow amplitude (A_q) of the wavefront to its temperature amplitude (A_f), with harmonic heat applied on one side of one homogenous semi-infinite wall body. It represents the sensitivity of the surface of a semi-infinite object to harmonic heat. In the presence of the same harmonic heat, the larger the heat storage coefficient, the small the surface temperature fluctuation is.

The polyurethane composite board has the smallest heat storage coefficient, which is only 0.75 times that of the EPS board. The heat storage coefficient of the rock wool board is 2.1 times that of the EPS board. The surface temperature of the polyurethane composite board fluctuates the most.

(9) Water absorption rate

The difference in the water absorption capacity is mainly determined by the chemical composition and internal structure of a material. (a) Polystyrene itself is not hygroscopic, so it absorbs only a small amount of water when soaked in water. The honeycomb walls of EPS granules are impermeable to water. Water can only infiltrate into the foamed plastic through micro-channels among the molten honeycombs. Therefore, the water absorption rate of the EPS board depends on the melting properties of raw materials during processing. The better the granules are molten, the larger the resistance to vapor diffusion and the smaller the water absorption rate is. (b) The XPS board with a closed-cell honeycomb structure has very low water absorption capacity. (c) The rigid polyurethane foamed plastic has a dense microcellular foamed structure with a large closed cell ratio, and its dense surface exposed to the air is not water-permeable. The water absorption rate of the polyurethane composite plate is close to that of the EPS board. (d) The phenolic board has a high water absorption rate because of the chemical composition and porosity of phenolic resin. The water absorption rate of the phenolic board is 2.5 times that of the EPS board. After water absorption and drying, the quality of the phenolic board will decline, which has great impact on its compressive strength.

The effect of the water absorption rate on the thermal conductivity is that the latter gradually rises with more and more water absorbed. The more water is absorbed, the lower the insulation performance is.

The water absorption capacity greatly affects the construction of insulation boards and performance indicators of supporting materials. The special interface treatment agent needs to be applied on the surface of the XPS board during construction, in order to facilitate bonding.

(10) Vapor permeability coefficient

The external thermal insulation system is required not only to resist the intrusion of external liquid water, but also to facilitate the transfer of water vapor. The vapor permeability coefficient of the XPS board is only 0.78 times that of the EPS board, which is not conducive to the outward transfer of water vapor.

(11) Tensile strength

The tensile strength perpendicular to the board surface indicates the degree of softness of a material. The XPS board has the highest tensile strength, equal to twice that of the EPS board, while the rock wool board has the lowest. Taking into account the impact of wind pressure in the structural design of the rock wool board insulation system, mechanical anchorage should be adopted. The phenolic board of low surface strength is prone to efflorescence, so appropriate surface treatment for reinforcement is needed prior to construction.

According to the above analysis, the properties of various insulation boards greatly differ from each other, especially their thermal deformation. The direct use of the thin-plastered EPS board structure and supporting materials may lead to major technical risks in the building insulation engineering. It is necessary to analyze the thermal stability of insulation boards and select more structures and supporting materials.

9.2.2 Study on Insulation Board Deformation

A number of studies have been conducted on the properties of insulation boards under the heat and humidity conditions in the literature, which provides a reference for insulation board applications.

9.2.2.1 Short-term Insulation Board Deformation at High Temperature

The deformation of the EPS board, XPS board, polyurethane composite board and phenolic board was tested by means of 48h hot curing at 70℃ in an oven. These insulation materials were cut into prisms of 40mm×40mm×160mm, with copper connectors bonded at two ends. The directions of specimens were marked. The initial length and 48h deformation were tested under the laboratory conditions.

It can be seen from Table 9-2-2 that the EPS boards shrank in a short time when heated. Their length changed a little, but their volumes were stable. The EPS boards were not significantly affected by the humidity.

The length and volume of the XPS board do not change regularly. The XPS board shrinks in the length direction. The volume of the XPS board 1 changed the most, but its length was not available, as shown in Figure 9-2-2. The XPS boards 2 and 3 shrank when heated, their length change rates are more than twice that of the EPS board, but their volumes changed steadily. The volume of the XPS board 2 changed much under the heat and humidity effects, as shown in Figure 9-2-3.

Figure 9-2-2 Length and Volume Changes of XPS Board 1 at 70℃

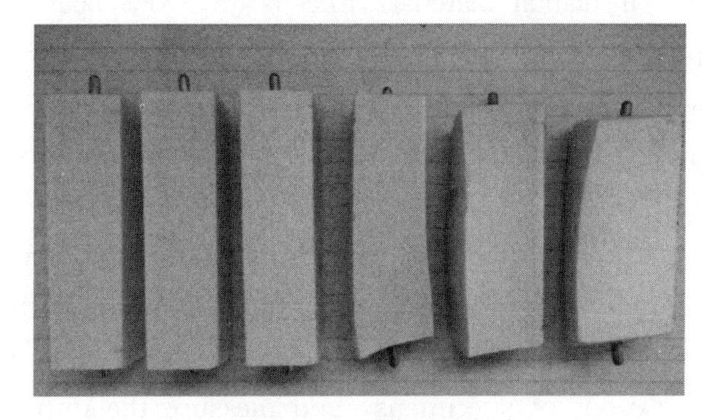

Figure 9-2-3 Length and Volume Changes of XPS Board 2 at 70℃

The shrinkage rate of the polyurethane composite board was 0.93% when heated, which was

Length Changes of Insulation Board at 70℃ in 48h **Table 9-2-2**

Insulation Board	Length Change(mm) at 70℃ in 48h	Length Change Rate (%)at 70℃ in 48h	Length Change(mm) at 70℃ in 48h (humidity:50%)	Length Change Rate (%)at 70℃ in 48h (humidity:50%)
EPS board	−0.4	−0.25	−0.33	−0.2
XPS board 1	No value	—	No value	—
XPS board 2	−0.6	−0.375	No value	—
XPS board 3	−0.8	−0.5	−1.04	−0.65
Polyurethane composite board	−1.48	−0.93	−0.24	−0.15
Phenolic board	−2.60	−1.6	−1.26	−0.79

3.72 times that of the EPS board. Under the heat and humidity effects, the length change rate of the polyurethane composite board was 0.15%. Its length changed steadily, but its volume changed little.

The shrinkage rate of the phenolic board was 1.6% when heated, which is 6.4 times that of the EPS board. Under the heat and humidity effects, the length change rate of the phenolic board was 0.79%, which was 3.8 times that of the EPS board. The volume of the phenolic board changed little.

9.2.2.2 Study on Long-term Thermal Deformation of Insulation Board

The technical center of Wacker Chemie AG conducted experiments on the deformation of insulation boards from different manufacturers under high-temperature conditions as well as high-temperature and high-humidity conditions.

Insulation material: EPS board, XPS board, polyurethane composite board, phenolic board and rock wool board. Curing: in an oven (70℃) and a box of constant temperature and humidity (70℃/RH95%).

Test procedures: Cut insulation materials into prisms of 40mm × 40mm × 160mm, with copper connectors bonded at two ends. Mark the directions of specimens, and measure the initial length and weight under the laboratory conditions. Put the specimens into the oven and box of constant temperature and humidity, and test their 28d deformation.

As can be seen from Figure 9-2-4:

(1) Under curing conditions of dry heating, the EPS boards from different manufacturers undergo the most stable deformation, the smallest change in length and the relatively stable quality.

(2) The deformations of polyurethane insulation boards from different manufacturers vary greatly. Two specimens were elongated significantly, and one specimen shrank.

(3) The XPS boards from different manufacturers change little but their deformations are not the same.

(4) The phenolic boards from different manufacturers shrink greatly.

(5) The deformation of rock wool boards is close to that of EPS boards, but the former is deformed steadily.

Insulation boards are greatly different in deformation under dry heating conditions, and separately subject to various volume deformations.

As shown in Figure 9-2-5:

(1) The EPS boards from different manufacturers are deformed little.

(2) The polyurethane insulation boards are elongated the most. There are differences in the elongation of polyurethane insulation boards from different manufacturers. Humidity has significant effects on the deformation of polyurethane composite boards.

(3) The deformations of XPS boards vary from each other, and some specimens are significantly affected by the humidity.

(4) The deformation of phenolic boards is obviously larger than that of EPS boards, and opposite to that in the heated state.

(5) The deformations of rock wool boards vary greatly. Changes in some specimens are opposite to those in the heated state, with obvious shrinkage.

It can be concluded that there are significant differences in the deformation of insulation boards under hot and humid conditions. Polyurethane composite boards, XPS boards and phenolic boards are greatly affected by humidity.

The above analysis shows significant differences in the deformation of insulation board under high temperature and humidity.

9.2.2.3 Determination of Thermal Expansion Coefficient of Insulation Board

Li Wenbo, et al. from Beijing University of Technology measured the thermal expansion coefficients of insulation materials. It is thought that there are differences in the thermal strain of mate-

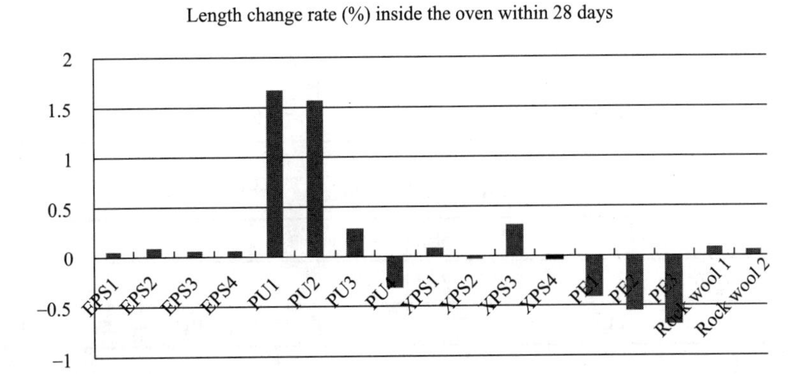

Figure 9-2-4 Length Change Rate of Boards under Dry Heating Conditions

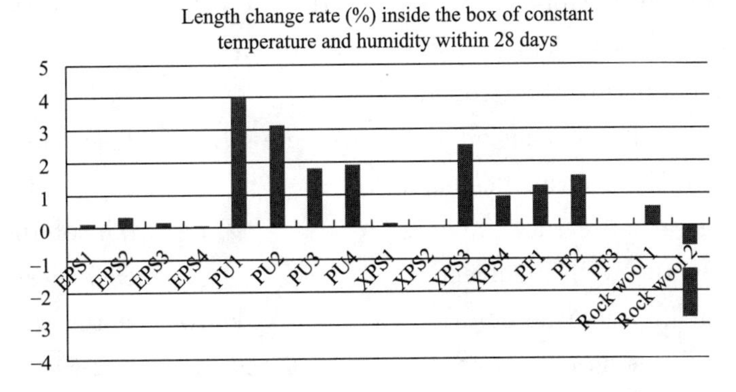

Figure 9-2-5 Length Change Rates of Boards from Different Manufacturers under Constant Temperature and Humidity

rials of an external thermal insulation system, stress is generated between materials of the external thermal insulation system with the temperature changing, and cracks are likely to occur in a long term, influencing the durability of the insulation system.

The strain test was conducted under the complicated conditions (including water spraying) to the XPS and EPS boards involved in the weathering resistance test, with the temperature ranging from 20℃ to 70℃.

As shown in the data of 10 groups of EPS boards in Table 9-2-3, the zero point of the surface strain of the EPS board fluctuates greatly at 20℃, with an average of 24$\mu\varepsilon$. When the temperature is 70℃, the peak strain is 2, 670$\mu\varepsilon$, and the average also reaches 2, 340$\mu\varepsilon$. This is associated with the properties of EPS boards, namely, low density and good insulation performance. The volumes of EPS boards change obviously, affected by the temperature.

It can be seen from the data of 10 groups of XPS boards in Table 9-2-4 that the strain of XPS boards is the most drastic among all test materials, with the peak value of 4, 571$\mu\varepsilon$. The average strain of XPS boards is 4, 036$\mu\varepsilon$ at 70℃, approximately twice that of EPS boards. With the ambient temperature fluctuating greatly, the volumes of XPS boards will undergo major changes. When the XPS board is used as a wall insulation material, such changes must be taken into account to select the appropriate insulation structure and supporting materials.

The test results show that the volumes of XPS and EPS boards change significantly with the temperature fluctuating. As the mortar of the protective layer is directly exposed to the insulation material and there are great differences in their thermal strain under temperature changes, the wall without effective measures is bound to crack,

Strain of EPS Board **Table 9-2-3**

Temperature	Strain($\mu\varepsilon$)										Average
20℃	−78	−53	−11	10	23	40	60	62	77	109	24
70℃	2324	2510	2555	1999	2411	2670	1932	2503	2388	2109	2340

Strain of XPS Board **Table 9-2-4**

Temperature	Strain($\mu\varepsilon$)										Average
20℃	−120	−76	−55	3	12	42	45	95	126	159	23.1
70℃	3480	4027	4571	4329	3875	4262	3553	4085	3961	4221	4036

affecting the normal use of the external thermal insulation system.

By analyzing the above test data, it can be concluded that insulation boards undergo significant expansion and contraction with the temperature and humidity changing. The expansion and contraction of EPS boards are the smallest and most stable. There are great differences among other insulation boards. As the wall is subject to complicated environmental impact, the deformation of insulation boards will be more complex.

9.2.3 Study on Deformation and Stress of Thermal Insulation System

9.2.3.1 Deformation of Insulation Board on Wall

Given that the energy-saving ratio of high-rise buildings is 75% and the concrete wall thickness is 180mm, the thickness of four types of insulation board was calculated, separately (Table 9-2-5). The anti-cracking protective layer is 3mm.

The elongation and stress of the external surface of each insulation board (1, 200mm × 600mm) were calculated at the low temperature of 20℃ and wall temperature of 70℃, respectively.

The elongation ΔL of the external surface of the insulation board:

$$\Delta L=\alpha\ (t_2-t_1)\ L$$

Where, α-thermal expansion coefficient of the insulation board, mm/ (m · K);

t_1 and t_2-surface temperature changes of the insulation board, ℃;

L—length of the insulation board, m.

With the elongation restricted, the thermal stress (σ) of the surface of the insulation board is:

$$\sigma=E\varepsilon$$

Where, E —elastic modulus of the insulation board, MPa;

ε—strain of the insulation board, dimensionless.

As shown by the calculation results (Table 9-2-6 and Figure 9-2-6):

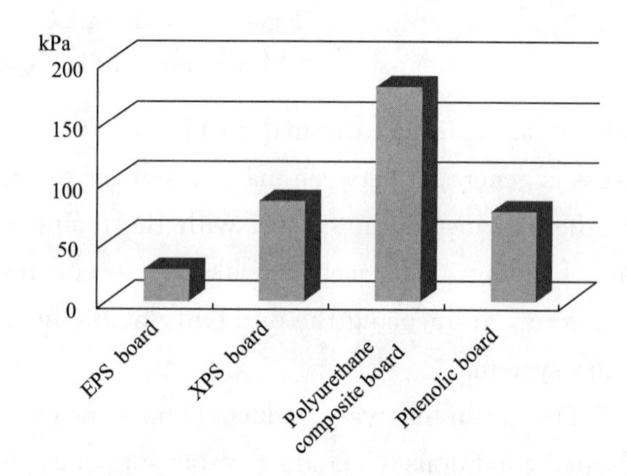

Figure 9-2-6 Histogram of Comparison of Deformation and Stress Estimates of Insulation Boards on Walls

The surface deformation of the polyurethane composite board is the largest (equal to 1.5 times that of the EPS board), and that of the XPS board and phenolic board is in between. The maximum thermal stress of the surface of the polyurethane composite board is 4.3 times that of the EPS board, that of the XPS board is 2.2 times, and that of the phenolic board is 2.4 times. If needed, the use of the XPS board, polyurethane composite board and phenolic board should be stu-

Thickness of Insulation Layer Table 9-2-5

Material	EPS Board	XPS Board	Polyurethane Composite Board	Phenolic Board
Thickness, mm	80	70	55	80

Comparison of Deformation and Stress Estimates of Insulation Boards on Walls Table 9-2-6

Insulation Board		EPS Board	XPS Board	Polyurethane Composite Board	Phenolic Board
Elongation	mm	3.6	4.2	5.4	4.8
	Multiple	1	1.2	1.5	1.3
Stress	kPa	27.3	60	117	65.6
	Multiple	1	2.2	4.3	2.4

died, and appropriate supporting materials should be added.

9.2.3.2 Stress between Insulation Board and Mortar

Since the thermal expansion coefficients of the insulation board and surface mortar are different, the thermal stress may be generated due to mutual constraints in the event of the volume contraction and expansion with the humidity and temperature changes. Insulation boards have difference linear expansion coefficients. When one insulation structure is adopted, the stresses between insulation boards and mortar can be compared by means of simplified mathematical calculation.

It is assumed that there is no stress when the free elongation is L_1 and L_2 ($L_1 \geqslant L_2$).

$$L_1 = \Delta T \cdot a_1$$

$$L_2 = \Delta T \cdot a_2$$

The final length is L ($L_1 > L > L_2$) if the insulation board and mortar are in contact with each other.

According to the mechanical equilibrium, the thermal stresses of the insulation board and mortar are equal to each other.

$$F_1 = (L_1 - L) \cdot E_1/(1-\gamma_1)$$

$$F_2 = (L - L_2) \cdot E_2/(1-\gamma_2)$$

$$F_1 = F_2$$

$$(L - L_2) \cdot E_2/(1-\gamma_1) = (L_1 - L) \cdot E_1/(1-\gamma_2)$$

Then:

$$L = \frac{L_1E_1(1-\gamma_2) + L_2E_2(1-\gamma_1)}{E_1(1-\gamma_2) + E_2(1-\gamma_1)}$$

Substitute them into the formula of mechanics:

$$F_1 = (L_1 - L)E_1/(1-\gamma_1)$$

$$= \Delta T \Delta a \frac{E_1E_2}{E_1(1-\gamma_2) + E_2(1-\gamma_1)}$$

Where, F—thermal stress, kPa;

ΔT—temperature variation, ℃;

E_1 and E_2—elastic modulus, MPa;

Δa—difference in the thermal expansion coefficient, mm/(m · K);

γ_1 and γ_2—Poisson's ratio, dimensionless.

(1) EPS board and surface mortar

The thermal expansion coefficient of the EPS board is 6×10^{-5}/℃, while that of the surface mortar is 1×10^{-5}/℃. The difference of both thermal expansion coefficients is $\Delta\alpha = 5 \times 10^{-5}$/℃. Given that the temperature change is $\Delta T = 50$℃, the elastic modulus of cement mortar is $E = 6$GPa, and the Poisson's ratio of cement mortar is 0.28, the average thermal stress of cement mortar is:

$$F = 25.3\text{kPa}$$

(2) XPS board and surface mortar

The thermal expansion coefficient of the XPS board is 7×10^{-5}/℃, while that of the surface mortar is 1×10^{-5}/℃. The difference of both thermal expansion coefficients is $\Delta\alpha = 6 \times 10^{-5}$/℃. Given that the temperature change is $\Delta T = 50$℃, the elastic modulus of cement mortar is $E = 6$GPa, and the Poisson's ratio of cement mortar is 0.28, the average thermal stress of cement mortar is:

$$F=83.1\text{kPa}$$

(3) Polyurethane composite board and surface mortar

The thermal expansion coefficient of the polyurethane composite board is 9×10^{-5}/℃, while that of the surface mortar is 1×10^{-5}/℃. The difference of both thermal expansion coefficients is $\Delta\alpha=8\times10^{-5}$/℃. Given that the temperature change is $\Delta T=50$℃, the elastic modulus of cement mortar is $E=6$GPa, and the Poisson's ratio of cement mortar is 0.28, the average thermal stress of cement mortar is:

$$F=178.4\text{kPa}$$

(4) Phenolic board and surface mortar

The thermal expansion coefficient of the phenolic board is 8×10^{-5}/℃, while that of the surface mortar is 1×10^{-5}/℃. The difference of both thermal expansion coefficients is $\Delta\alpha=7\times10^{-5}$/℃. Given that the temperature change is $\Delta T=50$℃, the elastic modulus of cement mortar is $E=$ 6GPa, and the Poisson's ratio of cement mortar is 0.28, the average thermal stress of cement mortar is:

$$F=75.3\text{kPa}$$

The above calculations (Figure 9-2-7) show that the stress varies greatly between different insulation boards and surface mortar. Due to the heat preservation and isolation effects of the insulation layer of the external thermal insulation system, the temperature of the part outside the insulation layer is too high or low in summer and winter. Thus, the thermal expansion coefficients of adjacent materials should not be quite different from each other. Otherwise, excessive thermal stress, hollowing, cracking and falling may be caused.

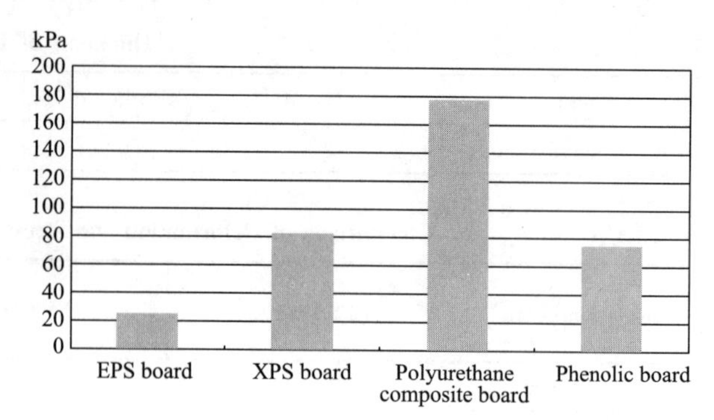

Figure 9-2-7 Stress between Insulation Board and Mortar

9.2.3.3 Impact of Transition Layer on Deformation of Insulation Board

The transition layer practice is stipulated in the *Products for External Thermal Insulation Systems based on Mineral Binder and Expanded Polystyrene Granule Plaster* (JG/T 158-2013) and Beijing's local standard Technical Specification for External Thermal Insulation Construction based on Rigid Polyurethane Foam Composite Board with Plastering Lightweight Mortar (DB11/T 1080-2014). If the adhesive polystyrene granule insulation mortar is used as a leveling transition layer, the surface temperature change of the insulation board should be calculated.

(1) Heat transfer principle

Calculation basis: the principle of steady and periodic non-steady heat transfer of building thermal engineering.

During the steady heat transfer, the amount of heat transferred is closely related to the effect of temperature difference, thermal conductivity of a material and heat transfer resistance of a structure. The heat flow density (Q) of a wall is:

$$Q=K\Delta T=K\ (t_e-t_i)\ =\frac{t_e-t_i}{R_i+\Sigma R_j+R_e}$$

Where,

$$K=\frac{1}{R_0}=\frac{1}{R_i+\Sigma R_j+R_e}$$

Heat flow density (Q_m) of the outer interface of one layer:

$$Q_m=\frac{t_m-t_i}{R_i+\Sigma_{j=0}^{m-1}R_j}$$

According to the principle that the heat flow density values of external surfaces of various layers are equal to each other, i.e. $Q=Q_m$:

$$t_m=t_i-\frac{R_i+\Sigma_{j=0}^{m-1}R_j}{R_0}\ (t_i-t_e)$$

$$(m=1,\ 2,\ \cdots,\ n+1)$$

Where, R_0—total thermal resistance of the outer wall, $m^2 \cdot K/W$;

R_i— thermal resistance of the internal surface of the outer wall, $m^2 \cdot K/W$;

R_j—thermal resistance of the material of the jth layer, $m^2 \cdot K/W$;

$\sum_{j=0}^{m-1} R_j$—sum of thermal resistance from the first to $(m\text{-}1)^{th}$ layer, calculated from indoors.

(2) Principle of periodic non-steady heat transfer

The periodic change in the external thermal effect over time is:

$$t_\tau = \bar{t} + A_t \cos\left(\frac{360\tau}{Z} - \phi\right)$$

Where, t_τ—medium temperature at the time τ, ℃;

$\bar{t}$— average temperature in one period, ℃;

τ— calculation time from the given time (such as zero o'clock in a day), h;

ϕ— initial phase of the temperature wave, deg. Given that the coordinate origin is the maximum temperature, $\phi=0$.

During the periodic heat transfer in the presence of harmonic heat, the material and the heat storage coefficient and thermal inertia of the material layer are involved.

In the building thermal engineering, the ratio of the outdoor temperature amplitude (A_e) to the internal surface temperature amplitude (A_{if}) of a flat wall under the effects of external harmonic heat is called the penetration attenuation of a temperature wave, hereinafter simply referred to as the total attenuation, expressed as ν_0:

$$\nu_0 = A_e / A_{if}$$

The attenuation of a temperature wave is exponential to the thermal inertia index of a material.

$$V_x = A_\theta / A_x = e^{\frac{D}{\sqrt{2}}}$$

The attenuation multiple is a ratio of the amplitude of harmonics of the outdoor air temperature to that of harmonics of the inner surface temperature of a flat wall, calculated from:

$$\nu_0 = 0.9 e^{\frac{\Sigma D}{\sqrt{2}}} \cdot \frac{S_1+\alpha_i}{S_1+Y_{1,e}} \cdot \frac{S_2+Y_{1,e}}{S_2+Y_{2,e}} \cdot \cdots \frac{S_n+Y_{n-1,e}}{S_n+Y_{n,e}} \cdot \frac{\alpha_e+Y_{n,e}}{\alpha_e}$$

Where, ΣD —total thermal inertia index of the flat wall, equal to the sum of thermal inertia indexes of all material layers;

S_1, S_2, …, S_n—heat storage coefficient of each material layer, $W/(m^2 \cdot K)$;

$Y_{1,e}$, $Y_{2,e}$, …, $Y_{n,e}$—heat storage coefficient of the external surface of each layer, $W/(m^2 \cdot K)$;

α_i—heat transfer coefficient of the internal surface of the flat wall, $W/(m^2 \cdot K)$;

α_e—heat transfer coefficient of the external surface of the flat wall, $W/(m^2 \cdot K)$;

e— base of the natural logarithm, e=2.718.

It is customary to use the delay time (ξ_o) to evaluate the thermal stability of an envelope. The delay time can be obtained based on the transformation relationship between the time and phase angle. The attenuation of a temperature wave is exponential to the thermal inertia index of a material.

$$V_x = A_\theta / A_x = e^{\frac{D}{\sqrt{2}}}$$

The attenuation multiple is a ratio of the amplitude of harmonics of the outdoor air temperature to that of harmonics of the inner surface temperature of a flat wall, calculated from:

$$\nu_0 = 0.9 e^{\frac{\Sigma D}{\sqrt{2}}} \cdot \frac{S_1+\alpha_i}{S_1+Y_{1,e}} \cdot \frac{S_2+Y_{1,e}}{S_2+Y_{2,e}} \cdot \cdots \frac{S_n+Y_{n-1,e}}{S_n+Y_{n,e}} \cdot \frac{\alpha_e+Y_{n,e}}{\alpha_e}$$

Where, ΣD—total thermal inertia index of the flat wall, equal to the sum of thermal inertia indexes of all material layers;

α_i—heat transfer coefficient of the internal surface of the flat wall, W/(m^2 · K);

$Y_{1,e}$, $Y_{2,e}$, …, $Y_{n,e}$—heat storage coefficient of the external surface of each layer, W/(m^2 · K);

α_i—heat transfer coefficient of the internal surface of the flat wall, W/(m^2 · K);

α_e—heat transfer coefficient of the external surface of the flat wall, W/(m^2 · K);

e— base of the natural logarithm, e=2.718.

$$\xi_0=\frac{1}{15}\left(40.5\sum D+\arctan\frac{Y_{ef}}{Y_{ef}+\alpha_e\sqrt{2}}-\arctan\frac{\alpha_i}{\alpha_i+Y_{if}\sqrt{2}}\right)$$

Where, Y_{ef}—heat storage coefficient of the external surface of the outer wall, W/(m^2 · K);

Y_{if}—heat storage coefficient of the internal surface of the outer wall, W/(m^2 · K).

(3) Calculation of temperature variation of insulation board of insulation mortar leveling layer

The calculation is based on the stable indoor temperature (t_i=20℃) in the north in summer, the surface temperature (t_e=70℃) of the dark finish, and the average temperature (28℃) of the surface of the outer wall.

Refer to Table 9-2-7 and 9-2-8 for the structure and relevant parameters of the wall with a thin plaster on the external surface of the insulation board in theadhesive polystyrene granule insulation mortar leveling layer.

When a 20mm transition layer of adhesive polystyrene granule insulation mortar is applied on the insulation board, the surface reaches the maximum temperature (60.6℃) after 1.15h; and when a 30mm transition layer is applied, the surface reaches the maximum temperature (57℃) after 1.52h (Table 9-2-9). It can be that the surface temperature and thermal stress drop sharply,

Construction Method of External Surface of EPS Board in 20mm Adhesive polystyrene granule Leveling Layer

Table 9-2-7

Structure	Thickness δ (mm)	Thermal Conductivity λ [W/(m · K)]	Thermal Resistance R(m^2 · K/W)	Heat Storage Coefficient S W/(m^2 · K)	Thermal Inertia Index D	Heat Storage Coefficient Y of External Surface
Reinforced concrete	180	1.740	0.10	17.06	1.76	17.06
EPS board	60	0.039	1.47	0.36	0.58	0.66
Insulation mortar	20	0.070	0.23	0.95	0.27	0.75
Anti-cracking mortar	3	0.93	0.003	11.37	0.04	1.14
Total	263	—	1.96	—	2.63	—

Construction Method of External Surface of EPS Board in 30mm Adhesive polystyrene granule Leveling Layer

Table 9-2-8

Structure	Thickness δ (mm)	Thermal Conductivity λ [W/(m · K)]	Thermal Resistance R(m^2 · K/W)	Heat Storage Coefficient S W/(m^2 · K)	Thermal Inertia Index D	Heat Storage Coefficient Y of External Surface
Reinforced concrete	180	1.740	0.10	17.06	1.76	17.06
EPS board	60	0.039	1.54	0.36	0.55	0.63
Insulation mortar	30	0.070	0.43	0.95	0.41	0.80
Anti-cracking mortar	3	0.93	0.003	11.37	0.04	1.22
Total	273	—	2.07	—	2.76	—

and the occurrence of high temperature is delayed, which greatly relieve the stress arising from temperature variations and ensuring good weathering resistance of the external thermal insulation system.

9.2.4 Conclusions

(1) Through the comparative analysis of the properties of various insulation boards, it is found that the performance indicators of insulation boards are quite different, especially the thermal deformation. The direct use of the thin-plaster EPS board construction method and supporting materials in the building insulation engineering may lead to major technical risks.

(2) There are significant differences in the deformation of insulation boards affected by the temperature and humidity. Under dry heating conditions, various types of insulation boards vary in the deformation, and even one type of insulation board shows different performances. All insulation boards are significantly different in the deformation under hot and humid conditions. The polyurethane composite board, XPS board and phenolic board are greatly affected by humidity. When the temperature fluctuates drastically, the volume of the XPS board undergoes great changes, so the suitable insulation structure and supporting materials should be selected in thermal insulation projects.

(3) The insulation boards on walls differ greatly in the deformation and thermal stress of the surface. There are also significant variations of stress between insulation boards and surface mortar. Due to the heat preservation and isolation effects of the insulation layer, the temperature of the part outside the insulation is too high or low in summer and winter, so the thermal expansion coefficients of adjacent materials should not be greatly different.

(4) With a leveling transition layer of adhesive polystyrene granule mortar, the surface temperature of the insulation board will be decreased greatly, and the occurrence time of high temperature will be delayed, thus reducing the deformation and stress of the insulation board as well as the cracking possibility of the insulation surface, relieving sharp changes in the stress arising from temperature fluctuations, and achieving high weathering resistance of the external thermal insulation system.

9.3 Engineering Application Guide for Composite External Thermal Insulation Technology based on Adhesive polystyrene granule

9.3.1 Basic Requirements

The composite external thermal insulation system based on the adhesive polystyrene granule shall meet the following requirements:

(1) It shall adapt to normal deformation of the base course wall, without cracking or hollowing.

Changes in Temperature and Stress of External Surfaces of EPS Boards in Different Adhesive polystyrene granule Leveling Layers **Table 9-2-9**

Leveling Thickness (mm)	Average Temperature (℃) of External Surface of Insulation Board	Temperature Amplitude (℃)	Maximum Temperature (℃) of External Surface of Insulation Board	Temperature Drop (℃) Compared with that without Rendering	Temperature Drop Amplitude	Stress Drop Amplitude	Delay Time (h)
10	27.3	37.2	64.5	5.5	7.9%	11%	0.64
20	26.8	33.8	60.6	9.4	13.4%	18.8%	1.0
30	26.3	30.7	57.0	13.0	18.6%	26%	1.37

(2) It shall be able to withstand long-term repeated effects of the dead load, wind load and outdoor climate, without harmful deformation or damage.

(3) It shall be reliably connected to the base course wall, to avoid falling off during an earthquake.

(4) It shall be able to prevent flame spreading.

(5) It shall be impermeable for water.

The thermal insulation performance of the composite external thermal insulation system based on the adhesive polystyrene granule shall meet the provisions of national and industry standards as well as local energy conservation design standards. The moisture-proof design shall conform to the *Thermal Design Code for Civil Building* (GB 50176).

All components of the composite external thermal insulation system based on the adhesive polystyrene granule shall have physical-chemical stability. All constituent materials shall be compatible and corrosion-resistant. In the case of possible biological infestation (mouse, insects, etc.), the composite external thermal insulation system based on the adhesive polystyrene granule shall be resistant to biological intrusion.

The service life of the composite external thermal insulation system based on the adhesive polystyrene granule shall not be less than 25 years, subject to proper use and normal maintenance.

9.3.2 Performance Requirements

9.3.2.1 System

1. The general properties of the adhesive polystyrene granule composite system should meet the requirements of Table 9-3-1.

2. The reaction-to-fire performance of the adhesive polystyrene granule composite system should meet the requirements of Table 9-3-2.

9.3.2.2 Composition

1. Polystyrene boards are divided into the expanded polystyrene board (EPS board) and polystyrene board (XPS board), of which the performance should meet the requirements of Table 9-3-3. The specifications of polystyrene boards should be determined by the seller and buyer. Allowable deviations should meet the requirements of Table 9-3-4.

General Properties of Adhesive polystyrene granule Composite System **Table 9-3-1**

Test Item		Performance Indicator		Test Method
		Paint finish	Face brick finish	
Weathering resistance	Appearance	No seepage cracks, efflorescence, hollowing or peeling		JGJ 144
	Tensile bonding strength (MPa)	≥0.1	—	
	Tensile bonding strength (MPa) between face brick and anti-cracking layer	—	≥0.4	
Water absorption (after 24h soaking in water) (g/m^2)		≤500		
Impact resistance	2nd floor and above	Class 3J	—	
	1st floor	Class 10J		
Wet flow density [$g/(m^2 \cdot h)$] of vapor permeation		≥0.85		
Freeze-thaw resistance	Appearance	No seepage cracks, efflorescence, hollowing or peeling		
	Tensile bonding strength (MPa) between anti-cracking layer and insulation layer	≥0.1	—	
	Tensile bonding strength (MPa) between face brick and anti-cracking layer	—	≥0.4	
Water impermeability		No water permeation on the inner side of the anti-cracking layer		
Thermal resistance		Conform to design requirements		GB/T 13475

Requirements for Reaction-to-fire Performance of Adhesive polystyrene granule Composite System

Table 9-3-2

Building Height H(m)		Reaction-to-fire			
		Cone calorimeter test	Combustion shaft furnace test	Window fire test	
Residential building	Non-curtain-wall type public building	Peak heat release rate (kW/m^2)	Remaining length after specimen combustion (mm)	Maximum temperature of measuring point of insulation layer on horizontal baseline (℃)	Combustion area (m^2)
$H \geq 100$	$H \geq 50$	≤5	≥800	≤200	≤3
$60 \leq H < 100$	$24 \leq H < 50$	≤10	≥500	≤250	≤6
$24 \leq H < 60$	$H < 24$	≤25	≥350	≤300	≤9
$H < 24$	—	≤100	≥150	≤400	≤12
Test method		GB/T 16172	GB/T 8625	GB/T 29416	

Performance Indicators of Polystyrene Board **Table 9-3-3**

Item	Unit	Indicator		Test Method
		EPS board	XPS board	
Apparent density	kg/m^3	≥18	22-35	GB/T 6343
Thermal conductivity	W/(m·K)	≤0.039	≤0.030	GB/T 10295 or GB/T 10294
Tensile strength perpendicular to board surface	MPa	≥0.10	≥0.15	JG 149
Dimensional stability	%	≤0.3	≤1.2	GB/T 8811
Bending deformation	mm	≥20	≥20	GB/T 8812.1
Compressive strength	MPa	≥0.10	≥0.15	GB/T 8813
Water absorption ratio (V/V)	%	≤3	≤2	GB/T 8810
Oxygen index	%	≥30	≥26	GB/T 2406.1 and GB/T 2406.2
Combustibility	—	Class B_2 or above	Class B_2 or above	GB 8624

Allowable Dimensional Deviations (mm) of Polystyrene Board **Table 9-3-4**

Item		Allowable Deviation	Item	Allowable Deviation
Length and width	<1000	±3.0	Thickness	+1.5 and −0.0
	1000-2000	±5.0	Two diagonal lines	≤5.0
	2000-4000	±8.0	Edge flatness	≤3.0
	>4000	−10.0, without limitation to positive deviation	Surface flatness	≤2.0

2. The performance of spared rigid polyurethane foam should meet the requirements of Table 9-3-5.

3. The polyurethane composite board should be kept for at least 28d at room temperature before delivery, and its main performance indicators should meet the requirements of Table 9-3-6 and 9-3-7.

4. The performance of the rock wool board should meet the requirements of Table 9-3-8 and 9-3-9.

5. The performance of the reinforced vertical-fiber rock wool board should meet the requirements of Table 9-3-10.

6. The performance of the phenolic board should meet the requirements of Table 9-3-11 and 9-3-12.

Performance Indicators of Sprayed Rigid Polyurethane Foam **Table 9-3-5**

Item	Unit	Indicator	Test Method
Density	kg/m^3	≥30	GB/T 6343
Thermal conductivity	W/(m・K)	≤0.022	GB/T 10295 or GB/T 10294
Compressive strength	MPa	≥0.15	GB/T 8813
Tensile bonding strength (with cement mortar, under normal temperature)	MPa	≥0.10	GB 50404
Dimensional stability (70℃, 48h)	%	≤1.0	GB/T 8811
Oxygen index	%	≥26	GB/T 2406.1 GB/T 2406.2
Combustibility	—	Class B_2 or above	GB 8624
Water absorption rate (V/V)	%	≤3	GB/T 8810

Performance Indicators of Polyurethane Composite Board **Table 9-3-6**

Item		Indicator	Test Method
Core	Apparent density, kg/m^3	≥32	GB/T 6343
	Thermal conductivity (average temperature: 25℃), W/(m・K)	≤0.024	GB/T 10294 or GB/T 10295
	Dimensional stability, %(70±2℃)	≤1.0	GB/T 8811
	Water absorption rate, %	≤3	GB/T 8810
	Combustibility	Class B1	GB 8624
	Compressive strength (deformation: 10%), MPa	≥0.15	GB/T 8813
Composite board	Tensile strength perpendicular to board surface, MPa	≥0.10 (without damage to the interface)	JG 149
	Combustibility	The average remaining length of specimens in each group should not be less than 35cm (remaining length of each specimen: more than 20cm). The average peak temperature of flue gas in each test should be no higher than 125℃. Combustion must not occur to the backs of specimens	GB/T 8625

Allowable Dimensional Deviations **Table 9-3-7**

Item	Indicator	Test Method
Length, mm	±3.0	Steel tape with divisions of 1mm
Width, mm	±3.0	
Thickness, mm	+2.0	
Diagonal difference, mm	3.0	Guiding rule and wedge feeler
Edge flatness, mm/m	±2.0	
Surface flatness, mm/m	2.0	

Performance Indicators of Rock Wool Board **Table 9-3-8**

Item	Unit	Index	Test Method
Density	kg/m^3	≥150	GB/T 5480
Thermal conductivity	W/(m・K)	≤0.040	GB/T 10295 or GB/T 10294

Continued

Item		Unit	Index	Test Method
Tensile strength perpendicular to board surface		kPa	≥7.5	GB/T 25975
Right angle deviation		mm/m	≤5	GB/T 5480
Flatness deviation		mm	≤6	GB/T 25975
Dimensional stability		%	≤1.0	GB/T 8811
Acidity coefficient		—	≥1.6	GB/T 5480
Compressive strength		kPa	≥40	GB/T 13480
Water absorption rate		%	≤1.0	GB/T 5480
Water absorption (partial immersion)	Short term (24h)	kg/m^2	≤1.0	GB/T 25975
	Long term (28d)		≤3.0	
Hydrophobicity		%	≥98	GB/T 10299
Combustibility		—	Class A	GB 8624

Allowable Dimensional Deviations (mm) of Rock Wool Board **Table 9-3-9**

Allowable Length Deviation	Allowable Width Deviation	Allowable Thickness Deviation
+10	+5	+3
−3	−3	−3

Performance Indicators of Reinforced Vertical-fiber Rock Wool Board **Table 9-3-10**

Item	Unit	Indicator		Test Method
		Rock wool strip of core	Reinforced vertical-fiber rock wool board	
Density	kg/m^3	≥100	—	GB/T 5480
Thermal conductivity	W/(m·K)	≤0.043	—	GB/T 10295 or GB/T 10294
Tensile strength perpendicular to board surface	kPa	≥7.5	—	GB/T 25975
Short-term water absorption	kg/m^2	≤1.0	—	GB/T 25975
Combustibility	—	Class A	Class A	GB 8624
Rock wool direction	—	—	Parallel to the thickness direction	Visual inspection
Protective layer thickness	mm	—	2-4	Measurement with ruler
Right angle deviation	mm/m	—	≤10	GB/T 5480
Flatness deviation	mm	—	≤5	GB/T 5480
Tensile strength of surface	kPa	—	≥150	Appendix B of DB11/T 463
Impact resistance	J	—	≥2	JG 149
Hydrophobicity	%	—	≥98	GB/T 10299

Performance of Phenolic Board **Table 9-3-11**

Item	Unit	Performance Indicator	Test Method
Apparent density	kg/m^3	≥45	GB/T 6343
Thermal conductivity	W/(m·K)	≤0.032	GB/T 10294 or GB/T 10295
Tensile strength perpendicular to board surface	MPa	≥0.08	GB/T 29906
Water absorption rate (V/V)	%	≤5	GB/T 8810
Dimensional stability	%	≤1.0	GB/T 8811
Compressive strength (compressive deformation: 10%)	MPa	≥0.10	GB/T 8813
Combustibility	—	Class B_1	GB 8624

Allowable Dimensional Deviations of Phenolic Board **Table 9-3-12**

Item	Unit	Allowable Deviation
Length	mm	±3.0
Width	mm	±2.0
Thickness	mm	+1.5 and 0.0
Diagonal difference	mm	≤3.0
Edge flatness	mm/m	≤2.0
Surface flatness	mm/m	≤1.0

7. The performance of the adhesive polystyrene granule mortar should meet the requirements of Table 9-3-13.

8. The performance of interface mortar should meet the requirements of Table 9-3-14.

9. The performance of moisture-proof primer should meet the requirements of Table 9-3-15.

10. The performance of the insulation board adhesive should meet the requirements of Table 9-3-16.

11. The performance of the rock wool board adhesive should meet the requirements of Table 9-3-17.

12. The performance of the insulation board interface agent should meet the requirements of Table 9-3-18.

13. The performance of the interface agent for the rock wool board should meet the requirements of Table 9-3-19.

14. The performance of anti-cracking mortar should meet the requirements of Table 9-3-20.

15. Glass fiber meshes are divided into two types: ordinary and reinforced. Ordinary type glass fiber meshes are suitable for paint finishes, while reinforced type for face brick finishes. Their performance should meet the requirements of Table 9-3-21.

16. The performance of the elastic primer should meet the requirements of Table 9-3-22.

17. The performance of the waterproof flexibly putty should meet the requirements of Table 9-3-23.

Performance Indicators of Adhesive polystyrene granule Mortar **Table 9-3-13**

<table>
<tr><th colspan="3" rowspan="2">Item</th><th rowspan="2">Unit</th><th colspan="2">Indicator</th><th rowspan="2">Test Method</th></tr>
<tr><th>Insulation mortar</th><th>Pasting mortar</th></tr>
<tr><td colspan="3">Dry apparent density</td><td>kg/m³</td><td>180-250</td><td>250-350</td><td>JG 158</td></tr>
<tr><td colspan="3">Compressive strength</td><td>MPa</td><td>≥0.20</td><td>≥0.30</td><td>GB/T 5486</td></tr>
<tr><td colspan="3">Softening coefficient</td><td>—</td><td>≥0.5</td><td>≥0.6</td><td>JG 158</td></tr>
<tr><td colspan="3">Thermal conductivity</td><td>W/(m·K)</td><td>≤0.060</td><td>≤0.075</td><td>GB/T 10295 or GB/T 10294</td></tr>
<tr><td colspan="3">Line shrinkage rate</td><td>%</td><td>≤0.3</td><td>≤0.3</td><td>JGJ/T 70</td></tr>
<tr><td colspan="3">Tensile strength</td><td>MPa</td><td>≥0.1</td><td>≥0.12</td><td rowspan="5">JG 158</td></tr>
<tr><td rowspan="4">Tensile bonding strength</td><td rowspan="2">With cement mortar block</td><td>Standard state</td><td rowspan="4">MPa</td><td rowspan="2">≥0.1</td><td>≥0.12</td></tr>
<tr><td>Soaking treatment</td><td>≥0.10</td></tr>
<tr><td rowspan="2">With EPS board</td><td>Standard state</td><td rowspan="2">—</td><td>≥0.10</td></tr>
<tr><td>Soaking treatment</td><td>≥0.08</td></tr>
<tr><td colspan="3">Combustibility</td><td>—</td><td>Class B1 or above</td><td>Class A (minimum A2)</td><td>GB 8624</td></tr>
</table>

Performance Indicators of Interface Mortar **Table 9-3-14**

Item		Unit	Indicator	Test Method
Tensile bonding strength	Standard state	MPa	≥0.5	JC/T 907
	Soaking treatment		≥0.3	

Performance Indicators of Moisture-proof Primer **Table 9-3-15**

Item		Unit	Indicator	Test Method
Drying time	Surface drying time	h	≤4	GB/T 1728
	Full drying time	h	≤24	
Adhesion	Dry base course wall	Grade	≤1	GB/T 9286
	Wet base course wall	Grade	≤1	
Alkali resistance		—	No blistering, wrinkling or peeling in 48h	GB/T 9265

Performance Indicators of Insulation Board Adhesive **Table 9-3-16**

Item		Unit	Indicator	Test Method
Tensile bonding strength (with cement mortar)	Initial strength	MPa	≥0.60	JC/T 992
	Water resistance (48h)		≥0.40	
Tensile bonding strength (with corresponding insulation board)	Initial strength	MPa	≥0.10 (XPS board: 0.15), with the damaged interface inside the insulation board	
	Water resistance (48h)		≥0.10 (XPS board: 0.15), with the damaged interface inside the insulation board	
Operable time		h	1.5-4.0	

Performance Indicators of Rock Wool Board Adhesive **Table 9-3-17**

Item		Unit	Indicator	Test Method
Tensile bonding strength (with cement mortar)	Standard state	MPa	≥0.6	JGJ 144
	Soaking treatment		≥0.4	
Tensile bonding strength (in rock wool direction)	Standard state	kPa	≥50 or until the rock wool board is damaged	Appendix C of DB11/T 463
	Soaking treatment			

Performance Indicators of Insulation Board Interface Agent **Table 9-3-18**

Item		Unit	Indicator				Test Method
			EPS board interface agent	XPS board interface agent	Polyurethane board interface agent	Phenolic board interface agent	
Tensile bonding strength (with corresponding insulation material)	Standard state	MPa	≥0.10, with EPS board damaged	≥0.15, with XPS board damaged	≥0.10, with polyurethane board damaged	≥0.10, with phenolic board damaged	JC/T 907
	Soaking treatment						

Performance Indicators of Interface Agent for Rock Wool Board **Table 9-3-19**

Item	Indicator	Test Method
Tensile bonding strength (in the rock wool direction, MPa)	≥0.15	Appendix C of DB11/T 463
Hydrophobicity (%)	>98	GB/T 10299

Performance Indicators of Anti-cracking Mortar **Table 9-3-20**

Item		Unit	Indicator	Test Method
Tensile bonding strength (with cement mortar block)	Standard state	MPa	≥0.7	JG 158
	Soaking treatment	MPa	≥0.5	
	Freeze-thaw cycle treatment	MPa	≥0.5	
Tensile bonding strength (with adhesive polystyrene granule mortar)	Standard state	MPa	≥0.1	
	Soaking treatment	MPa	≥0.1	
Operable time		h	1.5～4.0	
Compressive-to-flexural strength ratio		—	≤3.0	

Performance Indicators of Glass Fiber Mesh **Table 9-3-21**

Item	Unit	Indicator		Test Method
		Ordinary type	Reinforced type	
Mass per unit area	g/m^2	≥160	≥270	JG 158
Alkali-resistant breaking strength (warp and weft)	N/50mm	≥1000	≥1500	
Alkali-resistant breaking strength retention rate (warp and weft)	%	≥80	≥90	
Elongation at break (warp and weft)	%	≤5.0	≤4.0	
Zirconia and titanium oxide content	%	—	Conform to the standard JC/T 841	JC/T 841

Performance Indicators of Elastic Primer **Table 9-3-22**

Item		Unit	Indicator	Test Method
Drying time	Surface drying time	h	≤4	JG 158
	Full drying time	h	≤8	
Elongation at break		%	≥100	
Surface hydrophobicity		%	≥98	

Performance Indicators of Waterproof Flexible Putty **Table 9-3-23**

Item		Unit	Indicator	Test Method
Drying time (surface drying)		h	≤5	GB/T 23455
Initial cracking resistance (6h)		—	No crack	
Grindability		—	Suitable for manual grinding	
Water absorption		g/10min	≤2.0	
Water resistance (96h)		—	No blistering, cracking or peeling	
Alkali resistance (48h)		—	No blistering, cracking or peeling	
Bonding strength	Standard state	MPa	≥0.60	
	5 freeze-thaw cycles	MPa	≥0.40	
Flexibility		—	50mm diameter, without cracks	
Low-temperature storage stability of non-powder component		—	No change in 5h freezing at −5℃ or impact on application	

18. The coating must be compatible with the external thermal insulation system, and its technical performance indicators should comply with the relevant provisions on coatings for outer walls of building.

19. The performance of anchor bolts should meet the requirements of Table 9-3-24.

20. The hot-dip galvanized welded wire mesh should be produced by welding first and then hot-dip galvanizing, and its performance should meet the requirements of Table 9-3-25.

21. The performance of the face brick bonding mortar should meet the requirements of Table 9-3-26.

22. The performance of the pointing material should meet the requirements of Table 9-3-27.

23. In addition to conformity to relevant standards on face bricks of outer walls, such as the *Ceramic Tiles* (GB/T 4100) and *Ceramic Mosaic* (JC 456), the face brick should also meet the requirements of Table 9-3-28.

24. The performance of the flexible waterproof mortar should meet the requirements of Table 9-3-29.

Performance Indicators of Anchor Bolt **Table 9-3-24**

Item	Unit	Indicator	Test Method
Standard tensile strength (C25 concrete base course wall)	kN	≥0. 60	JG/T 366
Standard pull strength of anchor bolt disc	kN	≥0. 50	

Performance Indicators of Hot-dip Galvanized Welded Wire Mesh **Table 9-3-25**

Item	Unit	Indicator	Test Method
Wire diameter	mm	0. 90±0. 04	QB/T 3897
Mesh size	mm	12. 70±0. 64 in the warp direction and 12. 7±0. 25 in the weft direction	
Solder joint tension	N	>65	
Mass of galvanized layer of mesh surface	g/m^2	>122	

Performance Indicators of Face Brick Bonding Mortar **Table 9-3-26**

Item		Unit	Indicator	Test Method
Tensile bonding strength	Standard state	MPa	≥0. 5	JC/T 547
	Soaking treatment			
	Thermal ageing treatment			
	Freeze-thaw cycle treatment			
	20min drying treatment			
Transverse deformation		mm	≥1. 5	

Performance Indicators of Pointing Material **Table 9-3-27**

Item		Unit	Indicator	Test Method
Shrinkage		mm/m	≤3. 0	JC/T 1004
Flexural strength	Standard state	MPa	≥2. 50	
	Freeze-thaw cycle treatment		≥2. 50	
Water permeability (24h)		mL	≤3. 0	JG 158
Compressive-to-flexural strength ratio		—	≤3. 0	

Performance Indicators of Face Brick for External Thermal Insulation **Table 9-3-28**

Item		Unit	Indicator	Test Method
Dimensions	Single area	cm^2	≤150	GB/T 3810.2
	Side length	mm	≤240	
	Thickness	mm	≤7	
Mass per unit area		kg/m^2	≤20	GB/T 3810.3
Water absorption rate		%	0.5-6	
Freezing resistance		—	Without damage in 40 freeze-thaw cycles	GB/T 3810.12

Performance Indicators of Flexible Waterproof Mortar **Table 9-3-29**

Item	Unit	Indicator	Test Method
Compressive strength (3d)	MPa	≥10.0	GB 23440
Flexural strength (3d)	MPa	≥3.0	
Tensile bonding strength (7d)	MPa	≥1.4	
Coating impermeability pressure (7d)	MPa	≥0.4	
Specimen impermeability pressure (7d)	MPa	≥1.5	
Compressive-to-flexural strength ratio	—	≤3.0	

25. The performance indicators of the wire-mesh EPS board should meet the *EPS Board with Metal Network for Exterior Insulation and Finish Systems* (GB 26540).

26. The JS waterproof coating should meet the requirements of Type I specified in the *Polymer-modified Cement Compounds for Waterproof Membrane* (GB/T 23445).

27. The adhesive polystyrene granule composite system should be produced with Grade 42.5 ordinary Portland cement and its technical performance should comply with the Common Portland Cement (GB 175).

28. Sand should comply with the *Standard for Technical Requirements and Test Method of Sand and Crushed Stone (or gravel) for Ordinary Concrete* (JGJ 52), in which particles larger than 2.5mm should be sieved out. The mud content of sand should be less than 3%.

29. The accessories used in the adhesive polystyrene granule composite system, such as sealing paste, sealing strips and opening seals, should comply with the design requirements and related product standards.

9.3.3 System Structure

9.3.3.1 Basic Requirements

1. The adhesive polystyrene granule composite system is divided mainly as follows:

(1) External thermal insulation system based on the adhesive polystyrene granule insulation mortar (hereinafter referred to as the insulation mortar system);

(2) External thermal insulation system based on the adhesive polystyrene granule pasted EPS board system (hereinafter referred to as the pasted EPS board system);

(3) External thermal insulation system based on the EPS board in the outer mould, cast-in-situ concrete as well as adhesive polystyrene granule (hereinafter referred to as the cast-in-situ concrete and meshed/non-meshed EPS board system);

(4) External thermal insulation system based on the sprayed rigid polyurethane foam as well as adhesive polystyrene granule (hereinafter referred to as the sprayed polyurethane system);

(5) External thermal insulation system based on the anchor rock wool board as well as

adhesive polystyrene granule (hereinafter referred to as the anchored rock wool board system);

(6) External thermal insulation system based on the reinforced vertical-fiber rock wool board pasted with the adhesive polystyrene granule mortar (hereinafter referred to as the pasted and reinforced vertical-fiber rock wool board system);

(7) External thermal insulation system based on the bonded insulation board as well as adhesive polystyrene granule (hereinafter referred to as the bonded insulation board system).

2. There should be no anti-cracking partition joints in the anti-cracking layer of the adhesive polystyrene granule composite system.

3. If applied, the coating of the finish layer of the adhesive polystyrene granule composite system should comply with the following requirements:

(1) An additional layer of glass fiber mesh should be paved in the anti-cracking layer of an easy-to-collide part, such as the first floor, door or window of a building.

(2) A layer (300mm×200mm) of glass fiber mesh should be added in the 45° direction inside the anti-cracking layer of each corner of a door or window opening.

(3) The thickness of the anti-cracking layer should be 3-5mm, and not less than 6mm on the first floor of a building.

(4) The anti-cracking mortar surface should be coated with elastic primer.

(5) If needed, the waterproof flexible putty should be used in the leveling process.

(6) Light-colored finish paints should be used.

4. Face bricks on the finish layer of the adhesive polystyrene granule composite system should comply with the following requirements:

(1) The bonding height should conform to relevant regulations of the state and city.

(2) The thickness of the anti-cracking layer should not be less than 8mm if the hot-dip galvanized welded wire mesh is used, and 6mm if the reinforced glass fiber mesh is used.

(3) The hot-dip galvanized welded wire mesh or reinforced glass fiber mesh should be reliably connected to a base course wall via anchor bolts. The number of anchor bolts per square meter should not be less than 4.

(4) The face brick bonding mortar and pointing material should be flexible. In addition, the pointing material should be impermeable.

(5) Light-colored face bricks with dovetail grooves on the surfaces to be pasted should be used if needed.

(6) The face brick joints should not be less than 5mm wide, and the pointing depth should be 2-3mm.

(7) One 20mm wide horizontal face brick joint should be built once every six floors, and pointed with a flexible waterproof material.

(8) Waterproofing and drainage structures should be used in concave and convex parts of a wall, such as windowsills, cornices, decorative lines, appentice, balconies, and sinkholes.

(9) The top drainage slope of a horizontal external corner should not be less than 3%. Top face bricks should be paved on façade bricks, and the bricks at the bottom of a façade should be constructed on the face bricks of the bottom surface. In addition, a drip structure should be provided.

9.3.3.2 Insulation Mortar System

The insulation mortar system is a kind of no-cavity external thermal insulation system formed by on-site plastering. It passed the examination of the Ministry of Housing and Urban-Rural Development in November 2001, won the second prize of the National Green Building Innovation awarded by the Ministry of Housing and Urban-Rural Development in March 2005, and was included in the new product promotion catalog by the Ministry of Housing and Urban-Rural Development in 2002 and 2006, respectively. Five ministries of China have granted the national key new product certificate of this system. In addition, this system has been listed in the National Torch Plan.

This system has also been granted all China's independent intellectual property rights, with the invention patent "anti-cracking insulated wall and construction technology" (ZL 98103325.3) and utility model patent "plastic composite glass fiber mesh" (ZL 98207104.3).

1. System features

The insulation mortar system has high insulation performance, cracking resistance, fire resistance, wind pressure safety, constructability, as well as adaptability to changes in walls, doors, windows, corners, ring beams, columns, etc. Due to the high utilization rate of materials but small workload of base course wall repair, labor costs can be saved. This system also has the advantages of extensive applications, high technology maturity, good operability in construction, excellent cost performance, and it construction quality can be controlled easily. Finished with face bricks, the system also has good seismic performance.

2. Scope of application

The insulation mortar system is applicable to external thermal insulation of new buildings with the reinforced concrete base course wall and various masonry structures in the hot-summer cold-winter as well as hot-summer warm-winter regions. It should be integrated with the framed lightweight insulated and infilled wall or the outer insulated wall of an existing building (subject to energy conservation renovation) in cold and severe cold regions. Also, this system can be used with partition walls, staircase walls, elevator walls, household walls, etc.

3. Process principle

The insulation layer of the insulation mortar system is formed as a whole without gaps by mean of on-site plastering. The base course wall is treated with insulation mortar so that materials of different water absorption rates are adhered evenly. Hollowing of the insulation layer can be avoided by decreasing the reinforcement ratio and adding a large number of fibers. The paint finish is composed of the flexible anti-cracking mortar and alkali-resistant glass fiber mesh, which enhances the flexible deformation capacity and cracking resistance of the surface. The elastic primer is applied to effectively prevent the inflow of liquid water and discharge water vapor. The waterproof flexible putty is located on the surface of the insulation layer, thereby improving the flexibility. The exterior finish should be treated with acrylic water-soluble paint, which is compatible with the deformation of the insulation layer. When the face brick finish is adopted, the anti-cracking protective layer should be composed of the anti-cracking mortar and hot-dip galvanized wire mesh or reinforced glass fiber mesh, and anchored on the base course wall via nylon expansion bolts to ensure high seismic performance. The face brick bonding mortar and pointing material used in the finish layer have high bonding performance, flexibility and cracking resistance, as well as good waterproofing effects. The flexibility of materials changes layer by layer to fully release the thermal stress. The basic structure of the insulation mortar system is shown in Table 9-3-30.

4. Typical engineering case

The 5 # residential building (Figure 9-3-1) of Shimao Waitan Xincheng, covering an area of 78,000m^2, is located in the waterfront zone of Xiaguan District, Nanjing. It has a fully cast-in-situ shear wall structure, and is divided into three units, namely, 5-1, 5-2, and 5-3. The three units include 47, 50 and 53 floors, respectively. Their eave heights (excluding the machine rooms and water tank floor) are 142m, 151m and 160m, respectively. Insulation mortar systems are adopted above the 5th floor of the unit 5-1, 6th floor of the unit 5-2 and 7th floor of the unit 5-3. The external thermal insulation project was started on March 5, 2006 and completed on July 30, 2006. The thermal insulation design is based on the energy conservation rate of 50%.

Basic Structure of Insulation Mortar System

Table 9-3-30

Type	Structural Layer	Composition	Schematic Diagram
Paint finish	Base course wall (i)	Concrete or masonry wall	
	Interface layer (ii)	Interface mortar	
	Insulation layer (iii)	Insulation mortar or pasting mortar	
	Anti-cracking layer (iv)	Anti-cracking mortar + glass fiber mesh + elastic primer	
	Finish layer (v)	Waterproof flexible putty (if required in the design) + paint	
Face brick finish	Base course wall (i)	Concrete or masonry wall	
	Interface layer (ii)	Interface mortar	
	Insulation layer (iii)	Insulation mortar or pasting mortar	
	Anti-cracking layer (iv)	Anti-cracking mortar+hot-dip galvanized welded wire mesh or reinforced glass fiber mesh [fixed on the base course wall with anchor bolt (vi)]	
	Finish layer (v)	Face brick bonding mortar + face brick + pointing material	

Figure 9-3-1 5# Building of Phase I of Nanjing Shimao Waitan

9.3.3.3 Pasted EPS Board System

The pasted EPS board system, developed in early 2003, is an external thermal insulation system meeting China's energy conservation standard (65%) and other higher energy conservation standards. It passed the examination of the Ministry of Housing and Urban-Rural Development in December 2004, won the second prize of the National Green Building Innovation awarded by the Ministry of Housing and Urban-Rural Development in March 2005, and was listed in the National Key New Products and National Torch Plan.

The pasted XPS board system has been applied from early 2005, which is an innovation and development of the EPS board system. As technological difficulties in vapor permeation and interface bonding of XPS boards in the external thermal insulation field are overcome effectively, it is possible to safely apply XPS boards in external thermal insulation systems.

This system has also been granted all independent intellectual property rights, with the invention patent "polystyrene board composite insulation wall and construction technology" (ZL2004 10046100.4) and utility model patent "sandwich type composite external insulation wall" (ZL2005 20200307.2).

1. System features

The no-cavity full bonding method is adopted, ensuring high bonding strength and wind pressure safety. The thick plaster and compartments are used on the surface of the insulation board, so the fire resistance of this system meets the fire protection requirements of high-rise buildings. The flexibility of structural layers is changed from inside to outside to prevent cracking. Board holes and joints act as passages for vapor discharge. This system also has good adaptability during construction, thereby reducing the workload of base course wall cutting and leveling.

2. Scope of application

This system applies to external thermal insulation projects in various regions in accordance with relevant standards of building energy conservation. The base course wall may be constructed with concrete or masonry. This system can be adopted in buildings (height: 100m or less) that have low requirements for energy conservation and fire ratings.

3. Process principle

The EPS boards with transverse trapezoidal grooves are fully bonded and pasted through the adhesive polystyrene granule pasting mortar. Interface mortar is applied on both sides of EPS boards. There are 10mm joints among boards. After leveling of pasting mortar squeezed out and pasting of boards, two holes of each EPS board are filled with pasting mortar, to reinforce the connection between the EPS boards and bonding/leveling layer and improve the vapor permeability. Then a 10mm pasting mortar leveling layer is applied on the surface, so that the composite insulation layer is bonded in a no-cavity manner with the wall. The anti-cracking protective layer is composed of the flexible anti-cracking mortar and alkali-resistant glass fiber mesh, which enhances the flexible deformation capacity and cracking resistance of the surface. The elastic primer is applied to effectively prevent the inflow of liquid water and discharge water vapor. The waterproof flexible putty is located on the surface of the insulation layer, thereby improving the flexibility. The exterior finish should be treated with acrylic water-soluble paint, which is compatible with the deformation of the system. The flexibility of materials changes layer by layer, to fully release the thermal stress and thus prevent cracking. The basic structure of this system is shown in Table 9-3-31.

4. Engineering case

Beijing Baiziwan Residential Community (Figure 9-3-2) was designed by Beijing Xingsheng Engineering Design Co., Ltd., contracted by the Real Estate Development and Operation Department of Beijing Construction Engineering Group, and completed by the 12th Project Department of Beijing Liujian Group. It external thermal insulation system was constructed by Beijing Zhenli High-tech Co., Ltd. The outer wall of this project is a kind of cast-in-situ concrete shear wall, with the pasted EPS board system for external thermal insulation. Its heat transfer coefficient was designed based on the energy conservation rate (65%) stipulated in the Design Standard for Energy Efficiency of Residential Buildings (DBJ01-602-2004). This project, covering an area of 60,000m^2, was completed in September 2005. In December 2005, the National Center for Quality Supervision and Test of Building Engineering tested the energy conservation structure of Beijing Baiziwan Residential Community. The test results show that the heat transfer coefficient of the main structure is 0.59W/($m^2 \cdot K$), which complies with the national standards and building design requirements. The project was awarded the Beijing "Great Wall Cup" Prize.

Basic Structure of Pasted EPS Board System **Table 9-3-31**

Type	Structural Layer	Composition	Schematic Diagram
Paint finish	Base course wall (i)	Concrete or masonry wall	
	Interface layer (ii)	Interface mortar	
	Bonding layer(iii)	Pasting mortar	
	Insulation layer (iv)	EPS board	
	Leveling layer (v)	Pasting mortar	
	Anti-cracking layer (vi)	Anti-cracking mortar + glass fiber mesh + elastic primer	
	Finish layer (vii)	Waterproof flexible putty (if required in the design) + paint	
Face brick finish	Base course wall (i)	Concrete or masonry wall	
	Interface layer (ii)	Interface mortar	
	Bonding layer(iii)	Pasting mortar	
	Insulation layer (iv)	EPS board	
	Leveling layer (v)	Pasting mortar	
	Anti-cracking layer (vi)	Anti-cracking mortar + hot-dip galvanized welded wire mesh or reinforced glass fiber mesh [fixed on the base course wall with anchor bolt (viii)]	
	Finish layer (vii)	Face brick bonding mortar + face brick + pointing material	

Figure 9-3-2 Beijing Baiziwan Residential Community

9.3.3.4 Cast-in-situ Concrete and Non-meshed EPS Board System

The cast-in-situ concrete and non-meshed EPS board system is formed by concrete pouring of EPS boards (subject to interface treatment on two sides) with vertical dovetail grooves at a time. Its surface is leveled with pasting mortar. This system passed the examination of the Ministry of Housing and Urban-Rural Development in November 2001, and won the second prize of the National Green Building Innovation awarded by the Ministry of Housing and Urban-Rural Development in March 2005. The adhesive polystyrene granule composite insulation system was also listed in the National Key New Products and National Torch Plan of five ministries (including the Ministry of Science and Technology).

This system has also been granted all China's independent intellectual property rights, with the invention patent "anti-cracking insulated wall and construction technology" (ZL98103325.3) and the utility model patents "wholly-cast EPS composite insulation wall" (ZL01201103.7), "concrete and cast polystyrene foam external insulation board" (ZL01279693.X) and "plastic clip for integrated non-meshed cast EPS board of external thermal insulation" (ZL02282766.8).

1. System features

The main structure concrete and insulation layer are formed at the same time during construction. As there are dovetail grooves on EPS boards, the bonding area between the concrete and EPS boards is increased to roughly 120%. Both sides of EPS boards are coated with interface mortar, ensuring the bonding reliability. Adjacent EPS boards are secured with ABS engineering plastic clips, to effectively control the flatness of the poured part and prevent the overflow of mortar as well as the formation of thermal bridges. Due to the no-cavity structure and thick plaster, the fire resistance of this system meets the fire protection requirements of high-rise buildings. The flexibility of materials changes from the inside to outside. In addition, this system has excellent cracking resistance, wind pressure safety and seismic resistance, and can be constructed in a convenient and fast manner.

2. Scope of application

This system applies to cast-in-situ reinforced concrete outer walls constructed with large formworks in a variety of climatic zones, and complies with the energy conservation standards as well as the external thermal insulation needs of buildings with high fire rating requirements.

3. Process principle

This system is formed by simultaneous pouring of EPS boards and concrete on the construction site. Both sides of EPS boards with vertical dovetail grooves are coated with interface mortar. Boards are bonded with adhesives, secured with special plastic clips and tied to the embedded bars in the concrete wall. The adhesive polystyrene granule pasting mortar is applied as a fireproof and vapor-permeable transition layer on the surfaces of EPS boards, to improve the fire resistance and vapor permeability of this system. Meanwhile, pasting mortar is used for overall leveling, to repair the holes, edges and corners damaged during board construction, reduce local thermal bridges of special parts such as door and window openings, and improve the thermal insulation effects. The anti-cracking protective layer is composed of the flexible anti-cracking mortar and alkali-resistant glass fiber mesh, which enhances the flexible deformation capacity and cracking resistance of the surface. The elastic primer is applied to effectively prevent the inflow of liquid water and discharge water vapor. The waterproof flexible putty is located on the surface of the insulation layer, thereby improving the flexibility. The exterior finish should be treated with acrylic water-soluble paint, which is compatible with the deformation of the insulation layer. The flexibility of materials changes layer by layer to fully release the thermal stress. The basic structure of this system is shown in Table 9-3-32.

Basic Structure of Cast-in-situ Concrete and Non-meshed EPS Board System **Table 9-3-32**

Type	Structural Layer	Composition	Schematic Diagram
Paint finish	Base course wall (i)	Cast-in-situ concrete wall	
	Insulation layer (ii)	Dovetail-groove EPS board [subject to auxiliary fixing with plastic clip (vi)]	⑥ ① ② ③ ④ ⑤
	Leveling layer (iii)	Pasting mortar	
	Anti-cracking layer (iv)	Anti-cracking mortar + alkali-resistant glass fiber mesh + elastic primer	
	Finish layer (v)	Waterproof flexible putty (if required in the design)+ paint	
Face brick finish	Base course wall (i)	Cast-in-situ concrete wall	
	Insulation layer (ii)	Dovetail-groove EPS board [subject to auxiliary fixing with plastic clip (vi)]	⑥ ① ② ③ ④ ⑤ ⑦
	Leveling layer (iii)	pasting mortar	
	Anti-cracking layer (iv)	Anti-cracking mortar + hot-dip galvanized welded wire mesh or reinforced glass fiber mesh [fixed on the base course wall with anchor bolt (viii)]	
	Finish layer (v)	Face brick bonding mortar + face brick + pointing material	

4. Engineering case

The dormitory building (Figure 9-3-3) of Beijing Architectural Design and Research Institute, completed in October 2000, is the first high-rise residential building with the cast-in-situ concrete and non-meshed EPS board system in China. It consists of 22 floors in total, with the total height of 64.5m and construction area of 16,170m^2. Party A of this project is Beijing Architectural Design and Research Institute, and Party B is China Construction First Building (Group) Corporation (Huazhong) Co., Ltd.

Figure 9-3-3 Dormitory Building of Beijing Architectural Design and Research Institute

9.3.3.5 Cast-in-situ Concrete and Meshed EPS Board System

In the cast-in-situ concrete and meshed EPS board system, the obliquely embedded and meshed EPS boards are pretreated with interface mortar on two sides and poured simultaneously with the concrete wall to form a secured insulation layer. The surface layer is leveled with the adhesive polystyrene granule pasting mortar. This adhesive polystyrene granule composite system has been awarded the prizes of the National Key New Products and National Torch Plan by five ministries (including the Ministry of Science and Technology). It passed the examination of the Ministry of Housing and Urban-Rural Development in November 2001, and won the second prize of the National Green Building Innovation awarded by the Ministry of Housing and Urban-Rural Development in March 2005.

This system has also been granted all independent intellectual property rights, with the invention patent "anti-cracking insulated wall and construction technology" (ZL98103325.3) and the utility model patent "wholly-cast EPS composite insulation wall" (ZL01201103.7).

1. System features

The main structure concrete and insulation layer are formed at the same time during construction. Both sides of the meshed boards are coated with interface mortar, which enhances the bonding strength between these boards and concrete. The steel truss connectors of the meshed boards are poured into concrete to reinforce the connection between the system and base course wall. The adhesive polystyrene granule pasting mortar is applied as a fireproof and vapor-permeable layer, which effectively solves the problems of cracking and damage of the conventional cement mortar, reduces the surface load, and cut off thermal bridges caused by steel truss connectors. This system has high wind load resistance, fire resistance, insulation performance and weathering resistance, and can be constructed easily and fast. With double meshes, the stress can be fully dispersed and released to effectively prevent cracks.

2. Scope of application

This system applies to cast-in-situ reinforced concrete outer walls constructed with large formworks in a variety of climatic zones, and complies with the energy conservation standards as well as the external thermal insulation needs of buildings with high fire rating requirements.

3. Process principle

This system is formed by simultaneous pouring of single-sided meshed EPS boards and concrete on the construction site. The adhesive polystyrene granule pasting mortar is applied as a fireproof and vapor-permeable transition layer to improve the fire resistance and vapor permeability of this system. The anti-cracking protective layer is composed of the flexible anti-cracking mortar and alkali-resistant glass fiber mesh, which enhances the flexible deformation capacity and cracking resistance of the surface. The elastic primer is applied to effectively prevent the inflow of liquid water and discharge water vapor. The waterproof flexible putty is located on the surface of the insulation layer, thereby improving the flexibility. The exterior finish should be treated with acrylic water-soluble paint, which is compatible with the deformation of the insulation layer. The flexibility of materials changes layer by layer to fully release the thermal stress. The basic structure of this system is shown in Table 9-3-33.

4. Engineering case

Qingdao Lucion Changchun Garden Community (Figure 9-3-4) was invested and built by Shandong Lucion Real Estate Co., Ltd., and located at No. 1, Yinchuan East Road, Qingdao, with the construction area of approximately 990,000m^2. It consists of 99 buildings in total, which are composed of the cast-in-situ concrete, meshed EPS board and framed shear walls filled with aerated concrete blocks.

Basic Structure of Cast-in-situ Concrete and Meshed EPS Board System **Table 9-3-33**

Type	Structural Layer	Composition	Schematic Diagram
Paint finish	Base course wall (i)	Cast-in-situ concrete wall	① ② ③ ④ ⑤ ⑥ ⑦
	Insulation layer (ii)	Meshed EPS board [with mesh (vii) hooked by φ6 bar (vi)]	
	Leveling layer (iii)	Adhesive polystyrene granule mortar	
	Anti-cracking layer (iv)	Anti-cracking mortar + alkali-resistant glass fiber mesh + elastic primer	
	Finish layer (v)	Waterproof flexible putty (if required in the design) + paint	
Face brick finish	Base course wall (i)	Cast-in-situ concrete wall	① ② ③ ④ ⑤ ⑥ ⑦ ⑧
	Insulation layer (ii)	Meshed EPS board [with mesh (vii) hooked by φ6 bar (vi)]	
	Leveling layer (iii)	Adhesive polystyrene granule mortar	
	Anti-cracking layer (iv)	Anti-cracking mortar + hot-dip galvanized welded wire mesh or alkali-resistant glass fiber mesh [fixed on the base course wall with anchor bolt (viii)]	
	Finish layer (v)	Face brick bonding mortar + face brick + pointing material	

Figure 9-3-4 Pictures of On-site Construction of Lucion Changchun Garden Community

9.3.3.6 Sprayed Polyurethane System

The sprayed polyurethane system is an external thermal insulation system which complies with the energy conservation standard (65%) and applies to energy-efficient buildings. It is suitable for China's national conditions and climate characteristics.

This system has also been granted all China's independent intellectual property rights, with six invention patents such as the "polyurethane-based outer insulated wall and construction method" (ZL02153346.6), "construction method combining polyurethane spraying and bonding for internal/external corner and door/window opening and prefabricated polyurethane block thereof" (ZL03160003.4), "external/internal corner casting mould and construction method for external/internal corner casting of polyurethane insulated wall with mould" (ZL03137331.3), and "sprayed polyurethane insulated and finished wall and construction method" (ZL200510200767.X), as well as six utility model patents such as the "polyurethane-based outer wall" (ZL200420064725.9) and "wall corner structure involving polyurethane spraying and bonding" (ZL032825072) This system passed the examination of Beijing Municipal Construction Committee in April 2004, and won the second prize of the National Green Building Innovation awarded by the Ministry of Housing and Urban-Rural Development in March 2005. It is also listed in the National Key New Products and National Torch Plan.

1. System features

The sprayed polyurethane system is constructed by means of fast and efficient on-site mechanized spraying. Polyurethane is adaptive to the shape of a building, especially structures with complex nodes, such as overhanging components and attic windows. The base course wall is treated with moisture-proof polyurethane primer to improve the closed-cell ratio of the polyurethane insulation layer and homogenize the adhesion between this insulation layer and wall. The polyurethane surface layer is treated with polyurethane insulation layer to enhance the bonding between the polyurethane and leveling material. The polyurethane insulation layer is also leveled with the adhesive polystyrene granule mortar for high insulation performance, vapor permeability, cracking resistance and fire resistance.

2. Scope of application

This system can be used in external thermal insulation projects of buildings in different climatic zones and subject to various energy conservation standards. It is applicable buildings with high requirements for energy conservation and fire rating, with concrete and masonry base course walls.

3. Process principle

In the sprayed polyurethane system, a polyurethane insulation material is sprayed under high pressure in a vapor-free manner onto a base course wall surface to form an insulation layer on the construction site; the prefabricated polyurethane components are affixed on the edges and corners, to handle internal and external corners and control the thickness of the insulation layer; the base course wall is coated with moisture-proof polyurethane primer to improve the water resistance and vapor permeability; the polyurethane surface is treated to overcome the problems of bonding between organic and inorganic materials; the surface layer is leveled and additionally insulated with the adhesive polystyrene granule mortar; and the anti-cracking protective layer is composed of the anti-cracking mortar and plastic-coated alkali-resistant glass fiber mesh, and coated with the elastic primer to prevent liquid water; and the waterproof flexible putty and finish paint are applied on the surface. The basic structure of this system is shown in Table 9-3-34.

4. Engineering case

The 6# and 8# buildings (Figure 9-3-5) of Beijing Shanshui Huihaoyuan Community are located in Doudian, Beijing. They are pilot projects of energy-efficient (65%) residential buildings in Bei-

Basic Structure of Sprayed Polyurethane System **Table 9-3-34**

Type	Structural Layer	Composition	Schematic Diagram
Paint finish	Base course wall (i)	Concrete or masonry wall + cement mortar	
	Interface layer (ii)	Moisture-proof primer	
	Insulation layer (iii)	Sprayed rigid polyurethane foam + polyurethane interface agent	
	Leveling layer (iv)	Pasting mortar	
	anti-cracking layer (v)	Anti-cracking mortar and glass fiber mesh + elastic primer	
	Finish layer (vi)	Waterproof flexible putty (if required in the design) + paint	
Face brick finish	Base course wall (i)	Concrete or masonry wall + cement mortar	
	Interface layer (ii)	Moisture-proof primer	
	Insulation layer (iii)	Sprayed rigid polyurethane foam + polyurethane interface agent	
	Leveling layer (iv)	Pasting mortar	
	Anti-cracking layer (v)	Anti-cracking mortar+ hot-dip galvanized welded wire mesh or reinforced glass fiber mesh [fixed on the base course wall with anchor bolt (vii)]	
	Finish layer (vi)	Face brick bonding mortar + face brick + pointing material	

Figure 9-3-5 Beijing Shanshui Huihaoyuan Community

jing, with a framed brick-concrete structure, in which the inner and outer walls were constructed with 240mm clay hollow bricks, the eave height is 14m and the shape coefficient is less than 0.3. They were developed by Beijing Huihao Real Estate Construction Co., Ltd., supervised by Beijing Qianxing Supervision Co., Ltd. and generally contracted by Dingzhou Construction & Installation Engineering Co., Ltd., with a total construction area of 4,800m^2. The external thermal insulation system covers an area of 2,500m^2. Beijing Zhenli High-tech Co., Ltd. provided the complete set of "construction technology for the external thermal insulation based on the sprayed rigid polyurethane foam as well as adhesive polystyrene granule", and the guidance over the construction technology. Both buildings were completed in June 2003.

9.3.3.7 Anchored Rock Wool Board System

The anchored rock wool board system is composed of high-quality rock wool boards. The problems of application of rock wool boards in external thermal insulation systems are solved due to the advanced anchorage technology and the technical route of flexible and gradual change against cracking. Thus, the external thermal insulation system based on the rock wool board is promoted. In addition, this technology won the National Green Innovation Award.

This system has also been granted all China's independent intellectual property rights, with the invention patent "rock wool and EPS granule insulation mortar composite wall and construction technology" (ZL02100801.9) and the utility model patent "integrated cast concrete wall with rock wool composite external thermal insulation" (ZL02235565.0).

1. System features

The rock wool insulation system has good insulation performance, cracking resistance, fire resistance and durability. Meanwhile, the rock wool boards are effectively secured on the base course wall to improve the wind load resistance; and anchorage helps to increase the construction speed and simplify the construction process, thus shortening the construction duration and decreasing construction costs. This external thermal insulation system is worthy of promoting because of its environmental friendliness and appropriate cost.

2. Scope of application

This system is applicable to the external thermal insulation of outer walls (concrete or masonry) containing paint finishes, as well as energy-saving renovation of various existing buildings.

3. Process principle

In the anchored rock wool board system, rock wool boards as an insulation material are secured with anchors (such as plastic expansion bolts) and hot-dip galvanized wire meshes. Gaskets are applied between the hot-dip galvanized wire meshes and rock wool boards to form gaps between them and facilitate plastering of rock wool boards; and interface treatment is conducted to the surfaces of the secured rock wool boards, to enhance the waterproofing performance and surface strength and also overcome difficulties in bonding between the rock wool boards and adhesive polystyrene granule leveling layer. The surface layer is leveled with the adhesive polystyrene granule pasting mortar. The anti-cracking protective layer with good cracking resistance is composed of the anti-cracking mortar and plastic-coated alkali-resistant glass fiber mesh, and coated with the elastic primer to effectively prevent liquid water. The waterproof flexible putty and elastic paint are applied on the finish layer. The basic structure of this system is shown in Table 9-3-35.

4. Engineering case

The three-storey comprehensive office building project (Figure 9-3-6) of Tianjin Huachen, covering an area of 3, 400m^2, has a framed type concrete infilled wall structure. Its external thermal insulation area is 1, 700m^2. The outer wall is filled with 300mm aerated concrete and ceramsite blocks, of which the shape coefficient is less than 0. 3 and the window-to-wall area ratio is 0. 4. The reinforced concrete frame columns (500mm × 500mm) and crossbeams (700mm×270mm) are indented for 30mm on the outer surface, and filled with 300mm aerated ceramsite ceramic block.

The rock wool board insulation system in the project is composed of the "45mm rock wool boards + 20mm adhesive polystyrene granule insulation mortar + 5mm anti-cracking mortar and alkali-resistant glass fiber mesh".

In the previous construction process, rock wool boards may be damaged with steel hook type anchors, and both the scratchy rock wool fibers and the difficulties in pulling and fixing of pins of anchors affected the construction. Later, rock wool boards are first pre-fixed and then anchored via plastic expansion bolts mounted in the drilled holes, which reduce damage to rock wool boards and also difficulties in construction.

Basic Structure of Anchored Rock Wool Board System **Table 9-3-35**

Structural Layer	Composition	Schematic Diagram
Base course wall (i)	Concrete or masonry wall	①②③④⑤⑥⑦
bonding layer (ii)	Rock wool board adhesive	
Insulation layer (iii)	Rock wool board + hot-dip galvanized welded wire mesh[fixed on the base course wall with anchor bolt (vii)] + board interface agent	
Leveling layer (iv)	Pasting mortar	
Anti-cracking layer (v)	Anti-cracking mortar and glass fiber mesh + elastic primer	
Finish layer⑥	Waterproof flexible putty (if required in the design)+ paint	

Figure 9-3-6 Comprehensive Office Building Project of Tianjin Huachen

This project meeting the requirements stipulated in the technology contract has passed the acceptance.

This pilot project shows that the rock wool insulation system has the advantages of reasonable structure and convenient construction, involving no difficulty in rock wool fixing or cracking of the surface of the conventional rock wool insulation. As the thermal insulation, heat isolation and fire protection are integrated, the rock wool insulation system has a high promotion value and promising market prospects.

9.3.3.8 Pasted and Reinforced Vertical-fiber Rock Wool Board System

The pasted and reinforced vertical-fiber rock wool board system was developed gradually based on higher requirements for thermal insulation and fire protection of buildings after 2009. The four sides of reinforced vertical-fiber rock wool composite boards are coated, thereby solving the problems of low pull strength and easy loosening of the exposed rock wool boards. The Class A insulation material is used in this system, which meets the fire protection needs of high-rise buildings.

1. System features

As there are no cavities, the four sides of the glass fiber mesh are coated, and the reinforced vertical-fiber rock wool composite boards are bonded instead of anchorage, this system has good construction adaptability, and base course wall cutting or leveling is not required. The board size is appropriate, which is conducive to construction, false bonding prevention as well as environmental protection.

2. Scope of application

The system is suitable for buildings with high fire protection requirements, in which outer walls are constructed with concrete or masonry. It is more applicable to the energy-saving renovation of existing buildings.

3. Process principle

Taking into account their high strength and large dead load, the reinforced vertical-fiber rock wool composite boards in the system are fully bonded with the adhesive polystyrene granule pasting mortar of Class A2 combustibility, and metal brackets are used at the bottom of the building envelope to improve the connection safety and facilitate construction.

The face brick finish is composed of the Class A2 adhesive polystyrene granule pasting mortar, hot-dip galvanized welded wire mesh and anchors. Anchors are selected depending on the wall material. Reinforced vertical-fiber rock wool composite boards are fixed with anchors such as plastic expansion bolts and fasteners as well the hot-dip galvanized welded wire mesh. Thus, the tensile stress is evenly distributed on the hot-dip galvanized welded wire mesh and dispersed to the base course wall through anchor bolts, making the reinforced vertical-fiber rock wool composite board secured well on the outer wall surface, fully eliminating safety risks of brick bonding and also enhancing the system safety. The basic structure of this system is shown in Table 9-3-36.

4. Engineering case

Beijing Yuanyang Aobei Project (Figure 9-3-7) is located in Changping, Beijing, with a total construction area of 250,000m^2. The pasted and reinforced vertical-fiber rock wool board system, with an insulation area of 25,000m^2, is applied in the buildings DK6, DK7, DK14 and DK17 of Phase I, in which the finishes are made of lacquer.

Basic Structure of Pasted and Reinforced Vertical-fiber Rock Wool Board System **Table 9-3-36**

Type	Structural Layer	Composition	Schematic Diagram
Paint finish	Base course wall (i)	Concrete or masonry wall	①②③④⑤⑥
	Interface layer (ii)	Interface mortar	
	Bonding layer(iii)	Pasting mortar	
	Insulation layer (iv)	Reinforced vertical-fiber rock wool composite board	
	Anti-cracking layer(v)	Anti-cracking mortar and glass fiber mesh + elastic primer	
	Finish layer (vi)	Waterproof flexible putty (if required in the design)+ paint	
Face brick finish	Base course wall (i)	Concrete or masonry wall	①②③④⑤⑥⑦
	Interface layer (ii)	Interface mortar	
	Bonding layer(iii)	Pasting mortar	
	Insulation layer (iv)	Reinforced vertical-fiber rock wool composite board	
	Anti-cracking layer(v)	Anti-cracking mortar and hot-dip galvanized welded wire mesh or reinforced glass fiber mesh[fixed on the base course wall with anchor bolt (vii)]	
	Finish layer (vi)	Face brick bonding mortar + face brick + pointing material	

Figure 9-3-7 Beijing Yuanyang Aobei Project

9.3.3.9 Bonded Insulation Board System

The basic structure of the bonded insulation board system is shown in Table 9-3-37. The external surface is composed of EPS boards treated with the EPS board interface agent, while the internal surface consists of the XPS boards treated with the XPS board interface agent, rigid polyurethane foam, phenolic boards, or reinforced vertical-fiber rock wool composite boards. The bonding area between the insulation boards and base course wall must not be less than 40% of the insulation board area in the case of a paint finish, 70% of the insulation board area in the presence of a face brick finish. The tensile strength between the base course wall and insulation board bonding mortar should not be less than 0.3MPa, and the peeling area of the bonding interface must not exceed 50% of the total area. When the paint finish is adopted and the building height is more than 20m, insulation boards of parts affected greatly by the negative wind pressure should be boarded in the closed small cavity form (that is, the area of closed cavities surrounded by the insulation board bonding mortar is not larger than 0.3m^2), and the bonding area between the insulation boards and base course wall must not be less than 50% of the insulation board area; when the paint finish is used, insulation boards should not be more than 900mm long and 600mm wide; and when the face brick finish is applied, insulation boards should not be more than 600mm long and 450mm wide.

9.3.4 Engineering Design

1. The appropriate adhesive polystyrene granule composite system should be selected according to the engineering needs, and the system structure and composition must not be changed without permission.

2. The thermal and energy - saving design of

Basic Structure of Bonded Insulation Board System　　Table 9-3-37

Type	Structural Layer	Composition	Schematic Diagram
Paint finish	Base course wall (i)	Concrete or masonry wall	①②③④⑤⑥
	Bonding layer (ii)	Insulation board adhesive	
	Insulation layer (iii)	Insulation board (EPS board/XPS board/polyurethane composite board/phenolic board/reinforced vertical-fiber rock wool composite board)	
	Leveling layer (iv)	Pasting mortar	
	Anti-cracking layer (v)	Anti-cracking mortar and glass fiber mesh + elastic primer	
	Finish layer (vi)	Waterproof flexible putty (if required in the design)+ paint	
Face brick finish	Base course wall (i)	Concrete or masonry wall	①②③④⑤⑥⑦
	Bonding layer (ii)	EPS board adhesive	
	Insulation layer (iii)	Insulation board (EPS board/XPS board)	
	Leveling layer (iv)	Pasting mortar	
	Anti-cracking layer (v)	Anti-cracking mortar and hot-dip galvanized welded wire mesh or reinforced glass fiber mesh[fixed on the base course wall with anchor bolt (vii)]	
	Finish layer (vi)	Face brick bonding mortar + face brick + pointing material	

the neral binder and EPS granule composite system should meet the following requirements.

(1) The temperature of the internal surface of the insulation layer should be higher than 0℃, and that of the part involving a thermal bridge must not be below the dew point under the design indoor air temperature and humidity.

(2) The external thermal insulation system should cover the external opening of the door/window frame, parapet, balcony and overhanging components with thermal bridges, such as appentice, eave plates and outdoor units of air-conditioners. The XPS boards (not less than 30mm) may be pasted. The door/window frame should be mounted on the external edge of the outer wall.

(3) The gap between the outer door/window frame and door/window should be filled with closed-cell polyurethane. A waterproof partition should be mounted between the outer wall insulation material and outer door/window frame. Deformation joints of the wall should be filled with an insulation material of appropriate thickness and sealed as per the architectural design requirements.

(4) The impact of thermal bridges caused by mechanical fasteners should be taken into consideration.

3. The fireproof design of the adhesive polystyrene granule composite system should conform to relevant national standards. The thickness of the leveling layer should not be less than 10mm.

4. The adhesive polystyrene granule composite system should be sealed and waterproof to prevent water infiltration into the insulation layer and base course wall, or permeation into any part that may cause damage. Important parts should be detailed. Waterproofing should be conducted to the horizontal or oblique overhanging parts as well as parts extended to below the ground. Equipment or pipes mounted on the external thermal insulation system should be secured on the base course wall, with sealing and waterproofing measures.

9.3.5 Enginering Construction

9.3.5.1 General

1. A complete set of components of the adhe-

sive polystyrene granule composite system should be supplied.

2. The construction company undertaking external thermal insulation projects should have appropriate qualifications.

3. External thermal insulation projects should be constructed in accordance with the reviewed design documents and approved construction scheme, and the wall energy saving design must not be changed without permission during construction. If it is necessary to make changes, the design change documents should be submitted, examined by the original construction drawing/design review organization, and approved by the supervision and contracting units. The building energy efficiency must not decline in the design changes.

4. Special construction schemes should be prepared for external thermal insulation projects, and technical disclosure should be conducted. Construction workers should be trained and pass examinations.

5. External thermal insulation projects should meet the national and local provisions on the fire safety.

6. Except the cast-in-situ EPS board system, external thermal insulation projects should be constructed based on the qualified base course wall.

7. Base course wall inspection and treatment should be carried out before the construction of the insulation layer. The base course wall should be clean, solid and smooth, and also comply with the *Code for Acceptance of Construction Quality of Concrete Structure Engineering* (GB 50204) and the *Code for Acceptance of Construction Quality of Masonry Structure Engineering* (GB 50203).

8. Specimen walls should be prefabricated with the same materials and technologies on the site, and approved by the contracting, design, construction and supervision units prior to large-scale construction.

9. Prior to construction of external thermal insulation projects except cast-in-situ EPS board systems, external door and window openings should be accepted, the opening sizes and locations should meet the design and quality requirements, and the door/window frames or auxiliary frames should be installed. In addition, the embedded parts and connectors of firefighting ladders, downfalls, inlet pipes and air-conditioners should have been installed, and gaps should be reserved based on the thickness of the external thermal insulation system.

10. Protective measures against moisture, water and fire should be taken to materials used in external thermal insulation projects during construction. All materials should be stored in a classified manner, in accordance with the storage period and conditions in the operating instructions, and also prevented from rain, sunlight and fire. Moreover, materials should not be stored in the open air, and those in the open air must be covered with tarpaulins. The supporting EPS or XPS board interface agent should be sprayed before boards are delivered to the site.

11. On-site preparation should be performed in accordance with the operating instructions, involving accurate measurement. The prepared materials must be used within the specified period.

12. All machines and tool should be prepared and checked for the safety and reliability. Measurement tools should be accurately calibrated. Main machines and tools include the forced mortar mixer, vertical transport machinery, trolley, handheld mixer, electric hammer, sprayer, polyurethane spraying machine, special spray gun, pouring gun, pipe, commonly used plastering tools and dedicated test tools, theodolite, setting-out tool, water barrel, hand saw, scissors, roller brush, shovel, hand hammer, pliers, wallpaper knife, broom, electric basket, scaffold, etc.

13. The base course wall temperature and ambient temperature must not be below 5℃ dur-

ing construction and within 24h after completion of an external thermal insulation project. The direct exposure to sunlight should be avoided in summer. Construction must be suspended in the case of rain and Grade 5 wind or above.

14. The external thermal insulation project completed must be protected.

9.3.5.2 Key Construction Points of Insulation Mortar System

1. The construction procedures of the insulation mortar system should meet the requirements of Figure 9-3-8.

2. Base course wall treatment requirements:

(1) Base course wall treatment should be carried out prior to construction of the insulation layer. The base course wall surface should be flat, clean and dry, without quality problems such as dust, leak, oil, and hollowing. The protrusions larger than 10mm on the external surface of the wall should be removed. Holes through the wall and damaged parts on the wall surface should be cleaned and repaired with appropriate materials. Holes on the wall surface should be moistened and repaired.

(2) Hollowing and cracking parts of the outer wall of the existing building should be removed.

3. The base course wall surface should be coated evenly and fully with a thin layer of interface mortar. The masonry wall with a large water absorption rate should be wetted and dried in the shade before interface mortar is used.

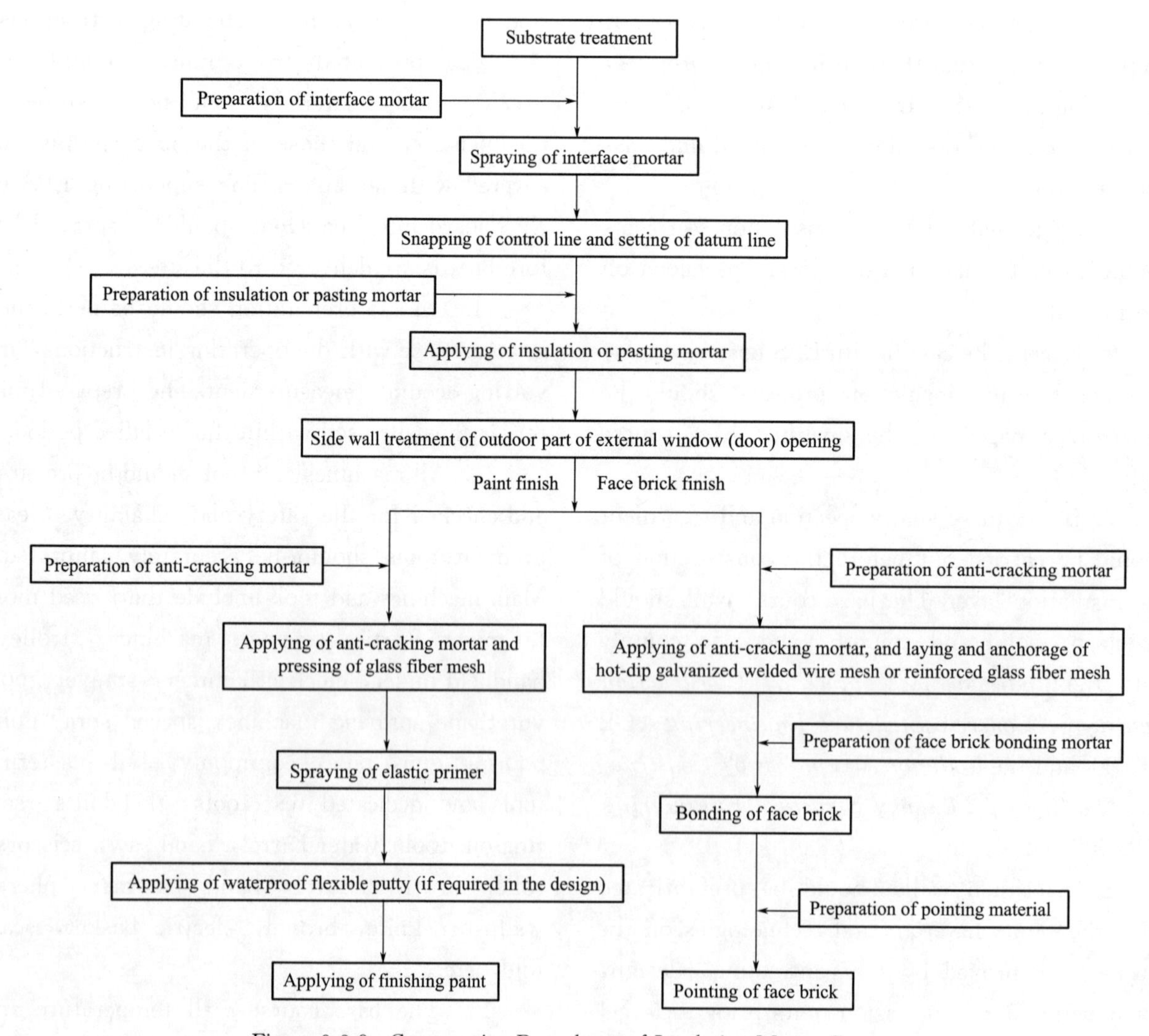

Figure 9-3-8 Construction Procedures of Insulation Mortar System

4. Requirements for snapping of the control line and setting of the datum line:

(1) Snap the horizontal and vertical control lines of external doors and windows on the wall surface, according to the requirements for the building facade design and external thermal insulation.

(2) Hang the vertical steel datum lines in large corners (internal and external corners) of the outer wall and other places (if necessary), and horizontal lines in appropriate positions of each floor.

5. Requirements for the applying of insulation or pasting mortar:

(1) Apply cement blocks at intervals of about 1.5m along the thickness control line. Determine the thickness of cement blocks based on that of the insulation layer. Use the cement blocks made of insulation or pasting mortar, or directly use EPS blocks.

(2) Apply the insulation or pasting mortar when interface mortar is basically hardened.

(3) Mix the insulation or pasting mortar according to the operating instructions. Check the mixing quality by measuring the wet apparent density or observing the operability, slip resistance and past state. Use the prepared insulation mortar within the permitted time.

(4) Apply the insulation or pasting mortar layer by layer (layer thickness: approximately 20mm) at intervals of more than 24h. Make the first layer of mortar compacted, and control the thickness of the final layer to be roughly 10mm. Flatten the mortar with a tie bar to be flush with cement blocks and conform to the flatness acceptance standards.

(5) Check the thickness of the insulation layer on the site for conformity to the design requirements, without negative deviations.

6. The side walls of outdoor parts of the external window (door) openings should be coated with XPS boards (thickness: 30mm or more). The 20mm wide gaps reserved between the external window (door) frames and XPS boards should be filled with flexible waterproof mortar, and treated with the JS waterproof coating.

7. When the paint finish is used, the anti-cracking layer and finish layer should be constructed in line with the following requirements:

(1) Apply anti-cracking mortar and press the glass fiber mesh 3-7d after leveling with the construction quality accepted.

(2) Make a drip line according to the design requirements before applying anti-cracking mortar.

(3) Apply a layer of 300mm×200mm glass fiber mesh in the 45° direction of each door/window opening before large-scale laying of glass fiber meshes.

(4) Lay glass fiber meshes (lap width: no less than 30mm) bottom-up along the outer wall immediately after applying anti-cracking mortar. Ensure that the glass fiber meshes are flat without wrinkles. When the fullness of mortar reaches 100%, press glass fiber meshes into anti-cracking mortar via a trowel, until they are fully covered by anti-cracking mortar.

(5) Lay double layers of glass fiber mesh on the wall of the first floor. The butt joints of glass fiber meshes in the first layer must not be located in internal and external corners, with the deviation of 200mm or more from internal and external corners. The anti-cracking mortar between two layers of glass fiber mesh should be full. Dry bonding is not permitted.

(6) After the applying of anti-cracking mortar, check the flatness and verticality of the mortar surface as well as the regularity of internal and external corners, and repair the nonconforming part with anti-cracking mortar. It is prohibited to make waist lines and window frame lines with ordinary cement mortar on the anti-cracking mortar surface.

(7) Following the initial setting of anti-cracking mortar, apply elastic primer evenly, without bottom exposure.

(8) Conduct local repair, and then conduct large-scale applying of waterproof flexible putty several times for leveling. Control the thickness of each layer of waterproof flexible putty to be approximately 0.5mm.

(9) Apply the finish paint in accordance with the Specification for Construction and Acceptance of Building Surface Decoration (JGJ/T 29).

8. When the face brick finish is used, the anti-cracking layer and finish layer should be constructed in line with the following requirements:

(1) Apply anti-cracking mortar and press the glass fiber mesh 3-7d after leveling with the construction quality accepted.

(2) Cut the hot-dip galvanized welded wire mesh based on the wall size.

(3) Drill anchor bolt holes in a plum shape in two directions (@500mm) on the wall. There should be at least three anchor bolt holes at hot-dip galvanized welded wire mesh seals of side window openings. The depth of holes in the structural wall should be more than 400mm, and anchor bolt sleeves should be inserted into holes.

(4) Lay hot-dip galvanized welded wire meshes top-down from left to right, with the lap width greater than 5 grids. The maximum number of lap layers is 3. Anchor bolts should be screwed or knocked into anchor bolt sleeves.

(5) Inspect the hot-dip galvanized welded wire meshes and apply anti-cracking mortar. Ensure that the hot-dip galvanized welded wire meshes are fully covered by anti-cracking mortar. The anti-cracking mortar layer should be 8-10mm and meets the flatness and verticality requirements.

(6) Install the reinforced glass fiber meshes in the door/window corners via anchor bolts, and make them pressed onto the hot-dip galvanized welded wire meshes.

(7) Apply a layer of anti-cracking mortar on the wall and then lay the reinforced glass fiber meshes, with the lap width no less than 100mm. Knock in anchor bolts in a plum shape in two directions (@500mm) and apply a layer of anti-cracking mortar (thickness: 6-8mm).

(8) Spray water for curing, and bond face bricks after about 7d.

(9) Bond the face bricks in accordance with the *Specification for Construction and Acceptance of Tapestry Brick Work for Exterior Wall* (JGJ 126). Ensure that the face brick bonding material is full and control the face brick finish to be 3-5mm.

(10) Promptly conduct pointing in accordance with the *Specification for Construction and Acceptance of Tapestry Brick Work for Exterior Wall* (JGJ 126), with the joint depth of 2-3mm. Then immediately scrub the face bricks around joints.

9.3.5.3 Key Construction Points of Pasted EPS Board System

1. The construction procedures of the pasted EPS board system should meet the requirements of Figure 9-3-9.

2. Base course wall treatment requirements:

(1) Base course wall treatment should be carried out prior to construction of the insulation layer. The base course wall surface should be flat, clean and dry, without quality problems such as dust, leak, oil, and hollowing. The protrusions larger than 10mm on the external surface of the wall should be removed. Holes through the wall and damaged parts on the wall surface should be cleaned and repaired with appropriate materials. Holes on the wall surface should be moistened and repaired.

(2) Hollowing and cracking parts of the outer wall of the existing building should be removed.

3. The base course wall surface should be coated evenly and fully with a thin layer of interface mortar. The masonry wall with a large water absorption rate should be wetted and dried in the shade before interface mortar is used.

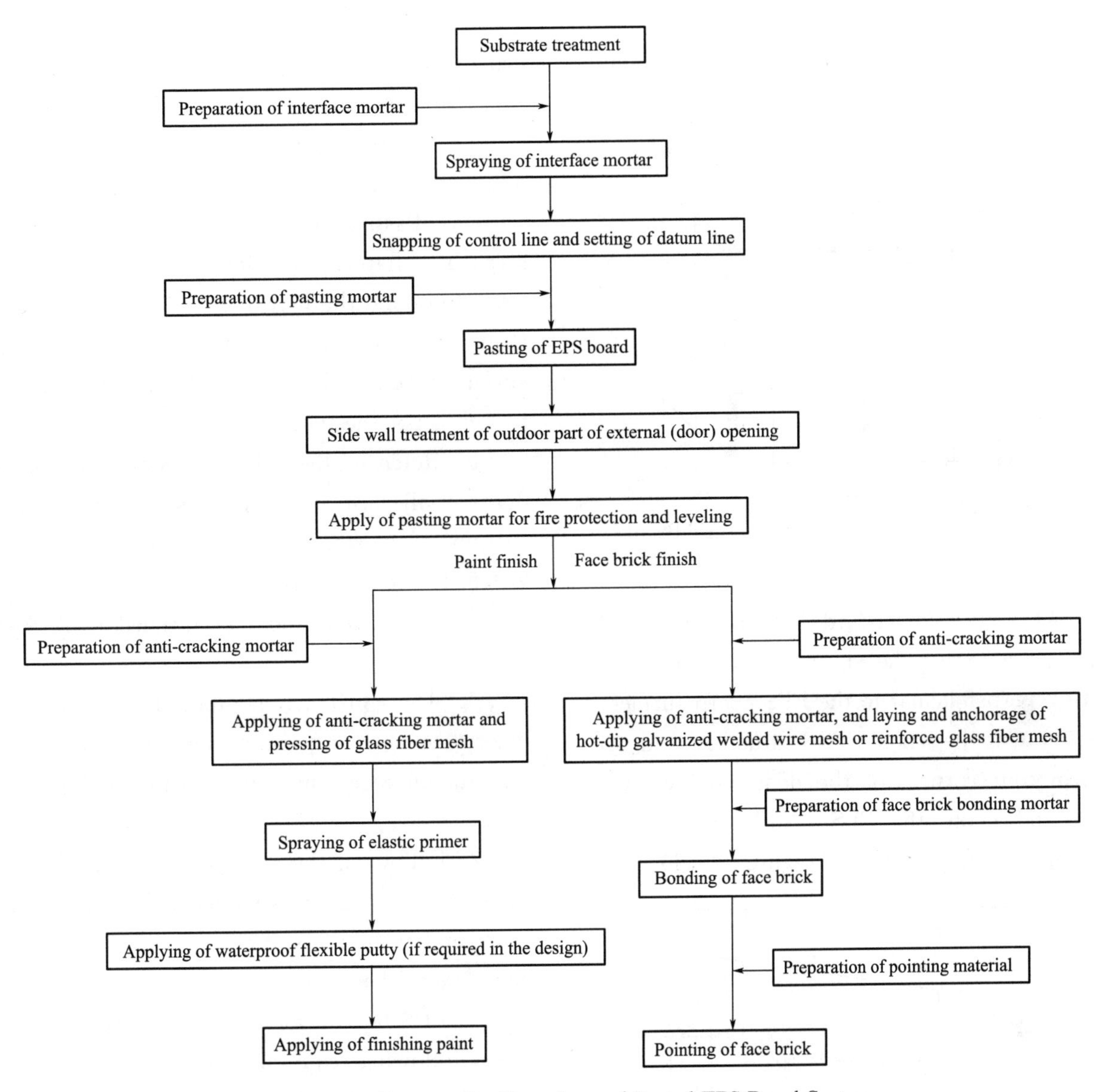

Figure 9-3-9 Construction Procedures of Pasted EPS Board System

4. Requirements for snapping of the control line and setting of the datum line:

(1) Snap the horizontal and vertical control lines of external doors and windows on the wall surface, according to the requirements for the building facade design and external thermal insulation.

(2) Snap horizontal lines at the starting positions of EPS board pasting around the building. Prior to EPS board construction, paste standard blocks in external corners, hang horizontal control lines on standard blocks, and set horizontal thickness control lines between two vertical through lines of one wall.

5. Requirements for EPS board pasting:

(1) Evenly apply a 5-10mm layer of pasting mortar on the bonding surface between the wall and EPS boards, and immediately paste EPS boards onto the wall, with the bonding layer thickness of approximately 10mm. Ensure that bonding mortar is squeezed out during EPS board pasting, and the mortar joints of EPS boards are 10mm wide. Apply pasting mortar in non-full joints as well as two holes of XPS boards.

(2) Horizontally paste EPS boards bottom-up from the starting positions also from edges and corners to the middle part. Make EPS boards interlaced and engaged in corners, and EPS board

gaps staggered on the wall surface (Figure 9-3-10). Promptly check the flatness of EPS boards with a guiding rule.

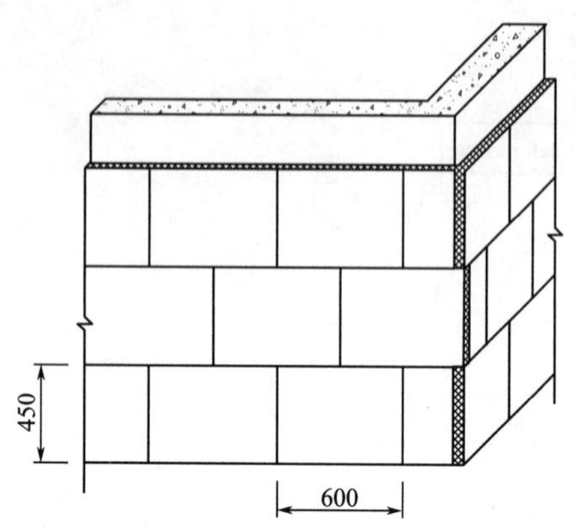

Figure 9-3-10 Schematic Diagram of EPS Board Layout in Corner (unit: mm)

(3) Cut non-standard EPS boards on the site, and note that the edges and openings are cut neatly, perpendicular to the EPS board surface.

(4) Cut the whole EPS boards to avoid splicing in four corners of the door/window openings, and make the EPS board gaps at least 200mm away from corners (Figure 9-3-11).

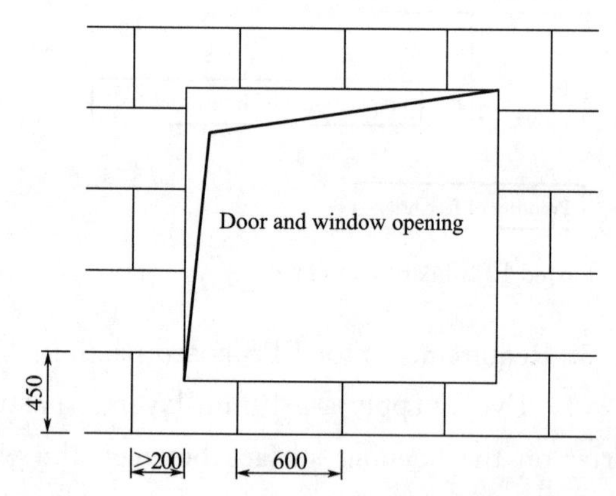

Figure 9-3-11 Schematic Diagram of EPS Board Layout in Door/Window Opening (unit: mm)

6. The side walls of outdoor parts of the external window (door) openings should be coated with XPS boards (thickness: 30mm or more). The 20mm wide gaps reserved between the external window (door) frames and XPS boards should be filled with flexible waterproof mortar, and treated with the JS waterproof coating.

7. Fire protection and leveling requirements for pasting mortar:

(1) Prior to leveling, snap the thickness control line of the leveling layer, and fabricate standard cement blocks with pasting mortar.

(2) Apply pasting mortar bottom-up from left to right for leveling.

(3) Pasting mortar may be applied in two steps. The flatness should reach ±5mm in the first step, and the mortar may be slightly higher than cement blocks in the second step. The wall surface should be leveled with a ruler and then repaired as per the flatness requirements.

8. Refer to the related construction procedures of the insulation mortar system during construction of the anti-cracking layer and finish layer.

9.3.5.4 Key Construction Points of Cast-in-situ Concrete and Non-meshed EPS Board System

1. The construction procedures of the cast-in-situ concrete and non-meshed EPS board system should meet the requirements of Figure 9-3-12.

2. Following the acceptance of steel bars of the shear wall, gaskets should be tied in a plum shape and at intervals of 600mm in two directions outside steel bars.

3. EPS boards should be installed according to the following requirements:

(1) Locate EPS boards outside the steel bars of the shear wall, and align their tongues and grooves.

(2) Install the boards of internal and external corners at first, and then assemble the boards among corner boards according to the specified sequence.

(3) Install plastic clips in a plum shape (at intervals of 600mm in two directions) at board gaps and in the middle of boards, and tie them to steel bars via wires.

4. Formworks should be installed as per the following requirements:

(1) The dimensions of corner and flat formworks should be determined based on the EPS board thickness. Large steel formworks should be used.

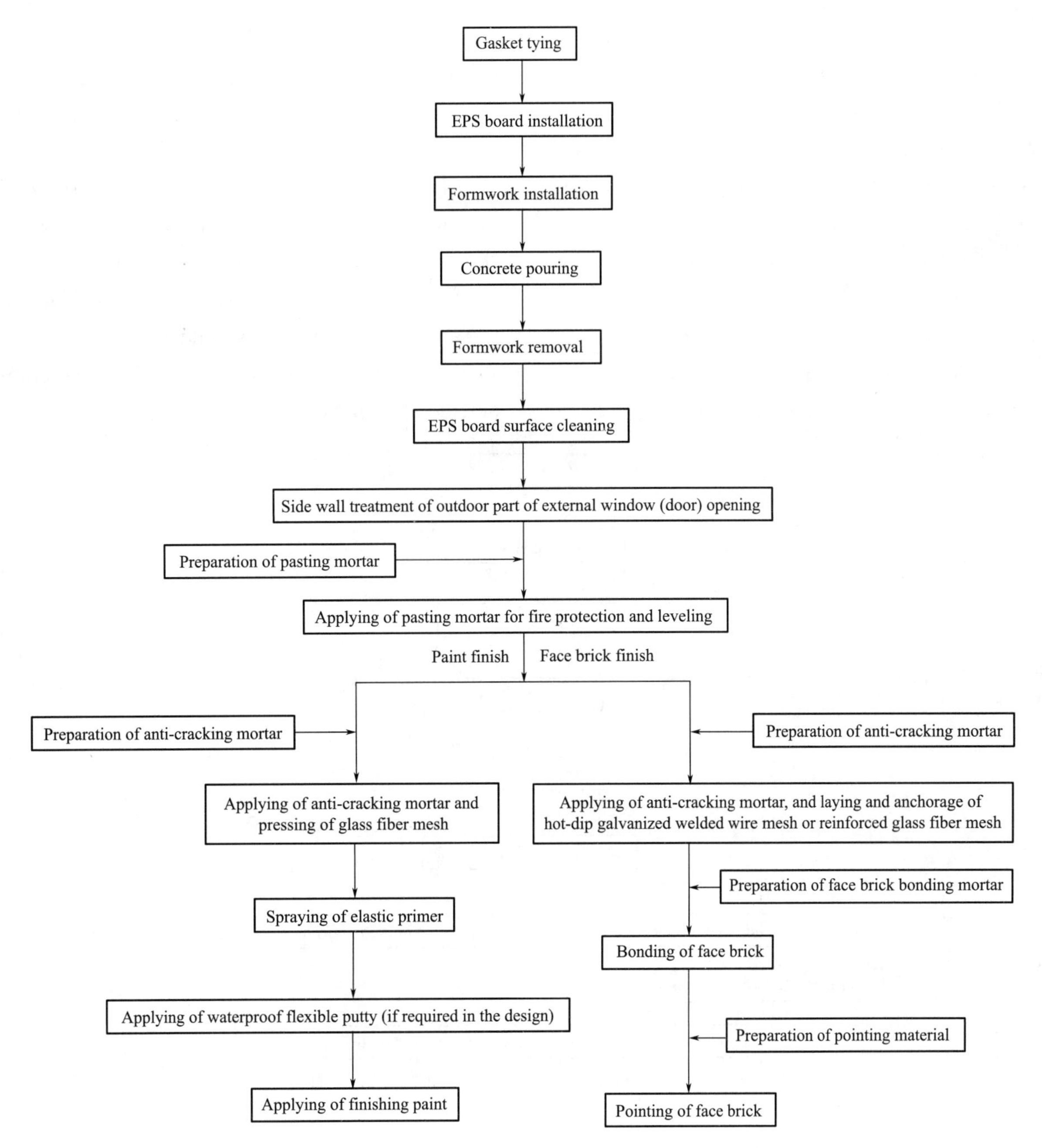

Figure 9-3-12 Construction Procedures of Cast-in-situ Concrete and Non-meshed EPS Board System

(2) The strength of concrete in the lower layer of a wall should be 7.5MPa before formworks are installed in an upper layer.

(3) Normally, corner formworks are first connected tightly and securely, with reliable positioning measures at the top and bottom. If necessary, additional supports should be used.

5. Requirements for concrete pouring:

(1) Metal "II" type shields should be used on the grooves of EPS boards as well as external formworks prior to concrete pouring.

(2) The vibrating rod should be moved horizontally at intervals of 400mm, and must be kept against EPS boards during vibration.

6. Requirements for formwork removal:

(1) External formworks should be removed first, followed by internal formworks.

(2) After removal of wall bushings, holes of the concrete wall should be filled with hard concrete, and margins larger than the EPS board thickness should be reserved outside and later filled with a high-efficiency insulation material.

(3) Damaged parts of the EPS board interface should be repaired after formwork removal.

7. Refer to the related construction procedures of the pasted EPS board system for the side wall treatment of outdoor parts of external window (door) openings.

8. Refer to the related construction procedures of the pasted EPS board system for fire protection and leveling of the pasting mortar layer.

9. Refer to the related construction procedures of the insulation mortar system during construction of the anti-cracking layer and finish layer.

9.3.5.5 Key Construction Points of Cast-in-situ Concrete and Meshed EPS Board System

1. The construction procedures of the cast-in-situ concrete and meshed EPS board system should meet the requirements of Figure 9-3-13.

2. Following the acceptance of steel bars of the shear wall, gaskets should be tied in a plum shape and at intervals of 600mm in two directions outside steel bars.

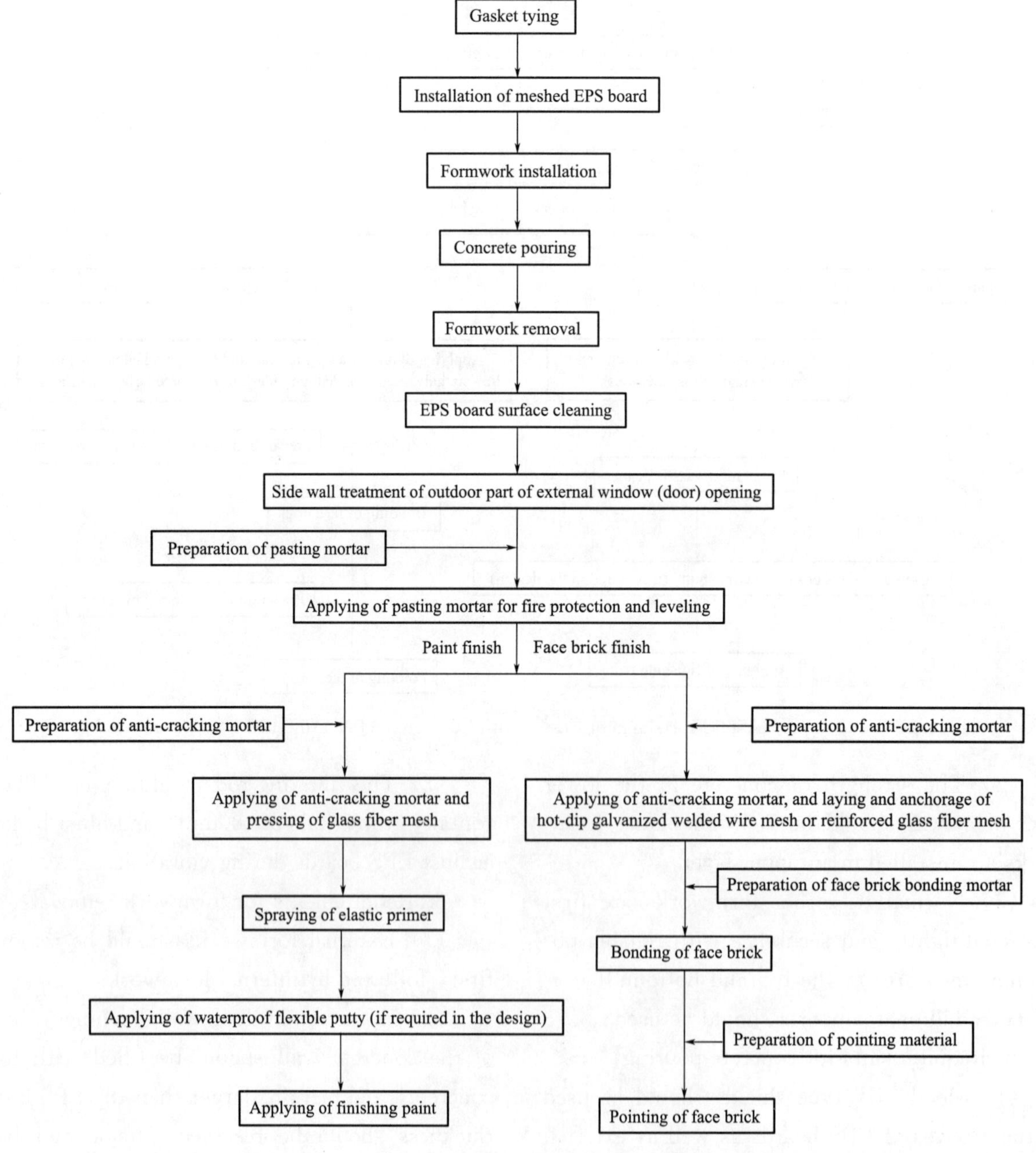

Figure 9-3-13 Construction Procedures of Cast-in-situ Concrete and Meshed EPS Board System

3. The meshed EPS boards should be installed according to the following requirements:

(1) Locate the meshed EPS boards outside the steel bars of the shear wall, and securely tie them to steel bars via mechanical anchors.

(2) Tie wire meshes at intervals of 150mm with annealed wires at board gaps, and make lap joints (width: 50mm). Use additional corner meshes and flat connecting meshes in the external corners of the outer wall and on the bottom edges of windows and balconies, with the lap width no less than 200mm.

4. Formworks should be installed as per the following requirements:

(1) Debris within the wall control lines should be cleaned before formwork installation.

(2) Corner formworks are first connected tightly and securely, with reliable positioning measures at the top and bottom. If necessary, additional supports should be used to prevent misalignment and leakage.

5. Requirements for concrete pouring:

(1) Metal "II" type shields should be used on the grooves of the meshed EPS boards as well as external formworks prior to concrete pouring.

(2) The vibrating rod should be moved horizontally at intervals of 400mm, and must be kept against the EPS boards during vibration.

6. Requirements for formwork removal:

(1) External formworks should be removed first, followed by internal formworks.

(2) After removal of wall bushings, holes of the concrete wall should be filled with expanded cement mortar, and those in the thickness direction of EPS boards should be filled with a high-efficiency insulation material.

(3) Damaged parts of the EPS board interface should be repaired after formwork removal.

7. Refer to the related construction procedures of the pasted EPS board system for the side wall treatment of outdoor parts of external window (door) openings.

8. Refer to the related construction procedures of the pasted EPS board system for fire protection and leveling of the pasting mortar layer.

9. Refer to the related construction procedures of the insulation mortar system during construction of the anti-cracking layer and finish layer.

9.3.5.6 Key Construction Points of Sprayed Polyurethane System

1. The construction procedures of the sprayed polyurethane system should meet the requirements of Figure 9-3-14.

2. Base course wall treatment requirements:

(1) Base course wall treatment should be carried out prior to construction of the insulation layer. The base course wall surface should be flat, clean and dry, without quality problems such as dust, leak, oil, and hollowing. The protrusions larger than 10mm on the external surface of the wall should be removed. Holes through the wall and damaged parts on the wall surface should be cleaned and repaired with appropriate materials. Holes on the wall surface should be moistened and repaired.

(2) Hollowing and cracking parts of the outer wall of the existing building should be removed.

3. Water must not be sprayed onto the base course wall surface. Instead, the base course wall surface should be evenly coated with moisture-proof primer, without bottom exposure.

4. Requirements for snapping of the control line and setting of the datum line:

(1) Snap the horizontal and vertical control lines of external doors and windows on the wall surface, according to the requirements for the building facade design and external thermal insulation.

(2) Hang the vertical steel datum lines in large corners (internal and external corners) of the outer wall and other places (if necessary), and horizontal lines in appropriate positions of each floor.

5. Requirements for polyurethane spraying:

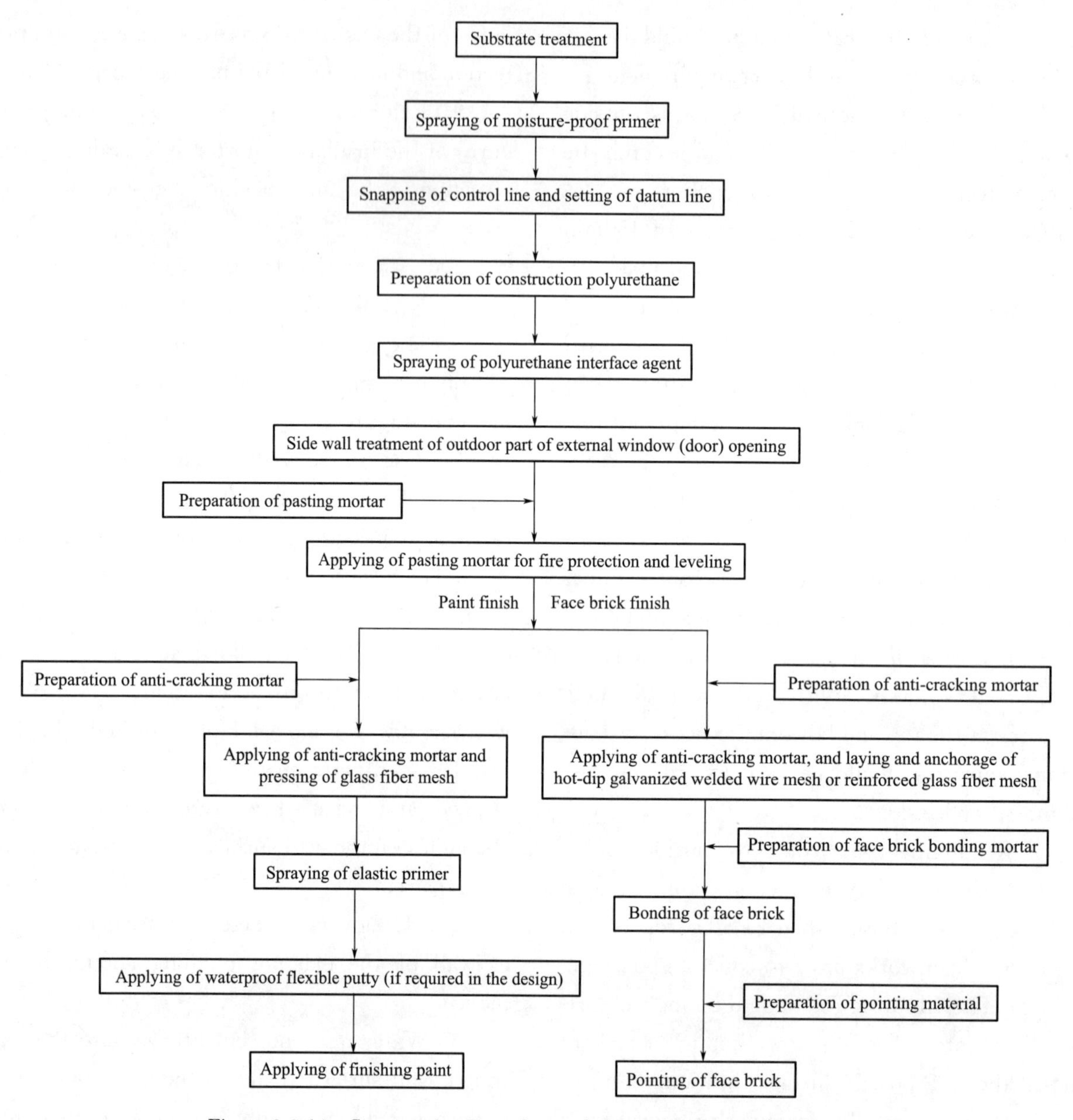

Figure 9-3-14 Construction Procedures of Sprayed Polyurethane System

(1) Fully cover doors, windows and other parts before spraying.

(2) Start the polyurethane spraying machine, and evenly spray rigid polyurethane foam onto the base course wall for foaming. The thickness of a single sprayed layer should not exceed 20mm, and the total spraying thickness should meet the minimum design requirements.

(3) After foaming of the first layer, vertically insert the insulation layer thickness signs at intervals of 400-600mm in two directions to the hard surface of the base course wall.

(4) Pay attention to wind protection in the polyurethane spraying process, and suspend construction if the wind speed exceeds 4m/s.

(5) Following 20min spraying, clean and repair the parts covered as well as the protruding parts exceeding the total thickness of the insulation layer, using tools such as the paper knife and hand saw.

6. Following the repair of the polyurethane surface or 4h after spraying, evenly spray the polyurethane interface agent onto the surface of the rigid polyurethane foam insulation layer, using the spray gun or roller brush.

7. The side walls of outdoor parts of external

window (door) openings are treated in the same method as those of the pasted EPS board system. The XPS boards pasted may be substituted with rigid polyurethane foam (minimum thickness: 30mm).

8. The sprayed rigid polyurethane foam layer should be fully cured for 48-72 hours before leveling. Refer to the related construction procedures of the pasted EPS board system for fire protection and leveling with pasting mortar.

9. Refer to the related construction procedures of the insulation mortar system during construction of the anti-cracking layer and finish layer.

9.3.5.7 Key Construction Points of Anchored Rock Wool Board System

1. The construction procedures of the anchored rock wool board system should meet the requirements of Figure 9-3-15.

2. Base course wall treatment requirements:

(1) Base course wall treatment should be carried out prior to construction of the insulation layer. The base course wall surface should be flat, clean and dry, without quality problems such as dust, leak, oil, and hollowing. The protrusions larger than 10mm on the external surface of the wall should be removed. Holes through the wall and damaged parts on the wall surface should be cleaned and repaired with appropriate materials. Holes on the wall surface should be moistened and repaired.

(2) Hollowing and cracking parts of the outer wall of the existing building should be removed.

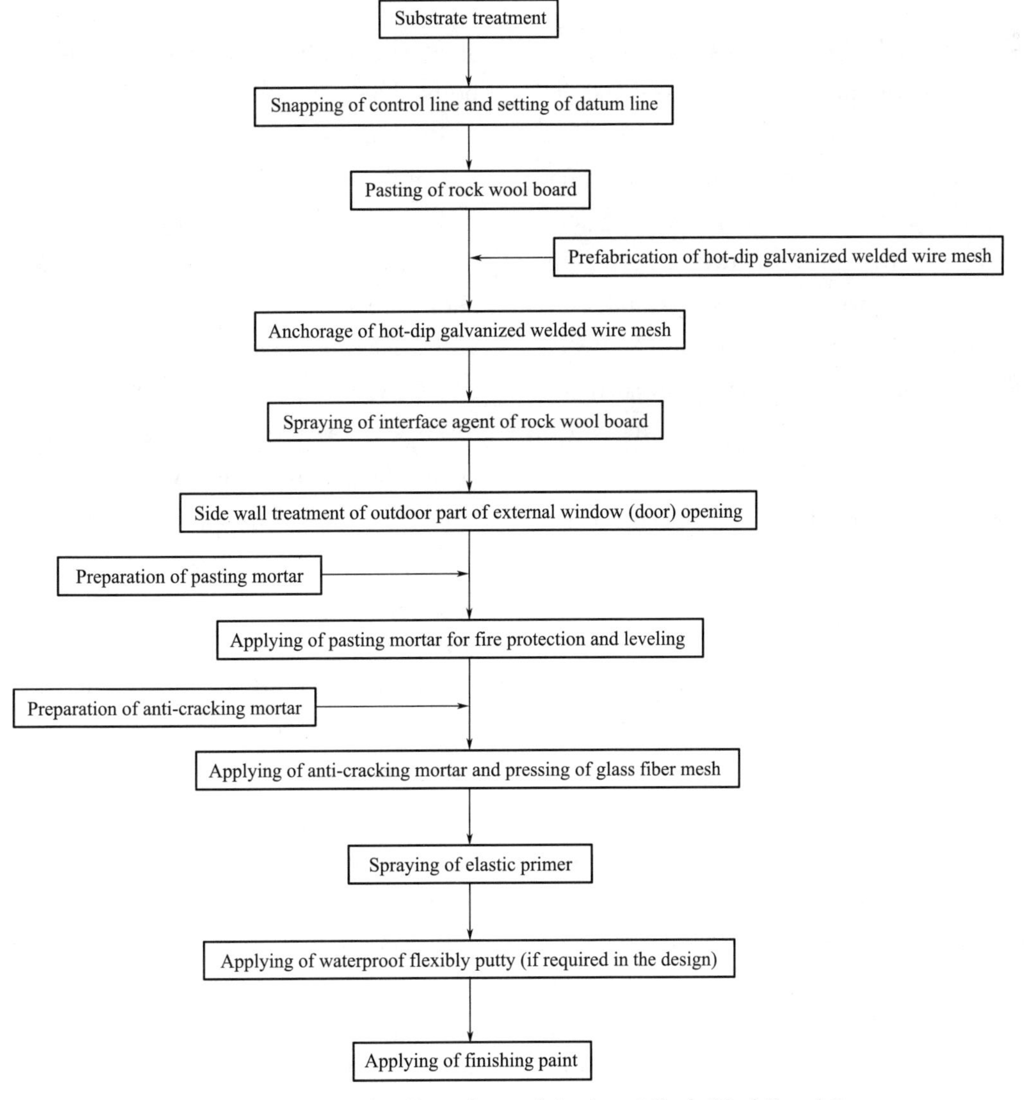

Figure 9-3-15 Construction Procedures of Anchored Rock Wool Board System

3. Requirements for snapping of the control line and setting of the datum line:

(1) Snap the horizontal and vertical control lines of external doors and windows on the wall surface, according to the requirements for the building facade design and external thermal insulation.

(2) Hang the vertical steel datum lines in large corners (internal and external corners) of the outer wall and other places (if necessary), and horizontal lines in appropriate positions of each floor.

4. Requirements for bonding of rock wool boards:

(1) Before bonding rock wool boards, check whether they are dry and whether their surfaces are flat and clean. The wet, unsmooth and contaminated rock wool boards must not be used.

(2) Snap horizontal lines with ink at the dispersion elevation of 20mm along the flattened outer wall. Take thermal insulation and waterproofing measures with XPS boards or rigid polyurethane foam boards at the lower ends of the first bottom row of rock wool boards and more than 200mm away from the dispersion part. Install metal brackets subject to anti-corrosion treatment via anchor bolts under the first layer of rock wool boards.

(3) Rock wool broads should be pasted bottom-up in the horizontal direction, with vertical joints staggered for 1/2 of the board length. They should be staggered and interlocked in corners to guarantee the verticality of such boards.

(4) Rock wool boards to be used in local irregular parts may be cut on the site, but cuts should be kept vertical to the board surfaces. The minimum size of rock wool boards for wall edges and corners should not be smaller than 300mm.

(5) The exposed sides (such as those at door/window openings, parapets and deformation joints) of rock wool boards should be fully covered with glass fiber meshes via the JS waterproof coating. The turning length of glass fiber meshes should not be less than 100mm. Once turned, glass fiber meshes should be promptly pasted onto rock wool boards.

5. Requirements for anchorage of hot-dip galvanized welded wire meshes:

(1) After rock wool boards are pasted, anchor bolt holes should be drilled in a plum shape according to the quantity requirements in the design. The depth of anchor bolt holes must not be less than 40mm. The number of anchor bolts should be calculated based on the local wind pressure.

(2) Depending on the positions of anchor bolt holes, supporting diffuser plates should be secured on the rock wool board surface via plastic clips.

(3) The hot-dip galvanized welded wire meshes should be laid and secured with anchor bolts. The side walls of doors and windows as well as the bottom of the wall should be wrapped with the prefabricated U-type hot-dip galvanized welded wire meshes, and corners with the prefabricated L-type hot-dip galvanized welded wire meshes. The wrapping meshes should be secured with rock wool boards via anchor bolts.

6. Once the hot-dip galvanized welded wire meshes are secured, the prepared interface agent should be promptly and evenly onto the rock wool board surface via the special spray gun. Both the rock wool board surface and hot-dip galvanized welded wire meshes should be fully coated with the interface agent.

7. Refer to the related construction procedures of the pasted EPS board system for the side wall treatment of outdoor parts of external window (door) openings.

8. Refer to the related construction procedures of the pasted EPS board system for leveling with pasting mortar.

9. Refer to the paint finish construction procedures of the insulation mortar system for construction of the anti-cracking layer and finish layer.

9.3.5.8 Key Construction Points of Pasted and Reinforced Vertical-fiber Rock Wool Board System

1. The construction procedures of the pasted and reinforced vertical-fiber rock wool board system should comply with the requirements of Figure 9-3-16.

2. Base course wall treatment requirements:

(1) Base course wall treatment should be carried out prior to construction of the insulation layer. The base course wall surface should be flat, clean and dry, without quality problems such as dust, leak, oil, and hollowing. The protrusions larger than 10mm on the external surface of the wall should be removed. Holes through the wall and damaged parts on the wall surface should be cleaned and repaired with appropriate materials. Holes on the wall surface should be moistened and repaired.

(2) Hollowing and cracking parts of the outer wall of the existing building should be removed.

3. The base course wall surface should be coated evenly and fully with a thin layer of interface mortar. The masonry wall with a large water absorption rate should be wetted and dried in the shade before interface mortar is used.

4. Requirements for snapping of the control line and setting of the datum line:

(1) Snap the horizontal and vertical control lines of external doors and windows on the wall surface, according to the requirements for the building facade design and external thermal insulation.

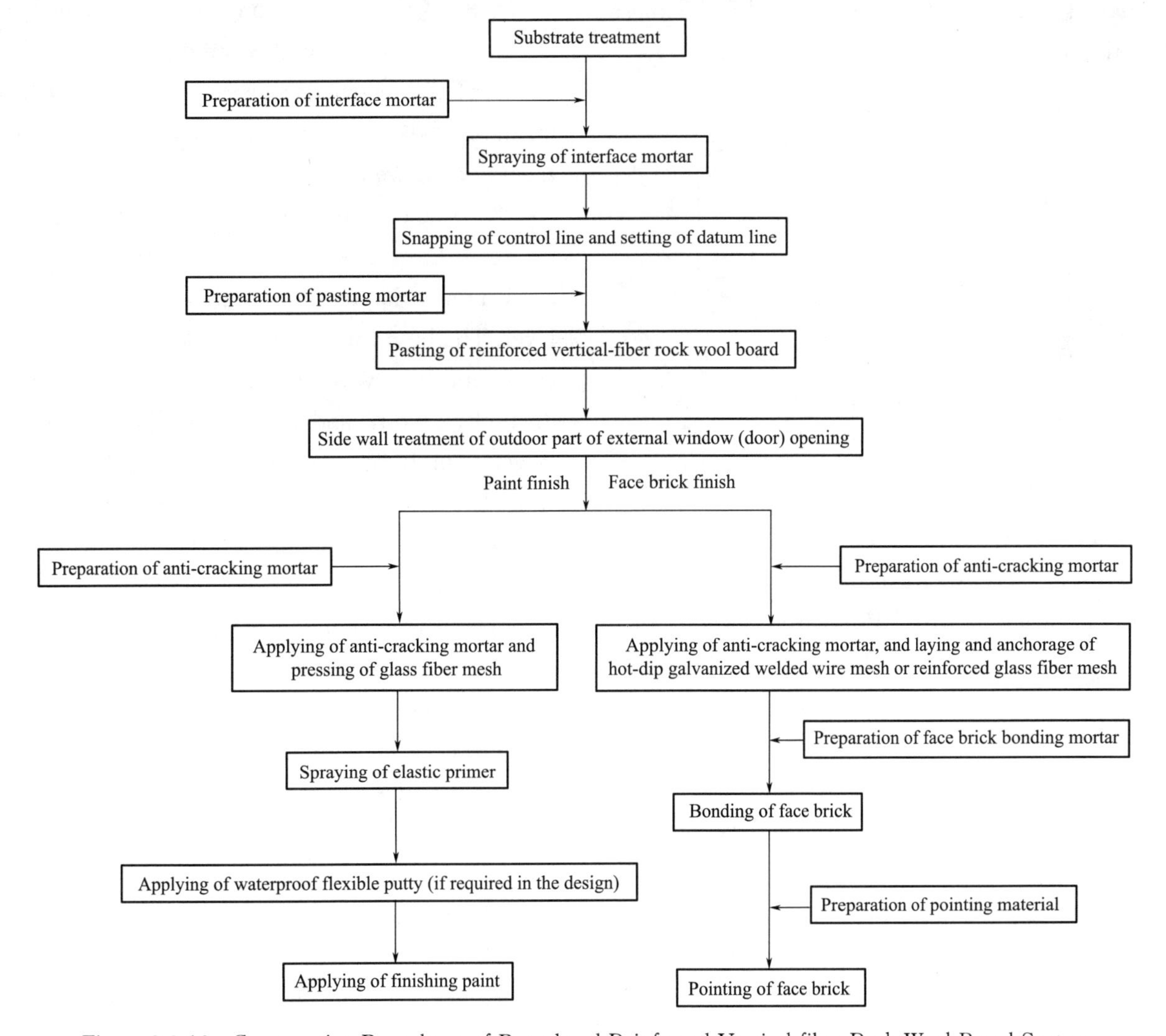

Figure 9-3-16 Construction Procedures of Pasted and Reinforced Vertical-fiber Rock Wool Board System

(2) Snap horizontal lines at the starting positions around the building for pasting of reinforced vertical-fiber rock wool composite boards. Prior to pasting the reinforced vertical-fiber rock wool composite boards, paste standard blocks in external corners, hang horizontal control lines on standard blocks, and set horizontal thickness control lines between two vertical through lines of one wall.

5. Requirements for pasting of the reinforced vertical-fiber rock wool board system:

(1) Install two L-shaped brackets (Figure 9-3-17) with the spacing of 300mm via fasteners or special anchor bolts on the lower side of each reinforced vertical-fiber rock wool composite board. Install the long ends of double U-shaped inserts into the slots of the brackets, and both ends of U-shaped inserts at the centers in the thickness direction of two insulation board layers.

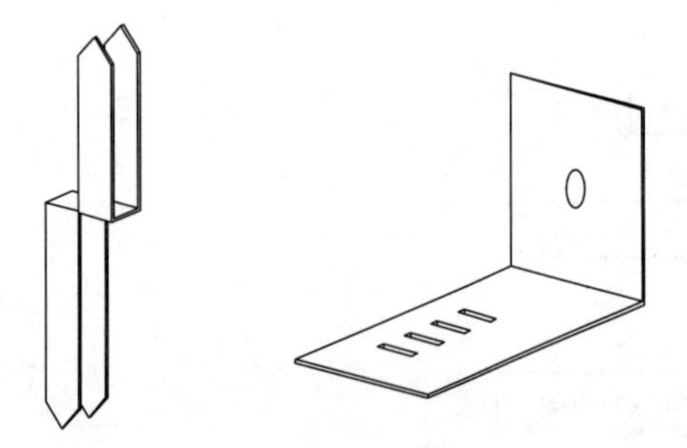

Figure 9-3-17 Double U-shaped Inserts and L-shaped Brackets

(2) Snap horizontal lines with ink at the dispersion elevation of 20mm along the flattened outer wall. Take thermal insulation and waterproofing measures with XPS boards or rigid polyurethane foam boards at the lower ends of the first bottom row of rock wool boards and more than 200mm away from the dispersion part.

(3) Apply a 5-10mm layer of pasting mortar on the wall surface and board bonding surface, and then paste the reinforced vertical-fiber rock wool composite board onto the wall, with the mortar joint width of 10mm. Level non-full joints with pasting mortar.

(4) Horizontally paste the reinforced vertical-fiber rock wool composite boards bottom-up from the starting positions and also from the edges and corners to the middle. They should be staggered and interlocked in corners, and staggered on the wall surface. The flatness should be promptly checked with the guiding rule.

6. Refer to the related construction procedures of the pasted EPS board system for the side wall treatment of outdoor parts of external window (door) openings.

7. Refer to the related construction procedures of the pasted EPS board system for fire protection and leveling with pasting mortar.

8. Refer to the related construction procedures of the insulation mortar system for construction of the anti-cracking layer and finish layer.

9.3.5.9 Key Construction Points of Bonded Insulation Board System

1. The construction procedures of the bonded insulation board system should meet the requirements of Figure 9-3-18.

2. Base course wall treatment requirements:

(1) Base course wall treatment should be carried out prior to construction of the insulation layer. The base course wall surface should be flat, clean and dry, without quality problems such as dust, leak, oil, and hollowing. The protrusions larger than 10mm on the external surface of the wall should be removed. If the base course wall is constructed with masonry or the flatness of the concrete wall does not comply with relevant standards, a cement mortar leveling layer should be applied on the external surface. Holes through the wall and damaged parts on the wall surface should be cleaned and repaired with appropriate materials. Holes on the wall surface should be moistened and repaired.

(2) Hollowing and cracking parts of the outer wall of the existing building should be removed.

3. Requirements for snapping of the control line and setting of the datum line:

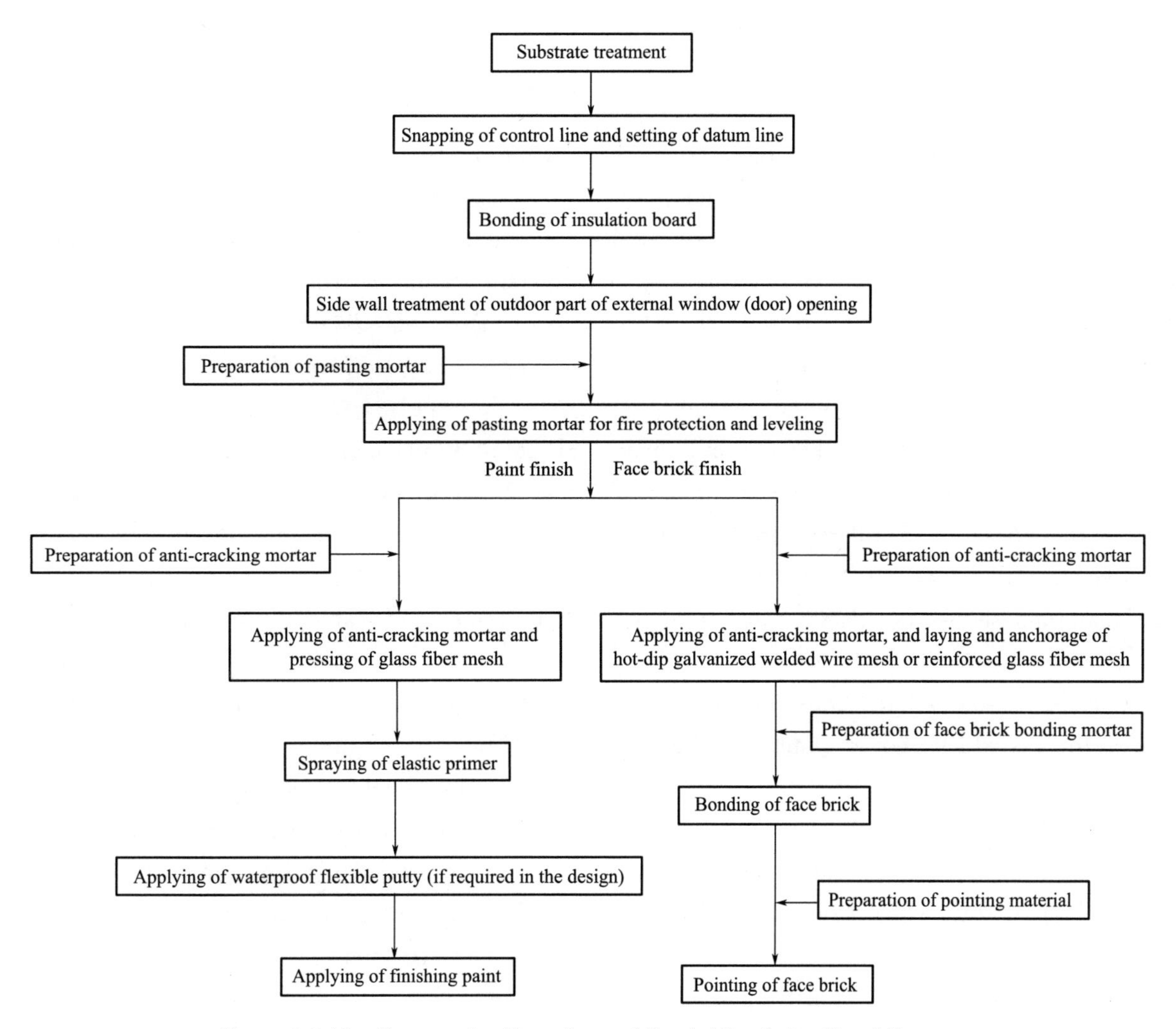

Figure 9-3-18 Construction Procedures of Bonded Insulation Board System

(1) Snap the horizontal and vertical control lines of external doors and windows on the wall surface, according to the requirements for the building facade design and external thermal insulation.

(2) Hang the vertical steel datum lines in large corners (internal and external corners) of the outer wall and other places (if necessary), and horizontal lines in appropriate positions of each floor.

4. Requirements for pasting of insulation boards:

(1) The supporting interface agent should be applied on the EPS boards, XPS boards and internal and external surfaces. The polyurethane composite boards and phenolic boards may be coated with the appropriate interface agent based on the board requirements.

(2) Insulation boards (EPS boards, XPS boards, polyurethane composite board, phenolic boards and reinforced vertical-fiber rock wool composite boards) should be bonded along the control lines, with vertical joints staggered. In addition, the insulation boards should be secured, and the effective bonding area must comply with the design requirements, without connecting cavities.

(3) Insulation boards should be staggered and interlocked in wall corners. Those used in four corners of door/window openings should be made by cutting the entire boards instead of assembly. The joints of insulation boards should be at least 200mm away from corners.

(4) Two L-shaped brackets (Figure 9-3-17)

should be installed with the spacing of 300mm via fasteners or special anchor bolts on the lower side of each reinforced vertical-fiber rock wool composite board. The long ends of double U-shaped inserts into the slots of the brackets, and both ends thereof at the centers in the thickness direction of two insulation board layers. Thermal insulation and waterproofing measures with XPS boards or rigid polyurethane foam boards should be taken at the lower ends of the first bottom row of rock wool boards and more than 200mm away from the dispersion part.

(5) If required, a fire barrier made of reinforced vertical-fiber rock wool composite board or other materials should be installed according to the design requirements.

5. Refer to the related construction procedures of the pasted EPS board system for the side wall treatment of outdoor parts of external window (door) openings.

6. Refer to the related construction procedures of the pasted EPS board system for fire protection and leveling with pasting mortar.

7. Refer to the related construction procedures of the insulation mortar system for construction of the anti-cracking layer and finish layer.

9.3.6 Acceptance of Construction Quality

9.3.6.1 General

1. The acceptance of the construction quality of the adhesive polystyrene granule composite system should be conducted in accordance with the *Unified Standard for Constructional Quality Acceptance of Building Engineering* (GB 50300) and the Code for Acceptance of Energy Efficient Building Construction (GB 50411).

2. The main composition of the adhesive polystyrene granule composite system should be witnessed and retested on the site as per the requirements of Table 9-3-38, and the number of samples should meet the *Code for Acceptance of Energy Efficient Building Construction* (GB 50411).

3. The quality inspection, concealment acceptance and inspection lot acceptance should be carried out promptly during construction, and the wall energy-saving sub-divisional works should be subject to acceptance after construction. The wall energy-saving works constructed simultaneously with the main structure should be accepted together with the main structure.

Witness, Sampling and Retesting Items of Main Composition of Adhesive polystyrene granule Composite System

Table 9-3-38

Material	Retest Item
EPS board	Apparent density, thermal conductivity, tensile strength and dimensional stability
Rigid polyurethane foam	Density, thermal conductivity and dimensional stability
Rock wool board	Density and thermal conductivity
Reinforced vertical-fiber rock wool board	Rock wool direction and protective layer thickness
Insulation mortar	Dry apparent density, thermal conductivity and tensile strength
Pasting mortar	Dry apparent density, thermal conductivity and tensile bonding strength
Interface mortar, interface agent, adhesive, and anti-cracking mortar	Tensile bonding strength
Moisture-proof primer	Adhesion
Glass fiber mesh	Mass per unit area, alkali-resistant breaking strength and alkali-resistant breaking strength retention rate
Hot-dip galvanized welded wire mesh	Wire diameter, mesh size and tensile strength of spot weld

4. The wall constructed with the same material, technology and method should be divided into inspection lots of 500-1,000m^2, and the area smaller than 500m^2 should be regarded as one inspection lot. The inspection lots may also be determined jointly by the construction and supervision (contracting) unit based on the principles of the consistency with the construction process and convenience for construction and acceptance.

5. The following documents should be submitted for the completion acceptance of the adhesive polystyrene granule composite system:

(1) Design documents, drawing reviews, design changes and negotiation records;

(2) Dedicated construction plan;

(3) Type test report, certificates of major materials, delivery test report, site re-inspection report and site acceptance records;

(4) Construction technology disclosure;

(5) Construction process records and construction quality inspection records;

(6) Other data to be submitted.

9.3.6.2 Main Control Items

1. The varieties, specifications and properties of materials used in the adhesive polystyrene granule composite system should meet the requirements of the design, this regulation and related standards.

Inspection method: visual inspection and measurement with rulers; examination of quality certificates; and check of the product certificates, delivery inspection reports and site re-inspection report.

2. The thickness of the insulation layer should meet the design requirements, with the negative deviation no greater than 3mm.

Inspection method: pin insertion..

3. The rigid polyurethane foam insulation layer should be free of sagging, collapse, breakage, core burning, etc. The foam structure should be uniform and delicate, without obvious shrinkage after 24h.

Inspection method: visual inspection.

4. The thickness of the leveling layer should meet the design requirements.

Inspection method: on-site measurement.

5. The bonding area of insulation boards in the bonded insulation board system should meet the requirements of this regulation.

Inspection method: on-site measurement.

6. The number, locations and anchoring depth of anchor bolts of the anchored rock wool board system should meet the design requirements.

Inspection method: visual inspection. The anchoring depth should be measured after anchor bolts are dismantled.

7. The safety and durability of the face brick finish must comply with the design requirements, without hollowing or cracking. The tensile bonding strength measured between the face brick finish and anti-cracking layer on the site should not be less than 0.4MPa.

Inspection method: knocking with a hammer and visual inspection. The tensile bonding strength between the face brick finish and anti-cracking layer should be tested in accordance with the *Test Standard for Bonding Strength of Building Facing Brick* (JGJ 110).

8. All structural layers must be firmly bonded with no delamination or hollowing. The surface layer must not be cracking. Both the bonding strength and the connection method should meet the design requirements.

Inspection method: visual inspection and on-site pull-out test.

9. The thermal insulation measures on peripheral sides of door/window openings of the outer wall or the water adjacent to a non-heated space as well as the thermal insulation measures on peripheral sides of bay windows should meet the design requirements.

Inspection method: visual inspection based on the design, and sampling and cutting inspection if necessary; and check of acceptance records of concealed works.

10. The isolation or insulation measures at thermal bridges of an outer wall should meet the design requirements.

Inspection method: visual inspection.

9.3.6.3 General Items

1. Except that of the cast-in-site concrete and EPS board composite system, the base course wall treatment of the adhesive polystyrene granule composite system should meet the requirements of the design and construction scheme.

Inspection method: visual inspection based on the design and construction scheme; and check of acceptance records of concealed works.

2. The internal and external surfaces of the EPS board should be fully coated with the supporting interface agent, without efflorescence; and the external surface of the rigid polyurethane foam and rock wool board should also be fully coated with the supporting interface agent.

Inspection method: visual inspection.

3. Glass fiber meshes and hot-dip galvanized welded wire meshes should be laid and overlapped in accordance with the requirements of the design and construction scheme, and must not be wrinkled or exposed. Anti-cracking mortar should be compacted, without hollowing.

Inspection method: visual inspection; and check of acceptance records of concealed works.

4. The anti-cracking layer should be smooth and clean, without delamination, hollowing or cracking.

Inspection method: visual inspection.

5. The finish layer must not be leaky. Sealing measures at connections and closures of the insulation, finish and other layers must comply with the design requirements.

Inspection method: visual inspection; and check of test reports as well as acceptance records of concealed works.

6. Wall defects generated during construction, such as wall bushings, scaffolding holes and thermal bridge isolation measures, should comply with the requirements of the construction scheme.

Inspection method: visual inspection based on the construction scheme.

7. The joints of insulation boards should be smooth and tight, and also conform to the requirements of the construction scheme.

Inspection method: visual inspection.

8. For special parts such as external corners prone to collision, door/window openings and connections between base course walls made with different materials, reinforcing measures against cracking and damage of the insulation layer should comply with the requirements of the design and construction scheme.

Inspection method: visual inspection; and check of acceptance records of concealed works.

9. The allowable deviation and inspection method of the surface layer of the wall with external thermal insulation should meet the requirements of Table 9-3-39.

Allowable Deviation and Inspection Method of Surface Layer of Wall with External Thermal Insulation

Table 9-3-39

No.	Item	Allowable Deviation (mm)	Inspection Method
1	Surface flatness	4	Inspection with 2m guiding rule and wedge feeler
2	Façade verticality	4	Inspection with 2m verticality measuring ruler
3	Uprightness of internal and external corner	4	Inspection with square
4	Straightness of dividing joint (decorative line)	4	Inspection with 5m line or through line (in the case of less than 5m) and steel ruler

9.4 Summary

1. The adhesive polystyrene granule mortar, which was originated abroad, has been developed in China. With the improvement of the building energy-saving level, the adhesive polystyrene granule insulation technology has been deeply researched and developed, a wealth of experience in application of external thermal insulation has been gained, and the theory on the external thermal insulation technology has been enriched and expanded, thus promoting the development of thermal insulation and energy conservation technologies in China.

2. Three major technical concepts on the external thermal insulation are the summary and theoretical sublimation of China's building insulation technology. Based on the principle that the safety and durability of the external thermal insulation system are main contradictions during applications of external thermal insulation and energy conservation technologies, the relationship between the location of the insulation layer and the temperature field of the outer wall is revealed, and the technical route for cracking control of the thermal insulation system is also put forward.

3. The physical and mechanical properties, cracking resistance, fire resistance, insulation performance and durability of the adhesive polystyrene granule mortar are studied in depth. The practice shows that the adhesive polystyrene granule mortar has excellent physical and mechanical properties, durability, fire resistance, and cracking resistance. The anti-cracking protective layer plays a key role in the cracking resistance of the entire thermal insulation system, so its anti-cracking structure is a guarantee for the cracking and weathering resistance of the adhesive polystyrene granule composite insulation system.

4. It is found in the comparison and analysis of insulation boards that their performance indicators vary greatly, especially their thermal deformation. The direct use of the thin-plastered EPS board practice and its supporting materials in building insulation projects may lead to major technical risks. There are significant differences in the deformation and thermal stress of insulation board surfaces. The thermal expansion coefficients of adjacent materials should not be far different from each other. If the adhesive polystyrene granule insulation mortar is used as a leveling transition layer, the surface temperature of insulation boards can be effectively reduced, and the occurrence time of high temperature can be delayed, which will decrease the deformation and stress of insulation boards, drastic changes in the stress as a result of temperature variations as well as cracking risks of the insulation surface, and achieving good weathering resistance of the external thermal insulation system.

5. The adhesive polystyrene granule composite insulation technology has significant technical advantages and applies to different climate regions and engineering forms. With the maturity of this technology, related industry standards and technical regulations have been formulated, and a large amount of engineering experience has been obtained, with appreciable economic and social benefits.

6. The engineering quality is critical to the success of a project and directly affects the future operation of the project as well as the reputation and benefits of a construction unit. In order to ensure the high-quality, high-efficiency, low-consumption, safety, sanitary and civilized construction on the site, the technology, quality, safety, material, progress and site management needs to be strengthened, and the construction quality and building energy efficiency should be improved constantly.

10. Comprehensive Resource Utilization of External Thermal Insulation System

10.1 Overview

In the field of building energy conservation, the external thermal insulation system, featuring the excellent thermal performance, high insulation performance, low overall investment and prolonged structural life, etc., has become a main technology for energy conservation and thermal insulation of outer walls of buildings in China. It mainly consists of organic polymer materials and cement for building energy conservation, but large quantities of energy and resources are needed and a lot of "three wastes" (waste gas, water and solids) are discharged during production. Meanwhile, there are a large number of solid wastes such as waste polystyrene plastics, waste polyester plastics, waste rubber tires, waste paper, fly ash and tailing sand in China, which not occupy much land but also cause serious environmental pollution. On February 9, 2010, the Ministry of Environmental Protection, the National Bureau of Statistics and the Ministry of Agriculture jointly issued the *First National Survey Communiqué of Pollution Sources*, announcing the national solid waste discharge data: 3.852 billion tons of industrial solid wastes were generated, 1.804 billion tons utilized comprehensively, 0.441 billion tons disposed of, 1.599 billion tons stored in the current year, and 49.1487 million tons dumped. Therefore, one important trend of external thermal insulation development is to develop external thermal insulation systems with solid wastes as part of raw materials, in order to reduce the energy consumption of production and meet the energy conservation and emission reduction requirements of the circular economy.

The comprehensive utilization of solid wastes in external thermal insulation products for buildings has been studied extensively and deeply. In 2002, the "external thermal insulation material based on the adhesive polystyrene granule" was recognized by Beijing Science and Technology Commission as one project under Beijing's Torch Plan, and listed in the National Key New Products by the Ministry of Science and Technology. The "external thermal insulation material based on the recycled polyurethane" won the certificate of 2004 National Key New Products, issued by the Ministry of Science and Technology. In order to promote the use of solid wastes such as fly ash, tailing sand, waste rubber particles and waste paper fibers in external thermal insulation systems, a large number of supporting mortar products have been developed in China's external thermal insulation industry, and widely applied in ten types of external thermal insulation systems, including the adhesive polystyrene granule insulation systems, sprayed rigid polyurethane foam insulation systems, adhesive polystyrene granule pasted EPS board systems, cast-in-situ concrete and non-meshed EPS board systems, and cast-in-situ concrete and meshed EPS board systems. The mass content of solid wastes (fly ash and tailing sand) in each dry-mixed mortar product is more than 30%, and that in each type of external thermal insulation system is more than 50%, thus achieving good economic and social benefits. Both the adhesive polystyrene granule mortar systems and composite systems thereof have been extensively applied. A large number of engineering practices have proved that the comprehensive utilization of solid wastes in external thermal insulation systems helps to improve the cracking and weathering resistance, without declining in the system quality.

10.2 Evaluation of Comprehensive Resource Utilization

10.2.1 Solid Contents of External Thermal Insulation System and Its Materials

The development of the traditional building material industry, depending mainly on the high consumption of resources, is a typical resource-dependent industry. According to calculations, mineral resources of 4.61 billion tons in total were consumed in the building material industry in 2007, including 2.41 billion tons in wall materials (52.3% of the total consumption) and 1.79 billion tons in cement (38.8% of the total consumption). That is, the resource consumption of these two sectors accounts for more than 90% of the total of the building material industry.

In addition to the high consumption of resources, the building material industry also involves the most applications of solid wastes among industrial sectors. Many industrial wastes can be used as substitutes and fuels to develop green building materials during production.

According to the Assessment Standard for Green Building (GB/T 50378-2014), it is encouraged, on the premise of satisfying performance, to use concrete blocks, cement products and recycled concrete made of aggregate produced with building wastes; to use cement, concrete, wall materials, insulation materials and other building materials made with industrial wastes, crop straws, construction wastes and sludge as raw materials; and also to use building materials made by treatment of domestic wastes. In order to guarantee the conformity of wastes to the quantity requirements, it is stipulated in this standard that the proportion (by weight) of building materials produced with wastes to all building materials of the same type should not be less than 30%. Gypsum blocks used as an internal partition material are taken as an example. The proportion of blocks made with industrial by-product gypsum (desulfurized gypsum, phosphogypsum, etc.) to all gypsum blocks used in buildings is more than 30%, which meets the requirements of this clause.

However, there are currently no standards on assessment of green building materials. The domestic building material industry should still endeavor to actively carry out the related work and formulate and implement relevant standards, in order to regulate and use green building materials and also promote the development of green buildings.

10.2.2 Energy Consumption and Waste Discharge in Production of External Thermal Insulation System and Its Materials

The unit energy consumption, waste gas emission and residue discharge of production should be used as major factors for assessment of green building materials. In recent years, the energy consumption of the building material industry rises year by year with the yield increasing, and the total energy consumption in the building material industry has been increased from 100 million tons of standard coal in 2001 to 209 million tons in 2008. The building material industry involves the kiln production as the main production means and coal as the main energy. Pollutants generated in the production process have significant impact on the environment, mainly including the dust, smoke, sulfur dioxide and nitrogen oxide. The amount of dust and smoke, accounting for the largest proportion, reached 4.612 million tons in 2008, equal to 36.7% of the total industrial emissions and 31% of the total emissions in China.

The high resource consumption and high pollution discharge of the building material industry must be changed. Therefore, it is a strategic goal of China's building material industry in the 21st century and an inevitable historical development trend to raise the awareness of the importance of sustainable development strategies, strive to develop green, ecological and environmentally-friendly building materials, fundamentally cha- nge the long-term high-investment, high-pollution and low-efficiency extensive production mode in China's building material industry, select the resource-saving, pollution-minimizing, quality-benefit and technology-based development method, and also organically combine the development of the building material industry with the ecological environment protection and pollution control.

10.3 Comprehensive Utilization of Solid Waste

10.3.1 Comprehensive Utilization of Solid Waste in Insulation Material

In accordance with the Eleventh Five-Year Plan, attention should be paid to environmental protection and resource conservation during improvement of the building energy efficiency, and energy-saving building products and technologies should be developed. In recent years, with the development of the world economy, especially the rapid economic growth of developing countries such as China and India, demands for strategic materials such as oil have grown fast, leading to changes in the relationship between oil supply and demand as well as the soaring of oil prices.

Main insulation materials used in external thermal insulation systems in China are organic materials such as the polystyrene board and polyurethane. A large amount of fossil energy is needed in the production process. Petrochemical products closely linked to oil prices, such as styrene monomers, are main raw materials in preparation of EPS granules, and their prices rise fast with the oil price increasing. Polyester polyol, one important component of polyurethane-based insulation materials, is mainly produced by means of synthesis of chemical materials, in which a lot of petroleum is consumed. Under the current conditions of energy shortage, extensive applications of the polystyrene board and polyurethane are equivalent to energy conservation on one hand and high energy consumption on the other hand.

China, a relatively resource-constrained country, should attach sufficient importance to environmental protection issues of the insulation material industry, actively develop insulation materials coordinated with the environment, and minimize the resource consumption and harm to the environment during raw material preparation (exploitation or transportation), production, use and future handling. The extensive use of wastes during production of insulation materials is conducive to the conservation of natural resources, and reduction of waste pressure on the environment as well as energy consumption in the production process.

10.3.1.1 Waste EPS Foam

The waste expandable polystyrene (EPS) foam, featuring light weight, shock absorption, low moisture absorption, easy formation and low price, is widely applied in shock-proof packages of electrical appliances, instruments, craftwork and other fragile valuables, as well as packages of fast food. Most of these packages are disposable, resulting in a large number of wastes. The EPS plastic with stable chemical properties, low density, large size, ageing resistance and corrosion resistance cannot be degraded by themselves, which causes increasingly serious environmental problems, likened to "white pollution". If waste EPS foam is recycled, the problem of "white pollution" will also be solved, so good economic and social benefits will be achieved. Currently, EPS foam is recycled mainly in the following four ways.

1. Lightweight insulation material

The waste EPS foam is first crushed and then mixed with concrete to produce lightweight bricks. The wall material made of EPS lightweight concrete is identified as a non-combustible building material, with good head and sound insulation effects. The EPS lightweight concrete can also be used as an antifreeze material during building and road construction. Additionally, the EPS scrap can be used in production of lightweight blocks, insulation mortar and lightweight mortar. The mixture of the clay and EPS scrap at an appropriate proportion can be cauterized at high temperature to produce hollow clay bricks with high strength and excellent insulation properties.

2. Fill type rigid polyurethane foam

Rigid polyurethane foam is generally produced by fast reaction molding of two components of low viscosity. As heat is released in the production process, the waste EPS scrap meets the filling requirements, namely, low cost, extensive sources, closed-cell structure, low water absorption rate, heat resistance, appropriate bonding strength and almost identical physical properties.

3. Depolymerization and regeneration

The defoamed waste EPS can be crushed into 3-5mm particles for preparation of styrene monomers.

The continuous bond in each monomer is the weak link of a styrene polymer, so the styrene monomer can be generated by means of heating at a high temperature, which is known as the depolymerization process:

$$\left[CH(C_6H_5)-CH_2 \right]_n \longrightarrow nCH(C_6H_5)=CH_2$$

At present, there are successful technologies both at home and abroad, but a lot of oil is needed. With the prices of raw materials soaring in recent years, the comprehensive production costs have exceeded ordinary production costs. According to preliminary calculations, the oil consumption for recycling of 1t EPS is 1.1 to 1.2 times that under normal conditions.

4. Adhesive polystyrene granule mortar

A large number of solid wastes such as EPS foam are used during production of the materials of adhesive polystyrene granule insulation systems. The adhesive polystyrene granule mortar is made by mixing the mineral binder, light EPS granule aggregate and water, and can be applied separately on a base course wall or compounded with other insulation materials to form a composite thermal insulation system. The light EPS granule aggregate, which accounts for more than 80% of this insulation material by volume, is completely made with the recycled EPS scrap. Only the 30mm insulation mortar layer can meet the heat transfer coefficient requirements in the energy conservation (50%) standards for southern areas. The composite thermal insulation system may be applied in northern areas, such as the adhesive polystyrene granule pasted EPS board system (total thickness of adhesive polystyrene granule mortar: normally 25-40mm). Given that the thickness of adhesive polystyrene granule mortar is 30mm, the waste EPS foam ("white pollutant") of 30,000m^3 is consumed per 1,000,000m^2 of the adhesive polystyrene granule insulation system. Assuming that the total construction area of new buildings is 2,000,000,000m^2 per year, the wall area accounts for 60% of the building area, the consumption of adhesive polystyrene granule insulation mortar accounting for 10% of the total consumption of new buildings, and the annual construction area of adhesive polystyrene granule insulation systems is 120,000,000m^2, the waste EPS foam ("white pollutant") of 3,600,000m^3 will be consumed. Solid wastes such as fly ash, tailing sand, waste paper fibers and waste rubber particles can be extensively used (comprehensive utilization rate: more than 50%) in supporting mortar of the adhesive polystyrene granule insulation systems, such as the interface mortar, insulation binder, anti-cracking mortar, face brick bonding mortar and pointing material. The mortar consumption is about 17kg per square meter of the paint finish of the adhesive polystyrene granule insulation system and 30kg per square meter of the face brick finish. Provided that the annual construction area of adhesive polystyrene granule insulation systems in China is 120,000,000m^2, 1-1.8 million tons of solid wastes such as fly ash and tailing sand can be utilized.

With the application of adhesive polystyrene granule insulation systems, especially the release of the industry standard "External Thermal Insulating Rendering Systems Made of Mortar with Mineral Binder and Using Expanded Polystyrene Granule as Aggregate" (JG 158-2004), the "white pollutant" waste EPS foam is highly demanded on the market. This has made and will continue to make great tremendous contributions to China's disposal of the "white pollution" of EPS foam.

10.3.1.2 Waste Polyester Plastic Bottle

Almost all bottles of carbonated beverages, mineral water, edible oil and similar products available currently on the market are made of polyester. According to the statistics, more than 1.2 billion polyester bottles are produced and consumed in China per year, equivalent to 6.3 thousand tons of polyester waste. The amount of polyester consumed each year in the world is 13 million tons, including 150 thousand tons for beverage bottles. Waste polyester bottle cannot be degraded by themselves if discarded, thereby causing serious environmental pollution and waste of resources. It is quite important and meaningful to effectively recycle waste polyester bottles.

Polyester polyol, a raw material of the recycled polyurethane insulation material, can be produced with waste polyester bottles. That is, polyester polyol produced by means of chemical treatment of solid wastes such as recycled polyester bottles can be combined as a component of polyurethane polyol into a polyurethane insulation material for external thermal insulation projects.

The utilization rate of waste polyester bottles in the polyol as an external thermal insulation material based on the recycled polyurethane is up to 30% (by mass). Taking Beijing as an example, when the thickness of the rigid polyurethane foam insulation layer is 40-50mm and the consumption of white polyurethane is about 0.6-0.7kg per square meter, about 10 polyester bottles are used per square meter (one per 500mL). In total, 10 million polyester bottles can be recycled per 1 million square meters of the external thermal insulation system based on the recycled rigid polyurethane foam, equivalent to the reduction of white volume by approximately 5000m^3.

10.3.1.3 Waste Polyurethane

Polyurethane is widely used in electromechanical, marine, aerospace, vehicle, civil engineering, light industry, textile and other sectors due to its excellent properties such as the foamability, elasticity, abrasion resistance, low temperature resistance, solvent resistance, and biological aging resistance. There are a variety of polyurethane products. The rapid development of the polyurethane industry had led to the increasing yield, but caused the production of a large number of wastes, including scraps in production and polyurethane materials scrapped as a result of ageing. Recycling of waste polyurethane has become an urgent problem to be solved. Currently, waste polyurethane is recycled mainly in physical and chemical methods.

1. Physical recycling method

The physical recycling method means that waste polyurethane is recycled by means of bonding, hot pressing and extrusion molding, or crushed into fine pieces or powder to be used as fillers. It is mainly divided into the bonding formation method, filler method, and hot pressing method.

(1) Bonding formation method

The bonding formation method consists of crushing the waste polyurethane foam into fine pieces, mixing them with the polyurethane adhesive, and forming the adhesive polyurethane foam under the specific temperature and pressure. The adhesive polyurethane foam can be used as a cushion or support material. This method applies to the recycling of various waste polyurethanes.

(2) Filler method

In generally, this method consists of crushing waste polyurethane into fine pieces or powder and filling them into a new PU raw material to make finished products. Using this method, waste polyurethane materials can be recycled, and the costs of finished products can be reduced effectively. This method applies to production of energy-absorbing foams and sound-insulating foams. As the waste polyurethane powder is added into raw materials for production of original components, the performance of parts will not be affected within appropriate ranges. In Japan, waste rigid PU foam has been used as a lightweight aggregate for mortar.

(3) Hot pressing method

Some polyurethane materials have thermal softening performance and properties within the temperature range from 100℃ to 220℃. When waste polyurethane foam is crushed and then hot-pressed within the above temperature range, adhesive will completely not be needed in the bonding process. The conditions of hot pressing are related to the type of waste polyurethane as well as the products of recycling.

2. Chemical recycling method

The chemical recycling method refers to the degradation of a polyurethane material into a material of low molecular weight under the action of a chemical degradation agent. The polymerization reaction of polyurethane is reversible under some conditions. That is, the product is gradually depolymerized into original reactants or other substances, and with distiller and other equipment, pure original substances such as monomer polyol, isocyanate and amine can be obtained. Depending on the degradation agents used, the chemical recycling method can be divided into alcoholysis, hydrolysis, alkaline hydrolysis, ammonolysis, aminolysis, pyrolysis, hydrocracking, phosphate degradation, etc. The decomposition products vary with methods. Generally, polyol mixture is produced in alcoholysis, polyols and polyamines in hydrolysis, amines, alcohols and carbonates in alkaline hydrolysis, polyols, amines and ureas in ammonialysis, mixture of gaseous and liquid fractions in pyrolysis, and oils and gases mainly in hydrocracking.

The chemical recycling method appeared relatively late, but new degradation methods are still emerging at present. Due to technical difficulties, large-scale industrialization of the chemical recycling method cannot be achieved in a short term.

10.3.2 Comprehensive Utilization of Solid Waste in Mortar Product

In accordance with the Eleventh Five-Year Plan as well as the principle and technical route of developing energy-saving building products, a large number of tests and researches have been conducted for comprehensive utilization of solid wastes in mortar of external thermal insulation systems, including fly ash, tailing sand, waste paper fibers and waste rubber particles. Currently, related technological results have been widely applied in a variety of external thermal insulation mortar and also supporting mortar.

10.3.2.1 Fly Ash

Fly ash is a kind of bulk industrial waste. Greenpeace, an international environmental protection organization, released the *True Cost of Coal-2010 Fly Ash Survey of China* in Beijing on September 15, 2010. It is pointed out in this report that, with the explosive growth of the installed capacity of thermal power from 2002, China's fly ash emissions have increased by 2.5 times in the past eight years, the fly ash emissions from thermal power generation have become the single largest source of pollution among industrial solid wastes in China, but this pollutant damaging the environment and public health has long been neglected. In 2009, China's output of fly ash reached 375 million tons, which is more than twice that of all urban domestic waste in the same year, and the fly ash volume was 424 million cubic meters, equivalent to the filling of one standard swimming pool every two and a half minutes, or the construction of one "Water Cube" per day. The accumulation of fly ash not only takes up land, but also leads to pollution of the air and groundwater concerned, which is harmful to the environment. However, fly ash also has potential pozzolanic activity. Due to its small particle size, fly ash can be used in the building materials industry. Currently, main applications of fly ash in China's building material industry include: subgrade filling materials, wall materials, fly ash cement and concrete admixtures. Studies are still needed on how to improve the utilization rate and level of fly ash and achieve the utilization of high added value.

Fly ash is an industrial waste which is discharged out of coal-fired boilers of thermal power plants and has pozzolanic activity. Its microscopic morphology and composition vary with the combustion temperature and type of coal, melting point of ash and conditions of cooling. There are several main forms of fly ash:

(1) Spherical particles: including floating beads, hollow sink beads, compound beads, dense sink beads, and iron-rich beads. They have regular shapes but different sizes. Due to the dense and smooth surface, they are main particles for dry discharge of fly ash. The first four types of particle have high activity, while iron-rich beads have poor activity.

(2) Irregular porous glass particles: mainly consisting of glass particles. They are spongy or honeycomb-like porous particles containing a lot of SiO_2 and Al_2O_3 in fly ash. With the large specific surface area and high activity, these particles have the absorption capacity.

(3) Obtuse-angle particles: mainly consisting of residues formed by no or partial melting of quartz particles in fly ash.

(4) Fine particles: main particles consisting mainly of debris and aggregates of various particles, with an agglomerate or flocculent structure. Main components include amorphous SiO_2 and a small amount of quartz debris.

(5) Carbon-containing particles: regular porous particles easy to break into porous materials.

The chemical composition of fly ash largely depends on the inorganic composition and combustion conditions of raw coal. More than 70% of fly ash is composed of silica, alumina, and iron oxide. Typical fly ash also contains oxides of calcium, magnesium, titanium, sulfur, potassium, sodium, and phosphorus. Another important chemical component of fly ash is unburned carbon, which has great impact on the application of fly ash. ASTM classifies fly ash into Grade C high-calcium fly ash and Grade E low-calcium fly ash based on the content of CaO.

Grade C high-calcium fly ash: generated by brown coal or sub-bituminous coal, $SiO_2+Al_2O_3+Fe_2O_3\geqslant 50\%$;

Grade F low-calcium fly ash: generated by anthracite or bituminous coal, $SiO_2+Al_2O_3+Fe_2O_3\geqslant 70\%$.

The main mineral phase of fly ash is the amorphous glass body. The glass body in high-calcium ash usually contains a large amount of cationic modifier, and has a low degree of polymerization but high activity; while that in low-calcium ash usually contains a small amount of cationic modifier, and has a high degree of polymerization but low activity. Main crystal phases present in high-calcium ash are anhydrite, tricalcium aluminate, melilite, merwinite, periclase and lime. The anhydrite, tricalcium aluminate and lime have hydraulicity, so high-calcium ash has self-hardening properties. However, the presence of excessive burnt lime leads to poor volume stability. Main crystal phases in low-calcium ash are mullite, quartz and magnetite with stable chemical properties, which are not engaged in chemical reactions in the cement-fly ash system.

Fly ash has three basic effects in mortar or concrete, namely, "morphological effect", "active effect" and "micro-aggregate effect".

(1) Morphological effect: generally refer to the effects of physical properties of mineral pow-

ders in mortar or cement concrete, such as the appearance, internal structure, surface properties and particle size distribution. Fly ash as a natural pozzolan material has more positive morphological effects than those negative. Its positive morphological effects comprehensively include the moisture reduction, compaction and homogenization of cement concrete; and its negative morphological effects include the loss of physical advantages as a result of the morphological inhomogeneity (such as coarse, porous, loose and irregular particles) and are harmful to the original structure and properties of mortar or concrete. In recent years, a large number of practical applications have proved that the positive morphological effects of fly ash have significant advantages, while those negative can be suppressed and overcome by appropriate means.

(2) Active effect: refer to the chemical effect of active components of fly ash in mortar or cement concrete. If fly ash is used as a gel component, this effect is naturally the most important basic effect. The active effect depends on the capacity and speed of reaction, the quantity, structure and properties of reaction products, etc. The active effect of low-calcium fly ash mainly refers to pozzolanic silicatization; and that of high-calcium fly ash also includes the activation of the glass phase with a large content of active alumina via lime and gypsum in fly ash to produce ettringite crystals, as well as subsequent changes in ettringite crystals. Main products of fly ash hydration are of course pozzolanic reaction products found on the glass bead surface. Test results show that such products are Type I or II C-S-H gel, similar to cement hydration products. Cross-linking of pozzolanic reaction products with cement hydration products plays an important role in promoting the strength (especially tensile strength). The secondary hydration reaction of glass-phase components in fly ash has auxiliary impact on cement hydration, which is demonstrated clearly mainly as a chemical activity effect in the late period of hardening. This indicates that the pozzolanic reaction of fly ash is latent, with little impact on the initial reaction of mortar or concrete. Instead, this action mainly occurs at the interface, thereby improving the interface between the cementing material and aggregate in mortar or concrete and enhancing its tensile and flexural strength.

(3) Micro-aggregate effect: meaning that fine particles of fly ash are evenly distributed in the matrix of cement mortar, just like fine aggregates. Compared with cement gel, clinker particles not only have high strength, but can be well combined with gels. The mixture of cement mortar with mineral powders can be used as a substitute of some cement clinker, so mineral powders have the micro-aggregate effect. This is conducive to the conservation of cement and energy. The excellent micro-aggregate effect of fly ash mainly results from the properties of micro-aggregates.

(i) Glass beads have high strength. The compressive strength of thick-walled hollow beads is higher than 700MPa.

(ii) The micro-aggregate effect significantly enhances the structural strength of hardened paste. The microhardness between fly ash particles and cement mortar is greater than that of cement paste.

(iii) Fly ash particles are dispersed well in cement mortar, which helps to improve the homogeneity of fresh mortar and hardened mortar and also fill and "refine" holes and pores in mortar.

To identify the effects and reasonable content of fly ash in mortar, researchers have conducted a large number of tests to the effects of fly ash in mortar and adaptability of fly ash to activators and water reducers in the cement-fly ash system, quicklime-sodium sulfate-fly ash system, slaked lime-sodium sulfate-fly ash system, cement-activator-fly ash system and cement-water reducer-fly ash system, respectively. The following conclusions are drawn:

(1) Cement-fly ash system

Fly ash with the water-reducing effect helps to improve the workability of mortar and reduce its compressive-to-flexural strength ratio. When the fly ash content is less than 60%, the flexural strength of the cement-fly ash mortar can be enhanced.

(2) Quicklime-sodium sulfate-fly ash system

The water demand of mortar decreases a little with the fly ash content rising. When the fly ash content rises from 60% to 90%, the water-cement ratio will decrease by 10% only.

The compressive strength and flexural strength of mortar in the quicklime-sodium sulfate-fly ash system rise over time during curing, but change little with the fly ash content. The compressive strength and flexural strength of quicklime-fly ash mortar are low in the early stage (3d curing), but rise obviously in the late period (28d curing), but their absolute values are not large. The compressive strength of quicklime-fly ash mortar is about 30% of that of cement mortar, and the flexural strength of the former is roughly 50% of that of the latter. In addition, quicklime has poor volume stability during operation.

(3) Slaked lime-sodium sulfate-fly ash system

The water demand of mortar in the slaked lime-sodium sulfate-fly ash system declines little with the fly ash content increasing. When the fly ash content is increased from 60% to 90%, the water-cement ratio will decrease by 5% only, and the water demand is approximately 10 times more than that of the quicklime-sodium sulfate-fly ash system with the same fly ash content. The compressive strength and flexural strength of mortar in the slaked lime-sodium sulfate-fly ash system rise over time during curing, but do not change greatly with the fly ash content increasing. The compressive strength and flexural strength of slaked lime-fly ash mortar are low in the early period (3d curing), and rise obviously in the late period (28d curing), but their absolute values are not large. The compressive strength of slaked lime-fly ash mortar is about 25% of that of cement mortar, and the flexural strength of the former is approximately 50% of that of the latter. Slaked lime-fly ash mortar with high workability also has great plasticizing effects in the slaked lime-sodium sulfate-fly ash system.

(4) Cement-activator-fly ash system

The water demand of the cement-fly ash system will increase to different degrees if fly ash activators are added. It will increase the most in the presence of calcium chloride. When the calcium chloride content is 3%, the water-cement ratio will rise by 15%.

With fly ash activators, the compressive strength and flexural strength of the cement-fly ash mortar will rise by 5% to 25% in the early period (3d curing), and their change rates increase with the activator amount. This indicates that sodium sulfate, calcium chloride, triethanolamine and calcium formate help to improve the early strength of cement-fly ash mortar, among which calcium chloride has the most significant effects; but they change little in the late period (28d curing). The mortar with calcium chloride and calcium formate has lower strength than that without activators. Both sodium sulfate and triethanolamine has no adverse effects on the late strength.

(5) Cement-sulfate-superplasticizer-fly ash system

If a super plasticizer is added, the water-cement ratio of the cement-fly ash system will decline greatly. With lignosulfonate, melamine and naphthalene sulfonate, the water-cement ratio drops by 8%, 18% and 16%, respectively. When lignosulfonate is added, the strength of the cement-fly ash system will fall by about 20% in the early period (3d curing) and rise slightly in the late period (28d curing); and when melamine and naphthalene sulfonate are used, the strength of the cement-fly ash system will rise by about 20% in the early period (3d curing) and 25% in the late period (28d curing). The air-entraining effect of lignosulfonate is a reason for insignificant increase in the strength.

(6) Cement-fly ash-sodium sulphate-slaked lime-silica ash cementation system

The generation of ettringite in this system is the reason for the increase the compressive strength. That is, the compressive strength of this system can be improved by increasing the content of sodium sulfate. However, excessive sodium sulfate may lead to efflorescence, so an appropriate amount of silica fume for effective prevention of efflorescence needs to be added together with sodium sulfate. The boundary phase of hydrated calcium silicate and hydrated calcium aluminate produced in the interfacial reaction between spherical glass substances in fly ash with calcium hydroxide and sodium hydroxide, instead of the calcium hydroxide boundary phase, is the reason for the increase in flexural strength. Sodium hydroxide also helps to raise the speed of interfacial reaction.

There is no reaction between crystal phases such as mullite, quartz and magnetite in fly ash and spherical glass bodies during cement hydration. Activators will accelerate the interfacial reaction of spherical glass bodies, thereby changing the state of the interface between hydrated calcium silicate gel and spherical glass bodies, changing the weak calcium hydroxide into hydrated calcium silicate, and improving the system strength, especially the flexural strength.

10.3.2.2 Tailing Sand

Tailing sand and tailings are wastes discharged by mining companies under specific technological and economic conditions and also potential secondary resources. Once permitted by technological and economic conditions, they can be further utilized effectively. According to the statistics, the total quantity of tailings produced by China's mines was 5.026 billion tons before 2000, including 2.614 billion tons of iron ore tailings, 2.109 billion tons of major non-ferrous metal tailings, 272 million tons of gold tailings, and 31 million tons of other tailings. Calculated based on the annual discharge (600 million tons) of tailings discharged by China's mines in 2000, the total quantity of tailings in 2006 is 8 billion tons.

Tailings account for about 1/3 of solid wastes in China, but their utilization rate is only about 8.2%. The discharge of tailings into river courses, valleys and lowlands may lead to the contamination of the water, soil and atmosphere, thereby causing the environmental pollution and even disasters. The storage of mine tailings also needs a large amount of farmland and forest land, which may lead to some environmental pollution.

Tailings containing a large number of nonmetallic minerals can be used as building materials and raw materials for glass production. With the national strengthening of environmental protection and land management, the land occupation of tailings will become an urgent problem that must be resolved. It is a best solution to only recycle valuable tailings. Instead, the fundamental disposal is to use tailings as building materials. Tailing sand of mines and tailings produced in concentration can be used as railway/highway slag and coarse concrete aggregates; and tailings of mines can also be used as building sand, non-sintered tailing bricks, blocks, square bricks, road bricks, raw materials of new wall materials, as well as excellent cement materials. High-silicon tailings can be used in glass production.

Large quantities of iron core tailings are produced in China, accounting for more than half of the total mine tailings. The iron core tailing sand with stable chemical properties and reasonable particle size distribution can be used during construction.

Researchers systematically studied iron ore tailing sand from Shougang Waterworks Concentrator subordinated to Mining Corporation of Shougang Group Co., Ltd. The water plant concentrator is located at the junction of Qian'an and Qianxi counties in Hebei Province, with the annual capacity of roughly 11 million tons of iron ore, 3.3 million tons of iron ore concentrate and

8 million tons of waste (containing a large amount of tailing sand).

The iron core tailing sand from Shougang Waterworks Concentrator was dried and screened to obtain dry tailing sand of different grades, namely, 10-20 meshes, 20-40 meshes, 40-70 meshes and 70-110 meshes. In accordance with the national standard "Sand for Construction" (GB/T 14684-2011), the particle size distribution, mud content, clay lump content, loose bulk density, compacted bulk density, solidity and alkali activity of aggregate were tested, and the tailing sand and ordinary washed river sand were compared, involving the cement mortar strength test, phase analysis and microstructure analysis.

(1) Particle size distribution

The particle size distribution of tailing sand was tested according to the method provided in Article 7.3 of the national standard "Sand for Construction" (GB/T 14684-2011). According to the test results in Table 10-3-1, by means of grading, tailing sand of specific fineness can be used to prepare the products meeting the particle size distribution requirements of this standard. Therefore, tailing sand is applicable to construction mortar.

(2) Mud content and clay lump content

The mud content and clay lump content of tailing sand were tested according to the methods provided in Article 7.4 and 7.6 of the national standard "Sand for Construction" (GB/T 14684-2011). According to the test results in Table 10-3-2, the mud content and clay lump content of tailing sand comply with Class I requirements specified in Article 6.2.1 of the national standard "Sand for Construction" (GB/T 14684-2011).

(3) Loose and compacted bulk density

The loose bulk density and compacted bulk density of tailing sand were tested according to the methods provided in Article 7.15 of the national standard "Sand for Construction" (GB/T 14684-2011). According to the test results in Table 10-3-3, the loose bulk density and compacted bulk density of tailing sand comply with the requirements of Article 6.5 of the national standard "Sand for Construction" (GB/T 14684-2011).

(4) Solidity

The solidity of tailing sand was tested according to the method provided in Article 7.13.1 of the national standard "Sand for Construction" (GB/T 14684-2011). According to the test results in Table 10-3-4, the tailing sand meets Class I requirements of Article 6.4.1 of the national standard "Sand for Construction" (GB/T 14684-2011).

Particle Size Distribution of Tailing Sand **Table 10-3-1**

Square Screen / Cumulative Screening Residue / Range	10-20 Meshes	20-40 Meshes	40-70 Meshes	70-110 Meshes
9.50mm	0	0	0	0
4.75mm	0	0	0	0
2.36mm	0	0	0	0
1.18mm	30	0	0	0
600μm	100	25	0	0
300μm	100	100	73	0
150μm	100	100	100	100

Mud Content and Clay Lump Content of Tailing Sand **Table 10-3-2**

Item	Indicator			
	10-20 meshes	20-40 meshes	40-70 meshes	70-110 meshes
Mud content (by mass), %	0.0	0.1	0.2	0.5
Clay lump content (by mass), %	0.0	0.0	0.0	0.0

Loose Bulk Density and Compacted Bulk Density of Tailing Sand　　Table 10-3-3

Item	Indicator			
	10-20 meshes	20-40 meshes	40-70 meshes	70-110 meshes
Loose bulk density, kg/m^3	1364	1380	1452	1528
Compacted bulk density, kg/m^3	1568	1588	1668	1720

Solidity of Tailing Sand　　Table 10-3-4

Item	Indicator			
	10-20 meshes	20-40 meshes	40-70 meshes	70-110 meshes
Mass loss, %, less than	2. 5	3. 3	3. 8	4. 2

(5) Alkali activity test (lithofacies method) of aggregate

The alkali active aggregate of tailing sand was tested according to the method provided in Appendix A of the national standard "Sand for Construction" (GB/T 14684-2011). Analyzed with a polarizing microscope, the main mineral phase of tailing sand is quartz, and no alkaline active aggregate was found. The photos taken by the polarizing microscope are shown in Figure 10-3-1 and 10-3-2.

Figure 10-3-1　Orthogonal Polarized Micrograph of Tailing Sand

(6) Cement mortar strength

The contrast test of tailing sand and washed river sand was conducted according to the method specified in the national standard "Test Method for Cement Mortar Strength (ISO method)" (GB/T 17671-1999). According to the test results of the cement mortar strength in Table 10-3-5, the compressive strength of the tailing is about 3MPa higher than that of the washed river sand, and the flexural strength of the former is 0. 5MPa

Figure 10-3-2　Single Polarized Micrograph of Tailing Sand

higher than that of the latter.

(7) Microstructures of tailing sand and washed river sand

According to the results of analysis with the polarizing microscope, the tailing sand is irregularly polygonal, mainly composed of quartz; and the washed river sand was regularly oval, mainly consisting of quartz and feldspar, and its microcrystalline quartz is an alkali active aggregate. The photos taken by the polarizing microscope are shown in Figure 10-3-1 and 10-3-3.

Figure 10-3-3　Cross-polarized Photo of River Sand

Cement Mortar Strength of Tailing Sand **Table 10-3-5**

Item	Compressive Strength				Flexural Strength			
	3d	7d	14d	28d	3d	7d	14d	28d
Washed river sand	30. 5	44. 5	56. 0	62. 0	5. 0	6. 6	7. 8	8. 2
Tailing sand 1	35. 5	49. 5	61. 0	66. 0	6. 0	7. 2	8. 3	8. 8
Tailing sand 2	34. 0	48. 5	60. 0	65. 0	5. 8	7. 0	8. 2	8. 7
Tailing sand 3	34. 5	50. 0	59. 5	64. 5	5. 7	6. 9	8. 2	8. 5

In summary, compared with the washed river sand, the tailing sand concerned is an excellent dry-mixed mortar aggregate, with stable supply and chemical composition, small mud content and few harmful components. However, tailing sand, varying with multiple factors such as the origin and deposit, must be systematically tested before comprehensive utilization, in order to minimize adverse effects on mortar products.

10. 3. 2. 3 Waste Paper Fiber

The waste paper fiber is obtained by crushing waste paper and has the capacity of water absorption. It can be used as a substitute of the expensive wood fiber for water retention, cracking prevention and workability improvement of mortar.

An appropriate number of wood fibers help to improve the workability of mortar and reduce its thermal conductivity, but the excessive use may affect the compressive strength of mortar. Based on the contrast study on wood and waste paper fibers, researches selected the fiber contents of 0. 1%, 0. 3% and 0. 5% in the test of the workability, thermal conductivity and compressive strength of insulation mortar. According to the test results in Table 10-3-6, the favorable content of water paper fiber is 0. 3%.

10. 3. 2. 4 Waste Rubber Particle

Waste rubber particles are obtained by crushing waste rubber tyres. Waste tyres are difficult to degrade but easy to burn. For stacking of a huge number of waste tyres, a lot of land will be occupied, and safety hazards and serious environmental pollution may be caused. According to the statistics, there are about 3 billion used tyres in the world, with a growth rate of 1 billion per year. China is a major country of tyre production and consumption. In 2004, China's output of new tyres was 239 million and that of used tyres was 120 million, both of which ranked second in the world. Also, the number of waste tyres increases at a rate of 12% per year. The efficient utilization of waste tyres has become a worldwide issue.

Dry-mixed anti-cracking mortar to be used in the anti-cracking protective layer of the face brick finish of the external thermal insulation system should have high workability, bonding strength and water resistance but low shrinkage rate, without cracking in the case of thick plastering. Researchers used the cement-fly ash system as the cementing material and tailing sand as the

Test Results of Insulation Mortar with Varying Fiber Content **Table 10-3-6**

Item	Content	Workability	Wet Apparent Density kg/m^3	Dry Apparent Density kg/m^3	Compressive Strength kPa	Thermal Conductivity W/(m • K)
Wood fiber	0. 1%	OK	400	215	240	0. 058
	0. 3%	OK	390	208	220	0. 056
	0. 5%	OK	375	185	195	0. 051
Waste paper fiber	0. 1%	OK	405	217	240	0. 058
	0. 3%	OK	396	210	226	0. 057
	0. 5%	OK	385	195	203	0. 054

aggregate, preferably redispersible polymer powder, cellulose ether, water-repellent agent and rheological additives to obtain an optimized formula for preparation of dry-mixed anti-cracking mortar. In order to prevent cracking in the case of thick plastering, waste rubber particles are used instead of part of tailing sand, to improve the flexibility and cracking resistance of mortar.

Studies have shown that, when waste rubber particles are used instead of gravel aggregates, the strength of cement-based materials will decline, but its workability, deformation capacity, cracking resistance and freezing resistance will be improved, and a series of excellent properties such as slip resistance, sound-proofing and thermal insulation will be achieved. Researchers used waste rubber particles instead of part of tailing sand in the test. According to the test results in Table 10-3-7, with the content of waste rubber particles increasing, the compressive-to-flexural strength ratio and bonding strength of dry-mixed anti-cracking mortar will decline, but its flexibility will rise. The optimal content of waste rubber particles is 5%. Table 10-3-8 lists the performance indicator and optimal formula test indicators.

10.3.2.5 Solid Waste Content in Mortar Products

This section summarizes the solid waste contents of supporting mortar products for 18 types of external thermal insulation system. Specific results are shown in Table 10-3-9 and 10-3-10. According to these two tables, the comprehensive utilization rate of fly ash in mortar products of external thermal insulation systems is more than 30%, that of tailing sand is 40% to 65%, and that of most solid wastes is roughly 60% and maximum 80.25%.

Test Results of Dry-mixed Anti-cracking Mortar with Different Contents of Waste Rubber Particles **Table 10-3-7**

Content (%) of Waste Rubber Particles	Bonding Strength (MPa)	Compressive Strength (MPa)	Flexural Strength (MPa)	Compressive-to-flexural Strength Ratio
1	0.98	20.8	6.3	3.30
3	0.84	18.6	5.9	3.15
5	0.81	15.4	5.6	2.75
7	0.74	12.3	4.3	2.86

Performance Indicators of Dry-mixed Anti-cracking Mortar Products **Table 10-3-8**

Item		Unit	Indicator	Test Result
Operable time		h	≥1.5	2.25
Bonding strength with cement mortar	Initial value	MPa	≥0.7	1.20
	After soaking in water	MPa	≥0.5	1.35
Compressive-to-flexural strength ratio		—	≤3.0	2.8

Solid Waste Contents of Dry-mixed Mortar Products for External Thermal Insulation Systems **Table 10-3-9**

Product Name	Solid Waste Content (%)			
	Fly ash	Tailing sand	Others	Total
Insulation binder	34	0	0.35	34.35
Roof insulation binder	40	0	0.30	40.3
Bonding insulation binder	39	0	0.50	39.5
Anti-cracking mortar I	23	57	0.25	80.25
Anti-cracking mortar III	15	42.5	5	62.5
Face brick bonding mortar	16	47	0.25	63.25
Face brick pointing material	25	57	0	81
Base course wall interface mortar	39	27	0.40	66.4
EPS board bonding mortar	30.5	38.5	0.30	69.3

Solid Waste Contents of Agent Products for External Thermal Insulation Systems **Table 10-3-10**

Product Name	Solid Waste Content (%)			
Interface agent	32	0	0	32
Waterproof flexible putty	40	0	0	40
EPS boar dinterface agent	33	0	0	33
XPS board interface agent	31	0	0	31
Polyurethane interface agent	0	34	0	34
Anti-cracking mortar I (two-component) (cement mortar anti-cracking agent)	17	55	0	72
Anti-cracking mortar III (two-component) (cement mortar anti-cracking agent)	12.5	55	0	67.5

10.3.3 Comprehensive Utilization of Solid Waste in External Thermal Insulation System

In the decades of domestic development, a large number of external thermal insulation systems involving the use of solid wastes have been developed, such as the adhesive polystyrene granule system (paint finish), adhesive polystyrene granule system (face brick finish), sprayed rigid polyurethane foam system (paint finish), sprayed rigid polyurethane foam system (face brick finish), adhesive polystyrene granule pasted EPS board system (paint finish), adhesive polystyrene granule pasted EPS board system (face brick finish), cast-in-situ concrete and non-meshed EPS board system (paint finish), cast-in-situ concrete and non-meshed EPS board system (face brick finish), cast-in-situ concrete and meshed EPS board system (paint finish), and cast-in-situ concrete and meshed EPS board system (face brick finish). After years of development, the application of a large number of solid wastes in external thermal insulation systems has also been mature. The solid waste contents of external thermal insulation systems are shown in Table 10-3-11. According to this table, the comprehensive utilization rates of the fly ash, tailing sand and waste EPS granule in the above ten external thermal insulation systems are greater than 50%.

Solid Waste Contents of External Thermal Insulation Systems **Table 10-3-11**

System Name	Solid Waste Content (wt%)			
	Fly ash	Tailing sand	Waste EPS granule	Total
Adhesive polystyrene granule system (paint finish)	30.48	19.47	3.68	53.63
Adhesive polystyrene granule system (face brick finish)	21.49	32.45	2.01	55.95
Sprayed rigid polyurethane foam system (paint finish)	22.72	25.05	5.91	53.68
Sprayed rigid polyurethane foam system (face brick finish)	16.36	37.17	0.47	56.34
Adhesive polystyrene granule pasted EPS board system (paint finish)	30.03	20.15	1.90	52.09
Adhesive polystyrene granule pasted EPS board system (face brick finish)	21.08	33.06	1.03	55.16
Cast-in-situ and non-meshed EPS board system (paint finish)	27.03	25.40	1.07	53.50
Cast-in-situ and non-meshed EPS board system (face brick finish)	18.07	37.80	0.49	56.36
Cast-in-situ and meshed EPS board system (paint finish)	25.66	24.11	1.02	50.79
Cast-in-situ and meshed EPS board system (face brick finish)	17.65	36.91	0.47	55.03

10.3.4 Comprehensive Evaluation

According to the national requirements for development of a circular economy and construction of a conservation-oriented society, the fly ash, tailing sand, waste rubber particle and waste paper fiber have been studied systematically, and the external thermal insulation system products involving the utilization of solid wastes have been developed. This not only solves the problem of raw material shortage arising from rapid development of China's building energy conservation and external thermal insulation industry, but purifies the environment due to the efficient and comprehensive utilization of a large number of solid wastes.

As large quantities of solid wastes such as fly ash and tailing sand are utilized, the costs of external thermal insulation systems are reduced. The fly ash activator can be added to make full use of the activity of fly ash and improve the properties of mortar products. Therefore, external thermal insulation systems have broad prospects of marketing and development. The extensive utilization of solid wastes complies with the national requirements for changing of wastes into building materials and development of the circular economy, and also provides the powerful technical support for comprehensive utilization of resources in China's external thermal insulation industry.

10.4 Analysis of Energy Consumption and Environmental Pollution in Insulation Material Production

At present, there are three main types of insulation technology used in construction: external thermal insulation, internal thermal insulation and sandwich insulation. In recent years, the external thermal insulation plays a dominant role on the market, involving a variety of insulation materials. The EPS board, XPS board, polyurethane, phenolic board, rock wool, inorganic insulation mortar and adhesive polystyrene granule mortar are widely applied currently.

China suffers from energy shortage, so the vigorous promotion of energy conservation and emission reduction is one of its current priorities. The external thermal insulation plays an essential role in building energy conservation. On one hand, the insulation material is conducive to energy conservation. On the other hand, a lot of energy and resources are consumed during production of insulation materials, which is not favorable for China's overall energy conservation and emission reduction. Several commonly used insulation materials are analyzed below.

10.4.1 EPS Board

The EPS board, which is made by pre-foaming, curing and foam molding of expandable polystyrene (EPS) granules as raw materials, has the advantages of light weight, good insulation performance, low water absorption capacity, high resistance to low temperature, etc. It is mainly applied in insulation layers of wall/roof insulation and composite insulation boards; insulation materials of vehicle/ship refrigeration equipment and refrigerators; and decoration, model carving, etc.

The production of foam products with EPS granules usually involves granule pre-foaming, curing, forming, product curing, hot curing, cutting, etc. There are small quantities of waste gas, wastewater and other contaminants in the production process. But scraps (accounting for 10% to 20% of the total materials) produced in the cutting process can be directly recycled by means of crushing.

Currently, the combustibility is a major factor constraining the development of EPS insulation technology. Benzene, toluene, ethylbenzene, p-xylene, o-xylene, m-xylene, styrene and other substances are produced in the polystyrene combustion process. Table 10-4-1 shows the types

Types and Concentrations (mg/m^3) of Thermal Decomposition Products of Polystyrene (PS) under Different Temperature Conditions **Table 10-4-1**

Thermal Decomposition Product	Temperature (℃)									
	80	100	120	140	160	180	200	220	240	260
Benzene	0.11	0.16	0.21	0.24	1.22	2.98	4.12	6.78	9.10	12.60
Toluene	0.08	0.14	0.20	0.22	0.73	1.24	2.28	3.42	6.82	9.22
Ethylbenzene	Not detected	Not detected	Not detected	0.18	0.38	0.66	1.06	1.31	2.56	5.81
P-xylene	Not detected	0.88	1.27	2.62	5.62	8.23	10.12	12.74	14.11	17.16
M-xylene	Not detected	Not detected	Not detected	Not detected	0.14	0.38	0.74	0.98	1.56	3.42
O-xylene	Not detected	Not detected	0.34	0.88	1.38	3.18	4.88	6.38	8.24	10.62
Styrene	Not detected	Not detected	Not detected	0.10	0.23	0.42	0.64	1.13	2.06	4.22

and concentrations of polystyrene decomposition products under various heating conditions. According to this table, polystyrene is decomposed at 80℃ (producing harmful gases such as benzene and toluene), melted at 140℃ and decomposed more rapidly at the temperature above 160℃. Its color changes as follows: colorless and transparent → light yellow → orange → brown → black. The highly toxic macromolecular organic styrene is produced at 140℃. Its amount increases with the temperature rising to 260℃, but the types of pyrolysis products do not change. Pyrolysis products are produced at different speeds: small-molecule organics are produced fast, with high concentration; macromolecular organics are produced slowly, with low concentration; the types and concentrations of macromolecular organics produced by pyrolysis increase with the temperature.

As water vapor is used instead of a foaming agent, there is not environmental pollution during the formation of EPS boards, so the major issue is recycling. In the case of improper disposal, a large amount of white pollution may be caused. Waste EPS boards can be widely used in the popular adhesive polystyrene granule insulation mortar available on the market at present, thereby solving the problem of white pollution.

According to the calculation method and limit in Tianjin's local standard DB12/046.84-2008, the comprehensive energy consumption per unit output of polystyrene foam products should not be more than 3,200kg (standard coal) per ton.

The EPS granule is a kind of resin prepared by suspension and polymerization of styrene with the liquid foaming agent. Its impact on the environment mainly lies in the production of styrene and use of liquid foaming agent (pentane). The liquid foaming agent (pentane) with the ODP of 0 has little impact on the environment. Currently, there are three styrene production routes in the world: (1) gas-phase catalytic dehydrogenation of ethylbenzene into styrene in the presence of steam, with ethylbenzene as the raw material, iron oxide-chromium or zinc oxide as the catalyst, using the multi-bed adiabatic or tubular isothermal reactor; (2) over-oxidation of propylene and ethylbenzene to prepare epoxypropane and styrene (by-product); and (3) extractive distillation and recycling of gasoline by means of steam cracking and pyrolysis. In total, 90% of styrene in the world is produced by the first method. Typical ethylbenzene dehydrogenation processes include the Badger and Roms method. Styrene production is mainly affected by the presence of toxic chemicals such as benzene, toluene, ethylbenzene, styrene, potassium hydroxide, and methanol. For example, the annual output of styrene of a new styrene device of a chemical plant can be designed as 500,000 tons of styrene. Styrene is produced by means of the traditional catalytic dehydrogenation of ethylbenzene. That is, the intermediate product ethylbenzene and a small amount of ethylbenzene are produced through the alkylation reaction of ethylene and excess benzene with the alkylation catalyst, and according to Badger's typical styrene technology, styrene is produced by dehydrogenation of ethylbenzene in the gaseous state at about 600℃, with iron-based oxides and other catalysts. Toxic chemical substances such as benzene, toluene, ethylbenzene, styrene, potassium hydroxide and methanol are also produced in the above process. As a lot of benzene is used as the raw materials, serious leakage may occur in the case of accidents at the benzene pipe. The exposure to a large amount of benzene steam generated because of leakage may easily lead to acute poisoning and even death.

The preparation of EPS granules by the styrene suspension method has minor effects on the environment, so the major issue is to prevent the leakage of styrene and discharge of wastewater. As the unreacted styrene is usually separated from wastewater by means of high-pressure distillation, the wastewater discharged actually meets the standards, leading to little pollution.

10.4.2 XPS Board

The extruded polystyrene (XPS) board with an independently closed bubble structure is a kind of rigid board made by continuous extrusion and foaming of polystyrene resin (as the raw material) through a special process, and an environmentally-friendly insulation material with excellent properties such as high compressive strength, no water absorption, moisture resistance, vapor impermeability, light weight, corrosion resistance, long service life and low thermal conductivity. It is widely applied in thermal insulation of walls, flat concrete roofs and steel structure roofs, moisture prevention and thermal insulation of low-temperature storages, floors, parking platforms, airport runways and highways, and control of ground expansion.

The main raw material for production of XPS boards is polystyrene resin with the average molecular weight of 170,000 to 500,000, and auxiliary materials include additives, foaming agents, etc.

Table 10-4-2 shows the production indicators of polystyrene resin production companies at home and abroad. The polystyrene production processes of different companies have respective shortcomings. For example, the consumption of styrene (main raw material) in polystyrene production of the company A is low, but the power consumption is high. In terms of pollutant discharge (mainly represented by the CODcr concentration in wastewater), the average CODcr concentration in wastewater discharged by the company A is 2561.4mg/L, and half of wastewater is discharged as clean water. If clean water is separated based on the separation requirements, the CODcr concentration of the wastewater discharged must be 4,000mg/L to 5,000mg/L, which has impact on the water inflow quality if such wastewater is discharged into the ABS plant for wastewater treatment.

As the Montreal Protocol comes into effect, the majority of European and American manufacturers have already completed the replacement of Freon foaming agents during production of XPS boards. The hydrocarbon foaming agent containing no halogenated carbon but with a high air replacement rate is used, thus avoiding damage to the ozone layer and guarantee the most of replacement in the initial stage of reaction, and leading to small changes in the thermal conductivity of materials after construction. In addition, the production technology with CO_2 as a foaming agent has been developed. Table 10-4-3 shows the material consumption of one plane using CO_2 as the foaming agent.

HCFC-22 and HCFC-142b are widely used as foaming agents in China's XPS foam industry. They are ozone depleting substances (ODS) and also strong greenhouse gases. Although the XPS board can be regarded as a circular economy

Production Indicators of Several Polystyrene Production Companies at Home and Abroad Table 10-4-2

S/N	Indicator	Unit	Company A	Company B	Company C	Company D
1	Production scale	10^4t/a	12	10	10	10
2	Residual styrene	ppm	<700	<300	<500	<500
3	Technology	—	Bulk polymerization	Bulk polymerization	Bulk polymerization	Bulk polymerization
4	Raw material consumption					
	Styrene	kg/t	906.25	915	965.5	934.0
	Polybutadiene rubber	kg/t	80.8	71.5	70.0	43.56
	Solvent and chemical	kg/t	22.95	31.2	21.2	34.75
	Unit consumption of total raw materials	kg/t	1010	1017.7	1056.7	1012.31
5	Water consumption					
	Fresh water	m^3/h	9.13	40	—	5
	Circulating water	m^3/t	—	67	58.5	37.5
6	Power consumption	kW·h/t	200	102	130	105.87
7	Energy consumption (fuel)	kcal/t	5.5×10^4	6.88×10^4	16.27×10^4	9.56×10^4
8	Wastewater amount	m^3/h	9.05	23.35	—	19
9	Average CODcr concentration of wastewater	mg/L	228.8	—	—	336.8

industry in China, there is huge pressure on environmental protection. It is a major problem encountered by the whole industry to stop using the unfavorable foaming agent and use the effective replacement technology. At present, the overwhelming majority of domestic XPS manufacturers have fully realized that HCFC-22 and HCFC-142b are greenhouse gases that damage the ozone layer and will certainly be banned in the future. In this sense, most XPS manufacturers are generally concerned about alternative technologies for HCFCs, but there are still doubts and confusions about the national policies, alternative technologies, technological maturity and feasibility, cost reasonableness, needs for transformation and repurchase of production equipment, impact of new alternative technologies on board properties, conformity to prevailing national standards, marketing of boards involving new technologies, etc.

The XPS board production process is divided into mixed feeding, melt mixing, extrusion molding, cooling, cutting, etc. The environmental pollution in the XPS board production process mainly occurs in the melting and plasticizing step. As the temperature in this step is above 200℃, the thermal decomposition of polystyrene occurs easily, especially the increase of the content of highly toxic styrene (Table 10-4-1). Part of the foaming agent is vaporized into the atmosphere, causing in the environmental pollution.

List of Raw and Auxiliary Material Consumption of Production Line for Annual Output of 60,000m³ XPS Boards

Table 10-4-3

XPS production line	Polystyrene	1200t
	CO_2	20t
	Flame retardant	70t

In essence, the main raw material for XPS board production is polystyrene, which is the same as that for EPS board production. The hazards resulting from combustion of the XPS board is basically equivalent to those of the EPS board. As a result of melting, plasticizing and extrusion, the XPS board cannot be recycled easily, which is different from that of the EPS board. Currently, there is no suitable way for recycling of waste XPS boards.

10.4.3 Polyurethane

The rigid polyurethane foam is a kind of organic polymer, which is prepared through the reaction of polyisocyanate (OCN-R-NCO) and polyol (HO-R1-OH) and consists of carbamate (R-NH-C--OR1) segments. The insulation effect of 50mm rigid polyurethane foam is equivalent to that of the 80mm EPS board and 90mm rock wool board. With the energy costs increasing and environmental protection requirements improved, the rigid polyurethane foam is applied more and more widely as an excellent insulation material in the energy conservation and thermal insulation of buildings.

Main raw materials used in production of rigid polyurethane foam are polyphenylmethane polyisocyanate and polyol, commonly known as black and white materials. The polyphenylmethane polyisocyanate (black material) has the average functionality of about 2.7 and viscosity of about 100-300mPa • s. The main preparation method is phosgenation, involving a large number of flammable, explosive and highly toxic substances. The production process is complex with many control points. In order to strictly control pollutant discharge in the isocyanate production process, China has formulated the access standards for the isocyanate industry. Specific requirements are listed in Table 10-4-4.

At present, main polyurethane materials used in building insulation include the on-site sprayed polyurethane, on-site cast polyurethane, prefabricated polyurethane insulation board, hollow brick filling polyurethane. The comparison of the environmental impacts of different construction methods are show in Table 10-4-5.

Major environmental problems of the polyurethane industry are caused by the foaming agent, especially the replacement of foaming agent in the sprayed polyurethane industry. Table 10-4-6 shows a few generations of foaming agent applied in the polyurethane industry. The parties related to the Montreal Protocol held the conference in Copenhagen in November 1992, in which the *Montreal Protocol on Substances that Deplete the Ozone Layer* was amended into the *Copenhagen Amendment*. Hydrogen containing chlorofluorocarbons were officially included in the list of controlled substances. China signed the amendment on April 22, 2003 and became a party thereof. The resolution to expedite the phase-out of hydrogen containing chlorofluorocarbons (HCFCs) was adopted in the Nineteenth Meeting of the Parties of the *Montreal Protocol* in September 2007, requiring in Article 5 that the production and consumption of

Raw Material and Power Consumption Standards of MDI Manufacturer **Table 10-4-4**

S/N	Name of Raw Material/Power	Specification (by percentage)	Unit	Unit Consumption
1	Aniline	100%	t/tMDI	≤0.75
2	Formaldehyde	100%	t/tMDI	≤0.15
3	CO	100%	NM3	≤195
4	Liquid chlorine	100%	t/tMDI	≤0.58
5	NaOH (including decomposition and neutralization)	100%	t/tMDI	≤0.165
6	Electricity	380V	KW • h/tMDI	≤450
7	Steam	4.0MPa	t/tMDI	≤1.1
		1.0MPa	t/tMDI	≤1.2

Environmental Impacts of Different Construction Processes — Table 10-4-5

S/N	Construction Process	Environmental Impact
1	On-site sprayed polyurethane	Mortar flashing affecting the environment around the construction site; large loss and significant volatilization of foaming agent; and harm to the body
2	Polyurethane insulation board	Small loss due to recycling of scraps; high concentration of isocyanate in the construction ship, with great harm to the body; and high concentration of cutting dust in the shop
3	On-site cast polyurethane	Small loss of raw materials and little environmental pollution
4	Hollow brick filling polyurethane	Small loss of raw materials and little environmental pollution

HCFCs should be frozen at the baseline level (2009 and 2010 average) in 2013, and reduced by 10% of the baseline level in 2015, 35% in 2020 and 67.5% in 2025, and the production and use of HCFCs (except a small number for maintenance) should be canceled in 2030.

Because of the constant changes in the global climate and acceleration of global warming, environmental protection has become an important issue. Though it is stipulated that the production and consumption of HCFC should be completely canceled in 2030, the replacement of 141b foaming agent is quite urgent with the development of the global economy.

There are many harmful gases and smoke during the combustion process of polyurethane. Main gases produced in polyurethane combustion are shown in Table 10-4-7.

10.4.4 Phenolic Insulation Board

The phenolic insulation board is a kind of closed-cell rigid foam board, which is made with phenolic resin (PF), flame retardant, smoke suppressant, curing agent, foaming agent and other additives according to a scientific formula. It is applicable to thermal insulation, heat isolation and sound absorption in the construction, chemical, petroleum, electricity, refrigeration, shipbuilding, aviation and other fields.

Main environmental problems in the production process of phenolic insulation boards are reflected in the discharge of wastewater. As shown in Figure 10-4-1, the wastewater discharged in production with general-purpose phenolic resin mainly consists of the clarified liquid separated in the polymerization stage and condensate and flushing

Comparison of Basic Parameters of Polyurethane Foaming Agents — Table 10-4-6

Name	Molecular Weight	Boiling Point ℃	Flash Point ℃	Gas-phase Thermal Conductivity at 25℃ mW/m. K	ODP	GWP
F-11	137.4	23.7	None	8.23	1	4600
141b	116.9	31.7	None	10.1	0.11	630
365mfc	148	40.0	−27	10.6	0	890
245fa	134	15.3	None	12.2	0	950
Cyclopentane	70	49.5	−37	12.0	0	11
Pentane	72	36	−56.2	15	0	11
Isopentane	72	27.8	−57	15	0	11
H_2O/CO_2	44	−78.4	None	16.6	0	1

Main Gases Produced in Polyurethane Combustion — Table 10-4-7

Polyurethane Type	Main Gas
Soft foam	HCN, acetonitrile, acrylonitrile, CO, CO_2, benzene, toluene and benzonitrile
Rigid polyester foam	HCN, formaldehyde, methanol, CO, CO_2, CH_4, C_2H_4 and C_2H_2
Rigid polyether foam	HCN, acetonitrile, acrylonitrile, CO, CO_2, pyridine and benzonitrile

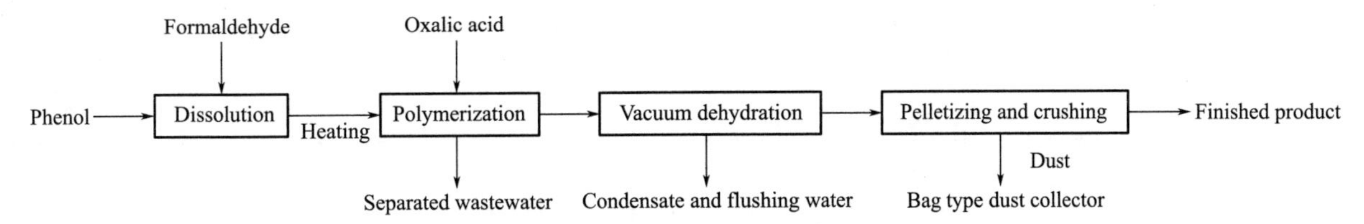

Figure 10-4-1 Phenolic Resin Process and Wastewater Discharge Flowchart

water generated in the vacuum dehydration and drying stage. Phenols and aldehydes in wastewater mainly come from the unreacted raw materials. In addition, there are usually alcohols in phenol and aldehyde reactions, with the concentration of about 1%. For example, methanol comes from the following sources: (1) methanol residues in formaldehyde; (2) methanol generated in the formaldehyde storage process; and (3) methanol added as a stabilizer. According to the statistics, the amount of wastewater (including process wastewater and flushing water) is 900-1,500kg during production of one ton of thermoplastic phenolic resin, and 1,200-1,800kg during production of one ton of thermosetting phenolic resin. The wastewater composition and concentration vary with the production process and operating conditions. Main pollutants in the wastewater include phenol, aldehyde and alcohol. Prior to resin recycling, wastewater contains phenols of 16-440g/L, aldehydes of 20-60g/L and alcohols of 25-272g/L.

At present, main technologies for wastewater treatment abroad include the phenol-formaldehyde condensation-recycling, phenol-formaldehyde condensation-incineration, biological oxidation, chemical oxidation, activated carbon adsorption, etc., or a combination of biological oxidation, chemical oxidation and activated carbon adsorption. The delayed condensation-biochemical method is often used in China. It is introduced briefly below.

The delayed condensation-biochemical method is a wastewater treatment method applied widely by domestic phenolic resin manufacturers. It is a combination of several single technologies, such as condensation (twice), neutralization, anaerobic reaction and aerobic reaction. The condensation technology is the foreign phenol-formaldehyde condensation technology. Phenol-formaldehyde condensation is involved twice in the delayed concentration. Currently, China's annual output of phenolic resin is 300,000 tons, involving the wastewater of about 520,000m^3, average phenol concentration of about 300g/L and average formaldehyde concentration of about 50g/L. Following the delayed condensation, 12,500 tons of low-molecular-weight phenolic resin can be recycled, equivalent to the reduction of phenol by 150,000 tons, formaldehyde by 25,000 tons and COD emissions by about 350,000 tons.

Most phenolic resin manufacturers in China have established wastewater treatment facilities. Among them, more than 80% of large-and medium-sized phenolic resin companies adopt the delayed concentration process for condensation of phenols and aldehydes in the wastewater, to recycle low-molecular-weight phenolic resin; and more than 50% of large-and medium-sized phenolic resin companies use the delayed condensation-biochemical method for comprehensive treatment of wastewater. Small-sized phenolic resin companies often use the entrusted treatment method as they produce a small amount of wastewater, namely, transfer of wastewater containing high-concentration phenols into the sewage treatment plant or hazardous waste treatment center. Some small-sized phenolic resin companies adopt the delayed condensation-biochemical method for treatment of wastewater containing phenols of high concentration. Table 10-4-8 shows the investigations on water pollutants of domestic general-purpose phenolic resin companies.

List of Investigations on Water Pollutants of Domestic General-purpose Phenolic Resin Companies Table 10-4-8

Wastewater Name	Water Content (m^3 per ton of products)	COD (mg/L)	Phenol (mg/L)	Formaldehyde (mg/L)	pH
Process wastewater	0.65-0.95	300000-380000	250000-320000	25000-55000	1.5-1.8
Flushing wastewater	0.60-0.90	380-3400	120-200	15-35	5.5-5.9

10.4.5 Inorganic Insulation Mortar

The inorganic insulation mortar is a kind of new insulation mortar applied on the inner and outer walls of a building, with inorganic light-weight particles as lightweight aggregates. Dry mortar is composed of the cementing material, anti-cracking additives and other fillers. The insulation performance of the insulation mortar is mainly achieved via the inorganic light aggregate thereof, which generally refers to the expanded perlite or vitrified microsphere.

Perlite is a kind of natural acidic glassy volcanic lava mineral (non-metallic), which is divided into perlite, rosin and obsidian. These three substances are different in the crystal water content. Its volume rapidly expands to 4-30 times at 1,000-3,000℃ to form the expanded perlite. Under normal circumstances, the expansion ratio should be 7-10 (black slate: more than 3), and the silica content should be approximately 70%.

The vitrified microbead is a kind of acidic glassy lava mineral (pitchstone ore), which is treated and produced with special processes and has a spherical fine particle structure containing internal pores and a vitrified and closed surface. It is a new inorganic lightweight insulation material with high performance, mainly composed of SiO_2, Al_2O_3 and CaO, with the thermal conductivity of 0.028-0.048W/(m·K), floatation rate and vitrification rate of more than 95%, water absorption rate of less than 50%, and melting point of 1200℃.

Both the vitrified microbead and expanded perlite are produced with perlite, but the former has higher requirements for the boiler temperature (generally above 1400℃).

China's perlite industry has developed rapidly in the northeast (accounting for more than half of the total output of China) since the beginning of production in 1966. With China's reform in wall materials and improvement of building energy conservation standards, inorganic insulation mortar products are mainly used in the south where heat isolation plays an important role.

The inorganic insulation mortar consists of the cementing material, anti-cracking additives and other fillers. The energy consumption in the production process is partly due to the production of lightweight aggregates and partly to that of cement. The environmental impact is mainly reflected by CO_2 emissions and dust produced in the production process.

According to the statistics of 177 new dry type cement production lines, the average heat consumption of clinker sintering is 828kCal/kg; the average power consumption per ton of clinker is 69.34kW·h; and the average comprehensive power consumption per ton of cement is 98.31kW·h, as shown in Table 10-4-9.

Comprehensive Energy Consumption and CO_2 Emissions of Cement Table 10-4-9

Product Name	Comprehensive Power Consumption (kW·h)	CO_2 Emissions (t)
Comprehensive energy consumption of clinker	69.34	0.89-1.22 (approximately 1.00)
Comprehensive energy consumption of cement	98.31	Approximately 0.12
Total	167.65	1.12

Note: The average power consumption per ton of cement is 100kW·h in the production process. Taking into account CO_2 generated in coal combustion, the CO_2 emission arising from the power consumption during production of one ton of cement is approximately 1.2t.

The mass-to-volume ratio of the binder and vitrified microbead is usually 1kg: 8.0L to 1kg: 4.5L. Improper proportioning will result in the nonconformity of the performance of the inorganic insulation mortar to the technical requirements of relevant standards. For a high cost performance, most of domestic inorganic thermal insulation mortar is prepared according to the following proportion: 1kg binder to 6L inorganic lightweight aggregate. The comprehensive energy consumption and CO_2 emission of cement are shown in Table 10-4-10.

10.4.6 Adhesive polystyrene granule Insulation Mortar

The adhesive polystyrene granule insulation mortar consists of the insulation binder and EPS granule, which are separately bagged (EPS granule volume: not less than 80%) and mixed with water according to the proportion in the production process. The calcium hydroxide, fly ash and amorphous silicon dioxide (accounting for more than 1/3 by weight) are used instead of a lot of cement. In the presence of an alkali activator, the C-S-H colloid is produced through the reaction of SiO_2 in fly ash with Ca $(OH)_2$. If a sulfate activator is used, the CAH colloid is produced through the reaction of SiO_2 in fly ash with Ca $(OH)_2$. When gypsum is added, stable ettringite will be formed. The C-H-S cementing system is generated on the surface of the fly ash particles and used for bonding particles in the system. The ettringite is filled into holes, making such holes denser in cement stones. Due to the gradual micro-expansion, the performance of cement is improved. The insulation mortar that is not produced with cement and gypsum has the following advantages. First, industrial solid wastes such as fly ash are comprehensively utilized. Secondly, the cement and energy consumption for production is reduced.

Comprehensively, the energy consumption is mainly caused by cement in the cementing material. The ratio of the cementing material to lightweight aggregate in domestic adhesive polystyrene granule insulation mortar is often 1kg: 8L to 1kg: 6.5L. The energy consumption and CO_2 emission of cement are shown in Table 10-4-11.

Comprehensive Energy Consumption and Comprehensive Utilization of Solid Waste in Production of Inorganic Insulation Mortar **Table 10-4-10**

Product Name	Consumption per 1,000m^2	Cement	Inorganic Lightweight Aggregate	Solid Waste Content Fly ash
Inorganic insulation mortar	13t	5.50t	5.00t	2.50t
Energy consumption	—	922.10kW·h	1465.00kW·h	—
CO_2 emissions	—	6.16t	1.76t	—

Notes: 1. The energy consumption of the electric furnace is about 293kW·h for production of one ton of inorganic lightweight aggregate. Taking into account CO_2 generated in coal combustion, the CO_2 emission arising from the power consumption during production of one ton of cement is approximately 0.352t.
2. The thickness of the insulation layer is 50mm.

Comprehensive Energy Consumption and Comprehensive Utilization of Solid Waste in Production of Adhesive polystyrene granule Insulation Mortar **Table 10-4-11**

Product Name	Consumption per 1,000m^2	Cement	Solid Waste Content (kg)	
			Fly ash	Waste EPS granule
Insulation mortar	7.5t	4.35t	2.55t	0.60t
Energy consumption	—	729.28kW·h	—	—
CO_2 emissions	—	4.87t	—	—

Note: The thickness of the insulation layer is 50mm.

10.4.7 Rock Wool

The rock wool insulation board is produced by melting of basalt and other natural minerals (main raw materials) into fibers at high temperature, and curing with an appropriate amount of adhesive.

The rock wool insulation board has the advantages of low thermal conductivity and high combustibility. It can be applied in energy conservation and thermal insulation works of outer walls of new, expanded and renovated residential and public buildings, including the external wall insulation, non-transparent curtain wall insulation as well as the fire barriers of EPS external thermal insulation systems.

Table 10-4-12 and Table 10-4-13 show the energy consumption ratings of rock wool boards in the *Energy Consumption Rating Quota of Rock Wool* (JC 522-1993), as well as the conforming energy consumption of rock wool production at different levels. The standard coal consumption should be less than 560kg/t, and the comprehensive power consumption should not exceed 400kW·h/t. Rock wool insulation boards to be used on the outer wall must have the appropriate tensile strength, with the density greater than 160kg/m^3 under normal circumstances. In accordance with relevant requirements at the minimum acceptable level, the standard coal consumption for production of per cubic meter of rock wool is 89.6kg, and the corresponding comprehensive power consumption is 64kW·h.

10.4.8 Comprehensive Evaluation

External thermal insulation systems used in south China are mainly based on the inorganic insulation mortar and adhesive polystyrene granule insulation mortar. Calculated based on the insulation layer thickness of 30mm, the energy consumption per 1,000m^2 is shown in Table 10-4-14.

In northern heating areas with high requirements for the thermal performance, the separate use of the inorganic insulation mortar and adhesive polystyrene granule insulation mortar cannot meet the design standards for building energy conservation. In this case, the composite insulation system may be used. Table 10-4-15 shows the comprehensive evaluation of energy consumption and harmful gas release of insulation materials, calculated based on the thermal insulation area of 1,000m^2.

The production of organic insulation materials requires the consumption of a large number of petrochemical products, while that of the inorganic insulation mortar and rock wool involves the consumption of a lot of energy and emission of a large amount of CO_2 and other gases. That is, a lot

Standard Coal Consumption Ratings for Rock Wool Production **Table 10-4-12**

National Super Level	National Level I	National Level II	Minimum Acceptable Level
300kg/t	380kg/t	450kg/t	560kg/t

Comprehensive Power Consumption Ratings for Rock Wool Production **Table 10-4-13**

National Super Level	National Level I	National Level II	Minimum Acceptable Level
310kW·h/t	330kW·h/t	350kW·h/t	400kW·h/t

Energy Consumption and CO_2 Emissions per 1,000m^2 of Inorganic Insulation Mortar and Adhesive polystyrene granule Insulation Mortar **Table 10-4-14**

Type of Insulation Material	Energy Consumption/kW·h	CO_2 Emissions/t
Inorganic insulation mortar	1432.26	4.75
Adhesive polystyrene granule mortar	437.57	2.92

Note: The thickness of the insulation layer is 30mm.

Comprehensive Evaluation of Insulation Materials **Table 10-4-15**

Type of Insulation Material		Insulation Layer Thickness /mm	Energy Consumption /kW·h	Coal Consumption /t	Oil Consumption /t	Toxic Gas Release
EPS		95	304	—	0.38	Benzene, toluene, ethylbenzene, p-xylene, m-xylene, o-xylene and styrene
XPS		70	1750	—	0.67	
Polyurethane (sprayed)		60	20	—	0.84	HCN, acetonitrile, acrylonitrile, CO, pyridine and benzonitrile
Phenolic insulation board		85	144	—	1.28	Phenol and its derivatives as well as dioxins
rock wool		115	644	828	—	—
Composite EPS board based on adhesive polystyrene granule	EPS	85	272	—	0.34	Benzene, toluene, ethylbenzene, p-xylene, m-xylene, o-xylene and styrene
	Adhesive polystyrene granule mortar	25	365	—	—	—

Notes: 1. The thickness of the insulation layer of the external thermal insulation system is based on Beijing's energy conservation standards (65%). The limit of K ([W/(m^2·K)]) of Beijing (Grade B cold zone) is 0.45, 0.60 or 0.70. In accordance with the *Design Standard for Energy Efficiency of Residential Buildings in Severe Code and Cold Zones* (JGJ 26-2010). Taking the limit of K is 0.45 [W/(m^2·K)], $K=1/(0.04+R+0.11)$.

2. The volume weight is 20kg/m^3 for the EPS board, 32kg/m^3 for the XPS board, 40kg/m^3 for the polyurethane, 60kg/m^3 for the phenolic insulation board, and 160kg/m^3 for the rock wool board. The energy consumption of rock wool production complies with the national Level II requirements of the Energy Consumption Rating Quota of Rock Wool (JC 522-1993).

of energy is consumed while energy is saved, which hinders the progress of energy conservation and emission reduction to a certain extent.

Waste polystyrene foam and solid wastes (such as fly ash) are used in the production of adhesive polystyrene granule mortar. On one hand, solid wastes are utilized comprehensively, which is conducive to the environmental protection. On the other hand, the energy consumption of buildings is reduced, which is consistent with the national development trend for energy conservation and emission reduction.

10.5 Prospects for Comprehensive Resource Utilization

A variety of solid wastes are produced extensively, causing indirect and long-term pollution. Direct pollution caused by solid wastes is far less than that of wastewater and waste gas, but solid wastes are converted in different ways into other pollutants that result in secondary and repeated pollution. At present, solid wastes are generally treated in a harmless way, namely, classified storage and disposal at the disposal site according to the national standards. The long-term stacking of solid wastes may lead to environmental pollution, and must be utilized comprehensively to eradicate the final pollution.

Solid wastes are used the most in the building material industry. Many wastes are used as alternative raw materials and fuels during production of building materials. To achieve the sustainable development of the building material industry, the conventional production method must be changed gradually, and the industry structure should be adjusted, resources and solid wastes should be utilized fully and reasonably depending on the advanced technologies, thereby saving the energy and reducing the environmental pollution in the production process. Green building materials are produced with the clean technology, involving no or low consumption of mineral resources and energy. That is, non-toxic, non-polluting and non-radiating building materials, which can be utilized within the specified application life and are beneficial to the environmental protection and physical health, are produced with a large number of industrial or urban solid wastes. The sustainable development goals of energy and resource conservation, environmental protection and comprehensive utilization can be achieved in the building material industry only by developing green building materials. Therefore, the energy conservation and emission reduction of the external thermal insulation industry will finally evolve into the development of green building materials, decease in the energy consumption for production of building materials and maximization of the comprehensive utilization of resources, in order to improve the utilization efficiency of resources, promote the circular economy and building a resource-saving and environmentally-friendly society.

11. Case Study of External Thermal Insulation Engineering Quality

11.1 EPS Board Insulation Project

The thin-plastered expanded polystyrene board (abbreviated as EPS board) system has the excellent insulation performance, waterproofing performance, wind pressure safety and impact resistance, in which wall cracking and leakage can be solved effectively. Featuring the convenience in construction and high cost performance, the system is the most popular and mature at home and abroad. It is widely applied in developed countries in Europe and America as well as various regions of China, especially in severe cold and cold regions. In the 1950s, Germany took the lead in researching the thin-plastered EPS board system and built the first external thermal insulation project that has withstood the test of more than 50 years. The thin-plastered EPS board system was first adopted in Beijing Yujing Garden Villa Group completed in the early 1990s in China, but is still in good conditions up to now. The thin-plastered EPS board system shows good performance during domestic engineering applications in over 20 years, but still has some problems. For example, EPS boards are blown off in the event of strong winds.

11.1.1 Cases of EPS Board Falling

The thin-plastered EPS board system is used in the following projects, but EPS boards fall off within a large scale, resulting in huge economic losses and also attracted the extensive attention

Case 1: 18# building of Kexueyuan Community on No. 1 Kexue Street, Beijing Road, Urumqi, China. all EPS boards on a side wall between the windows of the 1st floor and eaves of the 6th floor fell off, with a height of 15m and an area of about $200m^2$. Consequently, three cars were damaged, but fortunately without injuries (Figure 11-1-1).

Case 2: 5F and 6F of No. 16 building in Olympic Garden Community in Xinbei District, Changzhou City, Jiangsu Province. More than half of EPS boards on two outer walls fell off, making gray cement exposed, and the other half of EPS boards are separated from the walls and suspended in the air.

Case 3: 5# Unit, No. 17 Building, Chengsheng Garden Community, Urumqi. The EPS board insulation layer of approximately $40m^2$ is scattered on the ground (Figure 11-1-3).

11.1.2 Causes of EPS Board Falling

Falling of EPS boards of the thin-plastered EPS board system is mostly caused by nonconformity of the bonding method and area to the design requirements and technical standards. When the bonding area is too small, there is not enough bonding strength between the EPS boards and base course wall or connecting cavities are formed. In the event of nonconformity of bonding and sealing, the cavities are greatly affected by the negative wind pressure. The stress arising from the wind pressure is concentrated at board gaps, which may easily lead to the cracking of board gaps and finally large-scale falling of EPS boards.

11.1.2.1 Destruction Caused by Wind Pressure

The wind load of a building refers to the action of wind generated by the airflow on the building surface. It is associated with the wind properties (speed and direction), landform, surrounding environment, building height and shape, etc. The pressure arising from the wind load on the

Figure 11-1-1 Falling of EPS Board Insulation Layer

Figure 11-1-2 Suspension of EPS Board Insulation Layer

Figure 11-1-3 EPS Board Insulation Layer "Torn" by Wind

building is not distributed evenly. The wind load is divided into the positive and negative wind pressure. The positive wind pressure leads to the compressive force on a building surface, while the negative wind pressure causes the tensile force. The external thermal insulation system must be resistant to negative wind pressure in order not to fall under the negative wind pressure.

EPS boards are pasted with or without cavities. The EPS board system with connecting cavities is subject to the push-pull force from the inside to outside as the air pressure inside cavities is higher than the ambient air pressure in the area with negative wind pressure. The EPS boards of the system formed by means of pure point bonding may fall off within a large scale. The EPS board system without cavities has high wind pressure safety because of full bonding between the boards and base course wall, so this system does not fall off under normal circumstances.

When the negative wind pressure on the EPS board system is greater than the bonding strength between the EPS boards and base course wall or bonding mortar, the system may fall off. That is, the EPS board system is destroyed under the instantaneous negative wind pressure or in the case of a heavy wind (short-time). The common cases in which the EPS board system is blown off are related to the negative wind pressure, as shown in Figure 11-1-4 and 11-1-5.

Figure 11-1-4 Engineering Case (I) of Damage Caused by Negative Wind Pressure
(damage to the interface between the bonding layer and base course wall)

Figure 11-1-5 Engineering Case (II) of Damage Caused by Negative Wind Pressure
(damage to the interface between the point-bonded part and EPS board)

The negative wind pressure often occurs on two sides parallel to the wind direction and the leeward side. Two sides of a building are subject to the maximum negative wind pressure, so they are easy to damage under negative wind pressure. The wind load rises with the building height increasing. Therefore, particular attention must be paid to the impact of the wind load on the external thermal insulation system of a high-rise building. The prevailing wind direction of one area can be determined based on the wind rose diagram, and the area of negative pressure occurrence can be identified accordingly.

11.1.2.2 Connecting Cavity

Main methods of bonding between the EPS boards and base course wall in China's technical standards are strip bonding and point-and-frame bonding. In the case of poor flatness of the base course wall, point-and-frame bonding should be adopted. This method is used the mostly widely at present. However, the nonconformity of actual operations to the point-and-frame bonding requirements, such as "pure point bonding" with the frame omitted, will lead to connecting cavities, which is a kind of serious quality accident.

When pure point bonding is adopted, the deformation of EPS boards will not be constrained in the absence of bonding between four sides of EPS boards with the base course wall. From the perspective of force, this is equivalent to the change of a simply supported beam into a cantilever beam. When there is positive or negative wind pressure, the EPS boards will be deformed more greatly than those subject to point-and-frame bonding, which increases the possibility of large-scale falling. In the pure point bonding method, the closed cavities formed by means of point-and-frame bonding are transformed into large connecting cavities, and the consequent negative wind pressure is completely applied on the weak part involving a small bonding area, which will reduce the bonding strength and damage the bonded points one by one. As the freedom of the large-cavity external thermal insulation system in the direction perpendicular to the wall surface cannot be constrained by the protective layer, large-scale falling of EPS boards will be cause. If the insulation layer with connecting cavities are used in an area where the negative pressure is prone to occur, the push-pull force generated by the base course wall on EPS boards will be concentrated in the position with high negative pressure, and the area where the negative pressure is prone to occur may be damaged, resulting in cracking or falling (Figure 11-1-6).

Figure 11-1-6 Falling Caused by Pure Point Bonding

11.1.3 Measures for Prevention of EPS Board Falling

11.1.3.1 Closed Small Cavity Structure

In view that the external thermal insulation system with connecting cavities is easy to fall off under the negative wind pressure, it is stipulated in technical standards at home and abroad that the bonding area between the EPS boards and base course wall must be greater than 40%. The bonding surface between the EPS boards and base course wall must have a closed cavity structure, and be constructed mainly by means of point-and-frame bonding.

Currently, point-and-frame bonding is mostly used to EPS boards of 1200mm × 600mm or 900mm×600mm.

Although these large EPS boards can be constructed fast, the pressure at one end may result in the warping of the other end and the false bonding and hollowing of the other surface during construction of the insulation layer, so it is difficult to reach the fullness of 100%. The closed small cavity structure (Figure 11-1-7) with EPS boards of 600mm × 450mm, which is conducive to operations, can guarantee the effective bonding area and prevent connecting cavities. The bonding mortar consumption per EPS board of 1,200mm × 600mm is about 3.95kg and its bonding rate is approximately 41%, so the mortar consumption per square meter is 5.49kg. The bonding mortar consumption per EPS board of 600mm × 450mm is about 2.04kg and its bonding rate is approximately 56%, so the mortar consumption per square meter is 7.56kg. It can be concluded that the closed small cavity structure is not only beneficial to actual operations, but also involves a large bonding area. The closed small cavity structure has been included in local standards of Beijing and Shaanxi.

If the EPS boards of 1200mm × 600mm are constructed by means of point bonding without any frame (Figure 11-1-8), the bonding mortar consumption per board is 0.85kg, the bonding rate drops to roughly 8.70%, and the mortar consumption per square meter is only 1.73kg. Though materials are saved exponentially and the connection speed is increased greatly, it is a typical incomplete method, which is bound to cause quality accidents. Therefore, this method is prohibited.

11.1.3.2 No-cavity Structure

The "adhesive polystyrene granule pasted EPS board system" promoted by Beijing Zhenli has a no-cavity structure. It is included in the in-

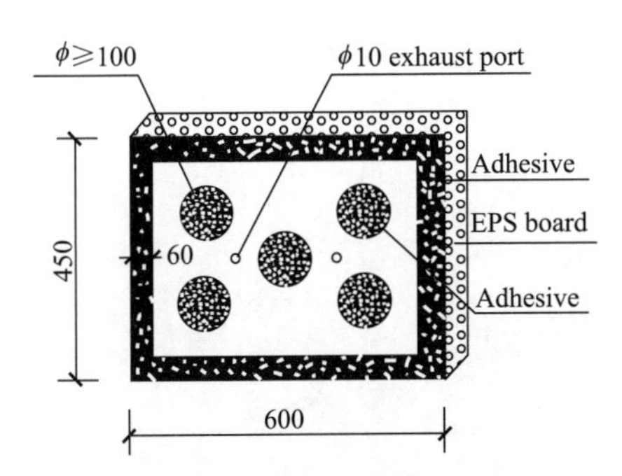

Figure 11-1-7 Bonding Diagram of Insulation Board with Closed Small Cavities

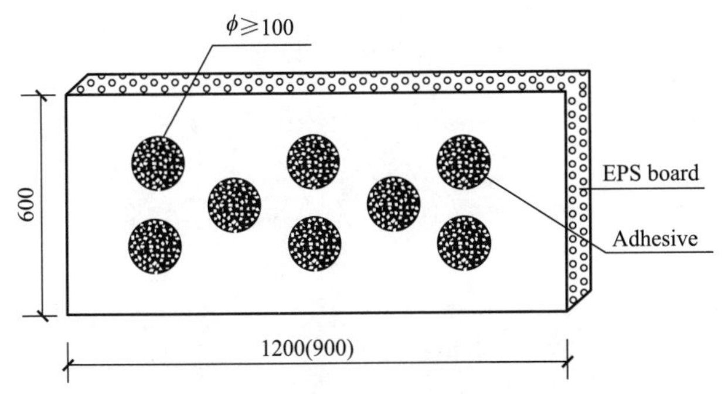

Figure 11-1-8 Schematic Diagram of Pure Point Bonding of Insulation Board

dustry standard "Products for External Thermal Insulation Systems based on Mineral Binder and Expanded Polystyrene Granule Plaster" (JG/T 158-2013) as well as local standards of Beijing, Shandong, Jilin and Shaanxi. This mature and reliable technology is an effective solution to EPS board falling.

The basic structure of the "pasted EPS board system" is shown in Figure 11-1-9. A 15mm layer of adhesive polystyrene granule pasting mortar is applied on the wall surface, transverse trapezoidal grooves are prefabricated, EPS boards pre-coated with the interface agent are pasted, and finally a 20mm leveling layer of adhesive polystyrene granule pasting mortar is applied on the EPS board surface, thereby forming a no-cavity composite insulation layer consisting of the "adhesive polystyrene granule pasting mortar+EPS board+adhesive polystyrene granule pasting mortar". The reserved 10mm wide board gaps are filled and leveled with the adhesive polystyrene granule pasting mortar extruded out during construction. The anti-cracking protective layer is composed of the anti-cracking mortar and plastic-coated alkali-resistant glass fiber mesh. On one hand, this structure is equivalent the use of pasting mortar anchors around each EPS board, thereby further enhancing the bonding performance and wind pressure safety of the entire system. On the other hand, the EPS board insulation layer with vapor permeability is able to decompose and absorb the concentrated stress arising from the expansion and contraction of EPS boards. The stress is transferred to the bonding layer and leveling layer, and then gradually released layer by layer to the surface, hence effectively preventing cracks. Furthermore, the six sides of EPS boards are completely covered by the adhesive polystyrene granule pasting mortar, which restricts the expansion and contraction of EPS boards to a certain extent. As the ageing, expansion and contraction of EPS boards are fully taken into account and the stress generated is limited, transferred, decomposed and absorbed by the bonding layer, leveling layer as well as adhesive polystyrene granule pasting mortar at board gaps, so cracking resulting from the post-shrinkage of EPS boards is solved effectively.

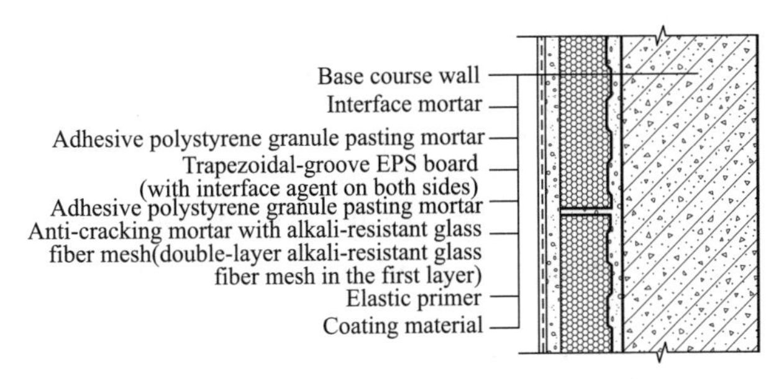

Figure 11-1-9 Basic Structure of Adhesive polystyrene granule Pasted EPS Board System

11.2 XPS Board Insulation Project

The extruded polystyrene board (hereinafter referred to as the XPS board) differs from the expanded polystyrene board (hereinafter referred to as the EPS board) in the fabrication process, performance indicator and application scope. Compared with the EPS board, the XPS board has higher density and strength, and lower thermal conductivity, water absorption rate, vapor permeability coefficient, bonding performance and dimensional stability. Thus, testing and research must be performed and targeted technical measures should be taken before the XPS board is applied for external thermal insulation. The construction method of the thin-plastered EPS board system are directly copied in some XPS board insulation projects, resulting in serious quality problems such as cracking, hollowing, falling and fire.

11.2.1 Cases of Quality Problems of XPS Board Insulation System

The XPS board insulation system is subject to serious surface cracking and falling under the repeated effects of thermal stress arising from changes in the external environment, as shown in Figure 11-2-1 and 11-2-2.

Figure 11-2-1 Finish Cracking of XPS Board Insulation System

Figure 11-2-2 Finish Cracking and Falling of XPS Board Insulation System

If there are connecting cavities in the bonding layer, the XPS board insulation system may be easier to damage under the negative wind pressure, resulting in large-scale falling, as shown in Figure 11-2-3.

The XPS board with a smooth surface and low water absorption rate cannot be bonded securely with the surface, so its finish layer may fall off, as shown in Figure 11-2-4.

Figure 11-2-3 Large-scale Falling of XPS Boards Caused by Connecting Cavity

Figure 11-2-4 Finish Falling Caused by Low Water Absorption Rate of XPS Board

11.2.2 Cause Analysis of Cracking and Hollowing of Thin-plastered XPS Board System

11.2.2.1 Drastic Stress of XPS Board

There are differences in thermal stress of the materials used in the thermal insulation system. if the temperature changes, stress is produced between external insulation materials, which will result in cracks and affect the durability of the system.

From 10 groups of stress data in Table 11-2-1, it can be seen that the surface strain of the EPS board fluctuates little at 20℃, with the average of approximately 24$\mu\varepsilon$. With the temperature rising, the strain increases greatly. When the temperature is 70℃, the peak strain is 2,670$\mu\varepsilon$, and the average is 2,340$\mu\varepsilon$.

From 10 groups of stress data in Table 11-2-2, it can be seen that the strain of the XPS board changes drastically. The peak strain is 159$\mu\varepsilon$ at 20℃. When the temperature is 70℃, the peak strain is 4,571$\mu\varepsilon$, and the average is 4,036$\mu\varepsilon$, close to twice that of the EPS board. The volume of the XPS board varies significantly in the case of violent changes in the ambient temperature. Such variations must be taken into consideration when the XPS board is used as a wall insulation material. The appropriate insulation structure and supporting materials should be used.

The above results show that the volumes of EPS and XPS boards undergo obvious changes with the temperature. If surface mortar is directly exposed to insulation boards, appropriate effective measures should be taken as a result of significant differences in the thermal strain of the surface mortar and insulation board, so that the normal use of the insulation system is not affected by cracks on the wall. Comparatively, the XPS board suffers from greater strain change and volume deformation, so it is less stable and affected in a more complex way than the EPS board. In the event of nonconformity to the specified technical requirements, the surface mortar will crack due to the thermal strain of XPS boards.

11.2.2.2 Significant Thermal Deformation of XPS Board

The shear stress is generated due to the thermal deformation of the interface under constraints. The significant relative deformation results from obvious differences in the temperature and thermal expansion coefficient of tow materials, and the essential condition for large stress is the major constraint. In the thin-plastered XPS board system, the temperature of the inner side of the XPS board varies little, basically stable at 20-30℃, and there is little thermal stress on each interface. The temperature of the outer side of the XPS boards fluctuates greatly (the temperature of the wall surface is approximately 70℃ in summer; the temperature of the outer side of the XPS board basically reaches 70℃ in summer as a result of no heat isolation through the anti-cracking layer and finish layer and below-20℃ in winter, and the temperature difference is 90℃ in a year and 50℃ in a day). The thermal expansion coefficients of the XPS board and surface mortar are greatly different from each other, but the former with high elastic modulus (more than 20MPa) forms strong constraints to the latter. Large stress will be concentrated at board gaps, which will result in the joint instability and cracking. Figure 11-2-5 shows the schematic diagram of thermal deformation of the insulation board in different seasons.

Strain of EPS Board **Table 11-2-1**

Temperature	Strain ($\mu\varepsilon$)										Average
20℃	−78	−53	−11	10	23	40	60	62	77	109	24
70℃	2324	2510	2555	1999	2411	2670	1932	2503	2388	2109	2340

Strain of XPS Board **Table 11-2-2**

Temperature	Strain ($\mu\varepsilon$)										Average
20℃	−120	−76	−55	3	12	42	45	95	126	159	23.1
70℃	3480	4027	4571	4329	3875	4262	3553	4085	3961	4221	4036

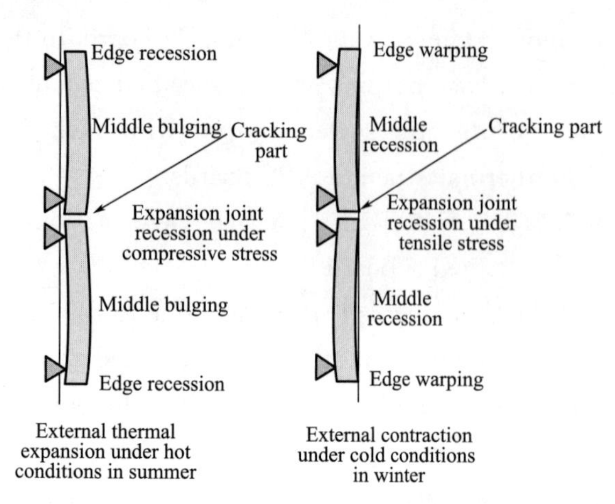

Figure 11-2-5 Schematic Diagram of Thermal Deformation of XPS Board in Different Seasons

11. 2. 2. 3 Poor Vapor Permeability of XPS Board

The XPS board has an independent cellular closed bubble structure, without gaps on the front and back sides, so it has almost no vapor permeability without moisture infiltration in the case of leakage, condensation, freezing and thawing.

The basic resin used in the EPS board is the same as that in the XPS board, but their production processes are different from each other. The EPS board is fabricated according to the following procedures: preparation of spherical polystyrene balls through the synthesis of polystyrene resin and other additives, pre-foaming of balls in a steam foaming machine, ageing, drying, mould installation, press molding by means of expansion and melting in the presence of hot steam, and cutting into the desired size. Although EPS balls are filled with gas and have a closed-pore structure, and the walls of balls are integrated by means of melting, there are open gaps among balls, which may become moisture erosion spaces or paths. Therefore, the EPS board has good vaporpermeability, and its permeation coefficient increases because of diffusion channels. The forming process ensures that the XPS board has a complete closed-cell structure. As there are no gaps among cells, the XPS board has an even cross section and continuous and smooth surface. The structures of XPS and EPS boards determine their significant differences in physical and chemical properties, especially the vapor permeability and bonding performance (Figure 11-2-6 and 11-2-7).

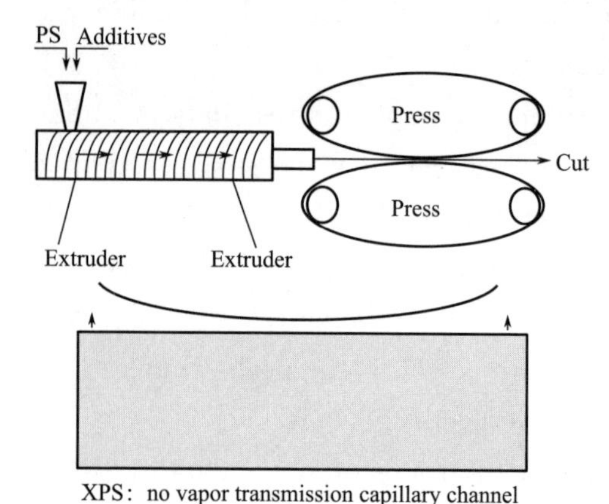

Figure 11-2-6 Schematic Diagram of Vapor Permeability of XPS Board

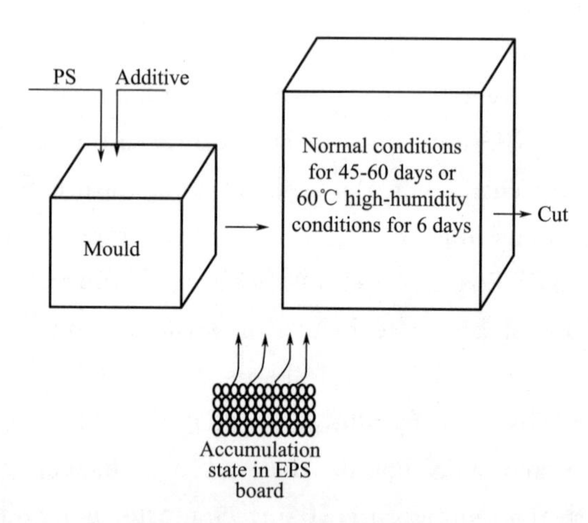

Figure 11-2-7 Schematic Diagram of Vapor Permeability of EPS Board

The XPS board has poor vapor permeability and low water absorption rate. Part of pores in polymer mortar are sealed with polymer emulsion, thereby increasing the difficulties in moisture transfer of vapor, making a lot of vapor accumulated between the polymer mortar and XPS board and finally leading to physical damage such as blistering. If these destructive effects are superimposed, more serious damage may be caused. It is found in microstructure analysis that the polymer film is swollen and softened due to water absorption, which reduces the bridging effect. Finally, the polymer film is degraded, resulting in fatal damage such as the efflorescence and adhesion loss (Figure 11-2-8).

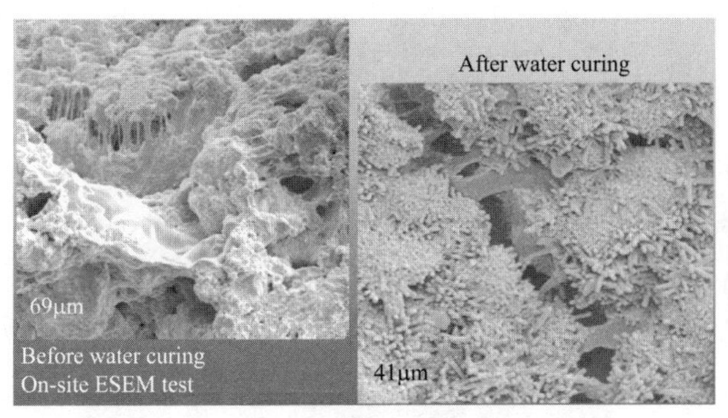

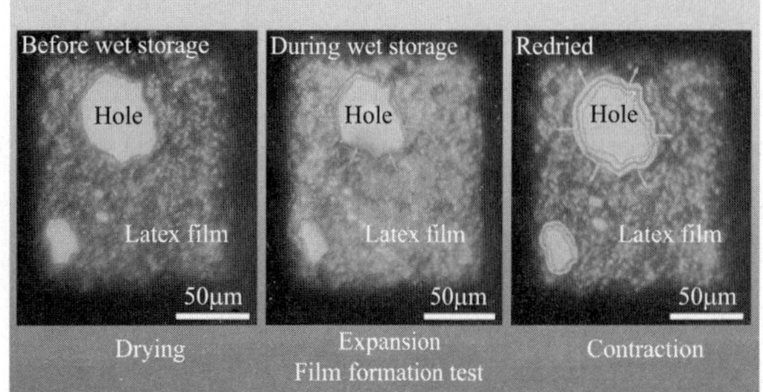

Figure 11-2-8 Microscopic Topography of Swelling of Polymer Film

11.2.2.4 Impact of Wind Pressure

1. Connecting cavity

The thin-plastered XPS board system is bonded mainly by means of point-and-frame bonding. However, pure point bonding with no frame is often adopted as a result of cheating in actual on-site operations, thus forming connecting cavities.

As four sides of the XPS board are not bonded with the base course wall in the case of pure point bonding, the board deformation is not constrained. From the perspective of force, this is equivalent to the change from a simply supported beam into cantilever beam. Under the positive and negative wind pressure, the deformation of the XPS board in the case of pure point bonding is far greater than that in the case of point-and-frame bonding, thereby increasing the cracking possibility and degree by a geometric multiple. At the same time, small cavities formed in point-and-frame bonding are changed into large through cavities. When one board is loose or cracking, the negative wind pressure generated by connecting cavities will be completely applied on the weak bonding points, affecting the bonding strength and break bonding points one by one. If there are large cavities in the external thermal insulation system, the freedom perpendicular to the wall surface cannot be constrained by the protective surface, which will lead to insufficient bonding area within a large range. When the insulation layer with connecting cavities is built in the area where the negative pressure is prone to occur, the tensile stress generated by the negative pressure will be concentrated in the position with the maximum negative pressure, thereby damaging the part where the negative pressure is prone to occur, and resulting in cracking or falling.

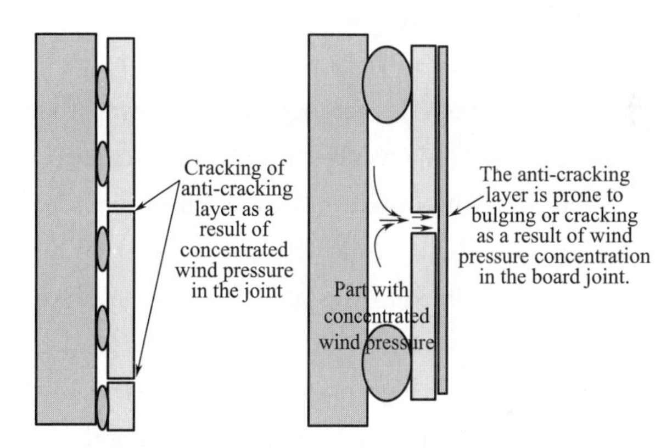

Figure 11-2-9 Schematic Diagram of Cracking of Thin-plastered System under Wind Pressure

Even if strip bonding or strict point-and-frame bonding is conducted, the XPS board with a smooth surface and high strength cannot be securely bonded with bonding mortar under normal circumstances. Instead, there will be through cavities, which are easier to damage under the wind pressure. The squeezing or suction force generated by the positive or negative wind pressure on the wall surface is released at board gaps, which may cause cracking of such joints. There is high negative wind pressure on the lee side at a height. The XPS boards bonded in such parts, even if reinforced with anchor bolts, may be sucked or torn under the negative wind pressure. Anchor bolts can only protect the parts covered by their discs.

2. Joint cracking

It is stipulated in the construction regulations on the thin-plastered insulation board system that gaps larger than 1.5mm should be reserved among boards and sealed with the polyurethane sealant or other materials. However, this time-consuming and labor-consuming procedure is often neglected, so board gaps are key parts involving quality problems. They may be destroyed as a result of board instability or thermal stress, or cracking under the wind pressure. As shown in Figure 11-2-9, the negative wind pressure is concentrated at board gaps, which may lead to hollowing, cracking and even falling of insulation boards.

11.2.2.5 Condensation Impact

Condensation in an outer wall refers to vapor liquefaction when the temperature is lower than the dew point temperature of air. It may occur on the surface of the outer wall or in the process of vapor transfer (i. e. inside the outer wall). When there is a difference in the partial pressure of vapor on two sides of the outer wall, vapor will be diffused from the high-pressure side to the low-pressure side. If the outer wall is not suitable for heat and moisture transfer, vapor will condense into water, causing condensation and moisture.

For the thin-plastered XPS board system, the temperature of the inner wall surface is far higher than the condensation temperature of air in summer, so condensation will not occur. As the temperature inside XPS boards is higher than the dew point temperature, there is also no condensation in winter. When the temperature of the external surface of the XPS board is lower than the dew point temperature and the air humidity is relatively high, condensation may occur on the surface mortar layer, generating condensate. As the thin surface mortar layer is able to absorb a small amount of liquid water, its strength and adhesion decline after long-term repeated immersion in liquid water. In addition, dry and wet deformation occurs in the drying process, which will lead to hollowing and falling. Figure 11-2-10 shows the damage caused by condensate to the thin-plastered EPS board system.

Figure 11-2-10 Damage Caused by Condensate to Thin-plastered XPS Board System

11.2.2.6 Dimensional Impact of XPS Board

At present, XPS boards of 1200mm × 600mm are used mostly in the point-and-frame bonding method. This helps to increase the construction speed. When one end of the XPS board is pressed, however, the other end may warp up, leading to false bonding or hollowing. The closed small cavity structure of the XPS board (600mm × 450mm) is conducive to the operations and guarantees the effective bonding area, without connecting cavities.

The analysis results of construction of boards with different specifications are shown in Figure 11-2-11 and 11-2-12. The "b" indicates the spacing between two hands, directly affecting the operating comfort. The greater the value of "b", the less evenly the force applied by workers is distributed, and the easier it is to cause false bonding. The most reasonable value of "*b*" is equal to the shoulder width of a worker. If the board is 1,200mm long, the spacing between both hands will be greater than the shoulder width, so the board may be bonded unevenly during construction. If the board is 600mm long, the spacing between both hands is basically identical to the shoulder width, which is conducive to operations and makes the pressing force distributed evenly thus effectively preventing false bonding. Meantime, the effective bonding area between the wall and small-sized boards is easier to control. Hence, the board specifications should be appropriate.

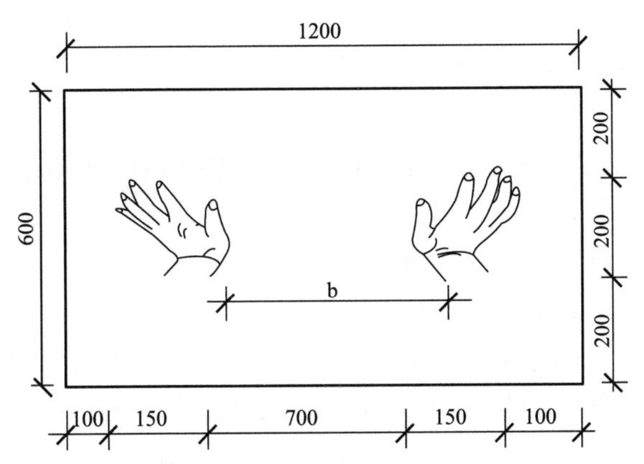

Figure 11-2-11 Schematic Diagram of Both Hands during Board Bonding (1200mm×600mm)

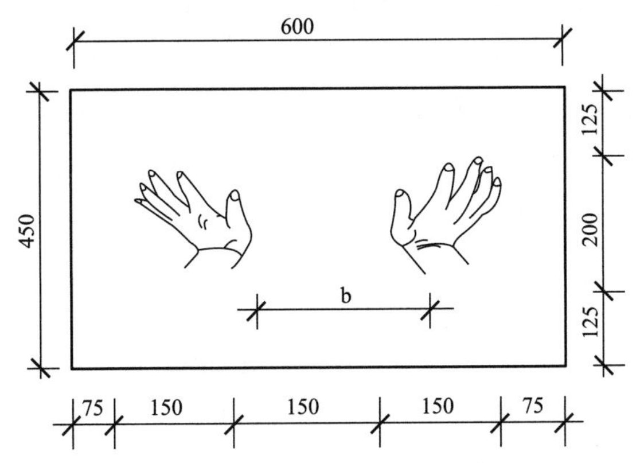

Figure 11-2-12 Schematic Diagram of Both Hands during Board Bonding (600mm×450mm)

The board width also affects the operations. Practices have proved that it is the most efficient to press and control boards when the palm length of a worker is greater than 1/2 of the board width. In this case, boards can be compacted easily. As the palm length is about 200mm, the board width should be less than 600mm, so it is reasonable to use the 450mm wide boards.

11.2.3 Troubleshooting in XPS Board Application

The *Products for External Thermal Insulation Systems based on Mineral Binder and Expanded Polystyrene Granule Plaster* (JG/T 158-2013) stipulates how to keep the stability of XPS board insulation systems, without cracking or falling, in order to guarantee the quality of external thermal insulation projects. Specific measures are as follows:

(1) A reasonable insulation structure should be selected. When the adhesive polystyrene granule pasted XPS board system is used, the thermal stress and deformation of XPS boards can be effectively resolved. When the XPS boards are fully bonded with the wall in a no-cavity manner, the damage under the negative wind pressure can be eliminated. The specific structure is shown in Figure 11-2-13. Two through holes (size: 40-60mm; center distance: 200mm) should be drilled along the center axis in the length direction (Figure 11-2-14), to facilitate full bonding, improve the vapor permeability and reduce hollowing.

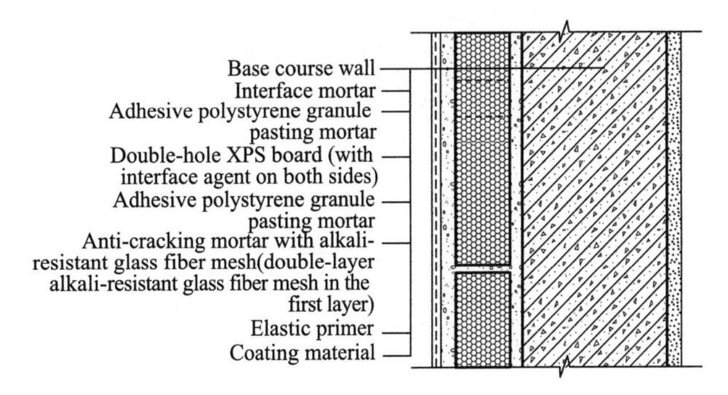

Figure 11-2-13 Basic Structure of Adhesive polystyrene granule Pasted XPS Board System

(2) The board size should be appropriate, preferably 450mm×600mm. As the XPS board is rigid, the excessive size will lead to false bonding.

(3) A leveling transition layer of adhesive

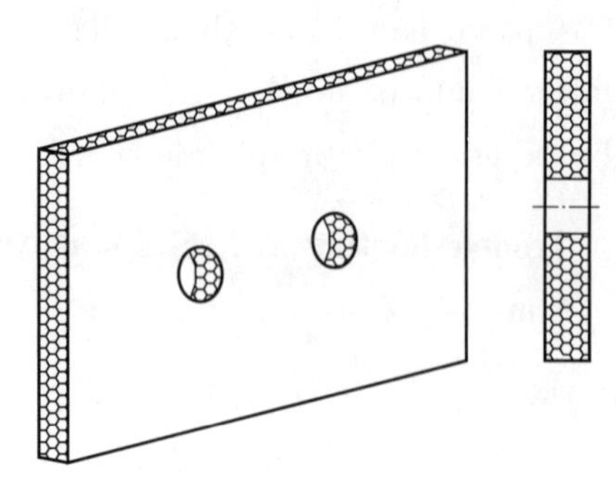

Figure 11-2-14 Double-hole XPS Board

polystyrene granule pasting mortar (thickness: not less than 20mm) should be applied outside the XPS board to achieve the following main functions:

(i) The XPS board insulation system will meet the weathering resistance requirements. According to the weathering resistance test results, the weathering resistance of the system is enhanced greatly with the thickness of the leveling layer increasing.

(ii) There are significant differences in the linear expansion coefficient and elastic modulus of the XPS board and polymer mortar of the anti-cracking protective layer. The adjacent materials affected by the temperature and humidity are deformed at different rates, resulting in stress concentration. When the stress exceeds the bonding strength of polymer mortar, there will be cracks in the system. The pasting mortar leveling layer (with the elastic modulus between those of the board and polymer mortar) applied outside the XPS boards has transitional impact, thereby avoiding direct exposure between the XPS boards and polymer mortar, reducing the difference in the deformation rate of adjacent materials as well as the thermal stress arising from such a difference, synchronizing the deformation of each structural layer, and guaranteeing the safety and durability of the entire system , without cracking, hollowing or falling.

(iii) The leveling layer made of adhesive polystyrene granule pasting mortar is equivalent to a moisture dispersion layer, with excellent functions of moisture absorption, conditioning and transfer. It is able to absorb the condensate produced in the presence of the vapor permeability difference of XPS boards or in the case of condensation, to avoid the damage caused by changes in three phases of the water accumulated, improve the bonding and breathing performance of the system, and guarantee the long-term safety and stability of external thermal insulation projects.

(iv) If the XPS board system is used, the construction quality should be controlled in strict accordance with the construction process and quality acceptance standards, in order to ensure the quality of the external thermal insulation project.

11.3 Cast-in-situ EPS Board Project

The cast-in-situ EPS board system, also known as the external thermal insulation system based on the external formwork, built-in EPS board and cast-in-situ concrete. It is fabricated by installing the non-meshed or meshed EPS boards on the inner side of the external formwork, integrating the external formwork with the non-meshed or meshed EPS boards after on-site concrete pouring, and applying a surface layer and finish layer outside the insulation layer. Depending on the EPS board structure, this system is divided into the cast-in-situ concrete and non-meshed EPS board system and the cast-in-situ concrete and meshed EPS board system. As the main structure and insulation layer are poured at the same time and the insulation is securely bonded with the base course wall, the system has high wind pressure safety, seismic resistance and low comprehensive costs and can be constructed fast. It is applicable to medium-and high-rise buildings. However, there are still quality problems in engineering applications, such as surface cracking. The cast-in-situ concrete and meshed EPS board system has more serious problems.

11.3.1 Cases of Quality Problems

11.3.1.1 Cast-in-situ Concrete and Non-meshed EPS Board System

Figure 11-3-1 is a picture of surface cracking of the cast-in-situ concrete and non-meshed EPS board system in Qingdao. It can be seen that there are obvious surface cracks in each direction as well as significant hollowing.

11.3.1.2 Cast-in-situ Concrete and Meshed EPS Board System

Surface cracking is common for cast-in-situ concrete and meshed EPS board systems. In one project located in Guang'anmen District of Beijing is taken as an example, with the cast-in-situ concrete and meshed EPS board system. There are a large number of cracks and peels after years of exposure to all weather conditions, as shown in Figure 11-3-2.

One high-rise building in Wangjing District of Beijing is also equipped with the cast-in-situ concrete and meshed EPS board system, involving both the paint finish and also the face brick finish. There are various types of cracks on the entire paint finish within a few years after the starting of use, but also tensile cracks on the face brick finish, as shown in Figure 11-3-3.

Figure 11-3-4 shows a photo of one project in Hohhot, Inner Mongolia, in which the cast-in-situ concrete and meshed EPS board system with a face brick finish is used. The face brick finish started to crack and fall off immediately after this project was put into use.

11.3.2 Cause Analysis of Quality Problem

11.3.2.1 Cast-in-situ and Meshed EPS Board System

1. The flatness and verticality are difficult to control.

As cast-in-situ concrete is constructed layer by layer, and the lateral pressure of the lower part of the cast-in-situ concrete is higher than that of the upper part, there is greater compressive force and deformation on the lower part of each layer. After the external formworks are removed, the resilience magnitude of the lower part of the

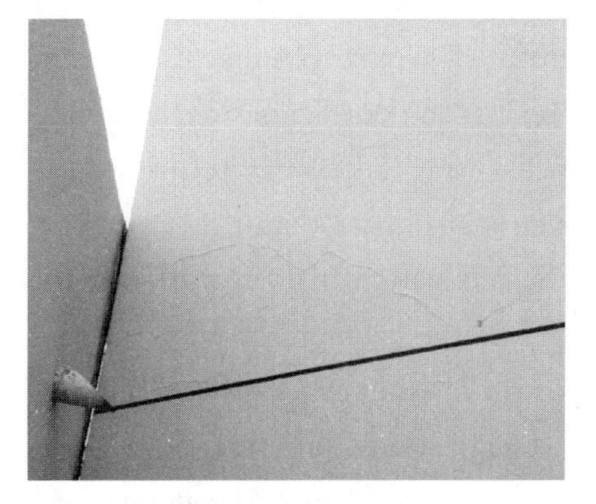

Figure 11-3-1 Surface Cracking of Cast-in-situ Concrete and Non-meshed EPS Board System

Figure 11-3-2 Surface Cracking and Peeling of Cast-in-situ Concrete and Meshed EPS Board System

Figure 11-3-3 Crack of Paint Finish and Tensile Crack of Face Brick Finish of Cast-in-situ Concrete and Meshed EPS Board System

Figure 11-3-4 Tensile Cracking and Falling of Face Brick Finish of Cast-in-situ Concrete and Meshed EPS Board System

EPS board is greater than that of the upper part, steps may be formed at connections of different layers, which results in poor flatness. Under normal circumstances, the flatness is controlled by making the lower part tighter than the upper part and adjusting the inclination of formworks in the typing process, but the problem of poor flatness cannot be solved thoroughly as a result of the limited effects and significant differences. In addition, the flatness of the cast-in-situ concrete surface is difficult to control, resulting in large vertical deviations in the height direction, partially 40-60mm. In order to ensure the final flatness and verticality, EPS boards are often ground, which leads to the unevenness of their thickness and the differences in thermal properties of the whole wall. In this case, the temperature of the protective surface is inconsistent, indicating that the protective surface is not deformed evenly but is cracking. Moreover, the bonding strength of polystyrene particles is affected and a lot of powder is produced in the grinding process, so the bonding strength between the surface mortar and EPS boards cannot be guaranteed. In addition, the protective surface is also used for leveling, which may result in the thickness unevenness and further cracking.

2. There is local damage and contamination.

The EPS board surface with low strength is unavoidably damaged when external formworks are erected and dismantled. For the external corners, lower supports of outer boards and bolt holes through the wall, thermal bridges may be formed as a result of leakage in the concrete pouring process. This will not only affect the thermal properties of the wall, but also may lead to local cracking.

11.3.2.2 Cast-in-situ Concrete and Meshed EPS Board System

The investigation and analysis results show

that a thick leveling layer of ordinary cement mortar or cement mortar containing a small amount of polymer is applied on the meshed EPS board surface in the cast-in-situ concrete and meshed EPS board system subject to cracking and face brick falling. This does not comply with the provisions of the industry standard "Technical Requirements of External Thermal Insulation with Expanded Polystyrene Panel for In-situ Concrete" (JG/T 228) that "an appropriate layer of lightweight fireproof insulation mortar shall be added on the EPS board or meshed EPS board surface", so there are risks of surface cracking.

1. The thick layer of ordinary cement mortar or polymer cement mortar is easy to crack.

Cement mortar itself tends to crack with various shrinkage deformations. A thick layer of cement mortar is easier to crack. When an appropriate amount of polymer is added, the flexibility of cement mortar can be modified to achieve the desired anti-cracking effects. Insufficient polymer will lead to poor flexibility and cracking of the mortar layer, while excess polymer will greatly increase the costs, thereby reducing the economic benefits and operability. Due to large grids of the wire mesh and high rigidity of wires, the effect of stress dispersion is insignificantly, so surface cracking will not be eliminated.

The thickness of ordinary cement mortar or polymer cement mortar in the cast-in-situ concrete and meshed EPS board system is up to 20-30mm, which may reduce the flexibility and lead to cracking of this layer. In addition, the flatness and verticality of the whole wall surface cannot be controlled accurately in the board pouring process, resulting in the thickness unevenness, local shrinkage and thermal stress inconsistency of the surface, and further causing cracks. As the layer of ordinary cement mortar or polymer cement mortar is located outside the meshed EPS board insulation layer, it is deformed greatly under the effects of outdoor temperature variations. The ordinary cement mortar or polymer cement mortar will crack as a result of fatigue deformation in a long time.

2. Extrusion cracking occurs under the excessive load.

The cast-in-situ concrete and meshed EPS board system has poor flatness and thick plaster. The load per square meter outside the meshed EPS boards is up to 80kg and even 100kg. The meshed EPS boards will creep in a long time, resulting in extrusion cracking of the whole hard surface under the gravity.

3. Cracking is caused by a unreasonable sandwich insulation structure.

When the meshed EPS boards are leveled with the ordinary cement mortar or polymer cement mortar (thickness: minimum 20mm), they are located between the leveling layer and reinforced concrete base course wall, similar to a sandwich insulation structure. It has been stated in Chapter 2 that the sandwich insulation structure is prone to crack.

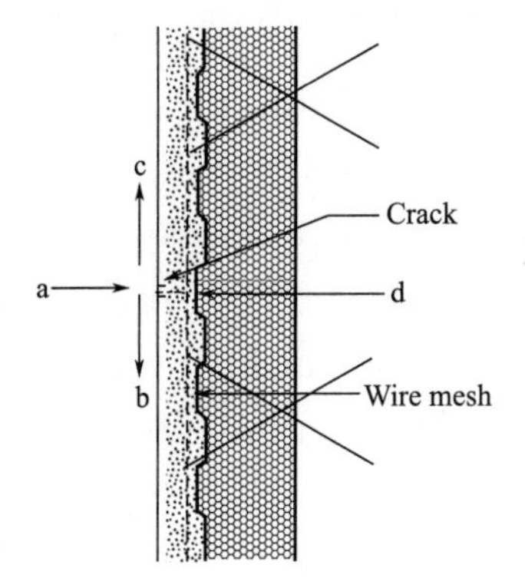

Figure 11-3-5 Location of Single-sided Wire Mesh in Mortar

After the meshed EPS boards are leveled with ordinary cement mortar or polymer cement mortar, it is impossible to seal cracks with the face brick finish. As the annual temperature difference of the external surface of the insulation layer is more than 80℃, the deformation is not the same on two sides of the insulation layer, and the surface mortar outside the meshes has too high strength, cracking will occur inevitably. The finishing face bricks may crack and even fall off under the tremendous deformation stress.

4. Cracking is caused by the unreasonable design of the single-sided wire mesh.

The positive and negative wind pressure, thermal expansion and contraction, wet expansion and dry shrinkage and seismic action lead to forces in two directions. The location of the single-sided wire mesh in mortar is shown in Figure. This type of reinforcement has good effects on resisting and dispersing the stress in the "a" direction, but limited effects in the other three directions "b", "c" and "d", which may lead to cracking. As a result of shrinkage of surface mortar and inconsistency of wire mesh locations in surface mortar, cracking often occurs in the surface layer. There is large deformation stress in surface cracks, the face brick pointing adhesive may crack during bonding of face bricks, and even face bricks may crack. In the case of water infiltration into cracks, wire meshes will be rusted directly and damaged more seriously.

11.3.3 Solution

In accordance with the industry standard "Technical Requirements of External Thermal Insulation with Expanded Polystyrene Panel for In-situ Concrete" (JG/T 228), an appropriate leveling layer of lightweight fireproof insulation mortar should be added on the surface of EPS boards or meshed EPS boards to effectively prevent surface cracks. The leveling layer should be made of adhesive polystyrene granule pasting mortar, with the thickness not less than 20mm.

11.3.3.1 Cast-in-situ Concrete and Non-meshed EPS Board System

The EPS board surface of the cast-in-situ concrete and non-meshed EPS system may be wholly leveled with adhesive polystyrene granule pasting mortar (thickness: not less than 20mm) based on the differences in flatness and verticality. In addition to the solution to steeping, overall flatness and verticality problems of EPS boards in different layers, the door/window holes, through holes reserved during construction and local damaged parts of EPS boards can be insulated and repaired. At the same time, the unavoidable "thermal bridges" can be flexibly cut off with adhesive polystyrene granule pasting mortar.

The stress is concentrated in and released from board gaps. Cracking is prone to occur due to the difference in the thickness of surface mortar in the case of stepping of board gaps. The leveling layer of adhesive polystyrene granule pasting mortar has the effect of homogenization, which can prevent cracking of board gaps and achieve good cracking resistance.

Figure 11-3-6 shows the comparison of external thermal insulation projects with the same structure, completed by one construction team on a construction site in Qingdao. The structure (*a*) was built by pouring EPS boards, grinding their uneven parts and then directly applying the anti-cracking mortar and glass fiber mesh. This structure is subject to serious cracking. The structure (*b*) was built by pouring EPS boards, leveling with the adhesive polystyrene granule pasting mortar, and then applying the anti-cracking mortar and glass fiber mesh. There are no cracks in this structure.

11.3.3.2 Cast-in-situ Concrete and Meshed EPS Board System

The cast-in-situ concrete and meshed EPS board system is usually leveled and plastered with polymer cement mortar (thickness: not less than 20mm) containing a small amount of polymer. As cracking is prone to occur, the face brick finish instead of paint finish is applied to cover cracks, thereby increasing the load of the meshed EPS board surface and making the system more unsafe. When a leveling layer (thickness: not less than 20mm) of adhesive polystyrene granule pasting mortar is used, the load can be greatly reduced and the thermal bridge can be blocked, thus achieving good effects of additional insulation, reducing the torque and enhancing the safety. At the same time, the adhesive polystyrene granule pasting mortar with good flexibility is not easy to crack, but is able to disperse and absorb part of stress to prevent the entire system from cracking.

(a)

(b)

Figure 11-3-6 Comparison of Cast-in-situ EPS Boards of One Project in Qingdao

(a) without leveling layer of adhesive polystyrene granule pasting mortar (cracking)

(b) with leveling layer of adhesive polystyrene granule pasting mortar (no cracking)

1. Elimination of thermal bridge

Table 11-3-1 shows the comparison of thermal resistance test results of the meshed EPS boards with a leveling layer of polymer cement mortar and those with a leveling layer of adhesive polystyrene granule pasting mortar. The steel truss connectors of the meshed EPS boards are welded with the wire meshes on the EPS board surface, and secured on the base course wall. Therefore, significant thermal bridges are inevitable between the leveling layer and wall, which will reduce the actual insulation effects of meshed EPS boards. The Institute of Physics of Beijing Building Research Institute of CSCEC verified that, based on the theory of three-dimensional heat transfer, each steel wire (Φ2) will lead to local thermal bridges within 30-50mm, and the insulation effect of the meshed EPS boards will decline by about 50%. The results of the thermal resistance test conducted b the Institute of Physics also verify the above viewpoints.

It can be seen from Table 11-3-1 that when the meshed EPS boards are leveled with the non-insulating polymer cement mortar, the surface heat can be transferred through oblique wires, thereby reducing the insulation effects of the insulation material; and when the leveling layer of adhesive polystyrene granule pasting mortar is used, the impact of thermal bridges caused by oblique wires can be blocked effectively, thereby improving the effects of wall insulation.

Thermal Properties of Meshed EPS Boards Leveled with Different Materials **Table 11-3-1**

Leveling Material	Polymer Cement Mortar	Adhesive polystyrene granule Insulation Mortar
Basic structure	30mm cement mortar as wall +50mm meshed EPS board (with 50mm×50mm wire meshes)+ 20mm polymer cement mortar leveling layer + 3mm anti-cracking mortar and alkali-resistant glass fiber mesh	30mm cement mortar as wall + 50mm meshed EPS board (with 50mm×50mm wire meshes)+ 20mm adhesive polystyrene granule pasting mortar + 3mm anti-cracking mortar and alkali-resistant glass fiber mesh
Thermal resistance	0.65($m^2 \cdot K$)/W	0.94($m^2 \cdot K$)/W
Heat transfer coefficient	1.25W/($m^2 \cdot K$)	0.93W/($m^2 \cdot K$)

2. Improvement of cracking resistance

When the external surface of the meshed EPS boards is leveled with adhesive polystyrene granule pasting mortar of high flexibility, the cracking resistance of the system can be improved to a certain extent. Meanwhile, a layer of anti-cracking mortar alkali-resistant glass fiber mesh is used on the surface of adhesive polystyrene granule pasting mortar to prevent cracking. With the alkali-resistant glass fiber meshes and the wire meshes on EPS boards, destructive forces in each direction can be fully eliminated and resisted, thereby achieving good cracking resistance and effectively prevent cracks.

3. Reduction of surface load

The dry density of polymer cement mortar differs greatly from that of adhesive polystyrene granule pasting mortar, but their bonding strength values vary a little. It can be seen from Table 11-3-2 that the adhesive polystyrene granule pasting mortar has higher shear resistance.

Table 11-3-3, based on Beijing's energy conservation standards (65%), shows the meshed EPS board thickness corresponding to the 20mm leveling layers of polymer cement mortar and adhesive polystyrene granule pasting mortar, as well as the torques of the leveling layers relative to the base course wall through steel truss connectors.

It is shown in Table 11-3-3 that the thickness of meshed EPS boards subject to leveling with polymer cement mortar is greater than that with adhesive polystyrene granule pasting mortar. Therefore, the torque of the former increases more significantly. This has adverse effects on the stability of the entire system. If used, the face brick finish will lead to the increase in the load and torque but decline in the stability.

If the leveling layer of adhesive polystyrene granule pasting mortar is used, the load will decrease by 30kg per square meter, the torque will drop significantly, but the system stability will be improved greatly, and the face brick finish can be applied safely. The wire meshes of EPS boards have resistance to external one-way impact but little impact on the destructive forces generated by the thermal expansion and contraction, positive and negative wind pressure, dry-wet cycles, seismic action and other factors in various directions. If an additional layer of wire mesh or alkali-resistant glass fiber mesh is used to form a two-way reinforcement structure, such destructive forces can be significantly eliminated and resisted. Before face bricks are bonded, one layer of hot-dip

Comparison of Dry Density and Bonding Strength of Adhesive polystyrene granule Pasting Mortar and Polymer Cement Mortar **Table 11-3-2**

Material \ Property	Dry Density (kg/m^3)	Bonding Strength (MPa)	Bonding Strength/Dry Density
Adhesive polystyrene granule pasting mortar	300	0.12	4.0×10^{-4}
Polymer cement mortar	1800	0.4	2.2×10^{-4}
Ratio of bonding strength/dry density of adhesive polystyrene granule pasting mortar and polymer cement mortar			1.82

Torque Calculations of Meshed EPS Board Leveling Layer **Table 11-3-3**

Item	Leveling Layer of Adhesive polystyrene granule Pasting Mortar	Leveling Layer of Polymer Cement Mortar
EPS board thickness, mm	75	90
Dry density, kg/m^3	300	1800
Mass of 20mm leveling layer, kg/m^2	6	36
Torque, N·m	4.5	32.4

galvanized welded wire mesh or alkali-resistant glass fiber mesh must be secured with plastic anchor bolts onto the leveling layer, in order to transfer the surface load to the base course wall.

11.3.4 Conclusions

Regardless of the cast-in-situ concrete and non-meshed EPS system and the cast-in-situ concrete and meshed EPS board system, a leveling layer of lightweight fireproof insulation mortar (preferably adhesive polystyrene granule pasting mortar) between the insulation layer and anti-cracking protective layer has appreciably beneficial effects. It cannot only improve the insulation performance, fire resistance, weathering resistance and stability of the system, but also can effectively control cracks.

The external thermal insulation project (Figure 11-3-7) of Jinzhou Baodi Manhattan Community was commenced in 2010 and completed in 2011. The cast-in-situ concrete and non-meshed EPS system is used in this project, with a total insulation area of more than 4 million square meters, and a 20mm leveling layer of adhesive polystyrene granule pasting mortar is located between the insulation layer and anti-cracking protective layer. The insulated wall completed in the early period of this project has been properly used for over three years. Cracking, hollowing, leakage and finish falling are not found.

Qingdao Lucion Changchun Garden Community consists of 99 buildings, with a construction area of about 990, 000m^2. The cast-in-situ concrete and meshed EPS board system is used for external thermal insulation (Figure 11-3-8). The meshed EPS boards were cast simultaneously with

Figure 11-3-7 Cast-in-situ Concrete and Non-meshed EPS Board System of Jinzhou Baodi Manhattan Community

Figure 11-3-8 Cast-in-situ Concrete and Meshed EPS Board Project with Face Brick Finish with Qingdao Lucion Changchun Garden Community

concrete, and the wire meshes were leveled with adhesive polystyrene granule pasting mortar, thereby improving the fire resistance, vapor permeability and cracking resistance of the system, effectively solving the problems of cracking and damage of the polymer cement mortar layer, reducing the surface load, and blocking the thermal bridges generated by steel truss connectors. The anti-cracking protective layer is composed of anti-cracking mortar and hot-dip galvanized welded wire meshes, and secured with plastic anchor bolts onto the base course wall, with good seismic performance. The dedicated face brick bonding mortar and pointing material used in the finish layer have the characteristics of high bonding strength, flexibility, cracking resistance and waterproofing performance. The results of use over years demonstrate that this project has stable quality, without cracking or falling.

11.4 Polyurethane Composite Board Insulation Project

The polyurethane composite board has the advantages of low thermal conductivity, high compressive strength and low water absorption rate, so it is widely applied in external thermal insulation systems of new residential and public buildings and also in energy-saving renovation of existing buildings. The polyurethane composite board is prefabricated in a factory by covering at least two main sides of the rigid polyurethane foam core with an appropriate layer of cement-based polymer mortar. The closed porosity of the polyurethane core is higher than 92%, and the gas pressure inside the closed pores changes with the ambient temperature. The macroscopic size of low-strength pore walls changes with the gas pressure inside pores, leading to low-temperature contraction or high-temperature expansion, so the system has poor volume stability. With the additional impact of the cement-based polymer mortar shell, the physical properties of the polyurethane composite board vary greatly from those of EPS and XPS boards. Instead of directly copying the construction method of the thin-plastered EPS board system, a reasonable structure should be found by comprehensively studying the dimensional stability of the polyurethane composite board as well as the structure and weathering resistance of the external thermal insulation system, in order to guarantee the engineering quality.

11.4.1 Study on Dimensional Stability of Polyurethane Composite Board

Affected by the ambient temperature, the dimensions of polyurethane insulation materials may change. The dimensional change rate is related to a number of factors, such as the raw material type, foam structure, core density, forming process and foaming agent type. With reference to the *Test Method for Dimensional Stability of Rigid Cellular Plastics* (GB/T 8811-2008), four representative specimens of polyurethane composite boards were selected, and the dimensional stability of the core and composite board were determined under different temperature conditions. The test results are shown in Table 11-4-1 to 11-4-4.

Test Results of Dimensional Stability at Room Temperature (23℃) **Table 11-4-1**

Specimen No.	Length Change Rate (%)		Width Change Rate (%)		Thickness Change Rate (%)	
	Core	Composite Board	Core	Composite Board	Core	Composite Board
a	0.1	0.1	0	0.2	0.9	0.4
b	0.03	0.03	0	0.1	0.1	0
c	0.1	0.1	0.1	0.1	0.1	0
d	0.1	0.03	0.1	0.1	0.1	0.2

Test Results of Dimensional Stability at 70℃ **Table 11-4-2**

Specimen No.	Length Change Rate (%)		Width Change Rate (%)		Thickness Change Rate (%)	
	Core	Composite Board	Core	Composite Board	Core	Composite Board
a	0.2	0.1	0.2	0.2	12.8	7.3
b	0.3	0.2	0.5	0.1	2.1	1.0
c	0.4	0.4	0.4	0.2	0.3	0.2
d	0.3	0.2	0.3	0.2	0.4	0.5

Test Results of Dimensional Stability at 80℃ **Table 11-4-3**

Specimen No.	Length Change Rate (%)		Width Change Rate (%)		Thickness Change Rate (%)	
	Core	Composite Board	Core	Composite Board	Core	Composite Board
a	0.4	0.4	0.6	0.3	8.0	11.1
b	0.5	0.4	0.2	0.2	2.5	1.2
c	0.5	0.3	0.3	0.2	0.3	0.2
d	0.3	0.2	0.3	0.2	0.4	1.0

Test Results of Dimensional Stability at-18℃ **Table 11-4-4**

Specimen No.	Length Change Rate (%)		Width Change Rate (%)		Thickness Change Rate (%)	
	Core	Composite Board	Core	Composite Board	Core	Composite Board
a	0.1	0.03	0.1	0.03	1.4	0.8
b	0.1	0.1	0.1	0.2	0.3	0.1
c	0.2	0	0.03	0.1	0.1	0.3
d	0.03	0.03	0.1	0.1	2.7	0.1

It can be seen from Table 11-4-1 to 11-4-4 that the polyurethane composite board has poor dimensional stability at low, normal and high temperature. The core dimensions vary significantly. The composite board with a protective layer, which is more stable than the core, still has low dimensional stability, especially in the thickness direction. The dimensional change rate of the polyurethane composite board in each direction is approximately 0.1% at the room temperature, reaches 0.2% to 0.5% at 70℃ to 80℃ and exceeds 10% in extreme cases. Under low temperature conditions (−18℃), it is roughly 0.1% at the room temperature and reaches 1% in extreme cases. Therefore, the dimensions of the polyurethane composite board are greatly affected by the temperature. Despite of minor constraints, they still change obviously with cement-based polymer mortar coating.

The insulation boards of external thermal insulation systems in cold areas are protected only with a thin plaster, so the temperature of the external surface of the insulation board is about50℃ in summer, and drops to approximately−10℃ in winter, with the annual temperature difference of about 60℃. The insulation board expands or contracts greatly with the temperature rising or dropping. When the insulation board is heated, compressive stress will be produced (Figure 11-4-1), and board gaps will be squeezed. The deformation of the insulation board (such as the polyurethane composite board) because of poor dimensional stability may finally lead to the hollowing of joints as well as damage to or falling of the bonding layer. When the insulation board contracts under cold conditions, the tensile stress will be produced (Figure 11-4-2) and board gaps will be widened. In the case of excessive deformation, board gaps will be broken at last, resulting in the water and air infiltration into the bonding layer and thus affecting the bonding strength. This is also a cause of board falling.

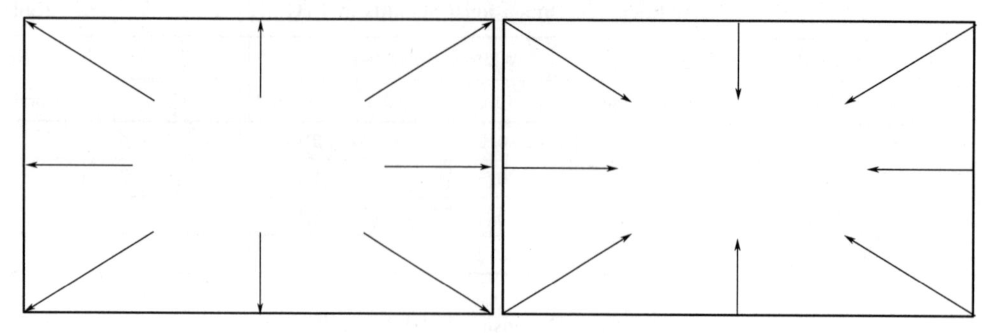

Figure 11-4-1 Schematic Diagram of Expansion of Insulation Boards under Hot Conditions

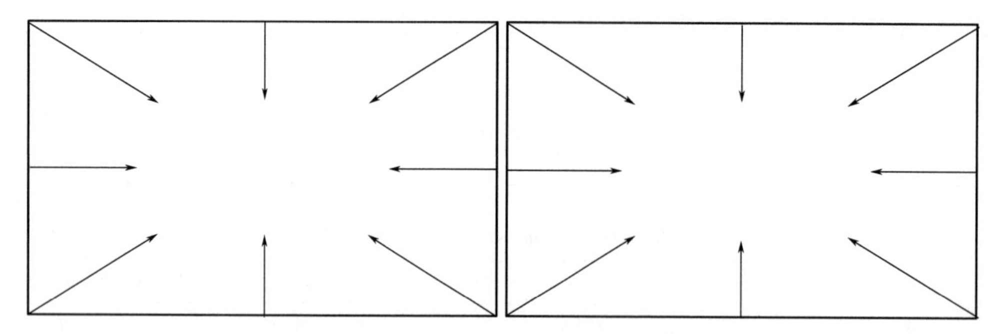

Figure 11-4-2 Schematic Diagram of Expansion of Insulation Boards under Cold Conditions

The polyurethane composite board has low thermal conductivity. Generally, the lower the thermal conductivity, the more difficult it is to transfer and disperse heat to the inside. In this case, the heat will be highly concentrated on the external surface of the insulation board, so there is a great difference between the internal and external surfaces. The greater the temperature difference between two sides, the more the insulation board is deformed, and the higher performance requirement should be satisfied by the external protective layer. When the transition layer is added between the insulation board and protective layer, and the thermal conductivity of the insulation layer falls between those of the insulation layer and protective layer, the surface temperature and deformation of the insulation board can be reduced, the impact of the insulation board on the protective layer can be eased, and the service life of the protective layer can be prolonged.

11.4.2 Structural Study and Analysis

11.4.2.1 Common Construction Methods

The polyurethane composite board can be used in external thermal insulation projects in four main typical methods: (1) board bonding and thin plastering (hereinafter referred to the thin plastering); (2) board bonding and hanging of double glass fiber mesh layers (hereinafter referred to the double-mesh hanging method); (3) board bonding and adhesive polystyrene granule mortar plastering (hereinafter referred to the adhesive polystyrene granule mortar transition method); and (4) board bonding and vitrified microbead mortar plastering (hereinafter referred to the vitrified microbead mortar transition method).

11.4.2.2 Comparison of Specimen Walls

The four typical construction methods of the polyurethane composite board were compared on one specimen wall. The temperature of each structural layer was tested in early September. The test results are listed in Table 11-4-5.

As can be seen from Table 11-4-5:

(1) The external surface of the anti-cracking protective layer has the maximum temperature in different construction methods.

(2) The external surface temperature of the polyurethane composite board in the thin plastering

Test Results of Each Structural Layer of Polyurethane Composite Board Insulation System Table 11-4-5

Construction Method	External Surface Temperature (℃) of Outer Wall	External Surface Temperature (℃) of Anti-cracking Protective Layer	External Surface Temperature (℃) of Polyurethane Composite Board	Internal Surface Temperature (℃) of Polyurethane Composite Board
Thin plastering method	67.25	70.00	70.70	29.95
Double-mesh hanging method	70.10	72.30	71.35	28.15
Adhesive polystyrene granule mortar transition method (transition layer: 10mm)	69.35	69.15	64.60	29.40
Vitrified microbead mortar transition method (transition layer: 10mm)	68.85	69.80	66.60	28.30

or double-mesh hanging method is close to the external surface temperature of the anti-cracking protective layer, with a difference of about 1℃. When a 10mm transition layer of lightweight insulation mortar is added on the external surface of the polyurethane composite board, the external surface temperature of the polyurethane composite board is more than 3℃ lower than that of the anti-cracking protective layer (as the thermal conductivity of adhesive polystyrene granule mortar is lower than that of vitrified microbead insulation mortar). When the thickness of the lightweight insulation mortar transition layer is increased, the external surface temperature of the polyurethane composite board will drop more greatly, and its deformation caused by temperature variations can be reduced effectively, thereby ensuring the stability of the overall structure.

(3) The internal surface temperature of the polyurethane composite board is relatively stable, and its daily difference is only about 4℃. Thus, the wall of the main structure is kept at a relatively stable temperature, which ensures the stability of the main structure.

According to the test results of these four methods on the specimen wall, board gaps are deformed obviously in a short time in the case of thin plastering, while obvious gaps are found in the case of double-mesh hanging (Figure 11-4-3). In the presence of the lightweight insulation mortar transition layer, no gaps are found among boards (Figure 11-4-4) or on the flat wall surface. Therefore, an appropriate lightweight insulation mortar transition layer on the polyurethane composite board surface is an indispensable structural layer to prevent cracking.

Figure 11-4-3 Wall Appearance for Test of Thin Plastering and Double-mesh Hanging

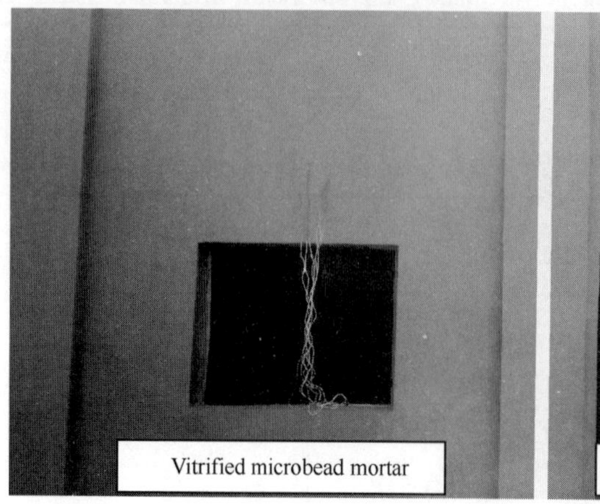

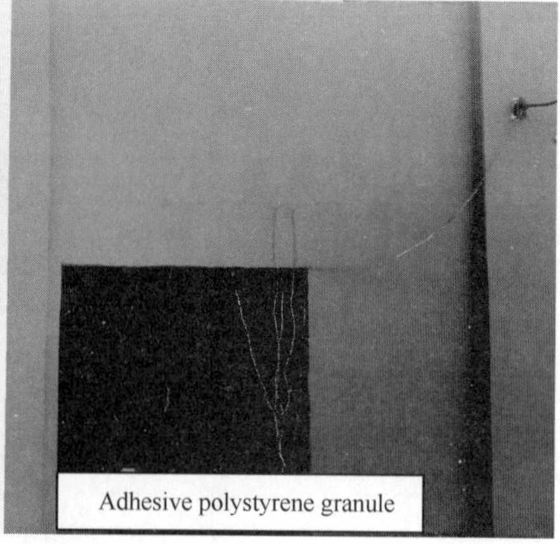

Figure 11-4-4 Wall Appearance for Test with Lightweight Insulation Mortar Transition Layer

11.4.2.3 Calculation of Surface Temperature of Each Structural Layer

According to the theory of steady-state heat conduction, the heat-conducting flow density of the external thermal insulation structure can be calculated from the equation (1):

$$q=\frac{t_1-t_{n+1}}{\frac{\delta_1}{\lambda_1}+\frac{\delta_2}{\lambda_2}+\cdots+\frac{\delta_n}{\lambda_n}} \tag{1}$$

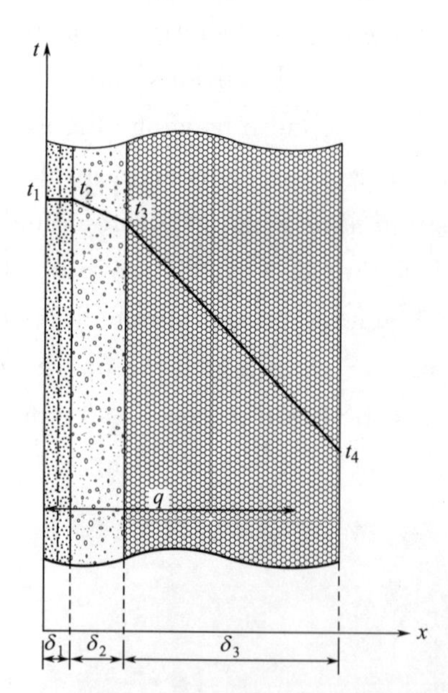

Figure 11-4-5 Schematic Diagram of Heat Conduction of External Thermal Insulation Structure based on Polyurethane Composite Board

The temperature of the contact surface between the kth and $(k+1)^{th}$ layers of the external thermal insulation structure can be calculated from the equation (2):

$$t_{k+1}=t_1-q\left(\frac{\delta_1}{\lambda_1}+\frac{\delta_2}{\lambda_2}+\cdots+\frac{\delta_k}{\lambda_k}\right) \tag{2}$$

Where,

q—heat flow density, W/m^2;

t_1—temperature of the external surface (high-temperature) of the external thermal insulation structure, ℃;

t_{n+1}—temperature of the internal surface (low-temperature) of the nth layer of the external thermal insulation structure, ℃;

δ_1, δ_2, …, δ_k, …, δ_n— material thickness of each structural layer, m;

λ_1, λ_2, …, λ_k, …, λ_n—thermal conductivity of the material of each structural layer, W/(m·℃);

t_{k+1}—temperature of the contact surface between the kth and $(k+1)^{th}$ layers, ℃.

As shown in Figure 11-4-5, it is assumed that the external surface temperature of the anti-cracking protective layer is t_1=70℃, the internal surface temperature of the insulation layer is t_4=29℃, and the thickness of the polyurethane composite board is 40mm. The external surface temperature (t_3) of the polyurethane composite board, calculated with the equations (1) and (2), is shown in Table 11-4-6.

As shown in Table 11-4-6: the external surface temperature of the polyurethane composite board can be reduced by 3℃ in the presence of a 10mm adhesive polystyrene granule leveling layer, 6℃ in the presence of a 20mm adhesive polystyrene granule leveling layer, and 9℃ in the presence of a 30mm adhesive polystyrene granule leveling layer. The thermal expansion coefficient of the polyurethane composite board is 9×10^{-5}m/℃. If a 20mm adhesive polystyrene granule leveling layer is applied on a 1.2m long polyurethane composite board, the elongation of the board surface will decrease by 9×10^{-5}m/℃×6℃×1.2=0.7mm. Thus, in the presence of a 20mm lightweight insulation mortar transition layer, the external surface temperature of the polyurethane composite board drops significantly (by more than 5℃), and the thermal deformation of its external surface decreased by more than 0.5mm, which is favorable for protection of the polyurethane composite board. With a 10mm lightweight insulation mortar transition layer, the external surface temperature of the polyurethane composite board drops by 3℃, and the thermal deformation of its external surface decreased by more than 0.3mm, which is obviously not enough to protect the polyurethane composite board.

11.4.3 Weathering Resistance Test

The weathering resistance test was conducted to the four typical polyurethane composite board structures. The results of 80 heat-rain cycles and 5 heat-cold cycles show that the four structures vary greatly in their own deformation as well as board gaps. Specific results are listed in Table 11-4-7.

Temperature Calculation of Each Structural Layer of Polyurethane Composite Board Insulation Structure Table 11-4-6

Structural Layer \ Construction Method		Thin Plastering	Double-mesh Hanging	Adhesive polystyrene granule Mortar Transition		
Anti-cracking protective layer	Thickness /mm	5	7	5		
	Calculated thermal conductivity /W/(m·℃)	0.93	0.93	0.93		
	External surface temperature (t_1)/℃	70	70	70		
Adhesive polystyrene granule mortar leveling layer	Thickness /mm	0	0	10	20	30
	Calculated thermal conductivity /W/(m·℃)	—	—	0.072	0.072	0.072
	External surface temperature (t_2)/℃	—	—	69.86	69.87	69.88
Polyurethane composite board insulation layer	Thickness /mm	40	40	40	40	40
	Calculated thermal conductivity /W/(m·℃)	0.029	0.029	0.029	0.029	0.029
	External surface temperature (t_3)/℃	69.84	69.77	66.12	63.02	60.39
	Internal surface temperature (t_4)/℃	29	29	29	29	29

Polyurethane Composite Board Structures after Weathering Resistance Test Table 11-4-7

Construction Method	Board Change	Board Gap Change	Board Gap Photo
Thin plastering method	Hollowing and shrinkage were found.	Significant	

Continued

Construction Method	Board Change	Board Gap Change	Board Gap Photo
Double-mesh hanging method	No abnormality	Large	
Adhesive polystyrene granule mortar transition method (transition layer: 10mm)	No abnormality	Minor	
Vitrified microbead mortar transition method(transition layer: 10mm)	No abnormality	Little	

As shown by the results of weather resistance test: the board constructed by the thin plastering method was damaged the most, and had the largest board gaps. The construction method with the lightweight insulation mortar transition layer is the most stable. Hollowing and shrinkage were found on the polyurethane composite board subject to thin plastering, and no abnormality in the other three construction methods. The board gap change in the thin plastering method is greater than that of the double-mesh hanging method, while the latter is greater than those of the adhesive polystyrene granule mortar transition method and vitrified microbead mortar transition method. There is almost no difference in the board gap change between the adhesive polystyrene granule mortar transition method and vitrified microbead mortar transition method, both of which involves small changes of board gaps. It can be concluded that the transition layer helps to reduce the adverse effects of the dimensional stability of the polyurethane composite board on the insulation system, as well as the thermal stress of the polyurethane composite board. Therefore, a transition is needed in order to improve the thermal stability of the polyurethane composite board system and effectively prevent hollowing, cracking, falling, etc.

11.4.4 Fire Resistance Verification and Analysis

11.4.4.1 Oxygen Index Test and Analysis of Polyurethane Composite Board

The oxygen index reflects the combustibility of a material. The oxygen index test results of four representative specimens are shown in Table

11-4-8.

Oxygen Index Test Results of Polyurethane Composite Boards Table 11-4-8

Specimen No.	Oxygen Index /%	Conformity to Class B_1 Requirements
a	21.5	Nonconforming
b	28.3	Nonconforming
c	29.7	Nonconforming
d	30.3	Nonconforming

As shown in Table 11-4-8, it is difficult for the polyurethane composite board to meet the requirements of Class B_1 combustibility. Among the four specimens, only the specimen "d" conformed to the requirements, and the oxygen index of the specimen "a" has not yet met the requirements of Class B_2 combustibility (26%). Polyurethane is an organic material that is quite easy to burn even if its combustibility meets the requirements of Class B_1. When a thin protective layer of cement-based polymer mortar is applied, the polyurethane composite board is still susceptible to fire attack. Therefore, the thin-plastered polyurethane composite board has great safety hazards in the absence of fireproof structural measures. An appropriate layer of fireproof transition layer is needed on the external surface of the polyurethane composite board in order to reduce safety hazards.

11.4.4.2 Combustion Shaft Furnace Test

The combustion shaft furnace test was carried out in accordance with the *Test Method of Difficult-flammability for Building Materials* (GB/T 8625-2005), in which the combustion power of methane gas was approximately 21kW, the flame of about 900℃ was applied for 20min, and temperature measuring points of the insulation layer, exposed to the fireproof protective layer, were set at intervals of 200mm along the centerline in the height direction of specimens, as shown in Figure 11-4-6. During this test, the flame power applied was a constant, and the area with 5# and 6# thermocouples was the fire test area.

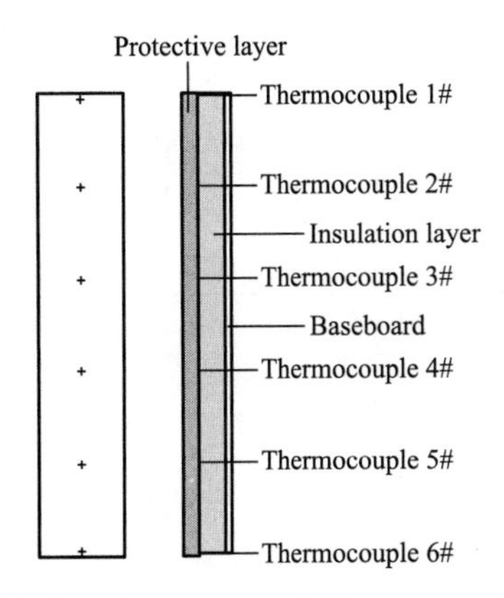

Figure 11-4-6 Specimen and Thermocouple Layout of Combustion Shaft Furnace Test

The fireproof protective layer was made of Class A adhesive polystyrene granule mortar, and specimens were fabricated in accordance with the requirements of Table 11-4-9. The results of the combustion shaft furnace test are listed in Table 11-4-10, and the specimen states after splitting are shown in Figure 11-4-7.

Structural Requirements for Combustion Shaft Furnace Test Specimens of Polyurethane Insulation Board

Table 11-4-9

Specimen No.	Thickness of Fireproof Protective Layer of Adhesive polystyrene granule Mortar (mm)	Thickness of Anti-cracking Layer + Finish Layer (mm)	Thickness of Polyurethane Insulation Board (mm)	Base course wall Thickness (mm)
PU-1	0	5	30	20
PU-2	10	5	30	20
PU-3	20	5	30	20
PU-4	30	5	30	20

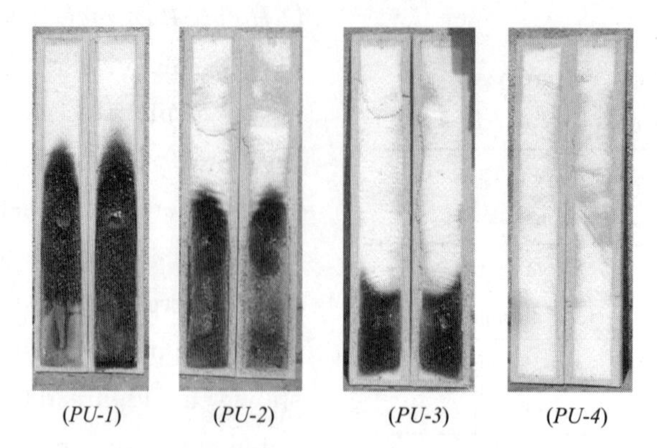

Figure 11-4-7 States of Split Specimens after Combustion Shaft Furnace Test

Results of Combustion Shaft Furnace Test of Polyurethane Insulation Board Specimens Table 11-4-10

S/N	Maximum Temperature (℃)Measured by Thermocouple						Residue Length (mm) of Combustion
	6# thermocouple	5# thermocouple	4# thermocouple	3# thermocouple	2# thermocouple	1# thermocouple	
PU-1	453.0	566.9	428.8	216.4	121.8	81.3	350
PU-2	92.2	386.5	330.1	95.4	91.3	74.4	500
PU-3	102.5	192.2	91.1	94.7	94.0	71.8	750
PU-4	96.3	95.0	45.1	95.9	34.5	31.2	1000

The results of the combustion shaft furnace test show that: with the thickness of the fireproof protective layer increasing, the temperature measured on the insulation layer tends to drop, and the residual length of combustion decreases. The residual length of combustion is only 350mm in the absence of a special fireproof protective layer; 500mm (one half of the original length) in the presence of a 10mm fire protective layer of adhesive polystyrene granule mortar; more than 750mm in the presence of a 20mm fire protective layer of adhesive polystyrene granule mortar; and 1,000mm (essentially with no loss) in the presence of a 30mm fire protective layer of adhesive polystyrene granule mortar. Taking into account other factors such as the economy and construction, the thickness of the fire protective layer of the polyurethane insulation board should not be less than 20mm.

11.4.5 Engineering Case Analysis

It is found through on-site inspections and questionnaire surveys that polyurethane composite boards are mainly constructed in the thin plastering method, but the adhesive polystyrene granule mortar transition method and the vitrified microbead mortar transition method are comparatively uncommon as the additional transition layer leads to the extension of the construction duration and increase in engineering costs. In terms of the engineering quality, the polyurethane composite board insulation project constructed in the same method as the thin-plastered EPS board project is more probable (more than 80%) to have quality problems than that constructed in the thin plastering method. The majority of thin-plastered polyurethane composite board projects contain obvious gaps as a result of hollowing (Figure 11-4-8). As the polyurethane composite board has poor dimensional stability and thin protective layer, external heat can be easily transmitted to the polyurethane composite board surface. Thus, the polyurethane composite board may be deformed as a result of significant rise of the temperature, thereby making the protective layer crack (Figure 11-4-9). The polyurethane composite board insulation project containing the thick enough (10-15mm) adhesive polystyrene granule mortar transition layer or vitrified microbead insulation mortar transition layer has better quality, with no gap or crack found (Figure 11-4-10). When the thickness of the transition layer exceeds 20mm, the project has stable quality and good surface flatness, involving no cracking, hollowing or board gaps. If the transition layer is too thin (less than 10mm), there will be obvious gaps on the polyurethane composite board (Figure 11-4-11).

Figure 11-4-8 Hollowing Photo of Thin-plastered Board

Figure 11-4-9 Surface Cracking Photo of Thin-plastered Board

Main causes of quality problems of the thin-plastered polyurethane composite board project, such as hollowing and cracking are as follows. The polyurethane composite board has poor dimensional stability but large deformation stress. In the case of thin plastering, heat is rapidly transferred through the thin protective layer to the external surface of the polyurethane composite board, leading to obvious changes in its dimensions. The

Figure 11-4-10 Photo of Thick Enough Transition Layer

Figure 11-4-11 Photo of Excessively Thin Transition Layer

deformation stress generated is released through board gaps and transmitted to the protective layer, so there is hollowing, warping, cracking and other anomalies at board gaps of the protective layer. In serious cases, boards may fall of. Meanwhile, the uneven deformation in the thickness of the polyurethane composite board leads to the flatness differences of materials and non-uniform stress on the surface, which will further cause cracking.

11.4.6 Conclusions

As the polyurethane composite board has poor dimensional stability, and the flexibility of the surface layer far different from that of the EPS board, the construction method similar to thin plastering of EPS boards should not be used. If an appropriate layer of lightweight insulation mortar transition layer is added on the surface layer, cracking and hollowing can be avoided effectively. Considering the fire safety, the optimal thickness of the transition layer should not be less than 20mm. Preferably, adhesive polystyrene granule mortar of high flexibility should be used, in order to greatly reduce the external surface temperature of the polyurethane composite board, prevent the surface cracking of the external thermal insulation system and prolong the service life of the external thermal insulation project.

11.5 Phenolic Board Insulation Project

The phenolic board is a thermosetting closed-cell rigid foam insulation board produced with the phenolic resin, curing agent, foaming agent, smoke suppressant, flame retardant, and the like. It has the advantages of low thermal conductivity, flame retardancy, low smoke and high temperature resistance, and also shortcomings of degradability for efflorescence, small bending deformation, poor dimensional stability, high water absorption rate and low tensile strength. The properties of the phenolic board are greatly different from those of EPS boards (Table 11-5-1). Therefore, quality accidents are likely to occur if the above thin plastering method is used without the study on the modification of phenolic board and the structure of the external thermal insulation system.

11.5.1 Quality Accident Analysis of Phenolic Board Insulation Project

11.5.1.1 High Water Absorption Rate

The chemical composition and porosity of a phenolic board determines its high water absorption rate. After water absorption and drying, the quality of the phenolic board will drop. The efflorescence is conducive to degradation to reduce the compressive strength, and also affects the bonding and insulation performance of the phenolic board. If moisture is absorbed by the phenolic board, degradation and efflorescence will be intensified Figure 11-5-1 shows the degradation and efflorescence occur with water infiltration into the insulation layer through the protective surface, which will further lead to large-scale falling and affect the engineering quality.

Comparison of Basic Properties of Phenolic and EPS Boards **Table 11-5-1**

Item	Unit	EPS Board	Phenolic Board
Apparent density	kg/m^3	≥18	≥45
Thermal conductivity	W/(m·K)	≤0.039	≤0.033
Tensile strength perpendicular to board surface	MPa	≥0.10	≥0.08
Compressive strength	kPa	≥100	≥100
Bending deformation	mm	≥20	≥4.0
Dimensional stability(70℃,2d)	%	≤0.3	≤1.0
Linear expansion coefficient	mm/m·K	0.06	0.08
Heat storage coefficient	W/(m^2·k)	0.36	0.36
Water absorption rate	%	≤3	≤7.5
Vapor permeability coefficient	ng/(Pa·m·s)	4.5	8.5
Elastic modulus	MPa	9.1	16.4
Poisson's ratio	—	0.1	0.24
Combustibility	—	Class B_2 or above	Class B_1 or above

Figure 11-5-1 Degradation and Efflorescence of Phenolic Board Eroded by Water

11.5.1.2 Low Strength of Phenolic Board

The phenolic board of low strength is prone to external damage. Figure 11-5-2 shows a project with phenolic boards integrated with finishing boards. Both the phenolic board and finishing board fell off within a large scale. With the temperature changing, finishing and phenolic boards are deformed. The strength of the phenolic board is not a guarantee of the bonding reliability. When there are connecting cavities, the phenolic board may fall off due to the suction effect under the negative wind pressure. If the structure has high bonding strength, phenolic boards will be torn or damaged.

Figure 11-5-2 Destruction of Phenolic Board by External Force

11.5.1.3 Poor Dimensional Stability

The phenolic board has poor dimensional stability and large Poisson's ratio. Its dimensional change rate is 3.3 times that of the EPS board, and its Poisson's ratio is 2.4 times that of the EPS board. Therefore, the phenolic board used in the external thermal insulation project is easy to deform under sudden cooling and heating conditions, which may lead to the volume instability and large cumulative deformation. Prior to plastic deformation, the transverse deformation of a phenolic board is greater than its longitudinal deformation, which can easily cause the thermal expansion, extrusion and falling of the phenolic board and the surface layer is likely to crack. According to the photo of one phenolic board insulation project the phenolic board has poor dimensional stability, but large Poisson's ratio and dimensional deformation in each direction, which results in the extrusion or falling of the board, cracking of the surface as well as liquid water erosion of the insulation layer.

Figure 11-5-3 Extrusion Caused by Poor Dimensional Stability of Phenolic Board

11.5.1.4 Large Elastic Modulus

The elastic modulus can be regarded as an indicator of elastic deformation of a material. The larger the elastic modulus, the greater the elastic deformation stress is. The phenolic board has a large elastic modulus and therefore produces more stress. An improper structure of the phenolic board insulation project may lead to the hollowing, cracking and falling of the finish layer as a result of excessive elastic deformation stress. The insulation structure may fall off in serious cases. Figure 11-5-4 shows the finish layer damaged as a result of the excessively large elastic modulus of the phenolic board used as an insulation layer.

Figure 11-5-4 Finish Layer Damaged by Large Elastic Modulus of Phenolic Board

11.5.1.5 Small Bending Deformation

The phenolic board has small bending deformation, high brittleness and fragility, and poor flexibility, so it is not able enough to absorb the internal stress and release the deformation. When used in external thermal insulation projects, the phenolic board is easy to break, thereby making the protective surface crack and external liquid water infiltrate to further damage the phenolic board. Figure 11-5-5 shows a phenolic board insulation project, clearly indicating that phenolic boards are broken and fall off with surface mortar. Meanwhile, the phenolic board of high brittleness is easy to damage during construction, which increases the difficulties in operation and leads to severe waste.

11.5.1.6 Structure Defect

Figure 11-5-6 presents a photo in which the phenolic boards in one phenolic board insulation project with structural defects fell within a large scale under the negative wind pressure. As the phenolic boards used in this project are subject to point bonding, there are obvious connecting cavities between the phenolic boards and base course wall, thus creating the conditions for the negative wind pressure. Under tremendous suction effects of negative wind pressure, phenolic boards will be phenolic blown off or torn away from the base course wall. Such damage is more obvious in the area where the negative wind pressure is prone to occur. It can be concluded that structural defects of the phenolic board insulation project probably lead to quality accidents.

Figure 11-5-5 Easy-to-damage Phenolic Board with Small Bending Deformation

Figure 11-5-6 Large-scale Blow-off of Phenolic Board due to Connecting Cavities

11.5.1.7 Quality Nonconformity

Special attention must be paid to the quality of phenolic boards with their respective defects in external thermal insulation project. The use of nonconforming phenolic boards is likely to cause quality accidents under the impact of freezing, thawing, wind pressure and thermal stress. The phenolic board or its protective layer will unavoidably fall off, just a matter of time. Figure 11-5-7 shows a photo of phenolic board falling of one external thermal insulation project in Liaoning. This project was originally an important "window" of one city in Liaoning Province, but now is "a piece of paper". Phenolic boards have fallen off within a large scale. It can be seen in the photo that bonding mortar blocks are still secured on the base course wall, and some phenolic boards have been broken. This indicates that phenolic boards have quality problems and cannot be effectively combined with bonding mortar. Another engineering photo (Figure 11-5-8) shows that the protective surface of the phenolic board insulation system fell off, and some phenolic boards have fallen down. In this project, the bonding area meets the construction specifications. The boards are bonded securely on the base course wall through anchor bolts and coated with the interface agent, but there are still quality accidents. Apparently, poor quality of phenolic boards is a major factor, as foams in inferior phenolic boards are not sintered fully, causing efflorescence and other adverse consequences. Thus, phenolic boards cannot be effectively combined with surface mortar. In the presence of small external forces, the protective surface may fall and phenolic boards may be torn and even fall off.

Figure 11-5-7 Large-scale Falling of Phenolic Board

Figure 11-5-8 Destruction of Protective Surface of Phenolic Board

Figure 11-5-9 provides a photo of one phenolic board insulation system in Beijing. A number of phenolic boards fell off and hollowing occurred on remaining parts within two years of application. The phenolic boards coated with the interface agent were still not effectively bonded with the surface mortar or bonding mortar. According to the introduction of a famous domestic insulation expert, there are no quality problems in the bonding mortar and surface mortar, or cracks on the remaining surface layer and finish layer. This fully proves that the quality of phenolic boards does not meet the insulation board requirements of external thermal insulation project. As shown in these photos, the bonding mortar was bonded securely with the base course wall and phenolic boards were fixed with anchor bolts, but large-scale falling still occurred. As anchor bolts apply the areas covered by bolt discs, some of them may be removed in the process of large-scale falling (Figure 11-5-10).

Figure 11-5-9 Large-scale Falling and Peeling of Phenolic Boards

Figure 11-5-10 Falling of Anchor Bolts with Protective Devices

11.5.2 Measures for Prevention of Quality Accident

11.5.2.1 Use of Thermal Stress Barrier

There is a great difference in the thermal conductivity of the phenolic board (low) and anti-cracking protective layer (high), which leads to obvious variations of the thermal stress between them. The anti-cracking protective layer should meet high performance requirements, so its materials must be able to withstand such thermal stresses in order to prevent hollowing, cracking or falling. This is impossible in actual external thermal insulation projects, but a transition layer can be added between the phenolic boards and anti-cracking layer to solve this problem. The transition layer is specified in the industry standard *Products for External Thermal Insulation Systems based on Mineral Binder and Expanded Polystyrene Granule Plaster* (JG/T 158-2013) and Beijing's local standard *Technical Specification for External Thermal Insulation Construction based on Rigid Polyurethane Foam Composite Board with Plastering Lightweight Mortar* (DB11/T 1080-2014), in order to reduce the thermal stress difference between adjacent materials.

The surface temperature changes of phenolic boards in summer in north China can be calculated based on the construction methods and related parameters in Table 11-5-2, the indoor temperature (t_i=20℃), the surface temperature (t_e=70℃) of the dark-colored finish, and the average surface temperature (28℃) and in the presence of a leveling transition layer of adhesive polystyrene granule pasting mortar.

Calculations prove that when a 20mm layer of adhesive polystyrene granule pasting mortar is applied, the temperature amplitude of the external surface of the phenolic board was 34.7℃, the surface temperature of the phenolic board was 26.8℃ on average, and maximum 61.5℃, which was 8.5℃ less than that (70℃) without the layer of adhesive polystyrene granule pasting mortar, and the delay time of the temperature wave was estimated to be 1.15h.

The elongation of the external surface of the phenolic board was $\Delta L=\alpha$ (t_2-t_1) L=3mm, and

Related Parameters of External Surface of Phenolic Boards Leveled with Adhesive polystyrene granule Pasting Mortar **Table 11-5-2**

Structural Layer	Thickness δ (mm)	Thermal Conductivity λ [W/(m·K)]	Thermal Resistance R [(m²·K/W)]	Heat Storage Coefficient S [W/(m²·K)]	Thermal Inertia Index D	Heat Storage Coefficient Y of External Surface [W/(m²·K)]
Reinforced concrete	180	1.740	0.10	17.06	1.76	17.06
Phenolic Board	60	0.033	1.47	0.36	0.58	0.66
Adhesive polystyrene granule pasting mortar	20	0.070	0.23	0.95	0.34	0.75
Anti-cracking mortar	3	0.93	0.003	11.37	0.04	1.14
Total	263	—	1.96	—	2.72	—

the resulting stress was $\sigma = E\varepsilon = 22659$Pa, and the decrease rate of the phenolic board with the adhesive polystyrene granule pasting mortar was $(27300-22659)/27300=17\%$ than that with no transition layer.

To sum up, when a transition layer of adhesive polystyrene granule pasting mortar is applied on a phenolic board and the surface temperature of a wall is 70℃, the external surface of the phenolic board will reach the maximum temperature (61.5℃) after 1.15h. If the thickness of the adhesive polystyrene granule pasting mortar is 20mm, the thermal stress will decline greatly, and the occurrence time of high temperature will be delayed for 1.15h, thereby alleviating violent changes in the thermal stress and achieving high weathering resistance of the external thermal insulation system.

11.5.2.2 Use of Moisture Dispersion Structure

The material of each layer should have appropriate vapor permeability in the structural design, in order to discharge vapor out of the housing and balance the moisture content of the wall.

A moisture dispersion structure can be used in the external thermal insulation system, to absorb a small amount of vapor and condensate produced because of differences in the vapor permeability, thus preventing liquid water in the system. For example, when the adhesive polystyrene granule pasting mortar layer with the functions of moisture absorption, conditioning and transfer is mounted as a moisture dispersion layer on the condensation side of the phenolic board, and water vapor condenses on the inner side of the anti-cracking protective layer at low ambient temperature in the discharge process, a small amount of condensate generated will be absorbed and dispersed in the structural layer, thereby preventing the destructive forces of changes in three phases of the accumulated water, improving the bonding and breathing performance of the system, and furthermore guaranteeing the long-term safety, reliability and apparent mass stability of external thermal insulation projects.

11.5.2.3 Setting of Waterproof Permeable Layer

One layer of polymer elastic primer can be applied on the anti-cracking protective layer in the external thermal insulation system, to greatly reduce the water absorption rate of the surface, and prevent the damage caused by the frost-heave stress to the external surface of a building in winter, with the vapor permeability coefficient unchanged essentially. Meanwhile, this can guarantee the vapor permeability of the surface material, and prevent the wall surface from being covered with non-permeable materials to hinder the moisture discharge and thus damage of the expansion stress generated in vapor diffusion resistance to the external thermal insulation system. The reasonable external thermal insulation structures and materials should be selected with waterproofing performance and vapor permeability, in order to improve the resistance of the external thermal insulation system to freezing, thawing, weathering and cracking, and extend the service life of the insulation layer.

11.5.2.4 Use of Partitioning Structure

Insulation boards are secured on a wall by means of "point-and-frame bonding" with auxiliary anchor bolts in the conventional thin-plastered structure. If fixed in this method, phenolic boards with performance defects are far from satisfactory.

Laboratory researches and pilot projects have proven that it is relatively reasonable to fix the adhesive polystyrene granule pasting mortar layer in the "full bonding + partitioning" method. The "Partitioning" means that the adhesive polystyrene granule pasting mortar is filled on four sides of phenolic boards, and then the phenolic boards are bonded and leveled with adhesive polystyrene granule pasting mortar until its six surfaces are fully covered. Surely, "full bonding" has a number of advantages. The "partitioning" has the following advantages: the impact of the deformation of the phenolic board on thermal insulation system can be eliminated; the original large-sized thermal insulation system is divided based on the size (normally 600mm×450mm) of one phenolic board to reduce the risk of overall collapse of the external thermal insulation system; longitudinal partitioning can be performed to prevent each phenolic board from extrusion by adjacent ones and prevent the flame from spreading in the transverse direction in the event of a fire; and transverse partitioning is equivalent to the use of "brackets" to support phenolic boards on each layer. The shear strength of adhesive polystyrene granule pasting mortar should be more than 50kPa. For the adhesive polystyrene granule pasted EPS board system, the 10mm "partitioning" structure fully meets the performance requirements. In the pasted phenolic board system, however, the board gaps of "partitioning" should be more than 20mm as the dead load of phenolic boards is 2.5 times that of EPS boards.

11.5.2.5 Selection of Reliable Phenolic Board

The reliable quality of phenolic boards is critical to guarantee the quality of phenolic board insulation projects. Therefore, phenolic boards must be controlled at various levels during construction of external thermal insulation projects.

11.5.3 Conclusions

The 20mm adhesive polystyrene granule pasting mortar is used as the thermal stress blocking layer and vapor dispersion layer, and phenolic boards are pasted with adhesive polystyrene granule pasting mortar, to effectively avoid the defects of phenolic boards and solve the quality problems caused by the unreasonable structural design of phenolic boards, such as hollowing, cracking and falling, so that phenolic boards can be safely applied in external thermal insulation projects. The basic structure of the system is shown in Figure 11-5-11. At present, the adhesive polystyrene granule pasted phenolic board system has been applied in projects with good practical effects. It should be further verified in the large-scale weathering resistance test and also long-time application in the future.

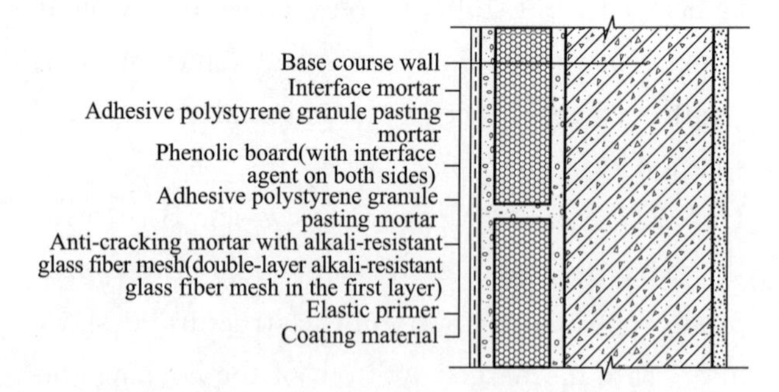

Figure 11-5-11 Basic Structure of Adhesive polystyrene granule Pasted Phenolic Board System

11.6 Rock Wool Board Insulation Project

With the improvement of fire protection requirements for external thermal insulation of buildings and also the issue of related policies in recent years, the rock wool board has attracted attention as a Class A non-combustible insulation material with excellent properties. It is applied extensively applied in actual projects. Because of inadequate testing and research as well as the rock wool board itself and related systems, however, quality accidents of rock wool board insulation projects occur from time to time.

11. 6. 1 Cases of Quality Problems of Rock Wool Board Insulation Project

The external thermal insulation project shown in Figure 11-6-1 was constructed by thin plastering of rock wool boards. The rock wool boards are bonded on the base course wall with the adhesive and secured with anchor bolts. This project failed to withstand external forces. When a strong wind occurred, some rock wool boards at a height of 55m fell off and the others left on the wall also cracked. The rock wool boards that were not bonded securely with the base course wall fell off (Figure 11-6-2) during a strong wind and were scattered on the ground after the wind (Figure 11-6-3).

Figure 11-6-1 Falling of Rock Wool Board from Outer Wall

Figure 11-6-2 Falling of Rock Wool Board from Outer Wall in Strong Wind

Figure 11-6-3 Scattered Rock Wool Boards after a Strong Wind

Figure 11-6-4 provides a photo of one project in Shenyang, in which rock wool boards are thin-plastered and coated with polymer cement mortar. It can be seen that there are obvious cracks, some of which are wide enough for moisture to infiltrate into the insulation layer, which seriously affects the insulation effect.

Figure 11-6-4 Surface Cracking of Thin-plastered Rock Wool Board Project

Figure 11-6-5 presents a photo of one project in Changzhou, Jiangsu, in which the insulation

Figure 11-6-5 Serious Surface Cracking of Thin-plastered Rock Wool Board Project

layer is composed of thin-plastered rock wool boards. A large number of cracks appeared immediately after this project was put into use. The white stripes in this photo are the paint applied during crack repair, forming an unsightly "bandages".

11.6.2 Cause Analysis of Quality Problems of Rock Wool Board Insulation Project

11.6.2.1 Defect of Rock Wool Board

Compared with other insulation boards, rock wool boards have low strength, high water absorption rate and poor hydrophobicity. They are significantly different in properties, and easy to peel off and expand by absorbing moisture. Such defects may lead to quality problems in external thermal insulation projects.

Rock wool boards available on the market are produced by the pendulum method. As rock wool fibers are distributed vertically, the comprehensive strength (up to40kPa) and interlayer bonding strength of rock wool boards are improved, the peel strength is 14kPa, and the tensile strength perpendicular to the wall surface is 7.5kPa. This is far from enough when rock wool boards are used in external thermal insulation projects.

The rock wool board that is mainly composed of transverse fibers has poor dimensional stability under natural conditions, especially hot and humid conditions. Fibers absorbing water are stratified and deformed severely, which is more obvious for interior rock wool boards (Figure 11-6-6).

Due to the air between fibers, the rock wool board is easy to fluff and bulge in the case of thermal expansion and contraction and under negative wind pressure (Figure 11-6-7). When rock wool boards are used in outer wall works, the surface is unable to resist the expansion stress, which will inevitably cause poor finishing, bulging, and gas (Figure 11-6-8).

11.6.2.2 Destruction Caused by Wind Pressure

Due to the loose structure, high density and low tensile strength perpendicular to the board surface, rock wool boards should be secured with an appropriate number of anchor bolts instead of direct bonding with the adhesive. If there is insufficient bonding strength to withstand the maximum negative wind pressure, rock wool boards will be inevitably blown off. Meanwhile, rock wool boards are prone to crack in a strong wind due to their low strength.

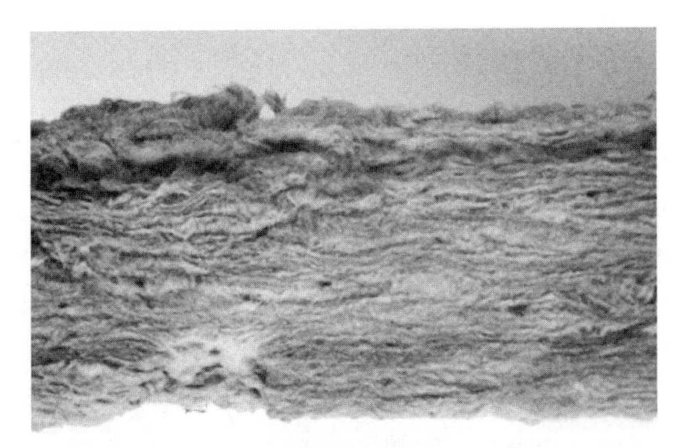

Figure 11-6-6 Obvious Expansion of Rock Wool Board after Water Absorption

Figure 11-6-7 Fluffing and Bulging of Rock Wood Board

Figure 11-6-8 Obvious Gaps of Rock Wool Board on Wall

If anchor bolts are directly secured on a rock wool board and there is tremendous negative wind pressure, only the parts covered by anchor bolt discs will be protected, and the rock wool board will be directly pulled out (Figure 11-6-9). Inferior anchor bolts will be damaged and anchor rods will be bent (Figure 11-6-10). When there is not enough bonding strength between the anchor bolts and base course wall, the anchor bolts will be directly pulled out.

In case that a rock wool board is secured by the wire mesh and anchor bolts but there are not enough anchor bolts, the rock wool board is not able to resist the wind pressure, but may fall off with the wire mesh (Figure 11-6-11). As steel bars are embedded in a concrete shear wall, the drill bit may be damaged by deformed steel bars in the drilling process, so the anchoring depth and number of anchor bolts could not be guaranteed.

Figure 11-6-9 Ineffective Anchor Bolt of Anchored Rock Wool Board

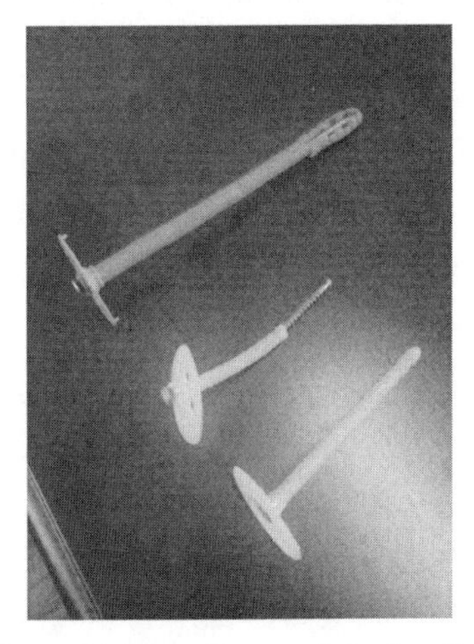

Figure 11-6-10 Damaged Anchor Bolt of Anchored Rock Wool Board

Figure 11-6-11 Large-scale Shedding of Rock Wool Board with Wire Mesh

11.6.2.3 Unreasonable Structural Design

The rock wool board insulation project involving surface cracking usually includes a thin plaster on the rock wool board surface, namely, a thin layer of polymer cement mortar and alkali-resistant glass fiber mesh to improve the cracking protection. However, the rock wool board of low strength is soft, flexible and prone to water absorption and stratification. The physical indicators (elastic modulus, linear expansion coefficient, etc.) and thermal properties of the rock wool board are greatly different from those of polymer cement mortar. If the rock wool board is plastered directly with a material of high density and rigidity, such as cement mortar or polymer cement mortar, cracking, hollowing and falling are likely to occur on its surface. Furthermore, fibers on the rock wool board surface are easy to break and fall off and cannot be bonded well with the plaster, so the plaster may crack and even fall off. In case that rock wool expands as a result of water absorption or heating, the plaster may be damaged. The high water absorption capacity of the rock wool board may also lead to moisture transfer from the surface mortar to rock wool board, and thus the plaster may crack as a result of quick loss of water but on effective curing.

11.6.3 Solutions

11.6.3.1 Better Structure Design

1. Reasonable and effective board fixing measures

As the rock wool board has low strength but large mass and rock wool fibers are easy to fall off, it is difficult to firmly fix the rock wool board on abase course wall with the adhesive. In other words, the adhesive only has the effect of temporary fixing. Instead, the rock wool board should be mainly secured with anchor bolts. The number and anchorage force of anchor bolts must comply with the requirements of wind load resistance. As the rock wool board is soft and prone to damage by external forces, only the parts around anchor bolt disc can be protected in the case of direct anchorage with anchor bolts. That is, the entire rock wool board cannot be protected. Therefore, reinforcing meshes should be paved on the rock wool board surface and secured with anchor bolts, to evenly distribute the anchorage forces of anchor bolts and effectively connect the rock wool board and base course wall. The wire mesh rather than the glass fiber mesh should be used for reinforcement. This is because both the rock wool board and glass fiber mesh are soft and therefore the anchorage forces of anchor bolts cannot be evenly distributed.

2. Flexible leveling transition layer

Direct plastering on the rock wool board surface will lead to cracking. When a leveling transition layer (thickness: more than 10mm) with the properties between those of the rock wool board and plaster is added between the rock wool board insulation layer and plaster layer, cracking can be prevented effectively. To overcome surface cracking and guarantee the stability and durability of the entire rock wool board insulation project, a lightweight leveling material should be used, which is not significantly different from the rock wool board in the density, specific strength, thermal conductivity and elastic modulus and beneficial for isolation of fire and heat. This is also conducive to the reliability of the rock wool board insulation project. The adhesive polystyrene granule mortar (adhesive polystyrene granule insulation mortar or adhesive polystyrene granule pasting mortar) satisfies these requirements: low dry density, with no excessive load on the rock wool board; heat-insulating and fireproofing performance; and high hydrophobicity and vapor permeability, conducive to the improvement of humidity of the external surface of the rock wool board insulation layer to keep a relatively stable status, and the reduction of vapor impact on the rock wool board insulation layer, thereby ensuring the stability of the entire insulation system.

3. Basic structure

The basic structure of the rock wool board insulation system designed reasonably is shown in Figure 11-6-12. Rock wool boards are mounted by means of point-and-frame bonding or strip bonding with the adhesive, and anchor bolts are secured on the wire mesh. The quantity of anchor bolts should be calculated based on the wind pressure. Gaskets should be used between the wire mesh and rock wool boards to prevent the wire mesh from direct attachment onto the rock wool boards and guarantee good bond stress between the leveling transition layer and wire mesh.

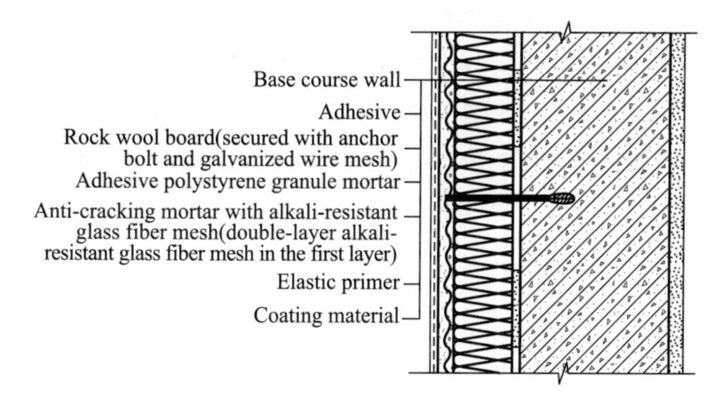

Figure 11-6-12 Basic Structure of Rock Wool Board Insulation System

4. Weathering resistance test

The wall used in the weathering resistance test was fabricated according to the basic structure shown in Figure 11-6-12. The finish layer is composed of four types of paint. After the weathering resistance test involving 80 heat-rain cycles and 20 heat-cold cycles, damage such as cracking, hollowing and falling was not found on the specimen surface. Blistering, efflorescence or peeling did not occur on the finish layer. Rock wool boards did not absorb water (Figure 11-6-13). After the test, the impact strength of the system was more than 10J, far higher than the value specified in the standards; and the pull strength between the leveling transition layer and rock wool boards was 0.1MPa, and damage was found in the adhesive polystyrene granule mortar layer.

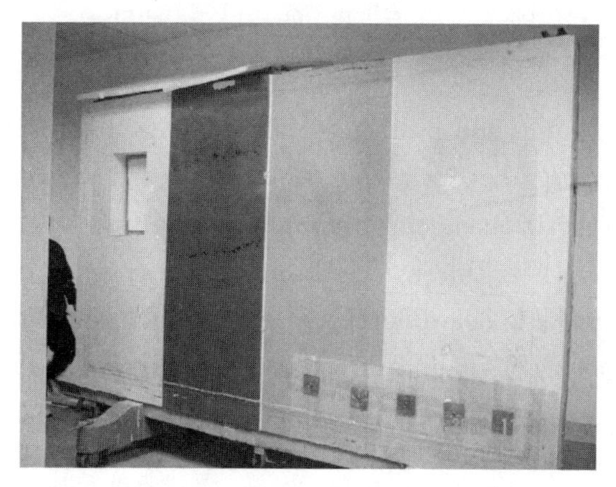
Figure 11-6-13 Weathering Resistance Test of Rock Wool Board Insulation System

11.6.3.2 Improvement of Rock Wool Board

It is difficult to produce rock wool boards with rock wool fibers perpendicular to the wall surface based on the prior art. However, reinforced vertical-fiber rock wool composite boards can be fabricated by cutting horizontal-wire rock wool boards into rock wool strips, reassembly rock wool strips into rock wool boards with rock wool fibers perpendicular to the wall surface, and making four surfaces in the length direction fully covered with mortar and glass fiber meshes (Figure 11-6-14). Thus, the distribution of rock wool fibers is changed, improving the tensile strength and dimensional stability of rock wool boards; the problem that rock wool fibers are easy to fall off is solved; the surface strength and water absorption capacity of rock wool boards is enhanced, avoiding the competition for moisture between the rock wool and plaster.

As four surfaces are coated with the glass fi-

Figure 11-6-14 Reinforced Vertical-fiber Rock Wool Composite Board

ber mesh and protective mortar, each reinforced vertical-fiber rock wool composite board is a relatively independent bearing element. Due to the glass fiber mesh, the bearing integrity of the composite board is greatly improved. If horizontal rock wool fibers are coated with the glass fiber mesh and protective mortar, the surface strength and waterproofing performance can be improved, but the tensile strength perpendicular to the wall surface cannot, which may lead to quality problems in external thermal insulation projects. The reinforcement of two main surfaces with the glass fiber mesh and protective mortar is not a fundamental solution to stratification. Instead, this structure is easy to stratify and fall off under cantilevering forces. Figure 11-6-15 shows the force conditions of several types of rock wool board on the wall. It can be seen that the reinforced vertical-fiber rock wool composite board is different from the ordinary rock wool board or rock wool board with two surfaces reinforced. In addition to the dead load and upward bonding force, the reinforced vertical-fiber rock wool composite board is also protected by the upward tension of the glass fiber mesh and that applied to the wall. Hence, this type of rock wool board has the highest safety and is the least likely to be destroyed.

When used in external thermal insulation projects, the reinforced vertical-fiber rock wool composite board can be bonded and anchored (Figure 11-6-16) or pasted (Figure 11-6-17). For the stability and the resistance to weathering and cracking, a 10-20mm adhesive polystyrene granule mortar leveling transition layer should be applied on the board surface. To prevent falling and improve the bonding to a base course wall, two L-shaped brackets (horizontal spacing: 300mm) are mounted with fasteners at the lower end of each reinforced vertical-fiber rock wool composite board, the long ends of two U-shaped inserts are mounted to the center positions of two insulation board layers in the thickness direction through appropriate jacks of the brackets (Figure 11-6-18). Both the L-shaped brackets and U-shaped inserts are made of metal.

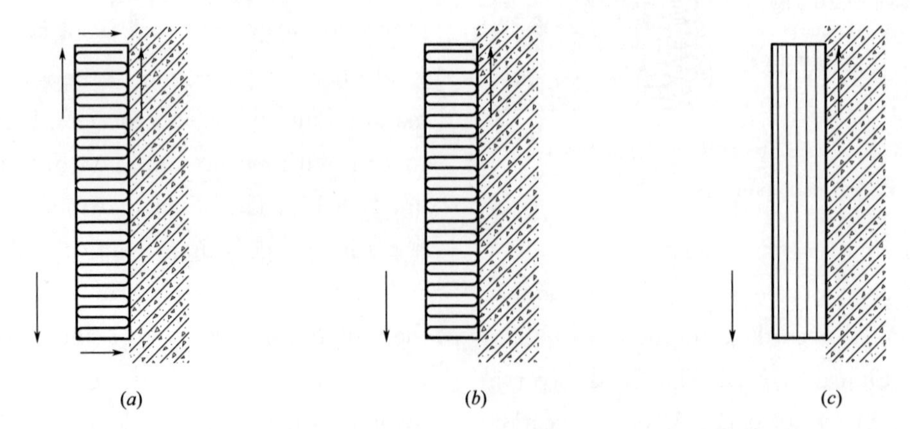

Figure 11-6-15 Comparative Analysis of Force Conditions of Rock Wool Boards

(*a*) reinforced vertical-fiber rock wool composite board with glass fiber meshes on four surfaces;

(*b*) rock wool board with two surfaces reinforced by glass fiber mesh;

(*c*) ordinary rock wool board

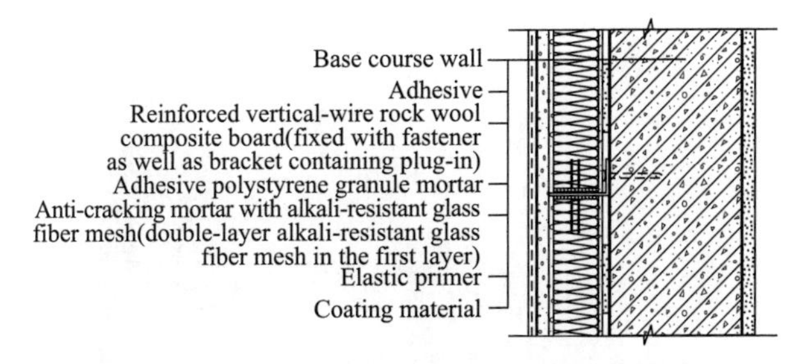

Figure 11-6-16 Bonding and Anchorage of Reinforced Vertical-fiber Rock Wool Composite Board

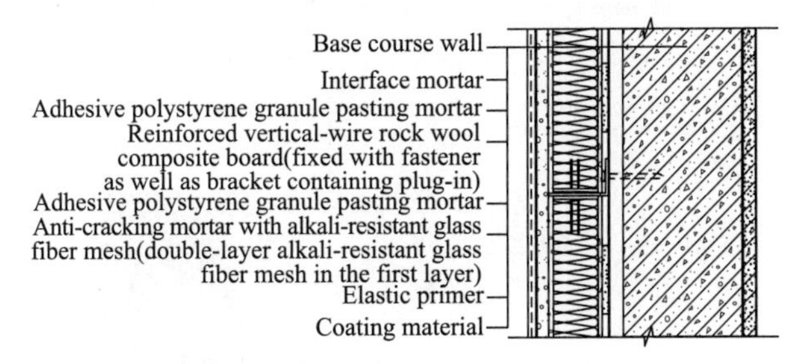

Figure 11-6-17 Bonding Structure of Reinforced Vertical-fiber Rock Wool Board System

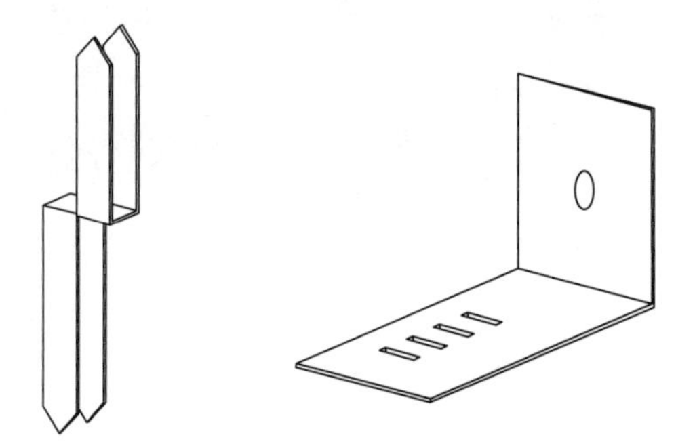

Figure 11-6-18 Double U-shaped Inserts and L-shaped Brackets

11.7 External Thermal Insulation Project with Rock Wool Fire Barrier

Fire accidents of external thermal insulation projects have stimulated the thoughts of all walks of life on the fire resistance of external thermal insulation system. Great important has been attached by the industry to this issue. The Ministry of Housing and Urban-Rural Development and the Ministry of Public Security jointly release the *Interim Provisions on Fire Protection of External Thermal Insulation System and Exterior Wall of Civil Buildings* (GTZ [2009] 46) on September 25, 2009. The national standard *Code for Fire Protection Design of Buildings* (GB 50016-2014) was released in 2014. It is stipulated in these documents that a fire barrier made of non-combustible insulation material shall be used depending on the building height in case that an external thermal insulation system is not made of Class A non-combustible insulation material. The rock wool board, as a non-combustible inorganic insulation material, is widely applied in fire barriers of external thermal insulation projects. In thin-plastered organic insulation board systems, however, bare rock wool boards are directly used as a fire barrier, and many quality problems occur as a result of significant differences between rock wool boards and adjacent insulation materials as well as defects of rock wool boards.

11.7.1 Engineering Case Analysis

Bare rock wool boards that have low strength are easy to deform and blister when heated and absorb water and fall off in a rain. Their surfaces are prone to damage. The properties of bare rock wool boards are greatly different from those of adjacent insulation materials. Therefore, there are potential quality hazards if bare rock wool boards are used as a fire barrier. Normally, bare rock wool boards are composed of horizontal fibers, which are soft and easy to peel off and cannot be effectively bonded with the base course wall and anti-cracking protective layer. This is why bare rock wool boards used as a fire barrier have quality problems such as falling and surface peeling. In addition, bare rock wool boards are able to absorb water and moisture, which may lead to the decline of insulation effects, and their expansion arising from water absorption may cause hollowing of the anti-cracking protective layer.

Figure 11-7-1 shows the obvious blistering and deformation of the fire barrier, caused by inconsistent expansion of bare rock wool boards and adjacent insulation materials.

Figure 11-7-2 shows the hollowing and falling

of the protective layer, caused by expansion of bare rock wool boards absorbing water.

Figure 11-7-1 Deformation and Blistering of Bare Rock Wool Board as Fire Barrier

Figure 11-7-2 Falling of Protective Layer Caused by Expansion of Bare Rock Wool Board Absorbing Water

As shown in Figure 11-7-3, the surface has been damaged seriously as a result of low strength of bare rock wool boards. In the case of loss of the hydrophobicity, the surface can be easily washed away by a rain (Figure 11-7-4), and moisture will infiltrate into adjacent insulation materials, resulting in damage to the insulation layer.

11.7.2 Solutions

11.7.2.1 Use of Reinforced Vertical-fiber Rock Wool Composite Board as Fire Barrier Material

It is a better choice to use the reinforced vertical-fiber rock wool composite board (Figure 11-7-5) instead of bare rock wool board as the fire barrier material.

Figure 11-7-3 Surface Damage Caused by Low Strength of Bare Rock Wool Board

Figure 11-7-4 Falling of Bare Rock Wool Board after Rain

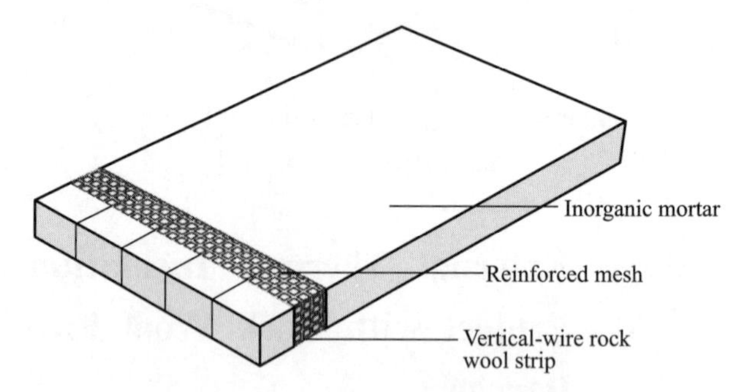

Figure 11-7-5 Reinforced Vertical-fiber Rock Wool Composite Board

The reinforced vertical-fiber rock wool composite board is a kind of prefabricated insulation board, in which the core is composed of rock wool strips (pieces) with rock wool fibers perpendicular to the wall surface and four surfaces in the length direction are coated with the glass fiber mesh and protective mortar. The rock wool strips or pieces are made by cutting rock wool boards or strips. The rock wool board or strip is a kind of board type insulation material (rock wool board) or strip type insulation material (rock wool strip) made by high-temperature melting and centrifugal injection of basalt or other natural igneous rocks (main raw materials) into mineral fibers, followed by adding of appropriate amounts of thermosetting resin adhesive, water repellent and other additives, pressing in a pendulum method, curing and cutting. The fibers of the rock wool board are parallel to the wall surface, while those of the rock wool strip are perpendicular to the wall surface. As the direction of rock wool fibers is changed and four surfaces are coated with the protective layer, the overall performance of the reinforced vertical-fiber rock wool composite board is improved. This insulation material has high tensile strength and integrity, and is easy to construct without pollution and securely bond with a base course wall. It is a solution to the problem of bare rock wool boards, such as settlement in the presence of water, stratification, falling and low tensile strength, and also to the problem of skin injury by rock wool fibers, with good effects of labor protection. Table 11-7-1 shows the comparison of workability between the reinforced vertical-fiber rock wool composite board and bare rock wool board.

11.7.2.2 Setting of Leveling Transition Layer

The impact of significant differences in the performance between the fire barrier material and adjacent insulation materials cannot be eliminated by one 3-5mm plaster layer (or anti-cracking protective layer). Instead, a 10-20mm leveling transition layer of adhesive polystyrene granule mortar can be added between the insulation layer and plaster layer (or anti-cracking protective layer). In this way, adverse effects arising from the differences in the performance between the fire barrier material and adjacent insulation materials can be eliminated, and deformation differences of fire protection parts can be absorbed. In addition, the plaster layer (or anti-cracking protective layer) is directly exposed to a uniform structural layer, thereby guaranteeing the stability of the entire system.

Workability Comparison between Reinforced Vertical-fiber Rock Wool Composite Board and Bare Rock Wool Board **Table 11-7-1**

Item	Reinforced Vertical-fiber Rock Wool Composite Board	Bare Rock Wool Board
Construction	This material of reliable quality is customized in the factory, and needs little on-site cutting. It is not harmful to construction workers and the environment, and can be constructed easily.	It is fabricated by on-site cutting in a manual way, so cutting effects cannot be guaranteed and the damage rate is high. This material is harmful to construction workers and the environment.
Appearance	The surface is flat with clear lines, so surface treatment is not needed. The outer wall constructed with this material has good visual effects.	This material is irregular as a result of difficulties in cutting. It rough surface cannot be treated easily, and the treatment effects are poor, which directly affects the appearance of the outer wall constructed with this material.
Structural performance	This material has high hydrophobicity due to the protective layer on its surface. It can be well bonded with a base course wall and surface layer. In addition, this material has good peel strength, impact resistance, durability and weathering resistance, and is not easy to fall off, be affected with damp, deform or damage.	This material has low strength and is easy to peel off. It cannot be well bonded with abase course wall or surface layer. After completion, quality problems often occur, such as falling and surface peeling. The hydrophobic layer on the surface of this material can be damaged easily. Moreover, this material is easy to absorb water and be affected with damp, resulting in poor insulation effects.

Figure 11-7-6 Schematic Diagram of Weathering Resistance Test of Reinforced Vertical-fiber Rock Wool Board System

11.7.3 Weathering Resistance Test

The large-scale weathering resistance test (Figure 11-7-6) was conducted to an external thermal insulation system, in which the fire barrier is made of reinforced vertical-fiber rock wool composite board, and the leveling transition layer is composed of adhesive polystyrene granule mortar. Cracking, blistering or peeling was not found on the surface, and the tensile strength of this system was 0.070-0.10MPa. Both the tensile strength and impact resistance complied with the standards. Therefore, it is feasible to use the reinforced vertical-fiber rock wool composite board instead of bare rock wool board as the fire barrier material and set an appropriate leveling transition layer of adhesive polystyrene granule mortar outside the insulation layer to withstand various climatic impacts.

11.7.4 Engineering Application

11.7.4.1 Insulation Board Bonding Method

When used in the insulation board bonding method, the reinforced vertical-fiber rock wool composite board as the fire barrier should be fully bonded onto a base course wall via adhesive polystyrene granule pasting mortar or adhesive and also secured with anchor bolts. The reinforced vertical-fiber rock wool composite board and all insulation board surfaces should be subject to leveling transition with adhesive polystyrene granule pasting mortar, as shown in Figure 11-7-7. The thickness of the reinforced vertical-fiber rock wool composite board should be the same as that of insulation boards.

During construction, insulation boards are pasted bottom-up on an outer wall to the fire barrier position, and then reinforced vertical-fiber rock wool composite boards are bonded from a corner at one end of the wall. Butt joints of reinforced vertical-fiber rock wool composite board should be spliced tightly and flush with insulation boards on the entire wall. The reinforced vertical-fiber rock wool composite boards bonded are additionally secured with anchor bolts (Figure 11-7-8) at intervals of 500-600mm. Then the overall leveling transition layer is constructed with adhesive polystyrene granule pasting mortar, followed by treatment of the anti-cracking protective layer. Figure 11-7-9 shows the effects after completion.

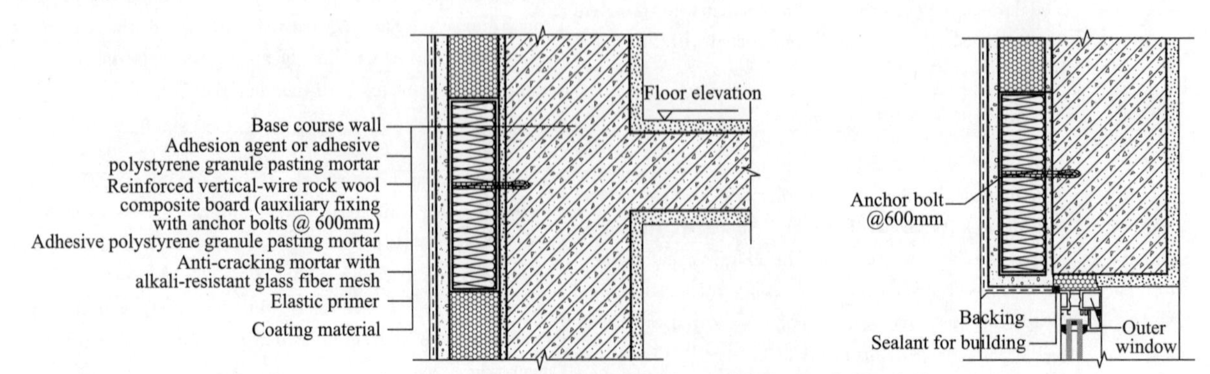

Figure 11-7-7 Basic Structure of Reinforced Vertical-fiber Rock Wool Composite Board as Fire Barrier in Insulation Board Bonding Method

Figure 11-7-8 Anchorage of Reinforced Vertical-fiber Rock Wool Composite Board

Figure 11-7-9 Effects after Completion of Anti-cracking Protective Layer

The EPS board bonding method was adopted in the Low-rent Housing Project of Wangsiying in Beijing and Wanliyuan Xianglan Jiayuan Project in Tianjin. The fire barrier is made of reinforced vertical-fiber rock wool composite board, and the insulation surface is subject to leveling transition with adhesive polystyrene granule pasting mortar. The Low-rent Housing Project of Wangsiying in Beijing covers a construction area of $202{,}000m^2$, the building height hereof is 60m, and the fire barrier of 13,602 linear meters based on reinforced vertical-fiber rock wool composite boards is used. The Wanliyuan Xianglan Jiayuan Project in Tianjin covers an insulation area of $45{,}000m^2$, and involves the barrier of 12,000 linear meters based on reinforced vertical-fiber rock wool composite boards. The insulation systems of both projects have high safety, reliably and weathering resistance, and no problems are found in more than two years after completion.

11.7.4.2 Cast-in-situ EPS Board Construction Method

Figure 11-7-10 shows the cast-in-situ EPS board construction method, in which the fire barrier is made of reinforced vertical-fiber rock wool composite board.

Reinforced vertical-fiber rock wool composite boards are connected to adjacent EPS boards and reinforced concrete through fire barrier clamps (Figure 11-7-11) at intervals of 500-600mm. The fire barrier clamps are made of metal or hard plastic. Reinforced vertical-fiber rock wool composite board of window openings are also secured with plastic clips (Figure 11-7-12) at intervals of 500-600mm.

When the lower EPS boards and reinforced vertical-fiber rock wool composite boards as the fire barrier are completed on the site, the sharp ends of the insert plates of fire barrier clamps should inserted into the upper sides of reinforced vertical-fiber rock wool composite boards, with

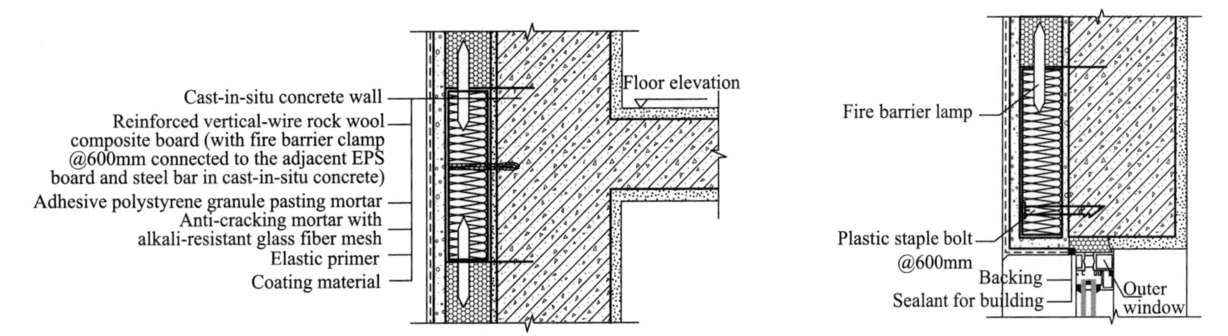

Figure 11-7-10 Basic Structure of Cast-in-situ EPS Board Construction with Fire Barrier Made of Reinforced Vertical-fiber Rock Wool Composite Board

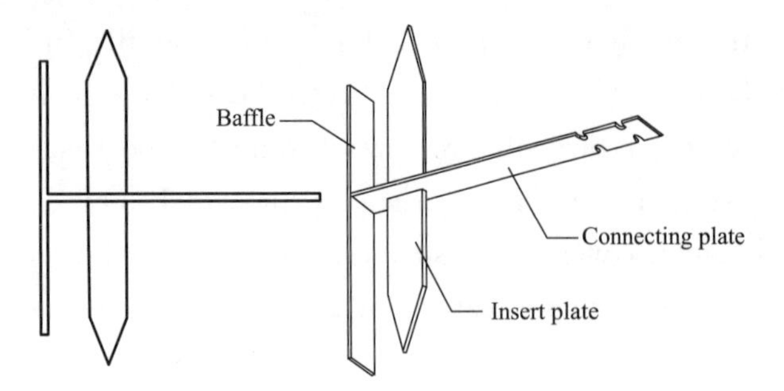

Figure 11-7-11 Fire Barrier Clamp

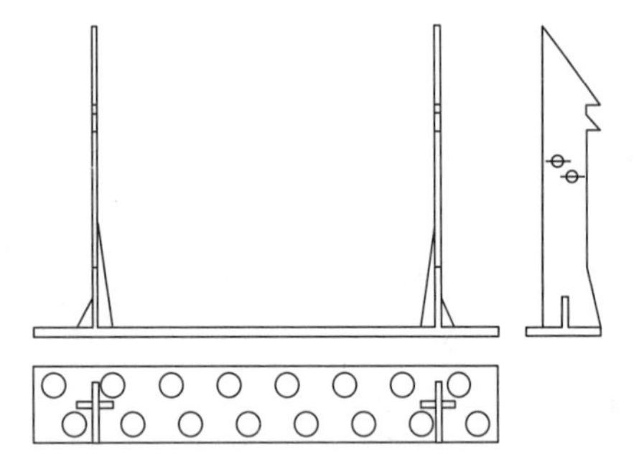

Figure 11-7-12 Plastic Clip

connecting plates facing steel bars, and baffles are securely attached onto reinforced vertical-fiber rock wool composite boards, as shown in Figure 11-7-13; then EPS boards of the upper layer are installed, with the other sharp ends of the insert plates of fire barrier clamps in EPS boards, and baffles are secured onto EPS boards, as shown in Figure 11-7-14; and finally the connecting plates of fire barrier clamps are tightly tied to steel bars, as shown in Figure 11-7-15. Figure 11-7-16 shows the wall effects after concrete pouring.

Figure 11-7-13 Installation of Fire Barrier Clamp

Figure 11-7-14 EPS Board Installation with Fire Barrier Clamp

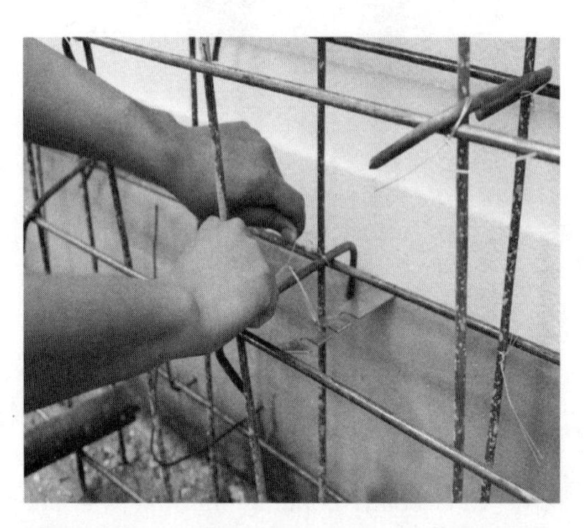

Figure 11-7-15 Typing of Fire Barrier Clamp

Figure 11-7-16 Wall Effects after Concrete Pouring

The external thermal insulation system based on the cast-in-situ composite EPS board is used in the collective housing project of the Administration for Industry and Commerce of Xinjiang Uygur Autonomous Region. The fire barrier is made of reinforced vertical-fiber rock wool composite board, and the insulation surface is subject to leveling transition with adhesive polystyrene granule pasting mortar. This project covers an insulation area of 40,000m^2, and involves the fire barrier of 3,500 linear meters based on the reinforced vertical-fiber rock wool composite board. It has good flatness, safety, reliability and weathering resistance in more than one year, and its fire barrier has no quality problem.

12. Conclusions

12.0.1 China has rich experience in external thermal insulation.

In the 1980s, China began to learn foreign experience in the external thermal insulation system based on the thin-plastered EPS board. With continuous and independent innovation, introduction and absorption, a number of external thermal insulation systems have been developed, involving differences in the material, structure and technology. At present, there are dozens of practices in five categories. The following systems are widely applied: pasted insulation board and thin plaster system, light-aggregate insulation mortar system, cast-in-situ concrete and built-in insulation board system, on-site sprayed or cast insulation system, and decorative insulation board system. These external thermal insulation technologies have their own characteristics and application ranges. With the gradual improvement of energy conservation requirements, the external thermal insulation has become the main form of energy-saving wall insulation in China, with the most mature technology and most extensive applications. It is particularly popular in heating areas, where the annual increase of application area is 100 million square meters and ranks top in the world. The external thermal insulation industry basically meets the development at present and in the near future. Practices have proved that the majority of external thermal insulation projects are successful and some projects completed in the early period of development of this technology are still in good conditions, making significant contributions to China's building energy conservation. However, some projects have quality problems, such as no insulation of local thermal bridges, poor safety of connection between the insulation layer and wall structure, infiltration of external moisture into the wall, resistance against outward permeation of internal vapor, low durability of the reinforced mesh, poor cracking resistance of the insulation surface, cracking, hollowing and even falling. The face brick finish is more prone to falling off. Additionally, some insulation materials are ignited on the construction site or burn on the wall, which has attracted attention of those parties concerned.

In other words, China has gained rich positive and negative experience in external thermal insulation projects over years.

12.0.2 The external thermal insulation is the most reasonable structure of outer wall instruction.

The outer wall subject to external thermal insulation is composed of multiple functional layers, including the load-bearing base course wall, insulation layer bonded on the wall, surface layer and finish layer, etc. There are several types of insulation with respective characteristics, namely, external thermal insulation, internal thermal insulation, sandwich insulation and self-insulation. Among them, the external thermal insulation technology with the insulation layer outside the outer wall is applied the most widely at home and abroad due to its many advantages.

As the external thermal insulation system was developed based on the needs of building energy conservation and thermal comfort and the energy conservation requirements are constantly improved, more and more importance is attached to the external thermal insulation system. The insulation layer is located outside the structural layers of a wall, so its thermal insulation effect can reduce the temperature variation and gradient inside structural layers, thereby stabilizing the temperature and decreasing the thermal stress and its change in structural layers. That is, the main structure is effectively protected with the external insulation layer, which can prolong the structural life. The structural life is the fundamental of a building. The extension of structural life will lead to inestimable economic and social benefits. The external thermal insulation helps to greatly reduce the heat loss of thermal bridges and prevent local condensation or mildew arising from the excessively low temperature on the internal surfaces of thermal bridges. The structural layer of a wall constructed with heavy materials has high thermal capacity. With the external insulation layer, a large amount of heat will be stored in the structural layer. Due to the powerful functions of automatic storage and radiation of heat, the temperature of the building envelope can be adjusted dynamically based on the ambient temperature, thereby making the room temperature stable and improving the comfort, physical health and living conditions, and also fully utilizing the building energy. If the external thermal insulation is adopted during energy-saving renovation of existing buildings, interference to residents will be reduced. In this sense, the external thermal insulation system is the most reasonable insulation structure which has a number of advantages.

12.0.3 The external thermal insulation system must be able to withstand the test of natural factors.

The external thermal insulation layer exposed to the outermost atmosphere of building is subject to long-term natural erosion. Natural conditions are constantly changing, including the seasons, day and night and climatic changes (sunny, cloudy, cold, hot, windy, rainy, freezing and thawing) and harsh sometimes. They repeatedly affect the external thermal insulation layer day by day. Accordingly, the temperature, humidity and stress inside the insulation layer vary continuously. When the stress reaches a critical value in a long time, the external protective layer and insulation layer may be damaged in an irreversible way. Initial cracks gradually expand due to freezing and thawing, which may cause large-scale damage. In addition, there are unpredictable factors. For example, earthquakes may occur in vast areas of China, damaging the external thermal insulation system and face brick finish. Fires may also occur during construction and use, involving the spreading of flame and burning of the insulation layer. Thus, effective and targeted protective measures must be taken during construction, so that the external thermal insulation system is able to withstand natural tests and challenges and its service life can be prolonged. All measures taken must be comprehensive but not single; economical but not costless; and durable but not short-lived. It is known that the damaged insulation layer must be dismantled before repair, which is especially costly for high-rise buildings. It is worthy to use the advanced and reliable technologies and take the measures specified in relevant standards in the construction process to achieve the safety and efficiency in dozens of years.

12.0.4 The moisture transfer inside a wall subject to external thermal insulation must be controlled.

The infiltration of moisture in nature into a wall subject to external thermal insulation has many influences. If the insulation material is eroded by rainwater or water vapor, the thermal insulation performance of the wall will decline and the building energy consumption will increase. Icing occurs in winter. As the ice volume is about 9% greater than that of water, the frost-heave stress will be generated, which may damage the wall. There are more serious results under the frost-heave stress inside a face brick system. Water may also lead to the decline of bonding strength between different layers, thereby reducing the weathering resistance of the system, accelerating the ageing of the insulation material and affecting the building durability. If the insulation layer is not applied properly, the system structure is unreasonable and vapor permeation is hindered, the insulation layer may absorb moisture, which will reduce the insulation performance and lead to condensation in winter and even black spots, mildew and water flow in serious cases. If contaminants formed in a humid environment in a long time are diffused along with the airflow, the indoor air will be affected. In the case of vapor erosion, the thermal comfort will drop. Thus, it is necessary to control the moisture inside the wall with external thermal insulation.

In the presence of water, essential conditions are provided for various destructive factors. Hence a layer of waterproofing elastic primer with a breathing function is needed on the anti-cracking polymer mortar surface, to prevent liquid water and discharge gaseous water. Provided that the vapor permeability coefficient is appropriate, the water absorption capacity of the external thermal insulation system and material should be reduced greatly to decrease the impact of water on the system and prevent the damage of the frost-heave stress to systems in cold areas and also that of an alkali environment to the bonding strength of polymer mortar under long-term humid conditions. This will keep a building in a relatively steady state and improve the safety and reliability of the external thermal insulation system as well as the long-term stability of the structural wall.

12.0.5 The thermal stress can be dispersed by a flexible transition layer to prevent cracking.

In order to ensure the service life of external thermal insulation systems, the large-scale weathering resistance test must be conducted prior to the application under normal conditions. The repeated actions of harsh climatic conditions in winter and summer should be simulated and the weathering of large-sized components should be accelerated. Severe temperature variations may lead to the uneven deformation of various layers of the external thermal insulation system and also changes in the internal stress. When the stress exceeds the limit, the part concerned will crack, thereby shortening the service life of the external thermal insulation system. The results of the large-scale weathering resistance test are well correlated with actual projects. The failure of external thermal insulation systems to pass the large-scale weathering resistance test proves their poor weathering resistance and inapplicability to projects. Problems of some methods can also be found in the weathering resistance test. In this case, actions need to be taken to improve the external thermal insulation technology.

For example, the test results of the XPS board insulation system demonstrate that, when XPS boards are fully bonded with adhesive polystyrene granule mortar and the 10mm board gaps are adopted, equivalent to six fully-bonded mortar reinforcement structures for each XPS board, the overall bonding strength and vapor permeability of the system will be enhanced, and the board deformation will be constrained. Meanwhile, the stress generated in the expansion and contraction of XPS boards can be dispersed or absorbed to reduce the possibility of cracking. XPS boards may be drilled for higher vapor permeability and overall bonding performance. Also, the test results show that, the sprayed polyurethane should be kept static for some time for full deformation and volume stabilization and then a layer of adhesive polystyrene granule mortar should be applied, to improve the weathering resistance of the sprayed polyurethane and achieve dual effects of leveling and cracking prevention.

The adhesive polystyrene granule mortar is a composite of organic and inorganic materials. Its linear expansion coefficient and elastic modulus fall between those of the EPS board and crack-resistance mortar. If a flexible transition layer of adhesive polystyrene granule mortar is used, the thermal stress will be dispersed, making the entire system change flexibly and gradually, facilitate the layer-by-layer release of deformation, reducing the difference in the deformation speeds of adjacent materials, greatly improving the weathering resistance of the system and thus solving the common problem of cracking of XPS board and polyurethane systems.

12.0.6 Fire protection of the construction site and entire thermal insulation system are critical to the fire safety of external thermal insulation.

At present, more than 80% of insulation materials in new buildings (including high-rise and super high-rise buildings) in China are made of organic combustible material such as the polystyrene foam and rigid polyurethane foam. Many construction fires are caused by welding sparks or improper use of flame. Therefore, the fire management on the construction site must be performed rigorously, and electric welding or opening flame should be prohibited on the construction site of insulation projects with organic materials.

Studies have shown that the thin-plastered EPS board system applied widely at home and abroad has poor fire resistance. Especially in the point-bonded EPS board system nonconforming to technical standards (bonding area: usually no greater than 40%), there are connecting air layers, which may quickly form "fire channels" in the event of a fire, resulting in fast spreading of flame. As a lot of smoke is generated during combustion, the visibility is greatly reduced, which lead to difficulties in escape and rescue. This system also has poor volume stability in the presence of a high-temperature heat source. In particular, the damage to the tile finish has more dangerous consequences and severe safety hazards. This issue is more prominent on high floors.

The fire safety is an importance conditions and basic requirement for the external thermal insulation technology. The current assessment on the fire safety of external thermal insulation is based on the fire test results. The large-scale model fire test is associated more closely with real fire conditions. The results of a number of fire tests prove that, with a reasonable structure, the insulation efficiency and fire safety of an organic material can be fully achieved at the same time. As long as the structure of the external thermal insulation system is reasonably designed, it has good reaction-to-fire performance and desired fire safety. The structure here refers to the bonded or secured structure (with or without cavities), fire compartment (partition or barrier), fireproof protective surface, surface of appropriate thickness, etc. The system with the fire compartment, fireproof protective surface of enough thickness and no cavity has excellent resistance to fire, and involves no flame spreading during the test. The fire barrier is able to prevent the spreading of flame. If a thin-plastered EPS board system of poor fire resistance is equipped with a fire barrier, the spreading of flame can be prevented effectively, and the applicable building height can be increased.

The combustibility of an insulation material is a technical indicator specified in the prevailing technical standards, and also an essential condition to ensure the fire safety during construction and use of external thermal insulation projects. It is not equivalent to the fire resistance of an external thermal insulation system. The test results demonstrate that the thin-plastered XPS board (Class B_1) has the tendency to spread the flame but the entire EPS board (Class B_2) system with reasonable fireproof structural measures has excellent overall fire resistance.

12.0.7 The negative wind pressure may lead to falling of the external thermal insulation system with cavities.

The external thermal insulation system is a

non-bearing structure secured onto a base course wall through the adhesive or anchor bolts. In the case of pure point bonding, there are connecting cavities in the system. When the negative wind pressure perpendicular to the wall surface is greater than the tensile strength of the material or interface, the external thermal insulation system may be blown off. This accident occurs frequently.

When the external thermal insulation system with a closed cavity structure is constructed as per the technical standards, its wind pressure safety complies with the requirements. However, improper construction (pure point bonding of insulation boards, too small bonding area, insufficient strength of the bonding material, and lack of necessary interfaces), lack of the sense of responsibility or poor supervision may lead to damage to the external thermal insulation system under the negative wind pressure. The external thermal insulation system without cavities or with small cavities has high wind pressure resistance, overall performance, stress transfer stability, safety, etc. In order to avoid the above problems, construction must be carried out in strict accordance with the technical standards or the no-cavity system should be used.

12.0.8 Appropriate safety measures must be taken to the face brick finish for external thermal insulation.

The face brick finish with good decorative effects, high impact or stain resistance and excellent color durability is popular with real estate developers and residents. A large number of new buildings are equipped with the face brick finish. Nevertheless, hollowing or falling occurs sometimes, so the face brick finish has potential safety hazards. It is generally thought that the face brick finish should not be used in external thermal insulation projects. On the contrary, it is still applied widely in actual projects, even on high-rise and super high-rise buildings. One problem to be resolved is how to guarantee the safety and quality of the face brick finish.

In order to meet the insulation requirements of the design standards for building energy efficiency, the insulation layer of the external thermal insulation system is usually made of soft organic foam, with the density, strength, rigidity and fire resistance far lower than those of the base course wall material. With the promotion of energy conservation and emission reduction and the improvement of design standards for building energy efficiency, the thickness of the insulation layer is gradually increased. At the same time, the face brick finish outside the external thermal insulation system is directly exposed to natural actions such as cold/heat, water/vapor, fire, wind pressure and earthquakes, the stress inside the system varies greatly, so appropriate and safe reinforcement measures need to be taken to keep the necessary safety of the building and its thermal insulation system and also prevent the quality accidents of the face brick finish, such as hollowing and falling. As the face brick finish has increasing safety problems, it is believed that this type of finish should be avoided in external thermal insulation systems in order to ensure the safety. If the face brick finish needed in actual projects, a series of appropriate safety measures should be taken carefully.

When porcelain tiles are pasted, the anti-cracking protective layer should be strengthened with reinforcing meshes, to combine the insulation layer of low density and strength with the face brick finish and form the transition from the insulation layer unsuitable for face brick bonding to a strong and flexible protective layer. The deformation of the tile adhesive in the external thermal insulation system should fall between those of the anti-cracking mortar and face brick finish. With the bonding strength guaranteed, the flexibility can be improved, so that the face brick finish can be securely combined with the thermal insulation system and external forces (especially thermal stress and seismic action) can be absorbed. A flexible pointing material should be used for free deformation of face bricks with the temperature. The face bricks with dovetail grooves on the sides to be bonded should be used, and release agents must not be used.

12. 0. 9 A flexible connecting structure should be adopted to reduce the seismic impact on the external thermal insulation system.

The external thermal insulation system should have appropriate deformation capacity to adapt to the displacement of a main structure. When a main structure is displaced under a large earthquake load, the external thermal insulation system in a seismic fortification zone should be free of excessive stress or unbearable deformation. Typically, functional layers of the external thermal insulation system are mostly made of flexible material. In the event of a small lateral displacement of the main structure, its impact can be absorbed via elastic deformation of the system. However, the external thermal insulation system is a composite system secured on structural wall by means of bonding or mechanical anchorage. When an earthquake occurs, the connection between different functional layers and that between the system and main structure should be able to reliably transfer the earthquake load and also bear the dead load of the system. To prevent the damage caused by the displacement of the main structure, connections must have appropriate adaptability.

The load capacity of the insulation material surface, bonding strength of the face brick adhesive and the flexible deformation capacity against violent movement under the seismic action must be taken into consideration during face brick bonding on the external thermal insulation surface. As thebase course wall and face brick finish are flexibly connected via an insulation material, they cannot be regarded as a whole under forces. Instead, they bear forces in different ways. Therefore, the face brick adhesive with appropriate flexibility and adaptability to the insulation material should be selected to form a system in which the deformation is changed flexibly and released layer by layer. The deformation capacity of the face brick adhesive should be lower than that of anti-cracking mortar but higher than that of the face brick finish, to eliminate the deformation of two materials that are different in the mass, hardness and thermal properties. Thus, the seismic force can be released separately by each face brick like scales, and the face bricks will not fall off as a result of deformation under the seismic action.

12. 0. 10 The adhesive polystyrene granule composite insulation system is an advanced technology suitable for China's national conditions as well as design requirements for building energy conservation in different regions.

The thin-plastered EPS board system and adhesive polystyrene granule mortar system was studied and applied in the 1950s and 1960s in Germany, but first studied and applied in the 1990s in Beijing. Different from the slow progress and non-extensive application of adhesive polystyrene granule mortar in foreign countries, Beijing Zhenli, on the basis of China's national conditions, persistent innovation and guidance of scientific researches over engineering applications, has changed a single insulation system into a "sandwich" system combining the insulation mortar and high-efficiency insulation boards, using the theory of material compounding and technological theory summarized from the tests, researches and engineering practices. In the "sandwich" system, comprehensive advantages of thermal insulation, heat isolation and fire protection are fully utilized, meeting the requirements of increasing standards of building energy conservation and fire protection for external thermal insulation. The adhesive polystyrene granule composite insulation technology with independent intellectual property rights and adaptability to severe cold and cold areas in the north and hot-summer cold-winter and hot-summer warm-winter areas in the south has been developed, including seven external thermal insulation systems based on the adhesive polystyrene granule, namely, pasted EPS board system, cast-in-situ EPS board system, pasted EPS board system, sprayed polyurethane system, anchored rock wool board system, pasted and reinforced vertical-fiber rock wool board system and insulation mortar system. They comprehensively reflect the scientific research achievements and engineering experience in the past 20 years, and also embody new achievements and progresses in current researches and applications, making important contributions to the technology progress of China's thermal insulation industry as well as the extensive and in-depth development of building energy efficiency.

For the technological theory, three technical concepts and five natural destructive forces have been creatively put forward, enriching and developing China's external thermal insulation technology; for the technological progress, a complete set of external thermal insulation technology meeting the building energy conservation requirements of different climatic zones has been developed independently and creatively, based on the European technology and fully considerations of the vast geographical area and complex climatic conditions in China, and this technology is far above the technological level of similar systems (products) abroad; and for the technological application, the adhesive polystyrene granule composite technology is a mature and suitable technology that is applied the most widely and tested the most extensively in China. So far, this technology has been successfully applied in more than 5,000 projects in over 20 provinces and cities in China, with the construction area more than 100 million square meters and good effects. Among engineering applications, the system and its composition have been tested tens of thousands of times by provincial and municipal test organizations, further proving that this technology is mature, reliable and trustworthy and will be used more extensively and increasingly.

12.0.11 One important direction of insulation development is to use solid wastes as raw materials.

At present, the large-scale application of wall insulation in China requires large quantities of organic insulation materials (EPS board and polyurethane) and mineral resources. A lot of energy is needed during production of such materials, so one important task of energy conservation and emission reduction is to develop energy-saving products and technologies for buildings. If industrial solid wastes are used as raw materials in external thermal insulation systems, natural resources can be saved, the environmental pollution arising from wastes can be reduced, and the energy consumption of production can be decreased. This is one essential direction for development of energy-saving technologies.

There are a large number of waste packages made of EPS foam, with stable chemical properties, low density, large volume, ageing resistance as well as corrosion resistance. They cannot be degraded by themselves. Stacking of fly ash and tailings needs the land and may result in the pollution of air and underground water; and waste tyres are difficult to degrade but easy to burn. All these industrial wastes are associated with the serious environmental pollution and safety hazards. However, the EPS granule has excellent insulation properties, fly ash has potential pozzolanic activity, waste paper fibers have the effects of water retention, cracking prevention and mortar workability improvement, and waste rubber particles help to improve the resistance of cement base course walls to deformation, cracking and freezing. All these wastes have the potential for development and utilization.

The systematic researches over years have proved that the application of large quantities of solid wastes in external thermal insulation systems is more and more mature. When appropriate quantities of EPS granules, fly ash, tailing sand, waste rubber particles and waste paper fibers are used, their special properties can be fully utilized for thermal insulation, bonding, leveling, connection, pointing, cracking prevention, putty application and pasting. The comprehensive utilization rate of solid wastes in external thermal insulation systems has exceeded 50%. In this way, solid wastes can be utilized efficiently and comprehensively, thereby reducing the environmental pollution, material costs and construction energy consumption, effectively solving the problem of raw material shortage arising from rapid development of China's building energy conservation and external thermal insulation industry, and also meeting the national requirements for the development of circular economy, construction of a conservation-oriented society and promotion of energy conservation and emission reduction strategies.

12.0.12 The external thermal insulation technology is developed constantly by means of testing, research and innovation based on positive and negative experience and lessons.

Similar to other objects, there are both successes and failures in the development process of external thermal insulation technology. It is important to summarize the successful experience and more important to transform the lessons from failures into successes. Beijing Zhenli collects and studies the quality problems of external thermal insulation projects in the research process, and also performs the theoretical analysis to investigate the causes and also the experimental research to find solutions for better effects. Seven representative engineering cases are used in this book. Even the thin-plastered EPS board system that is matured the earliest and applied the most widely may be blown off. If the construction procedures of the above system are mechanically copied, hollowing, cracking and falling may occur. Quality problems of external thermal insulation projects are generally caused for two reasons. First, the insulation boards and base course wall are not bonded securely, or there are connecting cavities, which cannot resist the impact of negative wind pressure. Secondly, the insulation boards have poor stability and the anti-cracking mortar has poor quality, resulting in low flexibility of the entire system and failure to resist the repeated effects of hot and wet loads. Hence the targeted studies have been carried out. First, insulation boards are fully bonded on a base course wall via adhesive polystyrene granule pasting mortar in a no-cavity manner. Secondly, a transition layer of adhesive polystyrene granule pasting mortar is added between the insulation boards and anti-cracking mortar. Engineering practices have demonstrated that such technical measures have excellent effects, including the enrichment and development of the external thermal insulation technology, and are listed into national industry standards and local standards, thus promoting the applications of external thermal insulation. In recent years, blind prevention of the use of organic insulation materials results in the blind development of rock wool production lines, hasty use of rock wool insulation systems, as well as quality accidents such as the blowing-off of rock wool boards, cracking of the insulation surface and decline in the insulation function. Beijing Zhenli has long researched and applied rock wool board systems, put forward scientific and reasonable solutions to the quality problems of rock wool board systems, and also creatively developed the reinforced vertical-fiber rock wool composite board. This type of rock wool board is made by cutting horizontal-wire rock wool boards into rock wool strips, reassembling rock wool strips into rock wool fibers, and installed perpendicular to the wall surface. The four sides in the length direction of this composite board are coated with the polymer mortar and glass fiber mesh. Specific innovations are as follows: (1) the rock wool board has higher tensile strength and dimensional stability and can be securely bonded on the base course wall; (2) the surface strength of the rock wool board is enhanced, thereby solving the problem of falling of rock wool fibers; (3) the waterproofing performance of the rock wool board is enhanced, preventing the failure in water absorption and insulation; and (4) the rock wool board can be constructed easily without pollution, preventing the air pollution and skin injury caused by rock wool fibers in a long time. The comprehensive application technologies (including the fire barrier, external thermal insulation system and composite EPS granule foam concrete lightweight wall with external thermal insulation) involving the reinforced vertical-fiber rock wool composite board is a new progress of the external thermal insulation technology, making new contributions to the building energy conservation industry.

References

[1] Jiang Zhigang and Long Jian. *Study on Heat Transfer Characteristics of Internal and External Thermal Insulation of Composite Outer Wall*. No. 2, 2006.

[2] Wang Jinliang. *Heat Transfer Analysis and Application Exploration of Internal and External Thermal Insulation of Composite Outer Wall*. Building Energy Conservation and Air Conditioning. December 2004.

[3] Research Institute of Standards & Norms of the Ministry of Housing and Urban-Rural Development. *Guidelines for External Thermal Insulation*. Beijing: China Building Industry Press. 2006.

[4] Yu Hongwei. *Analysis of Application Safety of External Thermal Insulation Technology in China*. Housing Science. October 2005.

[5] Chen Youtang. *Three-step Energy-saving Wall-Sandwich Insulation with Cast-in-situ Foam*. Tianjin Construction Department. No. 4, 2005.

[6] Li Zhilei and Gan Gang. *Numerical Simulation of Temperature Field of Building based on Radiation Heat Transfer*. Journal of Zhejiang University. 2004, 38 (7).

[7] South China University of Technology. *Building Physics*. Guangzhou: South China University of Technology Press. 2002.

[8] G. Weil, Die Beauspruchung der Betonfahrbahnplatten, Strassen und Tie-fbau, 17, 1963.

[9] Wen Zhouping and Wang Dan. *Research on External Thermal Insulation System of High-rise Building in Northeast China*. China Housing Facilities. No. 05, 2004.

[10] Wang Jiachun and Yan Peiyu. *Analysis of Application Effect of External Thermal Insulation System*. Bricks & Tiles. No. 7, 2004.

[11] Wang Hongxin. *Technical Requirements and Measures for Energy-saving and Heat-insulating Walls*. Energy Conservation Technology. 2002 (03).

[12] F. S. Barber, Calculation of maximum temperature from weather reports, H. R. B. bull, 168 (1957).

[13] Pang Liping, Wang Jun and Zhang Yanhong. *Study on Heat Transfer of Composite Insulation Wall*. Low-Temperature Architecture Technology. 2003 (04).

[14] Wang Jiachun, Yan Peiyu, and Zhu Yanfang. *Thermal Calculation and Analysis of Composite Wall with External Thermal Insulation*. Architecture Technology. 2004, 35 (10).

[15] Yang Chunhe. *Analysis of Temperature Field and Thermal Stress of Masonry Structure*. Jiangsu Architecture. 2005, (2).

[16] Li Hongmei and Jin Weiliang. *Numerical Simulation of Temperature Field of Building Envelope*. Journal of Building Structures, 2004, 25 (6).

[17] Yunus Ballim. A numerical model and associated calorimeter for predicting temperature profiles in mass concrete, Cement & Concrete Composites 26 (2004) 695-703.

[18] Yan Qisen and Zhao Qingzhu. Building Heat Process. Beijing: China Architecture & Building Press. 1986.

[19] Luo Zhele. *Dryvit Exterior Insulation and Finish System*. Chemical Building Materials. 1998, (6).

[20] S. Timoshenko and S. Woinowsky Krieger. *Theory of Plates and Shells*. Beijing: China Communications Press. 1977.

[21] Westergaard H M. Analysis of stress in concrete pavements due to variations of temperature. 6th Ann Meeting, Hwy. Res. Board, Washington, D. C., 201-215.

[22] Gu Tongzeng. *Necessity for Vigorous Development and Application of Single Insulation and Energy Conservation System of Wall*. China Building Materials, 2006 (2).

[23] Wang Wuxiang. *Research on Regenerative EPS Ultra-light Roof Insulation Materials*. Shandong Building Materials. 2003, (04).

[24] Peng Jiahui and Chen Mingfeng. *Development and Construction Technology of EPS Insulation Mortar*. Construction Technology. 2001, 30 (8).

[25] Peng Zhihui and Chen Mingfeng. *Research on External Insulation Mortar based on Waste Expandable Polystyrene (EPS) Foam*. Architecture Journal of Chongqing University. 2005, 27 (5).

[26] Bazant Z P, Panula L. Practical Prediction of Time Dependent Deformation of Concrete. Materials and Structures, Part I and II: Vol. 11, No. 65, 1978 pp307-328; parts III and IV : Vol. 11, No. 66, 1978, pp415-434; Parts V and VI: Vol. 12, No. 69, 1979, pp 169-173.

[27] Bazant Z P, Murphy W P. Creep and Shrinkage prediction model for analysis and design of concrete structures-model B3, Materials and Structures, 1995, 28, 357-365.

[28] Wang Jiachun and Yan Peiyu. *Energy Conservation Analysis of Composite Wall with External Thermal Insulation*. Insulation Material and Building Energy Conservation. 2004 (6).

[29] Zhu Bofang. *Thermal Stress and Temperature Control of Mass Concrete*. Beijing: China Electric Power Press. 1999.

[30] Zhang Yu. *Study on Cracking Causes and Countermeasures of Wall Insulation*. West China Exploration Engineering. 2005 (8).

[31] Teng Chunbo and Zhang Yuyao. *Analysis of Cracks in External Insulation Maintenance Wall of Composite Wall*. Heilongjiang Science and Technology of Water Conservancy. 2005 (5).

[32] Science and Technology Development Promotion Center of the Ministry of Housing and Urban-Rural Development and Beijing Zhenli High-tech Co., Ltd. *External Insulation Application Technology*. Beijing: China Architecture & Building Press. 2005.

[33] Sun Zhenping, Jin Huizhong, Jiang Zhengwu, Yu Long and Wang Peiming. *Structure, Technology and Characteristics of Exterior Insulation and Finish System*. Low-temperature Architecture Technology. 2005 (6).

[34] Kim, J. K., 2001. Estimation of compressive strength by a new apparent activation energy function. Cement and Concrete Research, 31 (2), 217-225

[35] Gutsch, A. W., 1998. Stoffeigenschaften jungen Betons-Versuche und Modelle. Doct. Th., TU Braunschweig.

[36] Xu Mengxuan and Han Yi. *Technical and Economic Analysis of External Thermal Insulation Technology*. Gas and Heat. 2005 (10).

[37] IPACS. Structual Behaviour: Numerical Simulation of the Maridal culvert. REPORT BE96-3843/2001: 33-8

[38] Timm, D. H. and Guzina, B. B., Voller, V. R., 2003. Prediction of thermal crack spacing. International Journal of Solids and Structures 40, 125-142

[39] Zhang Jun, Qi Kun and Zhang Minghua. *Calculation of Thermal Stress of Concrete Pavement Slab under Nonlinear Temperature Field in the Early Age*. Engineering Mechanics, 24 (11).

[40] Tu Fengxiang, et al. Persisting in the Developing Road of Energy Efficiency in Buildings with China's Characteristics. Beijing: China Architecture & Building Press. March 2010.

[41] Huang Zhenli and Zhang Pan. Utilization of Solid Waste in External Thermal Insulation of Building. Construction Technology. No. 6, 2007.

[42] Beijing Zhenli Energy Conservation & Environmental Protection Technology Co., Ltd. and Science and Technology Development Promotion Center of the Ministry of Housing and Urban-Rural Development. *Exploration of Wall Insulation Technology*. Beijing: China Architecture & Building Press. March 2009.

[43] Li Xin. *Main Approaches and Policies for Development of Green Building Materials in China*. 21st Century Building Materials. 2009, 1 (06).

[44] He Shaoming. *Common Misunderstandings of Insulation Materials in Building Energy Conservation*. China Value website. 2008.

[45] Cao Minggan and Cao Xiaorong. *Disposal and Recycling of Rigid Polyurethane Foam*. Plastics. 2005, 34 (1).

[46] Lin Huaying, Zhang Wei, Zhang Qiong, et al. *Analysis of Thermal Decomposition Products of Polystyrene by Gas Chromatography & Mass Spectrometry*. Chinese Journal of Health Laboratory Technology. 2009, 19 (9).

[47] Wang Yong. *Status Quo and Development Trend of Extruded Polystyrene (XPS) Foam in China*. China Plastics. 2010. 24 (4).

[48] Zhu Lvmin, et al. *Polyurethane Foam* [M] (3rd Edition). Beijing: Chemical Industry Press. 2005.

[49] Ma Deqiang, Ding Jiansheng and Song Jinhong. *Progress in Production of Organic Isocyanates* [J]. Chemical Industry Progress. 2007, 26 (5).

[50] Wang Chao. *HCFC-141b Alternative Solution for Solar Water Heater Industry* [A]. Proceedings of the 3rd Symposium on Polyurethane Foamer Alternative Technology [C]. Beijing: Shandong Dongda

Polymer Co., Ltd.

[51] Compilation team of *Water Pollutant Emission Standards for Phenolic Resin Industry*. *Instructions for Preparation of Water Pollutant Emission Standards for Phenolic Resin Industry*. March 2008.

[52] Zhou Yang. *Comparison of NSP System Energy Consumption in China, India and Japan*. China Cement. August 2007.

[53] Liu Jiaping. *Architectural Physics (Fourth Edition)*. Beijing: China Architecture & Building Press. 2009.

[54] Liu Nianxiong and Qin Youguo. *Building Thermal Environment* [M]. Beijing: China Architecture & Building Press. 2005.

[55] Su Xianghui. *Research on Heat and Moisture Coupling Migration Characteristics in Multilayered Porous Structures*. Master's thesis of Nanjing University of Aeronautics and Astronautics. 2002.

[56] Ji Jie. *Study of Impact of Moisture Transfer on Energy Consumption of Building Wall in Severe Cold Area*. Harbin University of Civil Engineering and Architecture. 1991.

[57] Carl Seifert. *Building Moisture Prevention* (translated by Zhou Jingde and Yang Shanqin). China Architecture & Building Press. 1982.

[58] Xu Xiaoqun. *Impact of Moisture Absorption and Desorption of Internal Surface of Building Envelope on Indoor Humidity and Wet Load*. Master's thesis of Central South University. 2007.

[59] Zhao Lihua, Dong Chongcheng and Jia Chunxia. *Study on Moisture Transfer of Wall with External Thermal Insulation*. Journal of Harbin University of Civil Engineering and Architecture. 2001.

[60] Zheng Maoyu and Kong Fanhong. *Impact of Insulation Layer on Water Vapor Permeability of Wall of Energy-efficient Building in North China*. Building Energy Conservation. 2007.

[61] Li Chaoxian. *Heat and Moisture Transfer of Multilayered Composite External Envelope*. Master's thesis of the Institute of Physics of Beijing Building Research Institute of CSCEC. 1984.

[62] Zhu Yingbao. Application of Insulation Material in Building Wall Energy Conservatoin. Beijing: China Building Material Industry Press. 2003.

[63] Sha Shengang. *Test Method and Experimental Analysis of Water Vapor Transmission Rate of Exterior Coating*. Wacker Chemical Investment (China) Co., Ltd. 2007.

[64] Shi Shulan. Microscopic Study on Properties of Polymer Mortar. Guomin Chemical Investment (China) Co., Ltd. 2005.

[65] Liu Zhenhe, Xu Xiangfei and Zheng Baohua. *Causes of Surface Cracking and Falling of Enclosure Wall*. Liaoning Province Building Science Research Institute. 2011.

[66] Sun Shunjie. *Study on Preparation and Properties of Organo-silicone Waterproofing Agent for Building Surface*. Beijing Technology Center of Longhu Science & Technology Co., Ltd. in Shantou Special Economic Zone. Beijing, 2007.

Postscript

Due to the need of building energy conservation, the external thermal insulation technology was introduced from developed countries into China in the late 20th century. It has a history of research and development for more than 20 years in China. Currently, high-efficiency insulation materials such as polystyrene, polyurethane and insulation mortar are widely applied, and the external thermal insulation technology is developed rapidly with flourishing prospects.

The development of the external thermal insulation technology in China relies on the efforts of experts, scholars and talents. They, based on a large number of investigations, tests, researches and engineering practices, carefully summarize and analysis China's basic experience in this field, and strive to find out rules to further improve China's theory and practice of external thermal insulation and develop world-class external thermal insulation projects.

The authors of this book are dedicated to the studies of this field. They have studied the effects of five natural forces on the wall insulation and attached great importance to the safety and life of the insulated wall. Through the numerical simulation with the mathematical model established in a finite difference method for analysis of the temperature field and thermal stress, walls with the external thermal insulation, internal thermal insulation, sandwich insulation and self-insulation were compared; the moisture transfer inside walls, and related hazards and countermeasures, as well as the results of large-scale weathering resistance tests, seismic tests and large-scale model fire tests were analyzed; the fire resistances of structures, destructive effects of the negative wind pressure on the cavity structure and measures for falling prevention of face bricks were studied; a series of scientific conclusions were drawn by analyzing engineering cases of falling of external thermal insulation layers in different regions; endeavors were made to classify the fire ratings of external thermal insulation; the concepts of overall fire protection of the external thermal insulation system was put forward; the fireproof design software was developed; and the ways of application of industrial wastes in wall insulation were explored. The authors have devoted themselves to difficult and persistent explorations. They should be appraised as backbones for the development of China's external thermal insulation technology. Due to the concerted efforts of these backbones, China's external thermal insulation technology is bound to further development.

The external thermal insulation technology integrates the research results of many disciplines and achievements of a number of experts. With the large-scale construction and continued deepening of energy conservation and emission reduction in China, there is still a long way to go in the field of external thermal insulation. The authors of this book, as active participants, have made unremitting efforts in theory and practice. They strive to summarize long-term test and research results, and explore the ways of controlling the impact of natural factors on external thermal insulation, hoping to make their own contributions during the development of building energy-saving technologies. Because of various constraints, however, there may be omissions and errors in this book. All criticisms and positive supplements will be appreciated. The rapid development of China's external thermal insulation technology has lead to the emergence of experts engaged in various disciplines as well as advanced technologies. If all achievements are summarized and published, China's external thermal insulation industry will present a bright and colorful scene.

In face of a new science and technology revolution and an industrial revolution characterized by energy conservation and emission reduction, we must immediately participate instead of waiting. China has a large number of talents and technological achievements associated with building energy conservation and external thermal insulation. We should pay close attention to energy-saving innovations and cutting-edge technologies, improve the industrial structure and develop industrial clusters of external thermal insulation, and cultivate more leading figures and enterprises, to change the external thermal insulation industry into an emerging strategic industry. We should endeavor to reach the forefront of the world's building energy-saving technologies, and occupy strategic commanding heights for future development, to significantly improve the building energy efficiency and generally construct long-life external thermal insulation projects. We should also build China into a powerful country in the external thermal insulation technology and industry, and achieve the sustainable and healthy development of China's cause of building energy conservation.

涂逢祥

Tu Fengxiang

March 5, 2011